Lecture Notes in Computer Science 9931

Commenced Publication in 1973
Founding and Former Series Editors:
Gerhard Goos, Juris Hartmanis, and Jan van Leeuwen

More information about this series at http://www.springer.com/series/7409

Feifei Li · Kyuseok Shim
Kai Zheng · Guanfeng Liu (Eds.)

Web Technologies and Applications

18th Asia-Pacific Web Conference, APWeb 2016
Suzhou, China, September 23–25, 2016
Proceedings, Part I

 Springer

Editors
Feifei Li
School of Computing
University of Utah
Salt Lake City, UT
USA

Kyuseok Shim
School of Electrical Engineering
Seoul National University
Seoul
Korea (Republic of)

Kai Zheng
Soochow University
Suzhou
China

Guanfeng Liu
Soochow University
Suzhou
China

ISSN 0302-9743   ISSN 1611-3349 (electronic)
Lecture Notes in Computer Science
ISBN 978-3-319-45813-7   ISBN 978-3-319-45814-4 (eBook)
DOI 10.1007/978-3-319-45814-4

Library of Congress Control Number: 2016949587

LNCS Sublibrary: SL3 – Information Systems and Applications, incl. Internet/Web, and HCI

Printed on acid-free paper

This Springer imprint is published by Springer Nature
The registered company is Springer International Publishing AG Switzerland

Message from the General Chairs and Program Committee Chairs

Welcome to APWeb 2016! This is the 18th Conference in the Asia Pacific Web Series. Since the first APWeb conference in 1998, APWeb has evolved over time to lead the frontier of data-driven information technology research. It has now firmly established itself as a leading Asia-Pacific focused international conference on research, development, and advanced applications on large-scale data management, Web and search technologies, and information processing. Previous APWeb conferences were held in Guangzhou (2015), Changsha (2014), Sydney (2013), Kunming (2012), Beijing (2011), Busan (2010), Suzhou (2009), Shenyang (2008), Huangshan (2007), Harbin (2006), Shanghai (2005), Hangzhou (2004), Xi'an (2003), Changsha (2001), Xi'an (2000), Hong Kong (1999), and Beijing (1998).

APWeb 2016 was held during September 23–25 in the beautiful and cultural city of Suzhou, China, a city proud of its history of 2500 years. The host organization of APWeb 2016 is Soochow University, one of the fastest developing universities in China.

As in previous years, the APWeb 2016 program featured the main conference with research papers, an industry track, tutorials, distinguished lectures, demos, and a panel. APWeb this year received 215 paper submissions to the main conference from North America, South America, Europe, Asia, and Africa. Each submitted paper underwent a rigorous review process by at least three independent referees from the Program Committee, with detailed review reports. Finally, 79 full papers, 24 short papers, and 17 demo papers were accepted and included in these proceedings. The conference this year had three satellite workshops:

- Second International Workshop on Web Data Mining and Applications (WDMA 2016)
- First International Workshop on Graph Analytics and Query Processing (GAP 2016)
- First International Workshop on Spatial-temporal Data Management and Analytics (SDMA 2016)

We were fortunate to have three world-leading scientists as our keynote speakers: Zhihua Zhou (Nanjing University, China), Cyrus Shahabi (University of Southern California, USA), and Yufei Tao (The University of Queensland, Australia). The Distinguished Lecture Series, co-chaired this year by Xiaokui Xiao (Nanyang Technology University, Singapore) and Xiaochun Yang (Northeastern University, China), invited active and high-impact researchers to discuss their work at APWeb. The two speakers this year were Chen Li (University of California, Irvine, USA) and Jianliang Xu (Hong Kong Baptist University, Hong Kong, China).

The success of APWeb 2016 would not have been possible without the hard work of a great team of people, including Workshop Co-chairs Rong Zhang (ECNU, China) and Wenjie Zhang (University of New South Wales, Australia); Tutorial Co-chairs, Wook-Shin Han (POSTECH, South Korea) and Weng-Chih Peng (National Jiao Tong University, Taiwan); Panel Chair, Ji-Rong Wen (Renmin University of China); Industrial Co-chairs, Luna Xin Dong (Google Research, USA) and Ying Yan (Microsoft Research, China); Demo Co-chairs, Zhifeng Bao (RMIT, Australia) and Xiangliang Zhang (KAUST, Saudi Arabia); Distinguished Lecture Series Co-chairs, Xiaokui Xiao (Nanyang Technological University, Singapore) and Xiaochun Yang (Northeastern University, China); Publication Chair, Guanfeng Liu (Soochow University, China); Social Media and Publicity Chair, Han Su (University of Southern California, USA); Sponsorship and Finance Chairs, An Liu and Lei Zhao (Soochow University, China); and Web Masters, Yan Zhao, and Yang Li (Soochow University, China).

We would also like to take this opportunity to extend our sincere gratitude to the Program Committee members and external reviewers. A special word of thanks goes to the Local Organization Chair, Zhixu Li (Soochow University, China) and his team of organizers and volunteers! Last but not least, we would like to thank all the sponsors, the APWeb Steering Committee led by Jeffrey Yu, and the host organization, Soochow University, for their support, help, and assistance in organizing this conference.

This year's APWeb conference was also the last APWeb conference in its current form. From next year, APWeb and WAIM will be officially combined into one conference, under the new name APWeb/WAIM Joint Conference on the Web and Big Data. These two conferences have many things in common, such as research topics, their regional focuses, and target audiences. In addition, they both share the same goal, to be a world-class research conference on World Wide Web, Internet, and Big Data research and applications with a clear focus on the Asia Pacific region. This year's APWeb was the end of a chapter that we are all proud of, and we are very excited about the new start!

We hope the participants enjoyed APWeb 2016 and Suzhou!

July 2016

General Chairs

Tamer Ozsu
Xiaofang Zhou

Program Committee Chairs

Feifei Li
Kyuseok Shim
Kai Zheng

Organization

General Co-chairs

Tamer Ozsu   University of Waterloo, Canada
Xiaofang Zhou   The University of Queensland, Australia and Soochow
                                University, China

Program Committee Co-chairs

Feifei Li   University of Utah, USA
Kyuseok Shim   Seoul National University, South Korea
Kai Zheng   Soochow University, China

Workshop Co-chairs

Atsuyuki Morishima   University of Tsukuba, Japan
Rong Zhang   ECNU, China
Wenjie Zhang   University of New South Wales, Australia

Tutorial Co-chairs

Wook-Shin Han   POSTECH, Korea, South Korea
Weng-Chih Peng   National Jiao Tong University, Taiwan
Yasushi Sakurai   Kumamoto University, Japan

Panel Chair

Ji-Rong Wen   Renmin University of China, China

Industrial Co-chairs

Luna Xin Dong   Google Research, USA
Ying Yan   Microsoft Research, China

Demo Co-chairs

Zhifeng Bao   RMIT, Australia
Xiangliang Zhang   KAUST, Saudi Arabia

Distinguished Lecture Series Co-chairs

Xiaokui Xiao   Nanyang Technological University, Singapore
Xiaochun Yang   Northeastern University, China

Research Students Symposium Co-chairs

Guoliang Li   Tsinghua University, China
Zi Huang   University of Queensland, Australia

Publication Chair

Guanfeng Liu   Soochow University, China

Social Media and Publicity Chair

Han Su   University of Southern California, USA

Sponsorship and Finance Chairs

An Liu   Soochow University, China
Lei Zhao   Soochow University, China

Local Organization Chairs

Zhixu Li   Soochow University, China
Yan Zhao   Soochow University, China

Web/Information Co-chair

Jun Jiang   Soochow University, China

Program Committee

Aimin Feng   Nanjing University of Aeronautics and Astronautics, China
Alex Thomo   University of Victoria, Canada
Alfredo Cuzzocrea   University of Calabria, Italy
Anirban Mondal   Xerox Research Centre, India
Aviv Segev   KAIST, Korea, South Korea
Baoning Niu   Taiyuan University of Technology, China
Bin Yang   Aalborg University, Denmark
Bingtian Dai   Singapore Management University, Singapore
Carson Kai-Sang Leung   University of Manitoba, Canada
Chaofeng Sha   Fudan University, China
Chen Lin   Xiamen University, China
Chengkai Li   University of Texas at Arlington, USA

Kun Ren	Yale University, USA
Leong Hou U	University of Macau, Macau, China
Liang Hong	Wuhan University, China
Lianghuai Yang	Zhejiang University of Technology, China
Lili Jiang	Max-Planck Institute, Germany
Ling Chen	University of Technology, Sydney, Australia
Man Lung Yiu	Hong Kong Polytechnical University, Hong Kong, China
Markus Endres	University of Augsburg, Germany
Meihui Zhang	Singapore University of Technology and Design, Singapore
Mihai Lupu	Vienna University of Technology, Austria
Mizuho Iwaihara	Waseda University, Japan
Mohammed Eunus Ali	Bangladesh University of Engineering and Technology, Bangladesh
Nicholas Jing Yuan	Microsoft Research, Beijing, China
Ning Yang	Sichuan University, China
Panos Kalnis	King Abdullah University of Science and Technology, Saudi Arabia
Peng Wang	Fudan University, China
Qi Liu	University of Science and Technology of China, China
Qingzhong Li	Shandong University, China
Quan Thanh Tho	Ho Chi Minh City University of Technology, Vietnam
Quan Zou	Xiamen University, China
Raymond Wong	Hong Kong University of Science and Technology, Hong Kong, China
Rui Zhang	University of Melbourne, Australia
Rui Chen	Hong Kong Baptist University, Hong Kong, China
Sanghyun Park	Yonsei University, South Korea
Sangkeun Lee	Oak Ridge National Laboratory, USA
Sang-Won Lee	Sungkyunkwan University, South Korea
Shengli Wu	Jiangsu University, China
Shinsuke Nakajima	Kyoto Sangyo University, Japan
Shuai Ma	Beihang University, China
Shuo Shang	China University of Petroleum-Beijing, China
Sourav Bhowmick	Nanyang Technological University, Singapore
Srikanta Bedathur	IBM Research India, India
Stavros Papadopoulos	Intel Labs, USA
Taketoshi Ushiama	Kyushu University, Japan
Tieyun Qian	Wuhan University, China
Ting Deng	Beihang University, China
Toshiyuki Amagasa	University of Tsukuba, Japan
Tru Hoang Cao	Ho Chi Minh City University of Technology, Vietnam
Vicent Zheng	Advanced Digital Sciences Center, Singapore
Wee Siong Ng	Institute for Infocomm Research, Singapore
Wei Wang	University of New South Wales, Australia
Wei Wang	Fudan University, China

Weining Qian	East China Normal University, China
Weiwei Sun	Fudan University, China
Weiwei Ni	Southeast University, China
Wen Zhang	Wuhan University, China
Wook-Shin Han	POSTECH, South Korea
Xiang Lian	University of Texas Rio Grande Valley, USA
Xiang Zhao	National University of Defence Technology, China
Xiangliang Zhang	King Abdullah University of Science and Technology, Saudi Arabia
Xiaochun Yang	Northeast University, China
Xiaofeng He	East China Normal University, China
Xiaowei Yang	South China University of Technology, China
Xike Xie	Aalborg University, Denmark
Xuanjing Huang	Fudan University, China
Xueqing Gong	East China Normal University, China
Yanghua Xiao	Fudan University, China
Yang-Sae Moon	Kangwon National University, South Korea
Yaokai Feng	Kyushu University, Japan
Yasuhiko Morimoto	Hiroshima University, Japan
Yi Zhuang	Zhejiang Gongshang University, China
Yijie Wang	National University of Defense Technology, China
Ying Zhao	Tsinghua University, China
Yinghui Wu	University of California, Santa Barbara, USA
Yong Zhang	Tsinghua University, China
Yoshiharu Ishikawa	Nagoya University, Japan
Yu Gu	Northeast University, China
Yuan Fang	Institute for Infocomm Research, Singapore
Yunjie Yao	East China Normal University, China
Yunjun Gao	Zhejiang University, China
Zakaria Maamar	Zayed University, United Arab Emirates
Zhaonian Zou	Harbin Institute of Technology, China
Zhengjia Fu	Advanced Digital Sciences Center, Singapore
Zhenying He	Fudan University, China
Zhiyong Peng	Wuhan University, China
Zhoujun Li	Beihang University, China
Zouhaier Brahmia	University of Sfax, Tunisia

Demo Track Program Committee

Chengbin Peng	KAUST, Saudi Arabia
Chuan Shi	Beijing University of Posts and Telecommunications, China
Feng Li	Microsoft Research, China
Huayu Wu	I2R, Singapore
Jianqiu Xu	NUAA, China
Jingbo Zhou	Baidu Research, USA

Ke Deng	RMIT University, Australia
Peng Zhang	University of Technology, Sydney, Australia
Pinghui Wang	Xian Jiaotong University, China
Qing Xie	Wuhan University of Technology, China
Rui Zhou	Victoria University, Australia
Sang-Wook Kim	Hanyang University, South Korea
Tao Jiang	Jiaxing University, China
Wei Lee Woon	Masdar Institute, United Arab Emirates
Xiangmin Zhou	RMIT University, Australia
Xiaoli Wang	Xiamen University, China
Xiaoying Wu	Wuhan University, China
Yong Zeng	Microsoft, USA
Yuchen Li	National University of Singapore, Singapore

Industry Track Program Committee

Liang Jeff Chen	Microsoft Research, China
Wee Hyong Tok	Microsoft Azure, China
Wenyuan Cai	Hypers, China
Sheng Huang	IBM China Research Lab, China
Chen Wang	RRX360, China

Contents – Part I

Research Full Paper: Social Media Data Analysis

Research Full Paper: Modelling and Learning with Big Data

Contents – Part II

Research Full Paper: Data Quality and Privacy

Research Full Paper: Query Optimization and Scalable Data Processing

Research Short Paper

Industry Full Paper

Demo Paper

Research Full Paper: Spatio-temporal, Textual and Multimedia Data Management

Probabilistic Nearest Neighbor Query in Traffic-Aware Spatial Networks

Shuo Shang[1(✉)], Zhewei Wei[2], Ji-Rong Wen[2], and Shunzhi Zhu[3]

[1] Department of Computer Science, China University of Petroleum-Beijing,
Beijing, China
jedi.shang@gmail.com
[2] Beijing Key Laboratory of Big-data Management and Analysis,
School of Information, Renmin University of China, Beijing, China
{zhewei,jrwen}@ruc.edu.cn
[3] Xiamen University of Technology, Xiamen, China
zhusz66@163.com

Abstract. Travel planning and recommendation have received significant attention in recent years. In this light, we study a novel problem of finding probabilistic nearest neighbors and planning the corresponding travel routes in traffic-aware spatial networks (TANN queries) to avoid traffic congestions. We propose and study two probabilistic TANN queries: (1) a time-threshold query like "what is my closest restaurant with the minimum congestion probability to take at most 45 min?", and (2) a probability-threshold query like "what is the fastest path to my closest petrol station whose congestion probability is less than 20 %?". We believe that this type of queries may benefit users in many popular mobile applications, such as discovering nearby points of interest and planning convenient travel routes for users. The TANN queries are challenged by two difficulties: (1) how to define probabilistic metrics for nearest neighbor queries in traffic-aware spatial networks, and (2) how to process the TANN queries efficiently under different query settings. To overcome these challenges, we define a series of new probabilistic metrics and develop two efficient algorithms to compute the TANN queries. The performances of TANN queries are verified by extensive experiments on real and synthetic spatial data.

1 Introduction

With the rapid development of GPS-enabled mobile devices (e.g., car navigation systems and smart phones) and online map services (e.g., Google-maps[1], Bing-maps[2] and MapQuest[3]), people can easily acquire their current geographic positions in interactive time and interact with servers to query spatial information regarding their trips (e.g., shortest path queries [12], nearest neighbor

---

[1] http://maps.google.com/.
[2] http://www.bing.com/maps/.
[3] http://www.mapquest.com.

© Springer International Publishing Switzerland 2016
F. Li et al. (Eds.): APWeb 2016, Part I, LNCS 9931, pp. 3–14, 2016.
DOI: 10.1007/978-3-319-45814-4_1

queries [9, 15], and route recommendation [10, 11]). In the meantime, trajectory sharing and search are pervasive nowadays. Travelers can easily upload their trajectories to some specialized web sites such as Bikely[4], GPS-Way-points[5], Share-My-Routes[6], Microsoft GeoLife[7]. By analyzing historical travel trajectories of commuters, it is possible to construct a traffic-aware spatial network [12–14], to describe traffic conditions and to help users plan travel routes and to avoid traffic congestions. The potential market of such location based services in the near future enables many novel applications. An emerging application is **traffic-aware nearest neighbor queries** (TANN queries) in spatial networks, which are designed to discover the nearby points of interest and to plan the corresponding convenient routes for users. We believe that this type of queries may bring significant benefits to users in many popular mobile applications such as travel planning and recommendation.

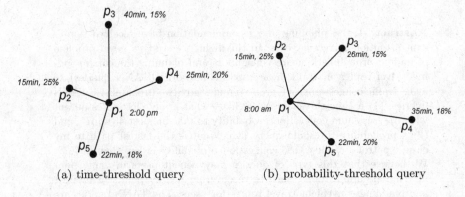

(a) time-threshold query   (b) probability-threshold query

Fig. 1. Two types of TANN query

In this light, we propose two novel probabilistic TANN queries: (1) a time-threshold query like "what is my closest restaurant with the minimum congestion probability to take at most 45 min?", and (2) a probability-threshold query like "what is the fastest path to my closest petrol station whose congestion probability is less than 20 %?". Examples of the TANN queries are shown in Fig. 1. Figure 1(a) describes the time-threshold TANN query, where p_1 is a user's current location and its departure time is 2:00pm, and p_2, p_3, p_4 and p_5 are his/her nearby restaurants. The time-threshold query retrieves a nearby restaurant with the minimum congestion probability to take at most 45 min. If only travel time is considered, p_2 is returned. However, p_2 has the highest congestion probability 25 % among all nearby restaurants. By taking both travel-time threshold and congestion probability into account, p_3 is returned. Similarly, in Fig. 1(b),

---

[4] http://www.bikely.com/.
[5] http://www.gps-waypoints.net/.
[6] http://www.sharemyroutes.com/.
[7] http://research.microsoft.com/en-us/projects/geolife/.

p_1's departure time is 8:00am and the probability-threshold query ask for a nearby petrol station whose congestion probability is less than 20 %. Although p_2 and p_5 are spatially close to p_1, their congestion probabilities exceed the user specified probability threshold. Petrol station p_3 is returned by taking both travel time and probability threshold into account.

The TANN queries are applied in spatial networks, since in a large number of practical scenarios, moving objects move in a constrained environment (e.g., roads, railways, and rivers) rather than a Euclidean space. The TANN queries are challenging due to two reasons. First, it is necessary to define a series of new metrics to describe travel time and congestion probability for the nearest neighbor queries in traffic-aware spatial networks. Second, it is necessary to develop efficient algorithms to compute the TANN queries under different query settings. To the best of our knowledge, there is no existing studies can process the TANN queries efficiently. In static spatial networks, nearest neighbor queries [6,7,15] do not take instant traffic conditions and traffic congestions into account, while in dynamic spatial networks, time-dependent and probabilistic routing queries [2,4] are based on different traffic models from ours, and traffic-aware unobstructed path queries [12–14] only focus on the fastest-path problems and their techniques can hardly be used in the TANN queries.

To overcome these challenges, we establish a traffic-aware spatial network $G_{ta}(V, E)$ according to the approaches in [12–14] to detect potential traffic congestions and to model instant traffic conditions practically. For each vertex $v \in G_{ta}(V, E)$, we maintain a set of traffic records to describe its traffic awareness. Based on $G_{ta}(V, E)$, we propose two novel probabilistic nearest neighbor queries (TANN) and develop efficient algorithms to compute them. To sum up, the main contributions of this paper are as follows:

- We define two novel TANN queries in traffic-aware spatial networks. It provides new features for advanced spatio-temporal information systems, and may benefit users in many popular mobile applications such as discovering nearby points of interest and planning convenient travel routes for users.
- We propose a series of new metrics to describe travel time and congestion probability for the probabilistic nearest neighbor queries in traffic-aware spatial networks (Sect. 2).
- We develop two algorithms to compute the time-threshold and probability-threshold TANN queries efficiently (Sect. 3).
- We conduct extensive experiments on real and synthetic spatial data to investigate the performance of the developed algorithms (Sect. 4).

The rest of the paper is organized as follows. Section 2 introduces traffic-aware spatial networks, spatial metrics, as well as problem definitions. The TANN query processing is detailed in Sect. 3, which is followed by the experimental results in Sect. 4. This paper is concluded in Sect. 6 after some discussions of related work in Sect. 5.

2   Preliminaries

2.1   Spatial Networks

A spatial network is modeled by a connected and undirected graph $G(V, E, F, W)$, where V is a vertex set and $E \subseteq V \times V$ is an edge set. A vertex $v_i \in V$ represents a road intersection or an end of a road. An edge $e_k = (v_i, v_j) \in E$ is defined by two vertices and represents a road segment that enables travel between vertices v_i and v_j. Function $F : V \cup E \rightarrow Geometries$ records geometrical information of the spatial network G. In particular, it maps a vertex and an edge to the point location of the corresponding road intersection and to a polyline representing the corresponding road segment, respectively. Function $W : E \rightarrow R$ is a function that assigns a real-valued weight to each edge. The weight $w(e)$ of an edge e represents the corresponding road segment's length or some other relevant property such as its travel time or fuel consumption [17]. Given two vertices p_a and p_b in a spatial network, the network shortest path between them is denoted by $SP(p_a, p_b)$, and its length is denoted by $sd(p_a, p_b)$. When weights model aspects such as travel time and fuel consumption, the lower bound of network distance is not necessarily the corresponding Euclidean distance; thus spatial indexes such as the R-tree [3] are not effective.

2.2   Traffic-Aware Spatial Networks

To detect potential traffic congestions and to model instant traffic conditions practically, we construct a traffic-aware spatial network $G_{ta}(V, E)$. The concept and establishment procedure of traffic-aware spatial networks are detailed in [12–14]. A traffic-aware spatial network $G_{ta}(V, E)$ is a spatial network, where each vertex $v \in G_{ta}.V$ has been assigned a threshold $v.k$ to describe its traffic processing capability. A moving object's processing time at v is $\frac{1}{v.k}$. If the gap between any two moving objects is less than $\frac{1}{v.k}$ minutes, a time-delay is triggered. The time-delay T_d for object o at vertex v is defined as follows.

$$T_d(o, v, T_a(o, v)) = \begin{cases} \sum_{o_i \in O_p} o_i.prob \cdot (\frac{1}{v.k} - T_a(o, v) \\ \quad + T_a(o_i, v)) + \sum_{o_j \in O_w} \frac{o_j.prob}{v.k} & \text{if } C_1 \\ \\ 0 & \text{if } C_2 \end{cases} \quad (1)$$

C_1: vertex v is occupied at object o's arrival time $T_a(o, v)$, and O_p is the set of moving objects being processed by v at $T_a(o, v)$. There are also $|O_w|$ objects waiting to be processed.
C_2: vertex v is non-occupied at $T_a(o, v)$, thus the time-delay is 0.

The total time cost of object o at vertex v is the sum of waiting time (time-delay) and processing time. The leaving time T_l of o from v is computed by

$$T_l(o, v) = T_a(o, v) + T_d(o, v, T_a(o, v)) + \frac{1}{v.k} \quad (2)$$

For each vertex $v \in G_{ta}(V, E)$, there are a set of traffic records attached to v to describe its traffic conditions. Each record is in the form of (object_id o, arrival_probability $o.prob$, expected arrival time $T_a(o, v)$, time-delay $T_d(o, v, T_a(o, v))$, processing time $\frac{1}{v.k}$).

If an object o' satisfies Eq. 3, it is labeled by "waiting object" (when object o arrives at v, o' is waiting for being processed) hence we put it into waiting-object set O_w. Otherwise, if o' satisfies Eq. 4, it is labeled by "processing object" (when o arrives at v, o' is being processed) hence we put it into processing-object set O_p.

$$T_a(o_i, v) < T_a(o, v) < T_a(o_i, v) + T_d(o_i, v, T_a(o_i, v)) \tag{3}$$

$$\begin{cases} T_a(o', v) + T_d(o', v, T_a(o', v)) < T_a(o, v) \\ T_a(o, v) < T_a(o', v) + T_d(o', v, T_a(o', v)) + \frac{1}{v.k} \end{cases} \tag{4}$$

3　Probabilistic TANN Query Processing

In this section, we propose two novel probabilistic nearest neighbor queries in traffic-aware spatial networks (TANN queries), and develop two efficient algorithms to compute them.

Given a traffic-aware spatial network $G_{ta}(V, E)$ and a moving object o, the congestion probability of object o at vertex v is defined by

$$o(v).prob_c = 1 - \prod_{o' \in O_p} (1 - o'(v).prob) \cdot \prod_{o'' \in O_w} (1 - o''(v).prob). \tag{5}$$

Here, O_p is the set of "processing objects" (objects are being processed when o arrives at v, refer to Eq. 4), and O_w is the set of "waiting objects" (objects are waiting to be processed when o arrives at v, refer to Eq. 3). Parameter $o'(v).prob$ is the probability of the scenario that when object o arrives at v, o' is being processed at v, and parameter $o''(v).prob$ is the probability of the scenario that when o arrives at v, o'' is waiting for being processed. In both scenarios, v is occupied by other objects and the movement of o is obstructed (a traffic congestion occurs). The travel time and congestion probability of path $P = \langle v_1, v_2, ..., v_k \rangle$ are defined by

$$P.time = \sum_{i=1}^{k-1} (w(v_i, v_{i+1}) + \frac{1}{v_{i+1}.k}) \tag{6}$$

$$P.prob_c = 1 - \prod_{i=1}^{k} (1 - o(v_i).prob_c) \tag{7}$$

where $w(v_i, v_{i+1})$ is the weight of edge (v_i, v_{i+1}), and $o(v_i).prob_c$ is the congestion probability of object o at vertex v_i.

3.1   Time-Threshold TANN Query

Time-Threshold TANN Query: Given a traffic-aware spatial network $G_{ta}(V, E)$, a query source q, a spatial object set O, a departure time t, and a time-threshold $\tau.t$, time-threshold TANN query finds a spatial object $p \in O$ with the minimum congestion probability and its travel time does not exceeds $\tau.t$, such that $\forall p' \in O \setminus \{p\}(P(q, p).time \leq \tau.t \land P(q, p').time \leq \tau.t \rightarrow P(q, p).prob_c < P(q, p').prob_c)$.

Network expansion is a conventional (e.g., Dijkstra's expansion [1]) approach to nearest neighbors in spatial networks, but it fails to address the time-threshold TANN problem due to the non-awareness of moving-object processing time and congestion probability. To overcome this weakness, we develop a novel algorithm called $TANN_{time}$ to compute the time-threshold TANN query efficiently. In $TANN_{time}$, we maintain a time label and a probability label for each vertex v in the expansion tree. The time label $P(s, v).time$ is defined by

$$P(q, v).time = P(q, v.pre).time + w(v.pre, v) + \frac{1}{v.k}, \tag{8}$$

where $v.pre$ is the parent vertex of v in the expansion tree (refer to Eq. 6). The congestion probability label of vertex v is defined by

$$P(q, v).prob_c = 1 - (1 - P(q, v.pre).prob_c) \cdot (1 - o(v).prob_c), \tag{9}$$

where $P(q, v).prob_c$ is the congestion probability of $P(q, v)$, and $o(v).prob_c$ is the congestion probability of moving object o at vertex v (refer to Eq. 7).

In the $TANN_{time}$ algorithm, we inherit the best-first search strategy from the Dijkstra's algorithm [1]. At each time, we select the vertex v with the minimum congestion probability label $P(q, v).prob_c$ for expansion. We also check the time label of v to guarantee that the value of $P(q, v).time$ dose not exceeds the time threshold $\tau.t$. Thus, the first scanned "qualified" spatial object is just the time-threshold closest spatial object to q, and other objects can be pruned safely.

The $TANN_{time}$ algorithm is detailed in Algorithm 1. The query input includes a set of spatial objects O, a source q, a departure time t, and a time threshold $\tau.t$, while the output is the spatial object $p \in O$ with the minimum congestion probability and the corresponding path $P(q, p)$. Initially, the set of scanned vertices is set to null, and the congestion probability label of each vertex $v \in V$ is set to 1 (line 1–2). The source q is labeled as scanned vertex and is put into O_s. Its time label is set to 0, and its congestion probability label is set to 0 (lines 3–4). In each iteration, we select the vertex c with the minimum congestion probability label $P(q, c).prob_c$ from the heap O_s (lines 5–7). If c is a spatial object, time-threshold closest spatial object is found and the spatial object and the corresponding travel path are returned (lines 8–13). We evaluate the distance label for each adjacent vertex of c. If the value of $P(q, c).time + w(c, v) + \frac{1}{v.k}$ exceeds the time-threshold $\tau.t$, v is pruned. Otherwise, if v's probability label exceeds the value of $1 - (1 - P(q, c).prob_c) \cdot (1 - o(v).prob_c)$, its probability label and time label are updated, and v is put into O_s (lines 14–20).

Algorithm 1. $TANN_{time}$ Algorithm

Data: $G_{ta}(V, E)$, O, q, t, and $\tau.t$
Result: $p \in O$ with the minimum congestion probability and $P(q, p)$

1   $P(q, p) \leftarrow \emptyset; O_s \leftarrow \emptyset$;
2   $\forall v \in V(P(q, v).prob_c \leftarrow 1)$;
3   $O_s.push(q)$;
4   $P(q, q).time = 0, P(q, q).prob_c = 0$;
5   **while** $O_s \neq \emptyset$ **do**
6   select $c \in O_s$ with the minimum $P(q, c).prob_c$;
7      $O_s.remove(c)$;
8   **if** c *is a spatial object* **then**
9         $p \leftarrow c$;
10   **while** $c.pre \neq null$ **do**
11         $P(q, p).push(c)$;
12         $c \leftarrow c.pre$;
13   **return** p *and* $P(q, p)$;
14   **for** *each vertex* $v \in c.adj$ **do**
15   **if** $P(q, c).time + w(c, v) + \frac{1}{v.k} < \tau.t$ **then**
16   **if** $P(q, v).prob_c > 1 - (1 - P(q, c).prob_c) \cdot (1 - o(v).prob_c)$ **then**
17             $P(q, v).prob_c \leftarrow 1 - (1 - P(q, c).prob_c) \cdot (1 - o(v).prob_c)$;
18             $P(q, v).time \leftarrow P(q, c).time + w(c, v) + \frac{1}{v.k}$;
19             $v.pre \leftarrow c$;
20             $O_s.push(v)$;

3.2   Probability-Threshold TANN Query

Probability-Threshold TANN Query: Given a traffic-aware spatial network $G_{ta}(V, E)$, a set of spatial objects O, a query source q, a departure time t, and a probability threshold $\tau.p$, the probability-threshold TANN query finds the spatial object $p \in O$ with the minimum travel time and its congestion probability does not exceed $\tau.p$, such that $\forall p' \in O \setminus \{p\}(P(q, p).prob_c < \tau.p \wedge P(q, p').prob_c < \tau.p \rightarrow P(q, p).time < P(q, p').time)$.

To compute the probability-threshold TANN query efficiently, we develop a novel $TANN_{prob}$ algorithm. Similar to the $TANN_{time}$ algorithm, for each vertex $v \in G_{ta}(V, E)$, we maintain a time label $v.time$ and congestion probability label $v.prob_c$ (refer to Eqs. 8 and 9).

Algorithm 2 introduces the $TANN_{prob}$ search. The query input includes a source q, a set of spatial object O, a departure time t, and a probability threshold $\tau.p$, while the output is the spatial object $p \in O$ with the minimum travel time and the corresponding travel path $P(q, p)$. Initially, the path $P(q, p)$ is set to null, and the set of scanned vertices is also set to null. The time label of each vertex $v \in V$ is set to $+\infty$ (line 1–2). The source q is labeled as a scanned vertex and is put into O_s (line 3). In each iteration, we select the vertex

---

Algorithm 2. $TANN_{prob}$ Algorithm

Data: $G_{ta}(V, E)$, O, q, t, and $\tau.prob$
Result: $p \in O$ with the minimum travel time and $P(q, p)$

1 $P(q, p) \leftarrow \emptyset$; $O_s \leftarrow \emptyset$;
2 $\forall v \in V(P(q, v).time \leftarrow +\infty)$;
3 $O_s.push(q)$;
4 **while** $O_s \neq \emptyset$ **do**
5   select $c \in O_s$ with the minimum $P(q, c).time$;
6       $O_s.remove(c)$;
7   **if** c is a spatial object **then**
8           $p \leftarrow c$;
9   **while** $c.pre \neq null$ **do**
10              $P(q, p).push(c)$;
11              $c \leftarrow c.pre$;
12   **return** p and $P(q, p)$;
13   **for** each vertex $v \in c.adj$ **do**
14   **if** $P(q, v).prob_c < \tau.p$ **then**
15   **if** $P(q, v).time > P(q, c).time + w(c, v) + \frac{1}{v.k}$ **then**
16                  $P(q, v).time \leftarrow P(q, c).time + w(c, v) + \frac{1}{v.k}$;
17                  $P(q, v).prob_c \leftarrow 1 - (1 - P(q, c).prob_c) \cdot (1 - o(v).prob_c)$;
18                  $v.pre \leftarrow c$;
19                  $O_s.push(v)$;

---

c with the minimum time label $P(q, c).time$ from O_s (lines 4–6). If c is a spatial object, probability-threshold closest spatial object is found and the spatial object and the corresponding path are returned (lines 8–12). If $P(q, v).prob_c$ exceeds the probability threshold $\tau.p$, v is pruned. Otherwise, if $P(q, v).time$ exceeds $P(q, c).time + w(c, v) + \frac{1}{v.k}$, $P(q, v).time$ is updated. We also update the probability and time labels of v and put v into O_s (lines 13–19).

4 Experimental Results

We conducted extensive experiments on real and synthetic spatial data to study the performance of TANN queries. The two data sets used in our experiments were the Beijing Road Network (BRN) and Synthetic Spatial Network (SSN), which contain 28,342 vertices and 7,000 vertices. In BRN, we used 100,000 real taxi trajectories to construct a traffic-aware spatial network according to the approaches in [12–14]. In SSN, 10,000 synthetic trajectories were used. All algorithms were implemented in Java and tested on a Linux platform with Intel Core i7-3520M Processor (2.90 GHz) and 8 GB memory. The experimental results were averaged over 10 independent trials with different query inputs. The main metrics were CPU time and the number of visited vertices. The number of visited vertices was selected as a metric because it describes the exact amount of data access. The parameter settings are listed in Table 1.

Table 1. Parameter settings

	BRN	ITN
Time Threshold $\tau.t$	20 − 50 (default 20)	10 − 25 (default 10)
Probability Threshold $\tau.prob$	10 % − 40 % (default 10 %)	10 % − 40 % (default 10 %)
Spatial-Object Density	1 % − 5 % (default 1 %)	1 % − 5 % (default 1 %)

For the purpose of comparison, two baseline algorithms were also implemented. Generally, baseline algorithms have two steps: (1) we make a range query from query source q, and $\tau.t$ and $\tau.prob$ are radiuses for the time-threshold and probability-threshold TANN queries; (2) we refine all spatial objects within the range and find the spatial object with the minimum congestion probability/travel time. The baseline algorithms are denoted by "baseline" in Figs. 2, 3, and 4.

4.1 Effect of Time-Threshold $\tau.t$

Figure 2 shows the effect of time threshold $\tau.t$ on the performance of the developed algorithms. For the baseline algorithm, the larger the time threshold is, the larger search space it may need. Thus the CPU time and the number of visited vertices are expected to be higher when the value of $\tau.t$ increases. For the $TANN_{time}$ algorithm, a larger $\tau.t$ may weaken its pruning effectiveness. In the meantime, a larger $\tau.t$ also leads to more "qualified" spatial objects and it may be easier to find the solution. By integrating these two reasons, the CPU time and visited vertices required by the $TANN_{time}$ algorithm increase very slow when $\tau.t$ increases.

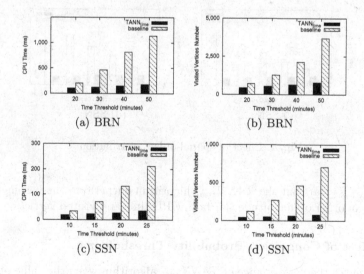

(a) BRN   (b) BRN

(c) SSN   (d) SSN

Fig. 2. Effect of time threshold $\tau.t$

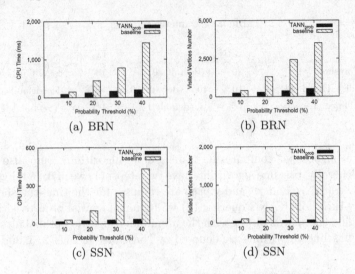

Fig. 3. Effect of congestion probability threshold $\tau.p$

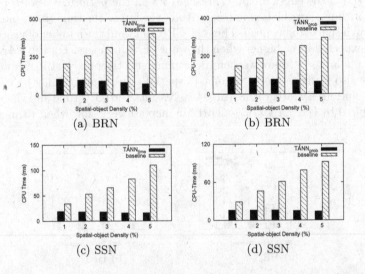

Fig. 4. Effect of spatial-object distribution

In Fig. 2, it is clear that the $TANN_{time}$ algorithm outperform the baseline algorithm by an order of magnitude (for both CPU time and visited vertices).

4.2 Effect of Congestion Probability Threshold $\tau.p$

Figure 3 shows the performance of $TANN_{prob}$ algorithm when the value of probability threshold $\tau.p$ varies. Similar to the $TANN_{time}$ algorithm, the CPU time and the number of visited vertices required by the $TANN_{prob}$ algorithm increase very slow when $\tau.p$ increases, and the $TANN_{prob}$ algorithm outperforms the baseline algorithm by almost an order of magnitude (for both CPU time and visited vertices).

4.3   Effect of Spatial-Object Density

Figure 4 considers the effect of varying spatial-object density under default settings of thresholds. For both $TANN_{time}$ and $TANN_{prob}$ algorithms, the denser the spatial object is, the smaller the required search space is, and thus the queries can be faster. For the baseline algorithms, since their search ranges are fixed, a denser spatial-object distribution will lead to more spatial objects to be refined; thus more CPU time is required. In Fig. 4, both $TANN$ algorithms outperform the baseline algorithms by a factor of 5–10 in terms of both CPU time and visited vertices.

5   Related Work

Nearest Neighbor (NN) queries aim to find objects close to a query source and constitute fundamental functionality in spatial data management. NN query processing may occur in different settings, including in Euclidean spaces [8,16], in spatial networks [6,7], and in higher dimensional spaces [5]. Despite the bulk of literature on nearest neighbor queries, none of the existing work can compute the TANN queries efficiently. In traditional static spatial networks, nearest neighbor queries [6,7,15] do not take instant traffic conditions and traffic congestions into account, while in dynamic spatial networks, time-dependent and probabilistic routing queries [2,4] are based on different traffic models from ours, and traffic-aware unobstructed path queries [12–14] only focus the fastest-path problems and their techniques can hardly be used in the TANN queries.

6   Conclusions and Future Directions

We proposed and studied a novel problem called traffic-aware nearest neighbor (TANN) query, including a time-threshold and a probability-threshold queries. We believe that this type of queries may benefit users in many popular mobile applications, such as discovering nearby points of interest and planning convenient travel routes for users. We define a series of novel probabilistic spatial metrics and develop two efficient algorithms to compute the TANN queries. The performances of TANN queries are verified by extensive experiments on real and synthetic spatial data. In the future, it is of interest to study the TANN query on new metrics that fully integrate travel time and congestion probability and to further extend the TANN queries to top-k queries.

Acknowledgements. This work is partly supported by the National Natural Science Foundation of China (NSFC.61402532, NSFC.41371386, and NSFC.61373147), the Beijing Nova Program, the Science and Technology Planning Project of Fujian Province (No.2016Y0079), and the Open Research Fund Program of Shenzhen Key Laboratory of Spatial Smart Sensing and Services (Shenzhen University).

References

1. Dijkstra, E.W.: A note on two problems in connection with graphs. Numerische Mathematik **1**, 269–271 (1959)
2. Ding, B., Yu, J.X., Qin, L.: Finding time-dependent shortest paths over large graphs. In: EDBT, pp. 205–216 (2008)
3. Guttman, A.: R-trees: a dynamic index structure for spatial searching. In: SIGMOD, pp. 47–57 (1984)
4. Hua, M., Pei, J.: Probabilistic path queries in road networks: traffic uncertainty aware path selection. In: EDBT, pp. 347–358 (2010)
5. Jagadish, H., Ooi, B., Tan, K.-L., Yu, C., Zhang, R.: iDistance: an adaptive B+-tree based indexing method for nearest neighbour search. ACM TODS **30**(2), 364–397 (2005)
6. Jensen, C.S., Kolarvr, J., Pedersen, T.B., Timko, I.: Nearest neighbor queries in road networks. In: ACM GIS, pp. 1–8 (2003)
7. Li, F., Cheng, D., Hadjieleftheriou, M., Kollios, G., Teng, S.-H.: On trip planning queries in spatial databases. In: Medeiros, C.B., Egenhofer, M., Bertino, E. (eds.) SSTD 2005. LNCS, vol. 3633, pp. 273–290. Springer, Heidelberg (2005)
8. Roussopoulos, N., Kelley, S., Vincent, F.: Nearest neighbor queries. In: SIGMOD, pp. 71–79 (1995)
9. Shang, S., Deng, K., Xie, K.: Best point detour query in road networks. In: ACM GIS, pp. 71–80 (2010)
10. Shang, S., Ding, R., Yuan, B., Xie, K., Zheng, K., Kalnis, P.: User oriented trajectory search for trip recommendation. In: EDBT, pp. 156–167 (2012)
11. Shang, S., Ding, R., Zheng, K., Jensen, C.S., Kalnis, P., Zhou, X.: Personalized trajectory matching in spatial networks. VLDB J. **23**(3), 449–468 (2014)
12. Shang, S., Liu, J., Zheng, K., Lu, H., Pedersen, T.B., Wen, J.: Planning unobstructed paths in traffic-aware spatial networks. GeoInformatica **19**(4), 723–746 (2015)
13. Shang, S., Lu, H., Pedersen, T.B., Xie, X.: Finding traffic-aware fastest paths in spatial networks. In: Nascimento, M.A., Sellis, T., Cheng, R., Sander, J., Zheng, Y., Kriegel, H.-P., Renz, M., Sengstock, C. (eds.) SSTD 2013. LNCS, vol. 8098, pp. 128–145. Springer, Heidelberg (2013)
14. Shang, S., Lu, H., Pedersen, T.B., Xie, X.: Modeling of traffic-aware travel time in spatial networks. In: 2013 IEEE 14th International Conference on Mobile Data Management, 3–6 June 2013, Milan, Italy, vol. 1, pp. 247–250 (2013)
15. Shang, S., Yuan, B., Deng, K., Xie, K., Zhou, X.: Finding the most accessible locations: reverse path nearest neighbor query in road networks. In: GIS, pp. 181–190 (2011)
16. Tao, Y., Papadias, D., Shen, Q.: Continuous nearest neighbor search. In: VLDB, pp. 287–298 (2002)
17. Yang, B., Guo, C., Jensen, C.S., Kaul, M., Shang, S.: Stochastic skyline route planning under time-varying uncertainty. In: ICDE, pp. 136–147 (2014)

NERank: Bringing Order to Named Entities from Texts

Chengyu Wang[1], Rong Zhang[1], Xiaofeng He[1(✉)], Guomin Zhou[2], and Aoying Zhou[1]

[1] Institute for Data Science and Engineering,
East China Normal University, Shanghai, China
chywang2013@gmail.com, {rzhang,xfhe,ayzhou}@sei.ecnu.edu.cn
[2] Zhejiang Police College, Hangzhou, Zhejiang Province, China
zhouguomin@zjjcxy.cn

Abstract. Most entity ranking research aims to retrieve a ranked list of entities from a Web corpus given a query. However, entities in plain documents can be ranked directly based on their relative importance, in order to support entity-oriented Web applications. In this paper, we introduce an entity ranking algorithm NERank to address this issue. NERank first constructs a graph model called Topical Tripartite Graph from a document collection. A ranking function is designed to compute the prior ranks of topics based on three quality metrics. We further propose a meta-path constrained random walk method to propagate prior topic ranks to entities. We evaluate NERank over real-life datasets and compare it with baselines. Experimental results illustrate the effectiveness of our approach.

Keywords: Entity ranking · Topic modeling · Topical tripartite graph · Meta-path constrained random walk

1 Introduction

Ranking problems have been extensively studied to bring order to varying types of objects, such as Web pages [1], products [2] and textual units [3]. With the number of entities increasing rapidly on the Web, the problem of Entity Ranking (ER) has drawn much attention. For example, ER tracks have been conducted in INEX and TREC since 2006 and 2009, to rank entities from Web corpora given a query topic [4].

In traditional ER tasks, the rank of entities is measured by the relevance between a query topic (e.g. *impressionist art in the Netherlands* [5]) and entities

A preliminary version of this paper has been presented in WWW'16 [6]. This work is partially supported by NSFC under Grant No. 61402180, the Natural Science Foundation of Shanghai under Grant No. 14ZR1412600, Shanghai Agriculture Science Program (2015) Number 3-2 and NSFC-Zhejiang Joint Fund for the Integration of Industrialization and Informatization under Grant No. U1509219.

F. Li et al. (Eds.): APWeb 2016, Part I, LNCS 9931, pp. 15–27, 2016.
DOI: 10.1007/978-3-319-45814-4_2

with contextual information. In this paper, we consider a different problem: *ranking entities in document collections based on the importance of entities*. For example, given news articles related to *Egypt Revolution* as input, ER aims to retrieve a ranked list of entities that are most relevant to *Egypt Revolution*, including people (e.g., *Hosni Mubarak, Mohamed Morsi*), locations (e.g., *Egypt, Cairo*), organizations (e.g., *Muslim Brotherhood*), etc[1].

The task of ER in this paper is vital for several Web-scale applications:

- **Entity-oriented Web Search:** It facilitates Web entity recommendation, rather than retrieving a list of Web documents that are relevant to the user query but contain abundant or irrelevant information.
- **Web Semantification:** It identifies important entities from Web documents and add semantic tags to the Web automatically.
- **Knowledge Base Population:** It potentially improves the performance of knowledge base population by extracting and ranking entities from the Web and linking them to knowledge bases.

The challenge of ER is that the ranking order of entities should be determined by the contents of the document collection, with no additional knowledge sources or user queries available. Additionally, the importance of entities is expressed implicitly in natural language text, which can not be measured directly. Therefore, it is difficult to extend traditional ER techniques to this scenario.

In this paper, we introduce a graph-based ranking algorithm *NERank* to solve this task. Given a document collection as input, we mine latent topics and model the semantic relations between documents, topics and entities in a graphical model called *Topical Tripartite Graph* (TTG). A ranking function is designed to estimate prior ranks of topics via three quality metrics (i.e., prior probability, entity richness and topic specificity). The prior ranks are propagated along paths in the TTG via a meta-path constrained random walk algorithm. The final rank of entities can be estimated when this process converges.

In summary, we make the following contributions in this paper:

- We formalize the ER problem. A graphical structure TTG is proposed to model the implicit semantic relations between documents, topics and entities.
- A ranking function is designed to calculate the prior ranks of topics based on three quality metrics. We introduce a meta-path constrained random walk algorithm to compute the ranks of entities by rank propagation.
- We conduct extensive experiments and case studies to illustrate the effectiveness of our approach.

The rest of this paper is organized as follows. Section 2 summarizes the related work. We define the ER problem in Sect. 3. The proposed approach is described in Sects. 4 and 5. Experimental results are presented in Sect. 6. We conclude our paper and discuss the future work in Sect. 7.

---

[1] See background info at:
https://en.wikipedia.org/wiki/Egyptian_Revolution_of_2011.

2   Related Work

Research efforts on ER have been put to address the problem of retrieving a ranked list of entities given a query. In the task of ER, entities can be of a certain type, for example, searching for experts in a specific domain [7]. The more general problem is ranking entities of various kinds. Recently, a lot of ER related research has been conducted in the context of INEX and TREC evaluation, started in 2006 and 2009, respectively.

Besides these ER tracks, ER provides a new paradigm to rank and retrieve information at an entity level in the field of Web search. Nie et al. [8] propose a link analysis model PopRank to rank Web "objects" (i.e., entities) within a specific domain, which considers the relevance and popularity of entities. For vertical search, Ganesan et al. [2] leverage online reviews to design several ER models based on user's preference for the purpose of product ranking and recommendation. Lee et al. [9] model multidimensional recommendation as an ER problem, and adopt Personalized PageRank algorithm [10] to rank entities for e-commerce applications.

External data sources are utilized to provide additional information for more accurate ER. Kaptein et al. [4] use the Wikipedia category structure as a pivot to identify key entities properly. They reduce the problem of Web ER to Wikipedia ER. Ilieva et al. [11] make use the rich attribute information in knowledge bases to improve the coverage and quality of ER. For short text analysis, Meij et al. [12] apply learning-to-rank models to extract key concepts in tweets and link to Wikipedia. However, none of the prior work considers the ER task in this paper. With entities in documents ranked correctly, a series of Web applications can be benefitted to provide entity-oriented service.

3   Entity Ranking Problem

According to the task setting of ER, we take a collection of documents (denoted as D) as input. Let $m \in M$ denote an entity mention in document $d \in D$, recognized by Named Entity Recognition (NER) techniques. Because entity mentions appeared in the plain texts are unnormalized, simply ranking on M will result in the "unnormalized ranking" issue. Consider the example in Table 1. Both

Table 1. Comparison between unnormalized and normalized ranking.

Unnormalized ranking		Normalized ranking	
Entity mention	Rank value	Normalized entity	Rank value
Egypt	0.25	Hosni Mubarak	0.35
Mubarak	0.2	Egypt	0.25
Hosni Mubarak	0.15	Cario	0.1
Cario	0.1	...	...

"Hosni Mubarak" and "Mubarak" refer to the former Egypt president "Hosni Mubarak". If they are unnormalized, they will receive separate, inconsistent and under-estimated rank values.

Therefore, we employ an entity normalization procedure to map each $m \in M$ to its normalized form $e \in E$. To accomplish the task of ER, we assign each entity $e \in E$ a rank $r(e)$ to represent the relative importance in D. For the illustration purpose, the high-level process of ER is presented in Fig. 1. We also provide a simple example to show the data processing steps of ER. Here, we present the definition of ER formally as follows:

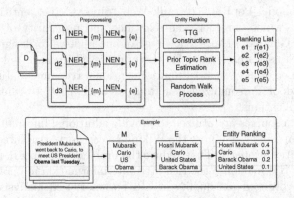

Fig. 1. Illustration of the ER process and a simple example.

Definition 1. *Entity Ranking. Given a document collection D and a normalized named entity collection E detected from D, the goal is to give each entity $e \in E$ a rank $r(e)$ to denote the relative importance such that (1) $0 \leq r(e) \leq 1$ and (2) $\sum_{e \in E} r(e) = 1$.*

The task definition in this paper is similar to the task *Ranked-concepts to Wikipedia* (Rc2W) [13] and the more general task *Ranked-concepts to Knowledge Base* (Rc2KB) in the entity annotator benchmark GERBIL [14]. We notice that both tasks are comprised of two sub-steps: (i) entity linking (which maps an entity mention to an entity in knowledge base) and (ii) ranking (which generates the ranked order of entities). While much of the previous work addressed the task of entity linking, we focus more on ER, which is not sufficiently studied. Another difference is that since existing knowledge bases still face the incompleteness issue, we do not require entities to be linked to Wikipedia or a knowledge base before they can be ranked.

4   Topical Tripartite Graph Modeling

The key for accurate ER is to mine the implicit semantic relations between documents and entities. Extracting language patterns that can help identify important entities from texts is difficult, due to the flexibility and complexity in expression of natural languages. However, by topic modeling, the gap between documents and entities can be bridged. In this section, we introduce the formal definition of TTG and show the construction process of the graph in detail.

Topical Tripartite Graph. The TTG is a tripartite graph to model the semantic relations among <document, topic> and <topic, entity> pairs. There are three types of nodes (i.e., documents D, topics T and (normalized) entities E), and two types of weighted, undirected edges (i.e., document-topic edges R_{DT} and topic-entity edges R_{TE}). Here, we give the formal definition of TTG as follows:

Definition 2. *Topical Tripartite Graph. A TTG w.r.t. document collection D is a weighted, tripartite graph $G_D = (D, T, E, R_{DT}, R_{TE})$. The nodes of the graph are partitioned into three disjoint sets: documents D, topics T and entities E. R_{DT} and R_{TE} are edge sets that connect nodes in $< D, T >$ and $< T, E >$ pairs, respectively.*

Additionally, weights of edges in TTG can be employed to quantify the degrees of relation strength. In this paper, we employ a weight $w_{dt}(d_i, t_j) \in (0, 1)$ for an edge $(d_i, t_j) \in R_{DT}$ and $w_{te}(t_i, e_j) \in (0, 1)$ for an edge $(t_i, e_j) \in R_{ET}$.

Graph Construction. The TTG construction process includes two parts: (1) *named entity recognition and normalization*, which discovers and normalizes named entities in document collections, and (2) *entity-aware topic modeling*, which mines the latent topics in documents and calculates the weights of edges in TTG. The general construction process of TTG is shown in Fig. 2.

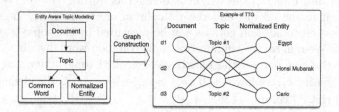

Fig. 2. Illustration of the TTG construction process.

Named Entity Recognition and Normalization. Entities in documents can be automatically recognized by the NER tagger, such as Conditional Random Fields [15]. Before we construct the TTG, entity normalization is necessary to transform entity mentions recognized by NER to normalized forms. Take the sentence "Mubarack was an former political leader in Egypt" as an example. It is processed by an NER tagger, shown as follows:

Mubarack#PERSON was a political leader in *Egypt*#LOCATION

After entity normalization, the tagging sequence becomes:

Hosni Mubarack#PERSON was a political leader in *Egypt*#LOCATION

In this paper, we employ the algorithm proposed by Jijkoun et al. [16] for entity normalization, which applies approximate name matching, identification of missing references and name disambiguation techniques. Due to space limitations, we omit the details here.

Entity Aware Topic Modeling. Topic models such as LDA [17] can model the latent topics in documents. However, LDA models a document using the "bag-of-words" model, without taking multi-word entities or unnormalized entity mentions into consideration. To better fit the ER task, we introduce an *entity aware topic modeling* approach, which models documents as a collection of textual units, consisting of *normalized entities* and *common words* (denoted as W). Additionally, we remove stop words and punctuations in these documents. Following the previous example, given sentence "Mubarack was a political leader in Egypt", we treat "Hosni Mubarack" and "Eqypt" as normalized entities, and "political" and "leader" as common words.

With entity normalization and preprocessing steps, LDA is employed to model the document-topic distributions Θ (represented as a $|D| \times |T|$ matrix) and the topic-textual unit distributions Φ ($|T| \times |E \cup W|$ matrix) given the document collection D. In Table 2, we present some topics we discovered in the document collection w.r.t *Egypt Revolution*. We also manually add a description of each topic to illustrate that this approach is effective to detect latent aspects in the document collection and model the relations between topics and entities.

The weights of edges in TTG are assigned based on distributions of entity aware topic modeling. If the probability of a topic is high in a document, it means the topic and the document have strong semantic associativity. Therefore, for document d_i and topic t_j, the weight is defined as $w_{dt}(d_i, t_j) = \theta_{i,j}$ where $\theta_{i,j}$ is the element in the ith row and the jth column of Θ. Similarly, the semantic relations between topics and entities can be measured by the topic-textual unit distribution. We remove the columns for topic-common word distributions in Φ, and denote the rest part of the matrix as $\hat{\Phi}$ (called topic-entity matrix). For topic t_i and entity e_j, we have $w_{te}(t_i, e_j) = \hat{\phi}_{i,j}$ where $\hat{\phi}_{i,j}$ is the element in the ith row and the jth column of $\hat{\Phi}$.

Table 2. Topics discovered in document collection w.r.t. *Egypt Revolution*.

Topic	Top normalized entities	Top common words	Description
#1	Egypt, Hosni Mubarak	political, military, revolution	Start of the revolution
#2	Mohamed Morsi, Egypt	President, constitution, vote	Presidential election
#3	Egypt, Israel, Iran	government, foreign, peace	Foreign countries' reaction
#4	Egypt, Cairo	economic, government, billion	Revolution's effect on economy
#5	Egypt	tourism, tourist, travel, sea	Revolution's effect on tourism

5   Entity Ranking Algorithm

In this section, we present our *NERank* algorithm. Based on TTG, we compute the prior ranks of topics in combination of three quality metrics. After that, a meta-path constrained random walk algorithm is proposed to calculate the ranks of entities by propagating the prior topic ranks to entities over the TTG.

5.1 Prior Topic Rank Estimation

Without additional knowledge sources, it is difficult to determine the relative importance of documents and entities. In contrast, entity aware topic modeling can provide prior knowledge about topics. For example, in Table 2, we can see that Topic #1 and Topic #2 are directly about major events in *Egypt Revolution* and Topics #3-#5 discuss different aspects related to *Egypt Revolution*, but are less relevant. To facilitate ER, we design the following quality metrics to calculate the prior ranks of topics.

Prior Probability. Different topics have different probabilities to be discussed in documents. Some topics are related to more documents in D (e.g. Topic #1 in Table 2), while others are only related to only a few articles (e.g. Topic #5). We define the prior probability $pr(t_i)$ of each topic $t_i \in T$ using document-topic distributions as $pr(t_i) = \frac{1}{|D|} \sum_{j=1}^{|D|} \theta_{j,i}$. Because $\sum_{j=1}^{|D|} \sum_{t=1}^{|T|} \theta_{j,i} = |D|$, $|D|$ is served as a normalization factor.

Entity Richness. Entity richness measures the "goodness" of a topic from an entity aspect. As entities play an important role in documents, the "richness" of entities is a useful signal to measure the quality of topics. Here, we compute the "richness" as the sum of all probabilities of entities given topic t_i, i.e., $\sum_{j=1}^{|E|} \hat{\phi}_{i,j}$. Therefore, the entity richness score for topic t_i is defined as: $er(t_i) = \frac{1}{Z_{er}} \sum_{j=1}^{|E|} \hat{\phi}_{i,j}$ where $Z_{er} = \sum_{m=1}^{|T|} \sum_{n=1}^{|E|} \hat{\phi}_{m,n}$ is a normalization factor.

Topic Specificity. Topic specificity measures the quality of a topic in an information theoretic approach. Based on the analysis on entities and common words in each topic, we obverse that some topics are specific about some events or latent aspects, while others only provide background information. We extract all probabilities of topic t_i in all $d \in D$ as a $|D|$-dimensional vector $< \theta_{1,i}, \theta_{2,i}, \cdots, \theta_{|D|,i} >$. The unnormalized "specificity" of topic t_i can be computed as $ts'(t_i) = \sum_{j=1}^{|D|} \theta_{j,i} \log_2 \theta_{j,i}$.

High "specificity" value means that there is no significant "burst" in topic distributions, which filters out topics that are only strongly related to few documents. However, if a topic rarely appears in any documents, it may receive a relatively high "specificity" score. In the implementation, we add a heuristic rule to avoid this problem: if the prior probability $pr(t_i)$ is smaller than a small threshold ϵ, we set $ts(t_i) = 0$. Hence, the topic specificity of t_i is defined as:

$$ts(t_i) = \begin{cases} 0 & pr(t_i) < \epsilon \\ \frac{1}{Z_{ts}} \sum_{j=1}^{|D|} \theta_{j,i} \log_2 \theta_{j,i} & pr(t_i) \geq \epsilon \end{cases}$$

where $Z_{ts} = \sum_{i=1}^{|T|} ts(t_i)$ is a normalization factor.

Ranking Function. Combined the three quality metrics together, we can generate a feature vector for each topic $t_i \in T$, i.e., $\boldsymbol{F}(t_i) =< pr(t_i), er(t_i), ts(t_i) >$.

Denote W as the weight vector where each element in W gives different importance for different features such that $\forall w_i > 0$ and $\sum_i w_i = 1$. Thus, the prior rank for topic t_i is defined as $r_0(t_i) = W^T \cdot F(t_i)$.

To learn the weights W for the features, we employ the max-margin technique introduced in [18]. Given two topics t_i and t_j, if t_i is a more important topic than t_j, judged by human annotators, we have $r_0(t_i) > r_0(t_j)$. This implies that the following constraint holds: $W^T \cdot F(t_i) - W^T \cdot F(t_j) \geq 1 - \xi_{i,j}$ where $\xi_{i,j} \geq 0$ is a slack variable. This learning problem can be modeled as training a linear SVM classifier with the objective function $\|W\|_2^2 + C \cdot \sum_{i,j} \xi_{i,j}$, where C is a tolerance parameter.

5.2   Meta-Path Constrained Random Walk Algorithm

With prior ranks of topics estimated, we aim to propagate ranks to other nodes in order to obtain entity ranks. Based on the graphical structure of TTG, we design a meta-path constrained random walk algorithm to rank entities.

In a TTG, we observe that (1) only topic nodes are connected with different types of nodes (i.e., documents and entities) and (2) we only have prior knowledge about ranks of topics. Thus, we define topic-centric meta-paths to constrain the behavior of random walkers. Denote $x \to y$ as the action where the random surfer walks from x to y. We define two types of meta-paths to embed the semantics of document-topic and topic-entity relations, shown as follows:

Definition 3. TDT Meta-path. *A TDT meta-path is a path defined over a TTG G_D which has the form $t_i \to d_j \to t_k$ where $t_i, t_k \in T$ and $d_j \in D$.*

Definition 4. TET Meta-path. *A TET meta-path is a path defined over a TTG G_D which has the form $t_i \to e_j \to t_k$ where $t_i, t_k \in T$ and $e_j \in E$.*

TDT meta-paths encode the mutual enforcement effect between ranks of documents and topics. The assumption is that "good" documents relate to "good" topics and vice versa. TET meta-paths update the ranks of entities and pass the rank back to topic nodes for the next iteration of random walk.

Because random walk algorithms in meta-paths are effective for inference based on previous research [19], we compute the ranks of entities by meta-path constrained random walk. To better fit the graphical structure of a topical tripartite graph, we require that the random surfer is only allowed to walk along TDT and TET meth-paths. To specify, the random surfer begins by selecting a topic node $t_i \in T$ with probability $r_0(t_i)$ (i.e., the prior rank of t_i) as the starting point. Next, the surfer makes the transfer along TDT and TET meta-paths. Denote α and β as tuning parameters where $\alpha > 0$, $\beta > 0$ and $\alpha + \beta < 1$. One iteration of the random walk process is shown as follows:

- With probability α, the random surfer walks through a TDT meta-path $t_i \to d_j \to t_k$. d_j is selected with probability $\frac{\theta_{j,i}}{\sum_{d_k \in D} \theta_{k,i}}$ for all $d_j \in D$. Next, t_k is selected with probability $\theta_{j,k}$ for all $t_k \in T$.

- With probability β, the random surfer walks through a TET meta-path $t_i \rightarrow e_j \rightarrow t_k$. e_j is selected with probability $\frac{\hat{\phi}_{i,j}}{\sum_{e_k \in E} \hat{\phi}_{i,k}}$ for all $e_j \in E$. Next, t_k is selected with probability $\frac{\hat{\phi}_{k,j}}{\sum_{t_m \in T} \hat{\phi}_{m,j}}$ for all $t_k \in T$.
- With probability $1 - \alpha - \beta$, the random surfer jumps to a topic node t_j. t_j is selected with probability $r_0(t_j)$ for all $t_j \in T$.

This random walk process can be repeated iteratively until the system reaches equilibrium. Each entity node e_i will receive a score $s(e_i)$, indicating the number of visits by random surfers. Thus, the rank of an entity e_i is computed as $r(e_i) = \frac{s(e_i)}{\sum_{e_j \in E} s(e_j)}$. In the appendix, we will prove that this process will converge after a sufficient number of iterations, and give the close-form solution of *NERank*.

6 Experiments

In this section, we conduct extensive experiments on news datasets to evaluate the performance of *NERank*. We also compare our method with baselines to make the convincing conclusion.

6.1 Datasets and Experimental Settings

We use two publicly available news datasets in our experiments (i.e., **Timeline-Data** [20] and **CrisisData** [21]), described as follows:

- **TimelineData** - The dataset has 4,650 news articles that are related to 17 international events, such as *BP oil spill, death of Michael Jackson*, etc. Each group of news articles belongs to a news agency, such as BBC and CNN.
- **CrisisData** - The dataset contains 15,534 news articles that report four recent armed conflicts, including *Egypt Revolution, Syria War*, etc. These articles are from 24 news agencies, obtained using Google search engine.

To generate document collections, we randomly sample 100 documents from news articles related to the same event at each time. In total, we have 34 document collections from **TimelineData** and 16 from **CrisisData**. We conduct separate experiments on all document collections in the following experiments.

6.2 Experimental Results and Analysis

Ground Truth. The document collections used in this paper are all related to news events. For ground truth, we first obtain the news summaries of each document collection from [20, 21], which are manually created by professional journalists. Based on the event summaries, we recruit a group of CS graduates to label the entities as "most important", "important", "relevant", etc. Following the evaluation framework in [22], we finally have a ranked list of 15

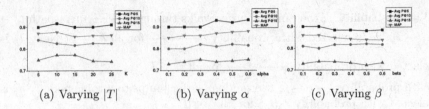

(a) Varying $|T|$ (b) Varying α (c) Varying β

Fig. 3. Evaluation results under different parameter settings.

entities w.r.t. a document collection by majority voting, which are regarded as "key" entities.

Evaluation Metrics. To evaluate different algorithms for ER, we compare the top-k entities generated by machines with the ground truth. We employ Precision@K ($K = 5, 10, 15$) and Average Precision as evaluation metrics. For multiple document collections, we take the average as results and report Average Precision@K and MAP in this paper. To compare *NERank* with baselines, we additionally use *paired t-test* to evaluate the level of statistical significance.

Parameter Settings. We tune parameters in *NERank*, namely, number of topics in LDA ($|T|$) and parameters for random walk (α and β). In Fig. 3, we present the experimental results when we vary only one parameter at each time. In Fig. 3(a), we fix $\alpha = \beta = 0.4$ and change the number of topics. It can be seen that although it is relatively hard to determine the number of topics, the performance of *NERank* is not sensitive to this issue. In Fig. 3(b) and (c), we set $|T| = 10$ and one parameter (α or β) to be 0.4 and vary the other. It shows that our algorithm is not sensitive to the change of parameters. Note that the weight vector W in the ranking function can be learned automatically and does not need to be tuned. We manually label 500 topic pairs to train the ranking model, and set $|T| = 10$ and $\alpha = \beta = 0.4$ in following experiments.

Method Comparison. To our knowledge, there is no prior work concerning ranking entities directly from document collections. However, there are abundant research on keyword extraction. In this paper, we take unsupervised keyword extraction methods as baselines. We first generate a ranked list of words using baselines and filter out common words in the list to produce the ranked list of entities. We also implement two variants of our approach, shown as follows:

- **TF-IDF** - rank entities based on TF-IDF scores.
- **TextRank** [3] - a graph-based iterative algorithm for textual unit ranking.
- **LexRank** [23] - a graph-based algorithm based on lexical centrality.
- **Kim et al.** [24] - a keyword extraction algorithm based on semantic similarity between words.
- **NERank$_{Uni}$** - the variant of *NERank* which sets prior topic ranks uniformly.
- **NERank$_{\alpha=0}$** - the variant of *NERank* which sets $\alpha = 0$ in random walk and thus ignores the semantic relatedness between documents and topics.

Table 3. Evaluation results for different methods. ($\star$: p-value$\leq$0.05)

Method	Average Precision@5	Average Precision@10	Average Precision@15	MAP
TF-IDF	0.85*	0.79*	0.73*	0.81*
TextRank	0.87*	0.83	0.73*	0.83*
LexRank	0.85*	0.8*	0.72*	0.8*
Kim et al.	0.87*	0.81*	0.76*	0.84*
NERank$_{Uni}$	0.80*	0.75*	0.71*	0.78*
NERank$_{\alpha=0}$	0.72*	0.61*	0.51*	0.62*
NERank	**0.92**	**0.87**	**0.79**	**0.89**

The results are shown in Table 3. We can see our method outperforms base-lines *TF-IDF*, *TextRank*, *LexRank* and *Kim et al.* and *TextRank* because these classical methods mostly capture the statistical characteristics of words and do not exploit the latent topics in document collections. The comparison between the variants and *NERank* shows that our topic rank function and meta-path constrained random walk algorithm are effective to boost the performance of ER. The results of paired t-test between *NERank* and baselines confirm that our method outperforms other approaches.

Case Study. We present the ER results of four events generated by our approach. Due to space limitation, we only present top-10 entities shown in Table 4. It can be seen that our approach can extract and rank entities from documents effectively.

Table 4. Top-10 entities of documents related to three events.

Entity	Egypt Revolution	Libya War	BP Oil Spill
1	Egypt	Libya	BP
2	Mohamed Morsi	Muammar Gaddafi	Gulf of Mexico
3	Hosni Mubarak	Tripoli	Barack Obama
4	Cario	NATO	Louisiana
5	Muslim Brotherhood	Benghazi	Coast Guard
6	Tahrir Square	Barack Obama	United States
7	Israel	Misrata	Tony Hayward
8	Middle East	United States	Deepwater Horizon
9	United States	National Transitional Council	Florida
10	Tunisia	Syria	Transocean

7 Conclusion and Future Work

In this paper, we formalize and address the problem of entity ranking. We design a TTG model to represent the semantic relations between documents, topics and entities. A meta-path constrained random walk algorithm is proposed to calculate the ranks of entities after estimating the prior ranks of topics by three quality metrics. The experimental results on two datasets demonstrate that the proposed approaches achieve accurate results. In the future, we will explore the task of joint entity linking and ranking for knowledge base population.

Appendix: Mathematical Analysis of *NERank*

We prove that the random walk algorithm of *NERank* will converge and derive the close-form solution. Let $\mathbf{T_n}$ denote the $|T| \times 1$ matrix which represents the ranks of topics in the nth iteration. Specially, $\mathbf{T_0}$ is the prior rank matrix for topics. Let $\mathbf{E_n}$ denote the $|E| \times 1$ entity rank matrix in the nth iteration. Based on the random walk process, the rank update of topics for TDT meta-path is formulated as: $\mathbf{T_n} = \mathbf{\Theta_R}^T \mathbf{\Theta} \cdot \mathbf{T_{n-1}}$ where $\mathbf{\Theta_R}$ is the row-normalized matrix of $\mathbf{\Theta}$. Similarly, for TET meta-path, we have $\mathbf{T_n} = \hat{\mathbf{\Phi}}_\mathbf{C} \hat{\mathbf{\Phi}}_\mathbf{R}^T \cdot \mathbf{T_{n-1}}$ where $\hat{\mathbf{\Phi}}_\mathbf{R}$ and $\hat{\mathbf{\Phi}}_\mathbf{C}$ are the row-normalized and column-normalized matrices of $\hat{\mathbf{\Phi}}$, respectively. The update rule in one iteration is formulated as:

$$\mathbf{T_n} = \alpha \cdot \mathbf{\Theta_R}^T \mathbf{\Theta} \cdot \mathbf{T_{n-1}} + \beta \cdot \hat{\mathbf{\Phi}}_\mathbf{C} \hat{\mathbf{\Phi}}_\mathbf{R}^T \cdot \mathbf{T_{n-1}} + (1 - \alpha - \beta) \cdot \mathbf{T_0}$$

For simplicity, we define $\mathbf{M} = \alpha \cdot \mathbf{\Theta_R}^T \mathbf{\Theta} + \beta \cdot \hat{\mathbf{\Phi}}_\mathbf{C} \hat{\mathbf{\Phi}}_\mathbf{R}^T$. By iteration, we have $\mathbf{T_n} = \mathbf{M}^n \cdot \mathbf{T_0} + (1 - \alpha - \beta) \cdot \sum_{i=0}^{n-1} \mathbf{M}^i \cdot \mathbf{T_0}$. Because $\lim_{n \to \infty} \mathbf{M}^n = \mathbf{0}$ and $\lim_{n \to \infty} \sum_{i=0}^{n-1} \mathbf{M}^i = (\mathbf{I} - \mathbf{M})^{-1}$, the limit of matrix series $\{\mathbf{T_n}\}$ is derived as:

$$\lim_{n \to \infty} \mathbf{T_n} = \lim_{n \to \infty} \mathbf{M}^n \cdot \mathbf{T_0} + (1 - \alpha - \beta) \lim_{n \to \infty} \sum_{i=0}^{n-1} \mathbf{M}^i \cdot \mathbf{T_0} = (1 - \alpha - \beta)(\mathbf{I} - \mathbf{M})^{-1} \mathbf{T_0}$$

where $\mathbf{I}$ is the $|T| \times |T|$ identity matrix. Therefore, the ranks of topics will converge in *NERank*. Because the rank of entities $\mathbf{E_n}$ can be computed by $\mathbf{E_n} = \hat{\mathbf{\Phi}}_\mathbf{R}^T \cdot \mathbf{T_n}$. Denote $\mathbf{E}^*$ as the close form solution vector for entity ranks. We have

$$\mathbf{E}^* = (1 - \alpha - \beta) \cdot \hat{\mathbf{\Phi}}_\mathbf{R}^T (\mathbf{I} - \alpha \cdot \mathbf{\Theta_R}^T \mathbf{\Theta} - \beta \cdot \hat{\mathbf{\Phi}}_\mathbf{C} \hat{\mathbf{\Phi}}_\mathbf{R}^T)^{-1} \cdot \mathbf{T_0}$$

where the rank of entity e_i (i.e., $r(e_i)$) is the ith element in $\mathbf{E}^*$.

References

1. Brin, S., Page, L.: The anatomy of a large-scale hypertextual web search engine. Comput. Netw. **30**(1–7), 107–117 (1998)
2. Ganesan, K., Zhai, C.: Opinion-based entity ranking. Inf. Retr. **15**(2), 116–150 (2013)

3. Mihalcea, R., Tarau, P.: Textrank: bringing order into text. In: EMNLP, pp. 404–411 (2004)
4. Kaptein, R., Serdyukov, P., de Vries, A.P., Kamps, J.: Entity ranking using wikipedia as a pivot. In: CIKM, pp. 69–78 (2010)
5. de Vries, A.P., Vercoustre, A.-M., Thom, J.A., Craswell, N., Lalmas, M.: Overview of the INEX 2007 entity ranking track. In: Fuhr, N., Kamps, J., Lalmas, M., Trotman, A. (eds.) INEX 2007. LNCS, vol. 4862, pp. 245–251. Springer, Heidelberg (2008)
6. Wang, C., Zhang, R., He, X., Zhou, A.: NERank: ranking named entities in document collections. In: WWW, pp. 123–124 (2016)
7. Balog, K., de Rijke, M.: Determining expert profiles (with an application to expert finding). In: IJCAI, pp. 2657–2662 (2007)
8. Nie, Z., Zhang, Y., Wen, J., Ma, W.: Object-level ranking: bringing order to web objects. In: WWW, pp. 567–574 (2005)
9. Lee, S., Song, S., Kahng, M., Lee, D., Lee, S.: Random walk based entity ranking on graph for multidimensional recommendation. In: RecSys, pp. 93–100 (2011)
10. Haveliwala, T.H.: Topic-sensitive pagerank. In: WWW, pp. 517–526 (2002)
11. Ilieva, E., Michel, S., Stupar, A.: The essence of knowledge (bases) through entity rankings. In: CIKM, pp. 1537–1540 (2013)
12. Meij, E., Weerkamp, W., de Rijke, M.: Adding semantics to microblog posts. In: WSDM, pp. 563–572 (2012)
13. Cornolti, M., Ferragina, P., Ciaramita, M.: A framework for benchmarking entity-annotation systems. In: WWW, pp. 249–260 (2013)
14. Usbeck, R., Röder, M., Ngomo, A.N., Baron, C., Both, A., Brümmer, M., Ceccarelli, D., Cornolti, M., Cherix, D., Eickmann, B., Ferragina, P., Lemke, C., Moro, A., Navigli, R., Piccinno, F., Rizzo, G., Sack, H., Speck, R., Troncy, R., Waitelonis, J., Wesemann, L.: GERBIL: general entity annotator benchmarking framework. In: WWW, pp. 1133–1143 (2015)
15. Finkel, J.R., Grenager, T., Manning, C.D.: Incorporating non-local information into information extraction systems by gibbs sampling. In: ACL (2005)
16. Jijkoun, V., Khalid, M.A., Marx, M., de Rijke, M.: Named entity normalization in user generated content. In: AND, pp. 23–30 (2008)
17. Blei, D.M., Ng, A.Y., Jordan, M.I.: Latent dirichlet allocation. J. Mach. Learn. Res. **3**, 993–1022 (2003)
18. Shen, W., Wang, J., Luo, P., Wang, M.: LINDEN: linking named entities with knowledge base via semantic knowledge. In: WWW, pp. 449–458 (2012)
19. Lao, N., Cohen, W.W.: Relational retrieval using a combination of path-constrained random walks. Mach. Learn. **81**(1), 53–67 (2010)
20. Tran, G.B., Alrifai, M., Nguyen, D.Q.: Predicting relevant news events for timeline summaries. In: WWW, pp. 91–92 (2013)
21. Tran, G., Alrifai, M., Herder, E.: Timeline summarization from relevant headlines. In: Hanbury, A., Kazai, G., Rauber, A., Fuhr, N. (eds.) ECIR 2015. LNCS, vol. 9022, pp. 245–256. Springer, Heidelberg (2015)
22. Zaragoza, H., Rode, H., Mika, P., Atserias, J., Ciaramita, M., Attardi, G.: Ranking very many typed entities on wikipedia. In: CIKM, pp. 1015–1018 (2007)
23. Erkan, G., Radev, D.R.: Lexrank: Graph-based lexical centrality as salience in text summarization. J. Artif. Intell. Res. (JAIR) **22**, 457–479 (2004)
24. Kim, Y., Kim, M., Cattle, A., Otmakhova, J., Park, S., Shin, H.: Applying graph-based keyword extraction to document retrieval. In: IJCNLP, pp. 864–868 (2013)

FTS: A Practical Model for Feature-Based Trajectory Synthesis

Jiapeng Li, Wei Chen, An Liu, Zhixu Li, and Lei Zhao[✉]

School of Computer Science and Technology, Soochow University, Suzhou, China
trajepl@gmail.com, wchzhg@gmail.com, {anliu,zhixuli,zhaol}@suda.edu.cn

Abstract. Driven by the GPS-enabled devices and wireless communication technologies, the researches and applications on spatio-temporal databases have received significant attention during the past decade. Hence, large trajectory datasets are extremely necessary to test high performance algorithms for these applications and researches. However, real-world datasets are not accessible in many cases due to privacy concerns and business competition. For this reason, we propose a practical model FTS to generate new trajectories in this work. We generate new trajectories based on features extracted from original dataset and validate the result by comparing the features of generated trajectories with the given dataset finally.

1 Introduction

With the ubiquitous of positioning technologies and the wide use of mobile devices, we benefit from various types of location based services(LBSs), such as online navigation, route planning and supply chain management, which inspire great efforts made on trajectory data analysis. In existing work, [1–3] study the problem of finding patterns from historical trajectories, [4,5] pay attention to design index structures and algorithms to support efficient trajectory search, [6,7] focus on trajectory clustering and [8,9] make contributions on privacy protection while searching trajectories.

Commonly, large datasets are necessary for these applications and researches, as the efficiency, effectiveness and scalability of the developed index structures and algorithms should be tested on large datasets. Without large datasets, the experimental results are not convincing and interpretable. However, in real life, it is difficult to obtain real-world datasets, which are very important to its owners, mainly due to privacy concerns and business competition. Having observed this, previous researchers make great efforts on data generation [10,11], where new trajectories are generated by the proposed models, and these trajectories are composed of new spatial locations. So, in a sense, it is difficult to keep the features of generated dataset consistent with the original data. The proposed model FTS can benefit lots of users, especially for people who are lack of large trajectory dataset. We can synthesize amounts of new trajectories, the features of which are similar to the given trajectory dataset.

© Springer International Publishing Switzerland 2016
F. Li et al. (Eds.): APWeb 2016, Part I, LNCS 9931, pp. 28–40, 2016.
DOI: 10.1007/978-3-319-45814-4_3

Despite the wide use of FTS, the synthesis of new trajectories turns out to be a challenging problem due to following reasons. (1) The data sparsity problem [12], which is inevitable in real life. Most of sample points fall into the main roads of a city, only a portion of the city is covered. Even paths with the same start and destination may vary on their trajectory locations due to the different sample rates. (2) It should be proved that the features of the synthesized data are similar to those of given dataset. However, the diversity of these features leads to the great difficulty to achieve the goal.

To tackle the above-mentioned problems, novel approaches are proposed in this study. First of all, we gather the features of historical trajectories, such as speed, acceleration, and U-turn. The details will be discussed in Sect. 4. Secondly, in order to avoid the data sparsity problem, we apply the grid map [12] to reconstruct historical trajectories. The map consists of $g \times g$ quad cells and all points in a grid cell are considered to be the same object. Then, a trajectory can be represented as a sequence of cells according to the locations of the trajectory. Next, we decompose all historical trajectories into sub-trajectories in their intersection, then we can use them to synthesize new trajectories. With the goal of making the features of new dataset consistent with those of original dataset, we propose a heuristic function during the synthesis, where all features of the given trajectories are taken into account. Thirdly, in real scenario, a trajectory is usually composed of several parts, which can be divided into some sub-trajectories with similar features. To maintain the features for each part, we design a novel approach Gene Partition (GP) to divide historical trajectories.

To sum up, we make the following contributions in this study.

- We propose a novel model FTS to synthesize trajectories based on the features of the original trajectory dataset.
- Novel functions and algorithms are proposed to synthesize trajectories, where all features are taken into account during the process. In addition, we use the Gene Partition (GP) to decompose the given trajectories, with the goal of reducing time cost and achieving higher performance.
- We conduct extensive experiments based on real datasets to study the performance of FTS by comparing the features of generated dataset with that of given dataset. The experimental results demonstrate the high performance of the proposed algorithms.

The rest of paper is organized as follows. The related work is presented in Sect. 2. In Sect. 3, we formulate the problem, which is followed by the feature extraction in Sect. 4. The baseline methods are introduced in Sect. 5. We optimize the proposed algorithms in Sect. 6 and report the experimental results in Sect. 7. This paper is concluded in Sect. 8.

2   Related Work

In this section, we present existing work related to our contribution and focus on the discussion of several representative generators.

Data generation has received great attention in the past decade, since data is the basis of many applications and it is hard to obtain real datasets for many people. There has been a lot of work on data generation [10, 13–17].

On the Generation of Spatiotemporal Datasets (GSTD) [13] is proposed to generate time evolving point or rectangular objects. The algorithm begins with three two-dimensional initial distributions, namely a uniform, a Gaussian and a skewed data distribution. The GSTD is extended in [10] to create more realistic trajectory data of moving objects for simulating real-world behaviors. Compared with GSTD, it takes more properties into account, such as preferred direction, clustered movement and obstructed movement.

Saglio et al. [14] develop another important generator for spatial-temporal data, i.e., Oporto: a realistic scenario generator for moving objects, which exploits the scenario with fishing ships moving in the direction of the shoals of fish and try to avoid storm areas. This generator is capable of creating unrestricted moving points and moving regions, which is composed of four main types of objects, i.e., harbors, fishing ships, spots and shoals. The result of Oporto is sufficient to cover the dynamics of the real scenarios, since the benchmark of which is the representative of real world applications.

Brinkhoff et al. propose a well-known network-based data generator in [15]. The basic idea of the generator is that moving real-world objects often follow a network, which is an important property of moving objects and is not considered by previous approaches. The behavior of this generator can be controlled by users with predefined functions and parameters. Especially, a moving object is created when its destination or the maximum time is reached.

Another important data generator is BerlinMOD: a benchmark for moving object databases [17], which presents a benchmark to design scalable and representative moving object data. The benchmark focuses on range queries and its ability has been demonstrated by presenting a preliminary extension handling various nearest neighbor queries. During the design of this benchmark, two different types of MOD are presented, i.e., object-based data and trip-based data.

3 Problem Statement

3.1 Preliminary

Due to the data sparsity problem, we apply the grid representation in this study, which is used in [12], where all locations in a grid cell are considered to be the same object, and each historical trajectory is denoted as a sequence of grid cells. In Fig. 1, the historical trajectory τ_2 can be represented as $(13, 10, 11, 12)$.

3.2 Problem Definition

Definition 1 Trajectory. *Let $p = (x, y, t)$ be a GPS sample point where x is the longitude, y is the latitude and t denotes the timestamp. A trajectory is a*

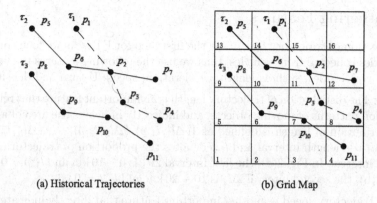

(a) Historical Trajectories   (b) Grid Map

Fig. 1. Grid representation

sequence of sample points, denoted as $\tau = (p_1, p_2, \ldots, p_n)$. The grid representation of τ is defined as $g(\tau) = (q_1, q_2, \ldots, q_m)$, where q_i is the grid representation of p_i, denoted as $q_i = g(p_i)$.

Definition 2 Crossing Trajectories. *Given two trajectories τ_1 and τ_2, $g(\tau_1) = (q_1, q_2, \ldots, q_n)$ and $g(\tau_2) = (q'_1, q'_2, \ldots, q'_m)$, τ_1 and τ_2 are considered as the crossing trajectories if there exists $q_i = q'_j (1 \leq i \leq n, 1 \leq j \leq m)$ such that $g(\tau_1) \cap g(\tau_2) \neq \phi$.*

Consider the example in Fig. 1, τ_1, τ_2 and τ_3 are historical trajectories, the grid representation of which are $g(\tau_1) = (14, 11, 7, 8)$, $g(\tau_2) = (13, 10, 11, 12)$ and $g(\tau_3) = (9, 6, 7, 4)$. As τ_1 and τ_2 share the common sample point p_2, and $g(\tau_1) \cap g(\tau_2) = g(p_2) = \{11\}$, thus τ_1 and τ_2 are crossing trajectories. Different from τ_1 and τ_2 sharing the same point p_2, τ_2 and τ_3 have no common locations, but they are also considered as crossing trajectories as $g(\tau_2) \cap g(\tau_3) = \{7\}$.

Definition 3 Trajectory Length and Speed. *Given a historical trajectory $\tau = (p_1, p_2, \ldots, p_n)$, the length and the speed of τ are denoted as $\mathcal{L}_\tau$ and $\mathcal{V}_\tau$ respectively, which are defined as follows:*

$$\mathcal{L}_\tau = \sum_{i=1}^{n-1} dis(p_i, p_{i+1})$$

$$\mathcal{V}_\tau = \mathcal{L}_\tau / \sum_{i=1}^{n-1} |t_{i+1} - t_i|$$

where $dis(p_i, p_{i+1})$ is the Euclidean distance from p_i to p_{i+1} and $|t_{i+1} - t_i|$ is the time interval between p_i and p_{i+1}.

Problem Formalization. Given a set of historical trajectories, we aim to generate a new trajectory set by synthesizing the sub-trajectories cut off from the given trajectories. Note that, we should make sure the features of generated trajectory set are similar to those of the given set.

4　Gathering Features

The collection of trajectory features is the first step for FTS. As we focus on the trajectories collected from vehicles, thus we use the following features to synthesize new trajectories, such as length, acceleration, speed, U-turn and density.

Length: The distribution of trajectory length is an important feature that reflects the motion patterns of moving objects, and indirectly indicates the area of a city. The distribution of length is defined as $\{(\Delta l_1, l_1.p), (\Delta l_2, l_2.p), \cdots, (\Delta l_n, l_n.p)\}$, where Δl_i is a length interval and $l_i.p$ denotes the proportion of trajectories fall into this interval. In Fig. 2(a), the first interval Δl_1 is $0 - 10\,\mathrm{km}$ and $l_1.p = 0.548$. In Fig. 2(b), the second interval Δl_2 is $10 - 20\,\mathrm{km}$ and $l_2.p = 0.06$.

Speed: Trajectory speed is another important feature that directly indicates the traffic conditions of a city. If most of historical trajectories have a high speed, it reflects the smooth traffic. Otherwise, it means the bad traffic conditions of a city, which may result from too many traffic jams and the bad road conditions.

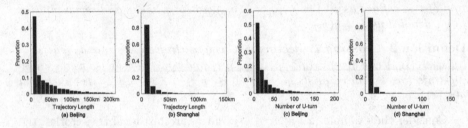

Fig. 2. Distribution of trajectory length and U-turn of 40K car trajectories in Beijing and Shanghai

U-turn: A U-turn is a GPS point where moving objects change direction sharply. The occurrence of U-turn is often caused by traffic lights or an individual change direction in an intersection. The more cross roads and more long trajectories, the larger number of U-turns. As seen in Fig. 2(c) and (d), most of car trajectories have more U-turns in Beijing. The corresponding distribution of U-turn is defined as $\{(\Delta u_1, u_1.p), (\Delta u_2, u_2.p), \cdots, (\Delta u_m, u_m.p)\}$, where Δu_i denotes a U-turn interval. In Fig. 2(c), Δu_5 is $40 - 50$ and $u_5.p = 0.004$.

Acceleration: Acceleration is the rate of change of velocity. The driver increases speed to a large value in a short time may indicate the smooth traffic conditions. However, if most of historical trajectories have a low acceleration, we can infer that the city may have many traffic jams and cross roads, which make the driver can not increase speed quickly. Consequently, acceleration is an important feature for trajectory synthesis.

Density: The density denotes the distribution of GPS points. Usually, this distribution is uneven. The center of a city is usually has a larger density than the suburbs, since the center usually has more commercial districts and residential areas. Obviously, the density is an important feature while describing the traffic conditions of a city.

5   Baseline Methods

5.1   Baseline 1: Random Generation

A straightforward method for FTS is to randomly select two crossing trajectories, and swap them in their intersection to generate new trajectories. Given two crossing trajectories τ_i and τ_j with $g(\tau_i) = (q_1, q_2, \cdots, q_n)$, $g(\tau_j) = (q'_1, q'_2, \cdots, q'_m)$, assume that there exists $q_l = q'_k$ ($q_l \in g(\tau_i), q_{k'} \in g(\tau_j)$) such that $g(\tau_i) \cap g(\tau_j) \neq \phi$, then we can synthesize new trajectories τ'_i and τ'_j with $g(\tau'_i) = (q_1, q_2, \cdots, q_k, q'_k, \cdots, q_{m'})$, $g(\tau'_j) = (q'_1, q'_2, \cdots, q'_k, q_l, \cdots, q_n)$, by swapping τ_i and τ_j in their intersection. Repeat the selection of crossing trajectories to make a synthesis, we can generate k trajectories.

5.2   Baseline 2: Sub-trajectory Synthesis

Even though it is easy to obtain new trajectories with above mentioned method, the characteristics of generated trajectories may be far from historical trajectories. With the goal of synthesizing trajectories that have similar characteristics with the given dataset, a novel method is proposed in the sequel, which contains two components.

Component 1: Given a set of historical trajectories $\mathcal{D}$, we get a set of sub-trajectories $\mathcal{C}$ by decomposing all crossing trajectories in their intersections. The formulation of Intersection Partition (IP) is given as follows.

Definition 4 *Intersection Partition (IP). Given a historical trajectory* $\tau = (p_1, p_2, \cdots, p_n)$, τ *is decomposed into a set of sub-trajectories* τ_i^j *that satisfy the following conditions:*

- $\tau_i^j = (p_i, p_{i+1}, \cdots, p_j)$, $\tau_i^j \in \tau$
- $\theta_{g(p_i)} \geq 2$, $\theta_{g(p_j)} \geq 2$
- $\theta_{g(p_k)} = 1$, $i < k < j$

where $\theta_{g(p_i)}$ denotes the number of trajectories pass through grid cell $g(p_i)$.

Continue the example in Fig. 1, the sub-trajectories generated by IP are presented in Table 1. τ_1 is decomposed in p_2 and p_3 as $\theta_{g(p_2)} = 2$ and $\theta_{g(p_3)} = 2$.

Definition 5 *Connected Sub-trajectories. For any two sub-trajectories* τ_i^j *and* $\tau_{i'}^{j'}$ *in* $\mathcal{C}$, *let* $g(\tau_i^j) = (q_1, q_2, \cdots, q_n)$ *and* $g(\tau_{i'}^{j'}) = (q'_1, q'_2, \cdots, q'_m)$, *they are defined as connected sub-trajectories if one of the following conditions is satisfied:* (1) $q_1 = q'_1$, (2) $q_1 = q'_m$, (3) $q_n = q'_1$, (4) $q_n = q'_m$.

Component 2: Synthesize new trajectories. As our goal is to generate a new dataset $\mathcal{S}$ having similar features to original dataset $\mathcal{D}$, which means that we should make the distributions of different traffic features of $\mathcal{S}$ are consistent with $\mathcal{D}$. Consequently, we should create a new trajectory that falls into the interval that has largest gap between $\mathcal{S}$ and $\mathcal{D}$ for each process. For example, in Fig. 2(a), if we want to generate a dataset having the same distribution of length

Table 1. Illustration of sub-trajectory

Trajectory	Sub-trajectory	Grid representation
τ_1	(p_1, p_2)	$(14, 11)$
	(p_2, p_3)	$(11, 7)$
	(p_3, p_4)	$(7, 8)$
τ_2	(p_5, p_6, p_2)	$(13, 10, 11)$
	(p_2, p_7)	$(11, 12)$
τ_3	(p_8, p_9, p_{10})	$(9, 6, 7)$
	(p_{10}, p_{11})	$(7, 4)$

to the given trajectories, we should synthesize a new trajectory with length $0 - 10\,\text{km}$ for the first process, i.e., a length interval Δl is selected before each generation. Similarly, we should select the speed interval Δv, U-turn interval Δu and acceleration interval Δa before synthesis. Then, we propose several heuristic functions to make a synthesis.

$$d_t = \frac{\theta_t}{\sum\limits_{\tau_i \in \mathcal{D}} |g(\tau_i)|}, \quad f_d(\tau_i) = \sum_{t \in g(\tau_i)} e^{\left|\frac{d_t^S - d_t^D}{d_t^D}\right|}$$

$$f_l(\tau_i, \Delta l) = e^{\left|\frac{\mathcal{L}_{\tau_i} - \Delta \bar{l}}{\Delta \bar{l}}\right|}, \quad f_v(\tau_i', \Delta v) = e^{\left|\frac{\mathcal{V}_{\tau_i} - \Delta \bar{v}}{\Delta \bar{v}}\right|} \quad (1)$$

$$f_u(\tau_i', \Delta u) = e^{\left|\frac{u_{\tau_i} - \Delta \bar{u} \cdot \frac{\mathcal{L}_{\tau_i}}{\Delta \bar{l}}}{\Delta \bar{u}}\right|}, \quad f_a(\tau_i, \Delta a) = e^{\left|\frac{a_{\tau_i} - \Delta \bar{a}}{\Delta \bar{a}}\right|}$$

$$f(\tau_i) = f_l(\tau_i', \Delta l) + f_v(\tau_i', \Delta v) + f_a(\tau_i', \Delta a) + f_u(\tau_i', \Delta u) + f_d(\tau_i)$$

where d_t represents the density of grid cell t. The median of selected intervals are denoted as $\Delta \bar{l}$, $\Delta \bar{v}$, $\Delta \bar{a}$ and $\Delta \bar{u}$ respectively.

Algorithm 1. Synthesize a New Trajectory

Input: a set of sub-trajectories $\mathcal{C}$
Output: a new trajectory
1: $\tau_c \leftarrow \tau_i$(with the minimum $f(\tau_i)$);
2: **repeat**
3:     **for** each connected sub-trajectories of τ_c **do**
4:         select a sub-trajectory with the minimum $f(\tau_i + \tau_c)$;
5:     **end for**
6:     $\tau_c \leftarrow \tau_c + \tau_i$;
7: **until** $f(\tau_i + \tau_c) > f(\tau_c)$
8: **return** τ_c;

For each process, we select a sub-trajectory τ_i connected with current synthesized trajectory τ_c according to Eq. (1). Repeat the action, then we can obtain a new trajectory. Note that, we should update the distribution of different features for S after each generation.

6   Optimization

Decomposing all historical trajectories in their intersections is not always reasonable, due to the following reasons. (1) It generates too many sub-trajectories, especially when there exist lots of crossing trajectories. As shown in Table 2, we randomly select taxi trajectories in Beijing from 10K to 50K, the number of sub-trajectories generated by IP increases fast. Note that the large number of sub-trajectories leads to high time cost for generation. (2) A trajectory is usually composed of several parts that consists of several sub-trajectories, even though these parts vary on traffic features, sub-trajectories in each part have similar features. Dividing a trajectories into too many sub-sections may lead to the lost of original features. As a consequence, we propose a novel method to decompose trajectories in the sequel.

Table 2. Number of generated sub-trajectories

	10K	20K	30K	40K	50K
IP	551564	1048207	1568904	2040944	2437631
GP	101624	195361	295424	378891	440082

Gene Partition (GP). In order to generate less sub-trajectories and avoid decomposing sub-trajectories with similar features, we should merge some sub-trajectories generated by IP. That is to say, given a sequence of sub-trajectories $\Gamma = (\tau_1, \tau_2, \cdots, \tau_n)$ generated by decomposing a historical trajectory with IP, we need to find a best label sequence $\omega = (\omega_1, \omega_2, \cdots, \omega_{n-1})$ such that sub-trajectories with similar features are merged and vice versa, where the state space of ω_i is $\{0, 1\}$, $\omega_i = 1$ means τ_i and τ_{i+1} should be merged and vice versa.

Different from traditional work Euclidean Distance [18], EDR [19] and LCSS [20] that directly use latitude, longitude and time-stamp to calculate the similarity between two trajectories, we use some traffic features instead, such as speed and acceleration. The features of a sub-trajectory τ_i is denoted as a vector $v_i = (v_i^1, v_i^2, \cdots, v_i^m)$, and $Sim(\tau_i, \tau_{i+1})$ is given as follows:

$$Sim(\tau_i, \tau_{i+1}) = \frac{\sum_{j=1}^{m} v_i^j \cdot v_{i+1}^j}{\sqrt{\sum_{j=1}^{m} (v_i^j)^2} \cdot \sqrt{\sum_{j=1}^{m} (v_{i+1}^j)^2}} \tag{2}$$

As our goal is to merge sub-trajectories with similar features, we set $\omega_i = 1$ if $Sim(\tau_i, \tau_{i+1}) > \sqrt{1 - Sim(\tau_i, \tau_{i+1})^2}$ (i.e., τ_i and τ_{i+1} own similar features, thus we should merge them). On the contrary, τ_i is set to 0, as τ_i and τ_{i+1} have different features.

$$\omega_i = \begin{cases} 1, & if\ Sim(\tau_i, \tau_{i+1}) > \sqrt{1 - Sim(\tau_i, \tau_{i+1})^2} \\ 0, & otherwise \end{cases} \tag{3}$$

7 Experiment Result

In this section, we conduct extensive experiments on real-world trajectory datasets to study the performance of the proposed algorithms, by comparing the features of the generated trajectory dataset with the original dataset $\mathcal{D}$. All algorithms are implemented on a Core i5-3470 3.20 GHz machine with 16 GB memory and a windows platform.

7.1 Dataset

In order to comprehensively display the performance of the proposed model FTS, we use a real-world trajectory dataset in Beijing, which contains 13245 taxi trajectories (1401413 GPS points) that collected from 1752 taxis during 1/10/2013-7/10/2013. Note that we use a 200×400 grid map on Beijing to generate 14500 new trajectories.

7.2 Performance Evaluation

We evaluate the proposed algorithms by comparing the trajectory features of the given dataset $\mathcal{D}$ with that of synthesized trajectories. Following methods are used: (1) Baseline 1: random generation (RG), (2) Baseline 2: intersection partition (IP), (3) Approach 3: gene partition (GP). Moreover, we use Original to denote the given trajectory dataset $\mathcal{D}$.

Distribution of Length. First of all, we compare the distribution of trajectory length between original dataset and generated datasets. Seen from Fig. 3, RG performs worst and it is more likely to generate long trajectories, as this approach has no restriction during synthesis. IP performs better than RG, since we have proposed several functions to control each synthesis. Note that, GP performs best, since it has the most similar distribution to the original dataset. This is because GP decomposes trajectories reasonably, then it is more likely to synthesize similar trajectories.

Distribution of Speed. Speed is of critical importance to reflect the traffic conditions of a city and describe motion behaviors of moving objects. Shown in Fig. 4, most of historical trajectories have speed 3m/s-8m/s in Beijing. The distribution of speed of trajectories generated by RG is different from the given trajectories, and IP performs better than RG. The trajectory speed of GP concentrate on 4m/s-8m/s, which is similar to the original dataset.

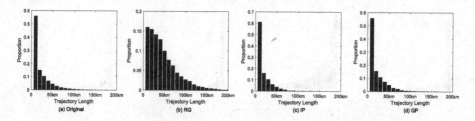

Fig. 3. Distribution of trajectory length in Beijing

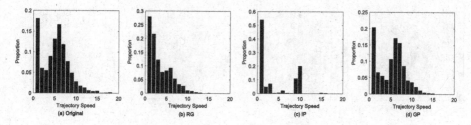

Fig. 4. Distribution of trajectory speed in Beijing

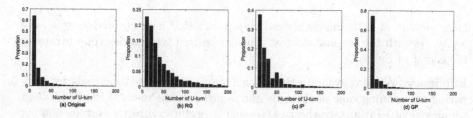

Fig. 5. Distribution of U-turn in Beijing

Distribution of U-turn. Trajectory U-turn is another important feature that directly indicates the traffic quality and how the moving objects travel. Seen from Fig. 5, GP has created the most similar trajectories to the given dataset, where most of synthesized trajectories have a small number of U-turn.

Distribution of Acceleration. We also investigate the performance of the propose approaches by comparing distribution of trajectory acceleration of corresponding datasets. Without surprise, in Fig. 6, IP performs better than RG and GP performs best. This is because we have designed several heuristic functions to select reasonable sub-trajectories for making a synthesis, and merge sub-trajectories with similar features. Consequently, we are more likely to synthesize new trajectories with similar features in IP and GP.

Distribution of Density. In order to display the generated trajectories intuitively, we plot all corresponding locations in Fig. 7. Note that, different from aforementioned features, by which we can distinguish the performance of different methods easily, all methods perform well in this feature. This is because the

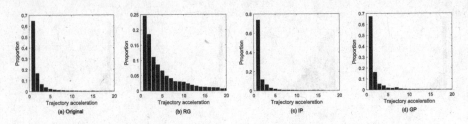

Fig. 6. Distribution of trajectory acceleration in Beijing

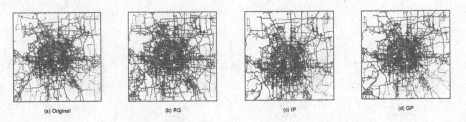

Fig. 7. Distribution of density in Beijing

large number of POIs in synthesized trajectories that have covered most area of a city. As with the original dataset, the synthesized trajectories concentrate on the center of Beijing.

Efficiency. We also evaluate the efficiency of the proposed methods by comparing the running time of them. Without surprise, the method RG needs least running time, as it just synthesizes crossing trajectories directly and the number of candidates is less than that of IP and GP. Furthermore, Table 3 shows GP is much more efficient than IP, since the number of sub-trajectories generated by GP is much less than that generated by IP shown in Table 2.

Table 3. Running time

Method	RG	IP	GP
Time cost (h:mm:ss)	0:04:02	3:1:24	1:21:32

8   Conclusions

We investigate a problem of synthesizing trajectories in this study. A novel model FTS has been proposed, which is composed of three components. (1) Gathering features of the original dataset, (2) Synthesizing new trajectories, (3) Validating the result. In order to make the features of generated dataset similar to the features of given dataset, we take many features into account during each generation. Furthermore, we design an optimized algorithm to reduce time cost and keep the consistency of data features. Experimental results based on real trajectory dataset show that the proposed algorithms can achieve high performance.

Acknowledgement. This work was supported by the National Natural Science Foundation of China under Grant Nos. 61572335, 61572336, and 61303019, the Natural Science Foundation of Jiangsu Province of China under Grant No. BK20151223, the Natural Science Foundation of Jiangsu Provincial Department of Education of China under Grant No. 12KJB520017, and Collaborative Innovation Center of Novel Software Technology and Industrialization, Jiangsu, China.

References

1. Lee, J.-G., Han, J., Whang, K.-Y.: Trajectory clustering: a partition-and-group framework. In: SIGMOD, pp. 593–604 (2007)
2. Jeung, H., Yiu, M.L., Zhou, X.F., Jensen, C., Shen, H.T.: Discovery of convoys in trajectory databases. Proc. VLDB Endowment **1**(1), 1068–1080 (2008)
3. Zheng, K., Zheng, Y., Yuan, N.J., Shang, S.: On discovery of gathering patterns from trajectories. In: ICDE, pp. 242–253 (2013)
4. Shang, S., Ding, R., Yuan, B., Xie, K., Zheng, K., Kalnis, P.: User oriented trajectory search for trip recommendation. In: EDBT, pp. 156–167 (2012)
5. Zheng, K., Shang, S., Yuan, N.J., Yang, Y.: Towards efficient search for activity trajectories. In: ICDE, pp. 230–241 (2013)
6. Li, Z., Lee, J.-G., Li, X., Han, J.: Incremental clustering for trajectories. In: Kitagawa, H., Ishikawa, Y., Li, Q., Watanabe, C. (eds.) DASFAA 2010. LNCS, vol. 5982, pp. 32–46. Springer, Heidelberg (2010)
7. Zeppelzauer, M., Zaharieva, M., Mitrovic, D., Breiteneder, C.: A novel trajectory clustering approach for motion segmentation. In: Boll, S., Tian, Q., Zhang, L., Zhang, Z., Chen, Y.-P.P. (eds.) MMM 2010. LNCS, vol. 5916, pp. 433–443. Springer, Heidelberg (2010)
8. Nergiz, M.E., Atzori, M., Saygin, Y.: Towards trajectory anonymization: a generalization-based approach. In: Proceedings of the 2008 International Workshop on Security and Privacy in GIS and LBS, pp. 52–61 (2008)
9. Terrovitis, M., Mamoulis, N.: Privacy preservation in the publication of trajectories. In: MDM, pp. 65–72 (2008)
10. Pfoser, D., Theodoridis, Y.: Generating sementic-based trajectories of moving objects. In: International Workshop on Emerging Technologies for Geo-Based Applications, pp. 59–76 (2000)
11. Pelekis, N., Ntrigkogias, C., Tampakis, P., Sideridis, S., Theodoridis, Y.: Hermoupolis: a trajectory generator for simulating generalized mobility patterns. In: Nijssen, S., Železný, F., Blockeel, H., Kersting, K. (eds.) ECML PKDD 2013, Part III. LNCS, vol. 8190, pp. 659–662. Springer, Heidelberg (2013)
12. Xue, A.Y., Zhang, R., Zheng, Y., Xie, X., Huang, J., Zhou, X.F.: Destination prediction by sub-trajectroy synthesis and privacy protection against such prediction. In: ICDE, pp. 254–265 (2013)

13. Theodoridis, Y., Silva, J.R.O., Nascimento, M.A.: On the generation of spatiotemporal datasets. In: Güting, R.H., Papadias, D., Lochovsky, F.H. (eds.) SSD 1999. LNCS, vol. 1651, pp. 147–164. Springer, Heidelberg (1999)
14. Saglio, J.-M., Moreria, J.: A realistic scenario generator for moving objects. In: Proceedings of the 10th International Workshop on Database and Expert Systems Applications, pp. 426–432 (1999)
15. Brinkhoff, T.: A framework for generating network-based moving objects. GeoInformatica **6**(2), 153–180 (2002)
16. Giannotti, F., Mazzoin, A., Puntoni, S., Renso, C: Synthetic generation of cellular network positioning data. In: Proceedings of the 13th Annual ACM International Workshop on Geographic Information Systems, pp. 12–20 (2005)
17. Duntgen, C., Behr, T., Guting, H.R.: BerlinMOD: a benchmark for moving object databases. VLDB J. **18**(6), 1335–1368 (2008)
18. Agrawal, R., Faloutsos, C., Swami, A.N.: Efficient similarity search in sequence databases. In: Proceedings of the 4th International Conference on Foundations of Data Organization and Algorithms, pp. 69–84 (1993)
19. Chen, L.: Similarity search over time series and trajectory data. Ph.D. dissertation (2005)
20. Vlachos, M., Kollios, G., Gunopulos, D.: Discovering similar multidimensional trajectories. In: ICDE, pp. 673–684 (2002)

Distributed Text Representation with Weighting Scheme Guidance for Sentiment Analysis

Zhe Zhao, Tao Liu[✉], Xiaoyun Hou, Bofang Li, and Xiaoyong Du

School of Information, Renmin University of China, Beijing, China
{helloworld,tliu,xiaoyunhou,libofang,duyong}@ruc.edu.cn

Abstract. With rapid growth of social media, sentiment analysis has recently attracted growing attention in both academic and industrial fields. One of the most successful paradigms for sentiment analysis is to feed bag-of-words (BOW) features into classifiers. Usually, weighting schemes are required to weight raw BOW features to obtain better accuracy, where important words are assigned more weights while unimportant ones are given less weights. Another line of researches for sentiment analysis focuses on neural models, where dense features are automatically extracted from texts by neural networks. In this paper, we take advantages of techniques in both lines of researches, where weighting schemes are introduced into the neural models to guide neural networks to focus on those important words. Neural models are known for their automatic feature learning abilities, however, we discover that when suitable guidance such as weighting schemes are applied, better features can be extracted for sentiment analysis. Experimental results show that our models outperform or can compete with state-of-the-art approaches on three commonly used sentiment analysis datasets.

Keywords: Distributed representation · Sentiment analysis · Weighting scheme

1 Introduction

Sentiment analysis aims at extracting users' subjective information from raw texts. It is valuable for companies and institutions to turn user-generated texts into knowledge for decision support. One of the most successful paradigm for sentiment analysis is to feed bag-of-words (BOW) features with weighting schemes into classifiers such as support vector machines [16]. In BOW, each word represents one dimension of features, and weighting schemes give those important words more weights [11]. Though simple, strong baselines are achieved on a range of sentiment analysis datasets.

However, traditional text features suffer from high dimensionality and data sparsity. They take each word as an atomic unit, which totally ignores the internal semantics of the words. Recently, many researchers have turned their attention to extracting dense text representation by neural networks (NNs) [5]. Typical neural networks include Convolutional NNs [7], Recursive NNs [17],

© Springer International Publishing Switzerland 2016
F. Li et al. (Eds.): APWeb 2016, Part I, LNCS 9931, pp. 41–52, 2016.
DOI: 10.1007/978-3-319-45814-4_4

Paragraph Vector (PV) [9], etc. Usually, these models learn dense text representations upon pre-trained dense word representations. One advantage of these models is that they can extract features automatically. In this sense, we can directly feed raw data into neural networks with no requirements of prior knowledge. However, a natural question is raised here: if prior knowledge is available, can we utilize the knowledge to help the training of the neural networks to achieve better results? This is the motivation of our models.

In this paper, a novel model is proposed which takes advantages of both neural models and traditional weighting schemes. Weighting schemes tell us which words are important and which words are not. Taking movie reviews for example, words like 'wonderful' and 'worst' reflect reviewers' sentiment tendencies, while words like 'of' and 'watch' contain little information about users' attitudes to movies. A lot of work has proven the effectiveness of weighting schemes on raw BOW features [2,8,11,15,19]. Inspired by their successes in BOW, we discover that weighting techniques are also very useful for guiding the training of neural networks. By knowing importance of words in advance, we can train neural models to focus on extracting features from words that is critical to the meanings of texts, while ignore those unimportant words. Since previous weighting schemes are originally applied to BOW features, we propose a novel weighting framework specially designed for neural models. As a result, very competitive results are achieved in both sentence-level and document-level datasets.

Section 2 reviews the related work of sentiment analysis. Section 3 discusses how to integrate weighting techniques into neural models. Section 4 proposes a novel weighting framework specifically designed for neural models. Experimental results are given in Sect. 5, followed by the conclusion in Sect. 6. Source code of the paper will be available at http://github.com/zhezhaoa/Weighted-Paragraph-Vector. Other researchers can reproduce our results easily and further improve models upon our new baselines.

2   Related Work

Bag-of-words Features. Text representation is of vital importance for sentiment analysis. One of the most popular features is bag-of-words (BOW). Early work that uses BOW feature on sentiment analysis is done by [16]. They found that binary weighted feature with SVM classifier gives promising results. Following their work, many researchers focus on applying various weighting schemes to original BOW text representations. One of the most successful attemptions is to introduce information of class preference into weighting schemes. [11] calculate the document frequency (DF) of words in positively and negatively labeled texts respectively, and use the ratios of them to weight words. [19] apply class preference weighting on bag-of-ngrams features and achieve very strong baselines on a range of sentiment analysis and text classification tasks. [8] add credibility weighting to determine which words are important for sentiment analysis. [2,15] give detailed discussions over different weighting schemes on BOW features.

Deep Neural Networks. Recently, neural networks (NN) have become increasingly popular in natural language processing (NLP). They can generate dense word and text representations and achieve state-of-the-art results on a range of tasks. One of the most fundamental work in this field is word embedding, where dense word representations are learned by modeling the relationships between target words and their contexts [1]. Almost all neural models for NLP are built upon word embeddings since neural networks require dense representation as input to make models computational. [7] propose to use Convolutional NN (CNN) for sentiment analysis, which can effectively capture n-gram features and achieve state-of-the-art results on many sentence-level datasets. Recursive NN (RecNN), proposed by [17], constructs neural networks on the basis of parse tree and is able to extract fine-grained information of sentences. Another family of neural networks for sentiment analysis is Recurrent NN (RNN). The hidden layers of the RNN store all previous information in theory. The hidden layer of the last word can be used as the representation of the whole texts. Most recently, combinations of neural networks are proposed to capture complex structures of the texts. Their training processes can be categorized in two steps. The first step is to extract sentence features from word features (word embeddings), and the second step is to extract document features from sentence features [3,10]. As a result, these models can not only capture word order and syntactic information, but also take relationships among sentences into considerations. The drawbacks of these deep neural networks are also obvious. These models are expensive in computational resources. Besides that, they are not as robust as traditional BOW approaches. The effectiveness of these models closely relies on careful hyper-parameters tuning and some sub-tasks such as pre-trained word embedding, parsing and sentence segmentation.

Neural Bag-of-words Models. Though deep neural models are very powerful in theory, only limited improvements are achieved in sentiment analysis when compared with traditional BOW features with weighting schemes. It seems that information like word order, sentence and document structure is not very crucial for sentiment analysis. Another line of neural models is neural bag-of-words models. Instead of constructing complex composition upon word embeddings, these models basically ignore order and syntactic information. Representative neural bag-of-words models include Deep Average Networks (DAN) [6] and Paragraph Vector (PV) [9]. DAN at first takes average of the words embeddings as the inputs and then constructs multiple neural layers upon them. PV embeds text by making it useful to predict the words it includes. These models enjoy the advantages of being simple and robust compared with other deep neural networks, and can still achieve competitive results on sentiment analysis tasks.

Existing neural bag-of-words models treat each word equally. To capture better features for sentiment analysis, in this paper we use weighting schemes to guide the training of Paragraph Vector models, thus the new models can pay more attention to words that reflect the polarities of texts. When suitable weighting schemes are used, significant improvements over traditional Paragraph Vector model are witnessed.

3 Weighted Paragraph Vector (WPV)

In this study, we introduce weighting schemes into a neural bag-of-words model, Paragraph Vector (PV). PV is just the direct extension of word2vec. Word2vec is a very popular word embedding training toolkit[1]. It achieves state-of-the-art results on a range of linguistic datasets, and at the same time only requires a fraction of time compared with previous word embedding models [13,14]. The main difference between PV and word2vec is that PV treats each text as a special word and uses this special word to predict the words in the text. PV includes two variants, PV-DBOW and PV-DM, which corresponds to two variants in word2vec, Skip-gram (SG) and continuous bag-of-words (CBOW). In the following subsections, we introduce weighting techniques into PV-DBOW and PV-DM respectively.

3.1 Weighted PV-DBOW

The original objective function of PV-DBOW is as follows:

$$\sum_{i=1}^{|T|}\sum_{j=1}^{|t_i|} logP(w_{ij}|t_i) + \sum_{i=1}^{|WN|} \sum_{-c\leq j\leq c, j\neq 0} logP(w_i|w_{i+j}) \tag{1}$$

$$where \quad P(a|b) = \sigma(e(a) \bullet e(b)) \prod_{k=1}^{K} E_{w_k \sim P_n(w)}(\sigma(-e(b) \bullet e(w_k)))$$

where $t_i=\{w_{i1}, w_{i2}, \ldots, w_{i|t_i|}\}$ denotes i_{th} text and $T=\{t_1, t_2, \ldots, t_{|T|}\}$ denotes the whole dataset. $|WN|$ is the number of training words in the whole dataset. Negative sampling [14] is used to define the 'conditional probabilities' of target word given its context: $e(\cdot)$ denotes the embedding of word (text) $\cdot$. $\bullet$ denotes inner product. σ is sigmoid function and $P_n(w)$ is uni-gram distribution raised to the n-th power. The first part of the objective uses text that contains the target word as context. The second part is just the objective of ordinary word embedding model, where words in local window are used as context. Traditional Paragraph Vector model treats each word in the text equally. In this sense, it can be viewed as the counterpart of BOW features where each feature is represented by the count of the word in the text.

Obviously, some words are more important for discriminating polarities of the sentiment. The main idea of our model is to make embedding of text pay more attention to those discriminative words, instead of neutral words which have little value for determining polarities of texts. We give each word a weight (a real value), which represents how important a word is for sentiment analysis. How to assign the weights to words will be discussed in Sect. 4. The objective of weighted PV-DBOW is as follows:

$$\sum_{i=1}^{|T|}\sum_{j=1}^{|t_i|} Weight(w_{ij})logP(w_{ij}|t_i) + \sum_{i=1}^{|WN|} \sum_{-c\leq j\leq c, j\neq 0} logP(w_i|w_{i+j}) \tag{2}$$

---

[1] http://code.google.com/p/word2vec.

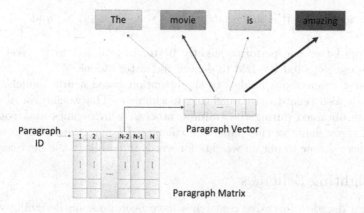

Fig. 1. Illustration of weighted PV-DBOW

where $Weight(w)$ reflects the importance of word w. By optimizing the above weighted objective, the trained text embedding is able to predict important word in larger probabilities while ignore those unimportant words. Figure 1 illustrates the framework of weighted PV-DBOW.

3.2 Weighted PV-DM

PV-DM uses average of the text embedding and word embeddings in local window to predict target word. The original objective of PV-DM can be written as follows:

$$\sum_{i=1}^{|WN|} logP(w_i|w_i^{context}) \tag{3}$$

$$where \ \ w_i^{context} = e(t_*) + \sum_{-c \leq j \leq c, j \neq 0} e(w_{i+j}) \ \ \ and \ \ w_i \in t_*$$

Like the way we introduce weighting into PV-DBOW. The objective of weighted PV-DM is as follows:

$$\sum_{i=1}^{|WN|} Weight(w_i)logP(w_i|w_i^{context}) \tag{4}$$

An alternative for introducing weighting information into PV-DM is to change the representation of target words' contexts. Since text embeddings should pay more attention to important words, we give text embeddings more weights in constructing contexts when target words are important. In this way, the text embeddings are more affected by those important words while less affected by those unimportant words. The objective function can be written as follows:

$$\sum_{i=1}^{|WN|} logP(w_i|w_i^{context}) \tag{5}$$

$$where \quad w_i^{context} = Weight(w_i)e(t_*) + \sum_{-c \leq j \leq c, j \neq 0} e(w_{i+j}) \quad and \quad w_i \in t_*$$

We find the latter one performs slightly better in practice. In the rest of the paper, we use weighted PV-DM to denote the latter model.

One may connect our models to the attention based neural models, where attention is also regarded as weighting mechanism. The weights in attention models are obtained during the training process, which comes at a cost [10]. However, the weights in this paper are calculated before the training process. We will discuss how to obtain weights for words in details in the next section.

4   Weighting Schemes

For several decades, intensive researches have been done on designing various weighting schemes on raw BOW features. Most of weighting schemes can not be directly applied to neural case, but the intuition behind them are very valuable, which guides us to design a new weighting framework for Paragraph Vector. As mentioned in above section, traditional PV treats each word equally and can be viewed as the neural counterparts of raw BOW feature. We use PV as baselines and gradually add new weighting techniques upon it.

4.1   IDF Weighting

Direct improvement over BOW baseline is inverse document frequency (IDF) weighting, where words that appear in every texts are given very low weights. Document frequency of word w is denoted by $df(w)$. A substitution for IDF weighting is sub-sampling technique used in [14], where high-frequency words are removed randomly. These two techniques perform almost the same in practice. For sake of space saving, in the following subsection we add new weighting techniques on the basis of IDF weighting. The weighting function of IDF is as follows:

$$Weight_1(w) = log(\frac{|T| + 1}{df(w)}) \tag{6}$$

4.2   Weighting with Class Preference

IDF and sub-sampling weighting do not take class preference into consideration. Intuitively, a word is important if it has uneven distribution over classes. In this paper, we only consider binary sentiment analysis. We use $Pos(w)$ and $Neg(w)$ to respectively denote the number of positively and negatively labeled texts that contain word w. [11,19] show that weighting word w with $log(Pos(w)/Neg(w))$ is very effective for BOW features. We adapt this term weighting function for neural case:

$$Weight_2(w) = \begin{cases} (Pos(w)/Neg(w))^k & Pos(w) > Neg(w) \\ (Neg(w)/Pos(w))^k & Pos(w) \leq Neg(w) \end{cases} \tag{7}$$

where k is a hyper-parameter and determined by validation set. Empirically, 0.5 is a good choice for k.

4.3   Credibility

The last block we add on our weighting framework is credibility. In fact, credibility is a supplementary of class preference weighting. The occurrence of a word in positive and negative texts can be modeled by binomial distribution. If a word occurs twice in positive texts and once in negative texts, we can not reject hypothesis that word has even distribution with reasonable significance levels. If a word occurs 200 times and 100 times in positive and negative texts respectively, we can conclude that the word has uneven distribution over classes with greater credibility, even though the ratio is the same with the previous case. Therefore the number of times a word occurs in the dataset is also an important factor for us to determine if a word is 'important' or not. The weighting for credibility used in this paper is as follows:

$$Weight_3(w) = log(log(Count(w) + 1) + 1) \tag{8}$$

where $Count(w)$ is used to denote the number of times word w occurs in the dataset.

Finally, we combine three components together and obtain the entire weighting framework for Paragraph Vector models.

$$Weight(w) = Weight_1(w)Weight_2(w)Weight_3(w)$$

$$= (Pos(w)/Neg(w))^{k*\lfloor Pos(w)>Neg(w) \rfloor} \\ *log(log(Count(w) + 1) + 1) * log(\frac{|T|+1}{df(w)}) \tag{9}$$

$$where \quad \lfloor \cdot \rfloor = \begin{cases} 1 & \cdot \, is \, True \\ -1 & \cdot \, is \, False \end{cases}$$

5   Experiments

5.1   Experimental Setup

Models in this paper are trained by stochastic gradient descent (SGD). Text embeddings obtained by models are regarded as texts features and are fed into logistic regression classifier [4]. Pre-processing of datasets and hyper-parameters setting follow the implementation by [12]. 10 percent of training data is selected as validation data to determine the number of training epochs and hyper-parameters in weighting schemes. The above settings are applied to all datasets unless otherwise noted. Experiments are conducted on two document-level datasets, IMDB[2] and RT-2k[3], and one sentence-level dataset, RT-s[4]. Detailed statistics of these datasets are shown in Table 1.

---

[2] http://ai.stanford.edu/~amaas/data/sentiment/aclImdb_v1.tar.gz.

[3] http://www.cs.cornell.edu/people/pabo/movie-review-data/review_polarity.tar.gz.

[4] http://www.cs.cornell.edu/people/pabo/movie-review-data/rtpolaritydata.tar.gz.

Table 1. Detailed statistics of datasets. CV denotes that dataset is evaluated by cross validation. l denotes the average length of texts and $|V|$ denotes the vocabulary size.

Dataset	Train(+,−)	Test(+,−)	l	$|V|$
IMDB	12500,12500	12500,12500	231	392K
RT-2k	1000,1000	CV	787	51K
RT-s	5331,5331	CV	21	21K

5.2 Effectiveness of Weighting Schemes

We gradually add weighting techniques upon Paragraph Vector baselines and observe the effectiveness of each weighting function over classification accuracies. From Tables 2, 3 and 4, we can observe that weighting schemes, especially weighting with class preference, are very effective for sentiment analysis. Since RT-2k and RT-s only contain limited texts, $Weight_2$ and $Weight_3$ are calculated by statistics of IMDB dataset. Adding additional texts from IMDB dataset to these two datsets is also helpful. We also conduct 'oracle' experiments where $Weight_2$ and $Weight_3$ are calculated by all data (including test data). Though knowing basic statistics of test data in advance is not realistic in many cases, 'oracle' experiments further demonstrate the effectiveness of class preference weighting in Paragraph Vector models.

When all weighting techniques are added, very competitive results are achieved on all three datasets. Though PV models are known for being able to extract features from texts automatically, weighting schemes are still very useful for these models.

Figures 2 and 3 illustrate the accuracies of PV-DM and Weighted PV-DM on IMDB dataset at different embedding dimensions and iterations level. We can observe that WPV achieves decent accuracies even when dimensions of text is very small, while the accuracies of PV declines sharply as we reduce the

Table 2. Illustration of effectiveness of weighting schemes on IMDB. Weighting techniques are gradually applied to PV models and competitive results are achieved finally.

Features	PV-DBOW	PV-DM
Baselines	87.6	88.5
+IDF	88.5(+0.9)	88.9(+0.4)
+Class Preference	89.5(+1.0)	90.0(+1.1)
+Credibility	90.0(+0.5)	90.2(+0.2)
+Oracle	91.4	91.5

Table 3. Illustration of effectiveness of weighting schemes on RT-2k. PV baselines are weak in this dataset. However, class preference weighting give rise to very significant improvements

Features	PV-DBOW	PV-DM
Baselines	84.9	86.0
+IDF	85.5(+0.6)	86.4(+0.4)
+Class Preference	87.9(+2.2)	88.3(+1.9)
+Credibility	88.5(+0.6)	88.8(+0.5)
+Additional texts	89.7(+1.2),	89.6(+0.8)
+Oracle	90.4	90.5

Table 4. Illustration of effectiveness of weighting schemes on RT-s. The corpus size of RT-s is very limited, additional text is necessary for this dataset.

Features	PV-DBOW	PV-DM
Baselines	71.5	70.1
+Additional texts	76.8	76.5
+Class Preference	77.8(**+1.0**)	77.7(**+1.2**)
+Credibility	78.4(+0.6)	78.3(+0.6)
+Oracle	79.4	79.2

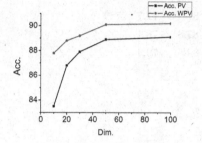

Fig. 2. Accuracies of PV-DM and Weighted PV-DM at different dimensions levels. Iter. is set to 20

Fig. 3. Accuracies of PV-DM and Weighted PV-DM at different iterations levels. Dim. is set to 50

text features dimensions. WPV also outperforms PV significantly at different iterations levels, and achieves optimal accuracy with less training epochs.

5.3   Comparison of State-of-the-art Models

IMDB is one of the most popular sentiment analysis dataset and has been studied by a large amount of researches. State-of-the-art results on IMDB are obtained by NBSVM and various neural networks. Different state-of-the-art models are compared in Table 5. From IMDB column we can observe that even though our models are essentially bag-of-words models, the results can compete with or even outperform the models that exploit complex information of texts. Since our models basically ignore word order and syntactic information, they are much faster in training time and require less computational resources compared with other state-of-the-art approaches. When our model is combined with NBSVM (Ensemble), state-of-the-art results are achieved on IMDB dataset.

The previous state-of-the-art models on RT-2k is NBSVM. Rare work reports the effectiveness of neural network on this dataset. We speculate the reasons are that the size of RT-2k is limited and the average length is too long. Original Paragraph Vector models only obtain weak baselines on this dataset. However, weighting schemes are very effective on this dataset and we can observe that new state-of-the-art results are achieved by our models.

Table 5. Comparison of models on three commonly used datasets. PV is implemented by source code[a] provided by [12]. Results of Ensemble is obtained by linear combination of WPV and NBSVM [12]. Results in rows 4–7 are from [19]. DCNN uses convolutional-convolutional neural network to extract text features upon word embeddings [3]. Results in row 9–11 are from [10], where recursive-recurrent neural network is used for text representations learning. RecNN: [18]. CNN: [7].

Models	IMDB	RT-2k	RT-s
PV	88.7	86.0	76.5
WPV	90.5	89.6	78.3
Ensemble	**92.6**	**90.5**	79.8
SVM-uni	87.0	86.3	76.2
SVM-bi	89.2	87.4	77.7
NBSVM-uni	88.3	87.8	78.1
NBSVM-bi	91.2	89.5	79.4
DCNN	89.4	-	-
RecNN-RNN	87.0	-	-
WNN	90.2	-	-
BENN	91.0	-	-
RecNN	-	-	77.7
CNN	-	-	**81.5**

[a]http://github.com/mesnilgr/iclr15

RT-s is a sentence-level dataset. Since the corpus size of RT-s is too limited, it requires additional data to achieve decent accuracy. The best result on RT-s is achieved by Convolutional neural network, which performs slightly better than our models. It is probably because that order information is considered in CNN. However, our models require only a small fraction of training time compared with CNN.

6   Conclusion

In this paper, we present a method for introducing weighting schemes into Paragraph Vector models. Weighting schemes tell which words are important, and can be easily used to guide the training of the neural models in our new method. We also propose a novel weighting framework specially designed for neural case. Three weighting techniques are gradually added to our weighting framework to better determine the importance of words for sentiment analysis. From Experimental results, we obtain following conclusions: (1) Though neural models are known for their ability to extract features automatically, prior knowledge like weighting schemes still show obvious effectiveness on neural models, just like weighting schemes work for raw BOW features. (2) Features extracted by our new method are very discriminative. We can obtain high-quality text features with less embedding dimensions and training epochs compared to traditional PV. Even 10-dimensional text feature can achieve good results. (3) Even though our models are essentially bag-of-words models, they can still rival the models that exploit complex compositionality over words and phrases. Very competitive results are achieved by our models on two document-level and one sentence-level sentiment analysis datasets.

Acknowledgements. This work is supported by National Natural Science Foundation of China (61472428, 61003204), the Fundamental Research Funds for the Central Universities, the Research Funds of Renmin University of China No. 14XNLQ06 and Tencent company.

References

1. Bengio, Y., Ducharme, R., Vincent, P., Janvin, C.: A neural probabilistic language model. J. Mach. Learn. Res. **3**, 1137–1155 (2003)
2. Deng, Z., Luo, K., Yu, H.: A study of supervised term weighting scheme for sentiment analysis. Expert Syst. Appl. **41**(7), 3506–3513 (2014)
3. Denil, M., Demiraj, A., Kalchbrenner, N., Blunsom, P., de Freitas, N.: Modelling, visualising and summarising documents with a single convolutional neural network (2014). abs/1406.3830
4. Fan, R., Chang, K., Hsieh, C., Wang, X., Lin, C.: LIBLINEAR: a library for large linear classification. J. Mach. Learn. Res. **9**, 1871–1874 (2008)
5. Goldberg, Y.: A primer on neural network models for natural language processing (2015). CoRR abs/1510.00726
6. Iyyer, M., Manjunatha, V., Boyd-Graber, J., Daumé III, H.: Deep unordered composition rivals syntactic methods for text classification. In: Proceedings of the 53rd Annual Meeting of the Association for Computational Linguistics: Long Papers, ACL 2015, vol. 1, pp. 1681–1691 (2015)
7. Kim, Y.: Convolutional neural networks for sentence classification. In: Proceedings of the 2014 Conference on Empirical Methods in Natural Language Processing, EMNLP 2014, pp. 1746–1751 (2014). A meeting of SIGDAT, a Special Interest Group of the ACL
8. Kim, Y., Zhang, O.: Credibility adjusted term frequency: a supervised term weighting scheme for sentiment analysis and text classification (2014). CoRR abs/1405.3518
9. Le, Q.V., Mikolov, T.: Distributed representations of sentences and documents. In: Proceedings of the 31th International Conference on Machine Learning, ICML 2014, pp. 1188–1196 (2014)
10. Li, J.: Feature weight tuning for recursive neural networks (2014). abs/1412.3714
11. Martineau, J., Finin, T.: Delta TFIDF: an improved feature space for sentiment analysis. In: Proceedings of the Third International Conference on Weblogs and Social Media, ICWSM 2009 (2009)
12. Mesnil, G., Mikolov, T., Ranzato, M., Bengio, Y.: Ensemble of generative and discriminative techniques for sentiment analysis of movie reviews (2014). abs/1412.5335
13. Mikolov, T., Chen, K., Corrado, G., Dean, J.: Efficient estimation of word representations in vector space (2013). CoRR abs/1301.3781

14. Mikolov, T., Sutskever, I., Chen, K., Corrado, G.S., Dean, J.: Distributed representations of words and phrases and their compositionality. In: Advances in Neural Information Processing Systems 26, pp. 3111–3119 (2013). 27th Annual Conference on Neural Information Processing Systems 2013

15. Paltoglou, G., Thelwall, M.: A study of information retrieval weighting schemes for sentiment analysis. In: Proceedings of the 48th Annual Meeting of the Association for Computational Linguistics, ACL 2010, pp. 1386–1395 (2010)

16. Pang, B., Lee, L., Vaithyanathan, S.: Thumbs up? Sentiment classification using machine learning techniques (2002). cs.CL/0205070

17. Socher, R., Huval, B., Manning, C.D., Ng, A.Y.: Semantic compositionality through recursive matrix-vector spaces. In: Proceedings of the 2012 Joint Conference on Empirical Methods in Natural Language Processing and Computational Natural Language Learning, EMNLP-CoNLL 2012, pp. 1201–1211 (2012)

18. Socher, R., Pennington, J., Huang, E.H., Ng, A.Y., Manning, C.D.: Semi-supervised recursive autoencoders for predicting sentiment distributions. In: Proceedings of the 2011 Conference on Empirical Methods in Natural Language Processing, EMNLP 2011, pp. 151–161 (2011). A meeting of SIGDAT, a Special Interest Group of the ACL

19. Wang, S.I., Manning, C.D.: Baselines and bigrams: Simple, good sentiment and topic classification. In: Proceedings of the 50th Annual Meeting of the Association for Computational Linguistics: Short Papers, vol. 2, pp. 90–94 (2012)

A Real Time Wireless Interactive Multimedia System

Hong Li, Wei Yang[✉], Yang Xu, Jianxin Wang, and Liusheng Huang

University of Science and Technology of China, Hefei, China
qubit@ustc.edu.cn

Abstract. Recent years, various interactive multimedia systems have been applied to relevant fields such as education, entertainment, etc. Researchers exploit sensors, computer vision, ultrasonic, and electromagnetic radiation to achieve human-computer interaction (HCI). This paper proposes an interactive wireless multimedia system which utilizes ubiquitous wireless signals to identify human motions around smart WiFi devices. Compared with related work, our system realizes interactions between human and computer without extra hardware devices. The system identifies human gestures around the smart devices (i.e., a laptop) equipped with the commercial 802.11n NIC, and it maps different gestures into distinguishable computer instructions. We build a proof-of-concept prototype using off-the-shelf laptop and evaluate the system in a laboratory environment with standard WiFi access points. The results show that our system detects human gesture with an accuracy over 95 % and it achieves an average gesture classification accuracy of 89 % for five different users.

Keywords: Gesture recognition · Human-computer interaction · WiFi

1 Introduction

Recent years witness a rising trend to incorporate gesture recognition system into various smart devices, including smart phones [1], laptops [2], gaming console [3]. These systems generally exploit the available sensors to enhance their functionality. The existing solutions adopt techniques such as computer vision [3], sensors [4–6], ultrasonic [2], and infrared to realize gesture recognition. These technologies are promising, however, they face some unavoidable disadvantages, including sensitivity to lighting conditions, requiring specialized hardware devices.

Given that the disadvantage of above techniques, WiFi-based gesture recognition [7–9] systems have been proposed to overcome the limitations of existing gesture recognition systems. These solutions are able to recognize in-air without extra equipments such as sensors or cameras. WiFi-based gesture recognition systems are based on analysis of the characteristics of signal patterns, including rising edge, falling edge, plateaus, caused by human motions. However, these system need sophisticated hardware devices to extract the desired signal features.

© Springer International Publishing Switzerland 2016
F. Li et al. (Eds.): APWeb 2016, Part I, LNCS 9931, pp. 53–65, 2016.
DOI: 10.1007/978-3-319-45814-4_5

For example, WiSee [8] and WiVi [9] adopt Universal Software Radio Peripheral (USRP) and the device-free radio-based activity recognition (DFAR) scheme [7] utilizes Software Defined Radio (SDR). Moreover, all these systems do not provide fine-grained interactions with a certain application in smart devices.

This paper presents a new method for controlling multimedia systems in smart devices by recognizing a set of human gestures under wireless environment. Our system does not need additional sensors, is resilient to environmental changes, and achieves recognizing human gestures in real time. The key insight is to leverage the effect of in-air gestures toward fine-grained channel state information (CSI) to recognize users' gestures. After that, the system maps the identified human gestures into intended instructions in smart devices to achieve system control. There are several challenges which must be solved in order to translate the above high-level idea into a practical system, including handling the noisy CSI time series due to multipath reflections, extracting and recognizing human gestures in CSI values, dealing with the variations of gestures as well as their attributes for different humans and even for the same human at different time.

To address these challenges, our system adopts Butterworth low pass filter to reduce the high frequency noise exists in CSI time series. The frequency of variations caused by the movements of hands lie at the low end of the spectrum, however, the frequency of noise lies at the high end of the spectrum. Butterworth low pass filter is a natural choice for eliminating these high frequency noise because its high fidelity in preserving both time and frequency resolution of WiFi signals. To extract and recognize subtle changes caused by human gestures in CSI time series, we introduce a unique signal pattern, for example, a preamble, to identify the beginning of the human gestures and counter the interference from irrelevant people. This also helps to enhance system's energy-efficiency which stems from the fact that detecting the preamble can be easily done by monitoring a simple threshold, rendering the system idle most of the time.

In summary, we make the following contributions in this paper:

(1) We present a proof-of-concept prototype on off-the-shelf laptops which extracts the physical layer CSI from the Intel 5300 NIC using a modified driver developed by Halperin *et al.* [10] to recognize a group of basic in-air gestures. Further, we use the identified gesture to control multimedia system in the smart WiFi devices.
(2) To evaluate the performance of this non-intrusive and device-free scheme, we test our system in our laboratory environment which covers an area of $50 \times 23\,\text{ft}^2$ with only one target user. The gesture set includes 7 gestures (6 normal gestures and 1 preamble gesture). Each gesture is performed by a target user for 30 times. Finally, we get 1050 gesture instances for 5 different users to evaluate our system. The experimental results show that our system can detect human gesture with an accuracy over 95 % using a single assess point within a distance of 1 ft around smart WiFi devices, and it achieves an average classification accuracy of 89 % for 5 different users in a multimedia player application case study.

2   Preliminary

Smart WiFi devices that support IEEE 802.11n/ac standards generally have multiple transceiver antennas. Hence, they support multiple-input multiple-output (MIMO) which provides several MIMO channels between transmit-receive (TX-RX) antenna pairs. Each TX-RX pair of transmitter and receiver consists of multiple subcarriers. These WiFi devices keep monitoring the MIMO channels to effectively acquire the signal strength, Signal to Noise Ratio (SNR), transmit power and rate adaptations. These devices quantify the detailed state of channel information in terms of channel state information (CSI). Recently, Halperin *et al.* proposed a new methods [10] to acquire fine-grained CSI values by modifying the commercial 802.11n NIC. It extracts the primitive signal variations from the physical layer. As the extracted signals is the resultant of constructive and destructive interference of multipath signal reflection. The variations caused by gestures are captured in the CSI time series for all subcarriers between every TX-RX antenna pair. Then the variations can be extracted to identify gestures. In frequency domain, the narrowband flat-fading channel with MIMO. A MIMO system at any time instant can be expressed as follows:

$$y = Hx + n, \tag{1}$$

where y is the received vector, x is the transmitted vector, n represents the noise vector and H denotes the channel matrix. CSI is an estimation of H. In Orthogonal Frequency Division Multiplexing (OFDM) system, CSI is represented at subcarrier level. CSI values in a single subcarrier can be formulated in the following equation:

$$h = |h|e^{j \sin \theta} \tag{2}$$

where $|h|$ and θ are the amplitude and phase respectively. Compare to Received Signal Strength Indicator (RSSI), CSI comprises fine-grained information. Hence, CSI can be utilized to sense subtle changes caused by human gestures.

3   System Conceptual Overview

In this section, we give the conceptual overview of our system including the Signal Processing, Gesture Set, and Multimedia Application Instruction.

(1) Signal Processing: This layer detects and extracts primitive CSI values in CSI streams, which reflect the signal diversity and space. These signal changes include rising edge, falling edge, pause. They are separately caused by moving the hand away from the receiver, moving the hand towards the receiver, and holding the hand still over the receiver. Other complicated gestures can be composed by combining these three variances.
(2) Gesture Set: Different CSI waveform patterns extracted from the primitive signals can be exploited to recognize higher level gestures. For example, an up-down hand gesture can be mapped to the primitive rising edge and

then falling edge. We define a set of gestures which can be represented by some primitive falling edges, rising edges and pauses. Considering all up-down, right-left or other gestures may have the similar effect on the signal variations and hence the same primitive sequence of a rising and then falling edge. We empirically choose the most suitable ways and positions to perform gestures which fit the applications and can be easy to distinguish and identify.

(3) Multimedia Application Instruction: We map the identified human gestures into a group of application instructions in this layer. We assume that each kind of gesture corresponding to a specific application instruction. As an example, for a multimedia system, a "pause" action can be performed with a push hand gesture, while a "speed up" action can be mapped to a right-movement gesture. In the next section, we give the details of system flow of extracting these different semantics and the relevant challenges.

4   System Design

In this section, we present the detail flow of our system and address the mentioned challenges. Our system flow covers three main procedures which corresponding to the system conceptual overview: Primitive Signal Processing, Gesture Recognition, and Gesture Mapping.

4.1   Primitive Signal Processing

The CSI values extracted from commodity WiFi Network Interface Cards (NIC) are inherently noisy because of the frequency changes in internal transmission rate, transmit power levels and even unavoidable Carrier Frequency Offsets (CFO) resulted from the hardware imperfections and environment variations [11]. To detect and extract human gesture information from CSI values, we must remove these innate noise. We empirically employ weighted moving average method for every 60 points to smooth the original signals. And then, the algorithm removes the DC component that accounts for the static reflections of the environment by subtracting the average value of CSI within a window containing 30 CSI values. Considering the high-fidelity of Butterworth filter, we first adopt a Butterworth low pass filter to remove high frequency noise which prevents us to identify human gestures. As the gesture movements while instructing applications around smart devices lie anywhere between 1 to 60 Hz, and the CSI sample rate is $F_s = 500$ samples/s, we set the cut-off frequency of Butterworth low pass filter with $w_c = \frac{2\pi * f}{F_s} = \frac{2\pi * 60}{500} \approx 0.75$ rad/s. To better compare the filtered signals with threshold, the system maps the filtered CSI values between their maximum and minimum interval. As can be seen in Fig. 1, we present a Up-Down gesture waveform as an example after the process of signal processing procedures. After that we obtain the filtered CSI time series. Assume that t represents the number of transmitting antenna and r represents the number of receiving antenna, then, we get a CSI matrix $\mathbf{M}_{t,r}$ with a dimension of $N \times T$,

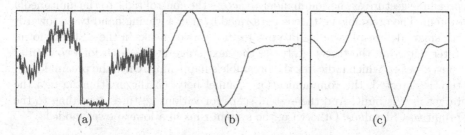

Fig. 1. Up-Down gesture waveform after different signal processing procedures. (a) Original signals (b) After applying weighted moving average method (c) Through low pass butterworth filter

where N is the number of CSI streams and T is the length of time. The value of N is related to the number of transmitting and receiving antennas. N can be calculated as $N = t \times r$ ($N = 9$ in our system), and we totally obtain $30 \times N$ CSI sucarriers. When a human gesture happens around our system, we experimentally observed that the CSI values change in all subcarriers. Hence, we find that the different subcarriers are correlated. In order to detect the human gestures in CSI time series, the system splits the CSI values of each subcarrier into R bins. We empirically set the bin size to be 100 CSI values to acquire the target CSI subcarrier which could be used for gesture detection. Then the algorithm calculates the variance of those bins. We compare the variances calculated for different bins of one subcarrier with the corresponding bins of other subcarrier, the subcarrier which has a larger number of higher variance bins is selected to be the target CSI subcarrier to detect gestures.

4.2 Gesture Recognition

After choosing the target CSI subcarrier. We use the target subcarrier to extract human gestures and their characteristics (i.e., frequency and waveform). It has the following two procedures: Detection and Recognition.

Detection: The gestures selected in our system are comprised of simple rising edges, falling edges, or pauses. To correctly detect the starting and finishing points of human gestures in target CSI subcarrier, we set thresholds to detect the occurrence of human gestures. The processed CSI values changes around the zero point, and the gesture waveforms lie both up and below the zero value. Hence, we set two thresholds to automatically detect human gestures. The positive threshold value is greater than zero which used for detect the gesture waveforms such as "Up". The negative threshold value is smaller than zero which facilitates the detection of the gesture waveform such as "Down". We empirically determined appropriate values of the two threshold. This method could efficiently detect the occurrence of target human gestures in real time. For the sake of saving energy and reducing the possibility of false detection. Our system sets a special preamble gesture as the commander to access the control right to the multimedia system. The preamble gesture is performed by user's waving hand twice towards the smart devices. It will lead to two regular convex peaks in the CSI waveform. After detecting the target gestures, the next stage is to search for two regular convex peaks, which indicates the preamble's happening. Once the preamble gesture is detected, the communication channel between the multimedia and the target user is built. And the system scans for various gestures according to the primitive CSI values. Otherwise, the system runs in a lower-power mode.

Recognition: Since different human gestures tend to cause different CSI changes in target CSI subcarrier, we can identify gestures by extracting CSI time series patterns caused by human gestures to achieve recognition. The system detects the onsets of target gestures by comparing CSI values with the defined thresholds. If the CSI value exceeds the value of the positive threshold or decreases to the value of negative value, it estimates the starting point of human gesture as s. We observe that on average the waveforms of a gesture spanned $t_{avg} = 500$ CSI values. Hence, we approximately get the finishing point as $e = s + t_{avg}$. Considering some gestures might have positive and negative waveforms such as "Up-Down". Then if the distance of two consecutive detected waveform less than d, the algorithm combines the two waveform to represent a same gesture. Finally, we set a guardian interval B which helps to extract the gesture waveforms. That means we add the guardian interval to both sides of the estimated gesture interval. Therefore, the gesture interval becomes $[s-B, e+B]$. Once the gesture onsets are determined, the algorithm extracts the CSI waveform between the gesture interval to identify gestures. We calculate the features from the acquired gesture waveform including zero-crossing rate, average value of gesture waveforms, first quartile and third quartile, variance, short time energy, short time average amplitude. We use the extracted features to form a feature vector to train FT, Naive Bayes (NB), and Random Forest Classifiers [12], respectively. We choose the classifier which has the best recognition performance to recognize gestures. Then the trained classifier can be used for recognizing the human gestures in real time.

4.3   Gesture Mapping

This section presents the direct mapping step based on the multimedia semantics. We map the application actions to their corresponding gestures as Table 1. After recognizing the human gestures using the pre-determined sequences, the system maps the identified gestures into their corresponding multimedia actions to control multimedia system. The gesture set in our interactive multimedia system generally covers 7 gestures which map to the most common 7 application actions in a multimedia system. Figure 2 shows the filtered waveforms of six gestures in our system. The developer can extend the gesture set and fully utilize the gesture attributes to enhance system's functionality. For example, the frequency attribute of gestures can be used to determine how fast the character should move in the multimedia system. We also note that multiple actions can

Table 1. Gestures and corresponding multimedia actions.

Human gestures	Multimedia actions
Up	Volume up
Down	Volume down
Up-Down	Play
Right	Speed up
Left	Slow down
Push	Stop

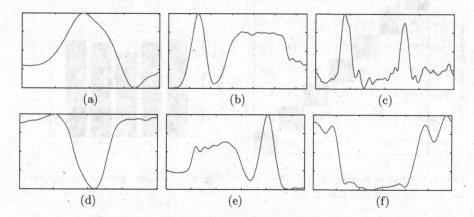

(a) (b) (c)

(d) (e) (f)

Fig. 2. The waveform of gestures in our system. (a) Up Gesture (b) Right Gesture (c) Wave Hand Twice (d) Down Gesture (e) Left Gesture (f) Push Gesture

be mapped to some other multimedia instructions such as the double right-hand could be mapping into speed up two times.

5 Evaluation

In this section, we analyze the performance of our proposed system in a typical laboratory environment. We first present the experimental setup in our environment and then we show the performance of our system.

5.1 Experimental Setup

The experimental setup includes two parts: hardware Setting and Data Collection. The details are illustrated below.

Hardware Setting. The system consists of two components: a laptop equipped with a commercial 802.11n WiFi card as a receiver and a commercially available WiFi access point (AP). We implement a proof-of-concept prototype of the

(a) Right-Left   (b) Up-Down   (c) Push

Fig. 3. The waveform of gestures in our system.

Gesture Classified

	Down	Up-Down	Left	Push	Right	Up	WaveHand
Down	0.85	0.15	0.00	0.00	0.00	0.00	0.00
Up-Down	0.08	0.92	0.00	0.00	0.00	0.00	0.00
Left	0.00	0.00	0.86	0.00	0.00	0.00	0.14
Push	0.06	0.00	0.00	0.94	0.00	0.00	0.00
Right	0.17	0.00	0.00	0.00	0.83	0.00	0.00
Up	0.00	0.00	0.00	0.07	0.00	0.93	0.00
WaveHand	0.00	0.00	0.00	0.00	0.00	0.00	1.00

Actual Gesture Prefomed

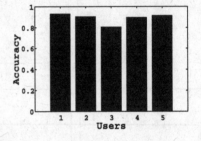

Fig. 4. Confusion matrix for the different gestures using NB classifier

Fig. 5. Average gesture recognition accuracy using NB classifier

system in a Think-pad E40 laptop with Intel 5300 WiFi card and test it using a TP-LINK TL-WDR4300 wireless router as an AP. Both the receiver and the AP have 3 working antennas. The distance between the receiver and the AP is around 8 ft. To obtain CSI values from regular data frames transmitted by the AP, we modified the firmware of the WiFi card as in [10] to report CSI values to upper layers. All the experiments were performed in the 5 GHz frequency band with 20 MHz bandwidth channels. The system acquires CSI measurements from the CSI tool and processes it in real-time using MATLAB.

Data Collection. Our laboratory environment covers an area of $50 \times 23\,\text{ft}^2$. There is only one target user in the experimental environment. The target user performs gestures near the receiver with a distance about 1 ft. Figure 3 shows the movement of hand gesture near the receiver. We collect gesture dataset from five student volunteers also mentioned as users 1–5. Users 1–5 performs each gesture for 30 times. We totally collect 1050 gesture instances for performance evaluation of the system.

5.2 Performance Evaluation

In this section, we present the system performance in various conditions such as different classifiers, different users as well as different time during a daytime.

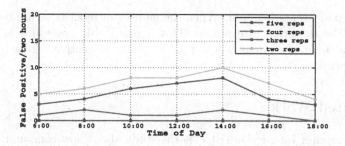

Fig. 6. False detection rate from a 12-h daytime trace

Different Classifiers. We test the performance of three classical classifiers (FT, NB, and Random Forest) in our experiment. We set feature vectors extracted from gesture instances of user 5 as the input samples for the three classifiers. These classifiers perform 10-fold cross validation. The result yields that the overall performance of the three classifiers are all above 90 %. However, the recognition accuracy of gesture "Right" in FT and Random Forest classifiers are 67 % and 50 %, respectively. In NB classifier, the recognition accuracy of all gestures are above 80 %. Figure 4 presents the confusion matrix for the seven gestures using NB classifier. Especially, the recognition accuracy of "WaveHand" reaches 100 %. It means our system can correctly identify the commander gestures of target user. Hence, we adopt NB classifier in our system for gesture recognition.

Different Users. To verify the system's resilience towards different users, the feature vectors of gesture instances collected from users 1–5 are used as the input of the selected NB classifier. We trained 5 user-specific NB classifiers for these 5 users. Every classifier performs 10-fold cross validation using each user's gesture instances. Figure 5 shows the average recognition accuracy of the gesture instances of these five users. Obviously, their average recognition accuracies are all above 80 %. The lowest extraction accuracy for user 3 shows that more gestures were falsely classified, which is due to the significant difference in his gesturing behavior compared to other users. The speed and magnitude of users' gestures also influence the recognition accuracy of our system. High gesturing speed will lead to short time span of gesture instance. And on the other hand, if users perform gestures in a larger magnitude, the amplitude of the signal change will be much greater than the original signal level. The accuracy of our system for such a user can be increased significantly by adjusting the thresholds of our algorithm for the given user.

Different Time. We test the robustness of our system during a daytime (6:00 AM to 6:00 PM) with a time span of 12 h. Figure 6 plots the number of false detection events every two hours as a function of time. The figure shows results for different number of repetitions in the preamble. The average number of false events is highest when the preamble contains only two repetitions. And with the number of repetitions increases, the false detection events significantly decline. Specifically, with four repetitions, the average false detection rate reduces to 0.67 events per hour. When the number of repetitions are more than four, the false detection rate is zero. This is reasonable because it is unlikely that typical human motions would produce five consecutive regularly convex CSI waveforms.

6   Related Work

Human-Computer Interaction (HCI) is the study about how computer technology influences human work and activities. These technologies generally cover from obvious computers to mobile phones, household appliances, car infotainment systems and even embedded sensors such as automatic lighting. Recent years, various techniques are used to HCI systems to improve user experience. The techniques include fundamental interaction styles such as direct manipulation, the mouse pointing device, and windows. Application types, like drawing, text editing, etc. And the Up-and-Coming Areas that will likely have the biggest impact on interfaces of the future, such as gesture recognition, multimedia, and 3D [13].

The fundamental interaction styles was first demonstrated by Ivan Sutherland in his PhD thesis about Sketchpad [14]. It enables the manipulation of objects using a lightpen, including grabbing objects, moving them, changing size, etc. Then the mouse pointing devices and other basic intersections were proposed by researchers. The application types of interactions, for example, the first drawing program presented by William Newmans Markup in 1975. Nowadays, researchers tend to integrate gesture recognition techniques to control devices. There are some products using the state-of-the-art techniques, for example, Xbox Kinect [15] adopts hybrid cameras to recognize human motions to realize human-computer interaction in multimedia systems. WiGest [16] extracts variations of the received signal strength indicator values to identify gestures.

Gesture recognition techniques have found a diverse set of applications, e.g., 3D in-air user-interface for mobile and laptops [17], remote control of home appliances [8], sterilized operation of medical devices and distraction-free management of in-car infotainment system [18,19]. The typical gesture recognition systems can be categorized into three types: computer vision based, sensors based, audio and radio based. Wahs *et al.* gave a comprehensive study in vision based techniques [18]. Recent arts include Xbox Kinect, LeaMotion [20] both utilize computer vision to recognize human gestures. Wearable or near-body sensing techniques such as Data glove, Sayre glove [21]. They use sensors in users' gloves to sense gestures of target users. Audio signals generated by mobile devices may be affected by human gestures. Researchers extract the resulting pattern to recognize human gestures [22,23], An alternative way extracts Doppler features from soundwaves reflected by human gestures relevant to interaction with computers [2]. WiSee [8] extends this approach to WiFi signals to identify 9 human gestures. After that, various WiFi based human motion work were proposed like WiVi [9], WiTrack [24], WiHear [25], Wikey [26], WiDraw [27], etc.

7   Conclusion

In this paper, we present a wireless interactive multimedia system that uses the fine-grained channel state information extracted from the physical layer to control the multimedia system by detecting and recognizing human gestures around a smart WiFi device. Our system does not need any extra hardware devices such as sensors, cameras, or sophisticated USRP platforms. We simply extract the CSI values by modifying the commercial 802.11n NIC. After applying typical signal processing methods, the system detects the variations in CSI time series caused by human gestures. Our system can realize interaction with multimedia systems in a smart WiFi device equipped with commercial 802.11n NIC (e.g., Intel 5300 NIC). We addressed the following system challenges including signal denoising, gesture extraction, interferences elimination. We evaluate the system in a typical laboratory environment using the gesture instances collected from 5 users. The results show that our system can accurately detect the target human gesture with an accuracy over 95 %, and it achieves recognize human gestures with an average accuracy of 89 % for five different users. This accuracy indicates that our system has the ability to use the ubiquitous wireless signals to sense human gestures to further control multimedia systems.

Acknowledgments. We would like to thank the anonymous reviewers for their valuable comments for improving the quality of the paper. This work was supported by the National Natural Science Foundation of China (No. 61572456) and the Natural Science Foundation of Jiangsu Province of China (No. BK20151241).

References

1. Toshibag55.   http://www.engadget.com/2008/06/14/toshiba-qosmiog55-features spursengine-visual-gesture-controls/
2. Gupta, S., Morris, D., Patel, S., Tan, D.: Soundwave: using the doppler effect to sense gestures. In: Proceedings of the SIGCHI Conference on Human Factors in Computing Systems, pp. 1911–1914. ACM (2012)
3. Shotton, J., Sharp, T., Kipman, A., Fitzgibbon, A., Finocchio, M., Blake, A., Cook, M., Moore, R.: Real-time human pose recognition in parts from single depth images. Commun. ACM **56**(1), 116–124 (2013)
4. Cohn, G., Morris, D., Patel, S., Tan, D.: Humantenna: using the body as an antenna for real-time whole-body interaction. In: Proceedings of the SIGCHI Conference on Human Factors in Computing Systems, pp. 1901–1910. ACM (2012)
5. Harrison, C., Tan, D., Morris, D.: Skinput: appropriating the body as an input surface. In: Proceedings of the SIGCHI Conference on Human Factors in Computing Systems, pp. 453–462. ACM (2010)
6. Kim, D., Hilliges, O., Izadi, S., Butler, A.D., Chen, J., Oikonomidis, I., Olivier, P.: Digits: freehand 3d interactions anywhere using a wrist-worn gloveless sensor. In: Proceedings of the 25th Annual ACM Symposium on User Interface Software and Technology, pp. 167–176. ACM (2012)

7. Scholz, M., Sigg, S., Schmidtke, H.R., Beigl, M.: Challenges for device-free radio-based activity recognition. In: Proceedings of the 3rd Workshop on Context Systems, Design, Evaluation and Optimisation (CoSDEO 2011), in Conjunction with MobiQuitous, vol. 2011 (2011)
8. Pu, Q., Gupta, S., Gollakota, S., Patel, S.: Whole-home gesture recognition using wireless signals. In: Proceedings of the 19th Annual International Conference on Mobile Computing & Networking, pp. 27–38. ACM (2013)
9. Adib, F., Katabi, D.: See through walls with wifi!, vol. 43. ACM (2013)
10. Halperin, D., Hu, W., Sheth, A., Wetherall, D.: Tool release: gathering 802.11 n traces with channel state information. ACM SIGCOMM Comput. Commun. Rev. 41(1), 53–53 (2011)
11. Wang, W., Liu, A.X., Shahzad, M., Ling, K., Lu, S.: Understanding and modeling of wifi signal based human activity recognition. In: Proceedings of the 21st Annual International Conference on Mobile Computing and Networking, pp. 65–76. ACM (2015)
12. Hall, M., Frank, E., Holmes, G., Pfahringer, B., Reutemann, P., Witten, I.H.: The weka data mining software: an update. ACM SIGKDD Explorations Newslett. 11(1), 10–18 (2009)
13. Myers, B.A.: A brief history of human-computer interaction technology. Interactions 5(2), 44–54 (1998)
14. Sutherland, I.E.: Sketch pad a man-machine graphical communication system. In: Proceedings of the SHARE Design Automation Workshop, pp. 6–329. ACM (1964)
15. XboxKinect. http://www.xbox.com/en-US/kinect
16. Abdelnasser, H., Youssef, M., Harras, K.A.: Wigest: A ubiquitous wifi-based gesture recognition system. In: 2015 IEEE Conference on Computer Communications (INFOCOM), pp. 1472–1480. IEEE (2015)
17. Pickering, C.A., Burnham, K.J., Richardson, M.J.: A research study of hand gesture recognition technologies and applications for human vehicle interaction. In: 3rd Conference on Automotive Electronics. Citeseer (2007)
18. Wachs, J.P., Kölsch, M., Stern, H., Edan, Y.: Vision-based hand-gesture applications. Commun. ACM 54(2), 60–71 (2011)
19. Melgarejo, P., Zhang, X., Ramanathan, P., Chu, D.: Leveraging directional antenna capabilities for fine-grained gesture recognition. In: Proceedings of the 2014 ACM International Joint Conference on Pervasive and Ubiquitous Computing, pp. 541–551. ACM (2014)
20. LeapMotion. https://www.leapmotion.com/
21. Dipietro, L., Sabatini, A.M., Dario, P.: A survey of glove-based systems and their applications. IEEE Trans. Syst. Man Cybern. Part C Appl. Rev. 38(4), 461–482 (2008)
22. Tarzia, S.P., Dick, R.P., Dinda, P.A., Memik, G.: Sonar-based measurement of user presence and attention. In: Proceedings of the 11th International Conference on Ubiquitous Computing, pp. 89–92. ACM (2009)
23. Scholz, M., Riedel, T., Hock, M., Beigl, M.: Device-free and device-bound activity recognition using radio signal strength. In: Proceedings of the 4th Augmented Human International Conference, pp. 100–107. ACM (2013)
24. Adib, F., Kabelac, Z., Katabi, D., Miller, R.C.: 3d tracking via body radio reflections. In: 11th USENIX Symposium on Networked Systems Design and Implementation (NSDI 14), pp. 317–329 (2014)

25. Wang, G., Zou, Y., Zhou, Z., Wu, K., Ni, L.M.: We can hear you with wi-fi! In: Proceedings of the 20th Annual International Conference on Mobile Computing and Networking, pp. 593–604. ACM (2014)

26. Ali, K., Liu, A.X., Wang, W., Shahzad, M.: Keystroke recognition using wifi signals. In: Proceedings of the 21st Annual International Conference on Mobile Computing and Networking, pp. 90–102. ACM (2015)

27. Sun, L., Sen, S., Koutsonikolas, D., Kim, K.H.: Widraw: enabling hands-free drawing in the air on commodity wifi devices. In: Proceedings of the 21st Annual International Conference on Mobile Computing and Networking, pp. 77–89. ACM (2015)

Mining Co-locations from Continuously Distributed Uncertain Spatial Data

Bozhong Liu[1,2(✉)], Ling Chen[2], Chunyang Liu[2], Chengqi Zhang[2], and Weidong Qiu[1]

[1] School of Electronic Information and Electrical Engineering, Shanghai Jiao Tong University, Shanghai, China
liu.bo.zhong@gmail.com
[2] Centre for Quantum Computation and Intelligent Systems, University of Technology Sydney, Sydney, Australia

Abstract. A co-location pattern is a group of spatial features whose instances tend to locate together in geographic space. While traditional co-location mining focuses on discovering co-location patterns from deterministic spatial data sets, in this paper, we study the problem in the context of continuously distributed uncertain data. In particular, we aim to discover co-location patterns from uncertain spatial data where locations of spatial instances are represented as multivariate Gaussian distributions. We first formulate the problem of *probabilistic co-location mining* based on newly defined prevalence measures. When the locations of instances are represented as continuous variables, the major challenges of probabilistic co-location mining lie in the efficient computation of prevalence measures and the verification of the probabilistic neighborhood relationship between instances. We develop an effective probabilistic co-location mining framework integrated with optimization strategies to address the challenges. Our experiments on multiple datasets demonstrate the effectiveness of the proposed algorithm.

1 Introduction

Co-location mining is an important application in spatial data sets. A co-location pattern is a subset of spatial features whose instances are frequently located close to each other. Spatial co-location patterns yield valuable knowledge for various applications such as Epidemiology [10], Ecology [5] and E-commerce [17]. Due to its importance, the problem of finding prevalent co-location patterns from spatial data sets has been explored extensively [10,14–16]. Traditional co-location pattern mining usually focuses on deterministic data sets, where instances of spatial features occur affirmatively at precise locations. However, it is not always the

This work was supported, in part, by the Australia Research Council (ARC) Discovery Project under Grant No. DP140100545, program of Shanghai Technology Research Leader under Grant No. 16XD1424400, program of New Century Excellent Talents in University under Grant No. NCET-12-0358, and public interest research of Institute of Forensic Science, Ministry of Justice, PRC under Grant No. GY2016G-6.

F. Li et al. (Eds.): APWeb 2016, Part I, LNCS 9931, pp. 66–78, 2016.
DOI: 10.1007/978-3-319-45814-4_6

case in practice. On one hand, the data' is inherently uncertain in many applications, especially in sensor environments and moving object applications [9]. On the other hand, artificial noise may be added deliberately for privacy protection [13]. Hence, data uncertainty is ubiquitous in real world and mining patterns from uncertain data has become an interesting and important task in the literature [3].

A few works on mining co-locations from uncertain spatial data have emerged recently, which consider data uncertainty from different aspects. Wang et al. [12] study mining co-location patterns from uncertain spatial data where instances are associated with *existential probabilities*. That is, whether an instance occurs or not is uncertain. However, if it occurs, its location is assumed to be deterministic. In contrast, Liu and Huang [8] explore the problem of co-location mining from uncertain data by recognizing the *location probabilities* of instances. Given an instance, their work considers several (typically 3-5) possible locations within a bounded range and assigns probabilities to indicate how likely the instance occurs at one of the locations. That is, the location of an instance is modeled as a discrete variable.

Considering the arrival of big data era, coupled with the continuous nature of spatial data, it is very likely that for each instance, a collection of possible locations may be gathered. For example, in the application of interesting constellation discovery in astrophysics, it is common to record the locations of stars in a long time period while the locations may vary every time the stars are observed. In this case, it is more reasonable to model the location of an instance as a continuous variable, instead of a discrete one. Therefore, in this paper, we focus on the problem of mining co-locations from uncertain spatial data where location of each instance is modeled as a continuous *multivariate Gaussian distribution*, which is widely used in modeling location uncertainty such as in spatial range querying [4] and localization in robotics [11]. To our knowledge, this is the first work that mines co-locations from Gaussian-based uncertain spatial data.

Mining co-location patterns from uncertain spatial data where locations are continuous variables is a challenging problem. Firstly, the existing framework of problem definition cannot be adopted directly because the existing interestingness measures cannot deal with locations modeled as probabilistic distributions. Secondly, the mining process will be computationally expensive and complicated. For example, when locations of instances are represented as probabilistic distributions, expensive integration will be involved to examine whether two instances are probabilistic neighbors.

To address the challenges, we first re-define the interestingness measures to cope with continuously distributed spatial data, based on which the problem of *probabilistic co-location mining* is formulated (Sect. 3). To compute the newly defined prevalence measure, it is essential to find out the probability that an instance supports/participates a feature set. We propose proper and effective schemes to compute the probability efficiently (Sect. 4). After handling the definition and computation of interestingness measures, a framework for probabilistic co-location mining is put forward (Sect. 5). Observing the bottleneck of the mining process lies in the discovery of probabilistic neighbors of instances, we further devise an optimization strategy to skip verifying the neighborhood relationship between particular instance pairs (Sect. 6). The main contributions of this paper are summarized as follows.

1. We formulate the problem of *probabilistic co-location mining* from Gaussian-based uncertain spatial data, based on newly defined prevalence measures.
2. We develop a framework for mining probabilistic co-locations from Gaussian-based spatial data, with effective strategies to address the computation of prevalence and the verification of probabilistic neighborhood relations.
3. We conduct experiments on multiple data sets to examine the effectiveness of the proposed methodologies.

2   Problem Definitions

In this section, we first review definitions related to co-location mining from deterministic data. Next, we formally define our problem based on redefined prevalence measures in the context of Gaussian-based uncertain spatial data.

2.1   Co-location Patterns in Deterministic Data

Given a deterministic spatial data set $\mathcal{F}$, measures related to characterizing the interestingness of a subset of features have been defined in [10].

Definition 1. *Given a subset of features* $F = \{f_1, f_2, \ldots, f_k\}$, $E = \{e_1, e_2, \ldots, e_k\}$ *is a **Row Instance (RI)** of F, denoted as $RI(F)$, if $\forall i \in [1, k]$, e_i is an instance of f_i and $\forall i, j \in [1, k], \|e_i - e_j\| \leq \tau$, where $\|e_i - e_j\|$ refers to the distance between two events.*

Definition 2. *The **Table Instance (TI)** of a subset of features $F \subseteq \mathcal{F}$, denoted as $TI(F)$, is the collection of all its row instances $\{RI_1(F), \ldots, RI_m(F)\}$.*

Definition 3. *Given a subset of features $F = \{f_1, \ldots, f_k\}$, the **Participation Ratio** of a feature $f_i \in F$, denoted as $PR(f_i, F)$, is the fraction of events of feature f_i that participate in the table instance of F. That is,*

$$PR(f_i, F) = \frac{|\{e_j | e_j \in \widehat{TI}(\{f_i\}), e_j \in \widehat{TI}(F)\}|}{|\{e_j | e_j \in \widehat{TI}(\{f_i\})\}|}, \tag{1}$$

where $\widehat{TI}(\cdot)$ is the union of elements in TI set. Hence, the denominator refers to the total number of events of feature f_i and the numerator refers to the distinct number of events of feature f_i that appear in the table instance of F.

Definition 4. *The **Participation Index** of a subset of features $F = \{f_1, \ldots, f_k\}$, denoted as $PI(F)$, is defined as*

$$PI(F) = \min_{i \in [1, k]} PR(f_i, F). \tag{2}$$

Definition 5. *Given a user-specified threshold min_{PI}, a subset of features $F \subseteq \mathcal{F}$ is a prevalent **Co-location Pattern** if $PI(F) \geq min_{PI}$.*

2.2   Co-location Patterns in Gaussian-Based Data

In this paper, we model the location of an event as a continuous variable. In particular, given an event e_i, the location of e_i is represented as a d-dimensional Gaussian distribution, $e_i = (x_i^{(1)}, x_i^{(2)}, \ldots, x_i^{(d)})^T$, with its mean location center μ_i and the corresponding covariance matrix Σ_i. The probability distribution function is given by $P_{e_i}(x) = \frac{1}{(2\pi)^{\frac{d}{2}} |\Sigma_i|^{\frac{1}{2}}} \exp\left[-\frac{1}{2}(x - \mu_i)^T \Sigma_i^{-1}(x - \mu_i)\right]$.

In the context of uncertain data, the distance between two events becomes a probabilistic distribution. Therefore, we define the probabilistic neighborhood relationship between a pair of events as follows.

Definition 6. *Given a distance threshold τ ($\tau \geq 0$) and a probabilistic neighborhood threshold θ ($0 < \theta < 1$), two events e_i and e_j are **probabilistic neighbors** if the probability that the distance between them is no greater than τ is no less than θ. That is, $\Pr[\|e_i - e_j\| \leq \tau] \geq \theta$.*

Based on the definition of probabilistic neighbors, we can define a *clique instance* of a co-location as follows, corresponding to the concept of row instance in the context of deterministic data.

Definition 7. *Given a subset of features $F = \{f_1, \ldots, f_k\}$, a set of events $E = \{e_1, \ldots, e_k\}$ is a **Clique Instance** of F, denoted as $CI(F)$, if $\forall i \in [1, k]$, e_i is an event of f_i, and $\forall i, j \in [1, k]$, $R_{\tau, \theta}(e_i, e_j) = 1$.*

Given a subset of features, we can find a collection of clique instances from the input spatial data. We record the set of clique instances, $\{CI_1(F), CI_2(F), \ldots, CI_m(F)\}$, in a *clique instance table*, denoted as $CIT(F)$.

Recall that in deterministic data, the participation ratio of a feature in a subset of features is computed as the fraction of events of this feature that participate in the collection of row instances of the feature set. However, in the context of uncertain data, whether an event participates in a clique instance is probabilistic. Let $PR(f_i.e_j, F)$ be the probability that the jth event of feature f_i participates in the collection of clique instances of F (we will explain how to compute this value in the next section). Then, the probabilistic participation ratio of a feature in a feature set can be defined as follows.

Definition 8. *Given a subset of features $F = \{f_1, \ldots, f_k\}$, the **Probabilistic Participation Ratio** of a feature f_i, denoted as $PPR(f_i, F)$, is defined as:*

$$PPR(f_i, F) = \frac{1}{|f_i|} \sum_{j=1}^{|f_i|} PR(f_i.e_j, F). \tag{3}$$

Definition 9. *The **Probabilistic Participation Index** of a subset of features $F = \{f_1, \ldots, f_k\}$ is defined as*

$$PPI(F) = \min_{i \in [1, k]} PPR(f_i, F). \tag{4}$$

Based on the newly defined measures, we formalize the problem of *probabilistic co-location mining* from Gaussian-based uncertain spatial data as follows:

Problem Definition. Given a set of spatial features $\mathcal{F}$, a set of events $\mathcal{E}$ on $\mathcal{F}$ where each event is associated with a location random variable represented as a d-dimensional Gaussian distribution, a distance threshold τ, a neighborhood probability threshold θ, and a minimal probabilistic participation index threshold min_{PPI}, the objective is to discover the complete set of *probabilistic co-location patterns* where for each pattern $F \subseteq \mathcal{F}$, $PPI(F) \geq min_{PPI}$.

3   Probabilistic Participation Ratio Computation

In this section, we discuss how to compute the probabilistic participation ratio of an event of a feature in a feature set, i.e., $PR(f_i.e_j, F)$.

Note that, since the neighborhood relationship between two events is probabilistic, each clique instance of a feature set is also associated with a probability representing how likely the set of events constitutes a clique instance.

Definition 10. *Let* $CI(F) = \{e_1, \ldots, e_k\}$ *be a clique instance of a subset of features* $F = \{f_1, \ldots, f_k\}$, *we associate a probability with the clique instance as*

$$\Pr[CI(F)] = \underbrace{\iint \cdots \int}_{k} \Psi(\boldsymbol{x_1}, \ldots, \boldsymbol{x_k}) \cdot P_{e_1}(\boldsymbol{x_1}) \ldots P_{e_k}(\boldsymbol{x_k}) d\boldsymbol{x_1} \ldots d\boldsymbol{x_k}, \quad (5)$$

where $\Psi(\boldsymbol{x_1}, \ldots, \boldsymbol{x_k}) = \begin{cases} 1 & if \ \forall \boldsymbol{x_i}, \boldsymbol{x_j}, \|\boldsymbol{x_i} - \boldsymbol{x_j}\| \leq \tau \\ 0 & otherwise \end{cases}$.

Given a collection of uncertain spatial data, each feature set F can be associated with a clique instance table $CIT(F)$ where each clique instance is accompanied with a probability obtained by Eq. (5). To find the probability of an event of a feature $f_i.e_j$ participates a feature set F, we consider the following two situations. (1) If the event participates in only one clique instance of the feature set (e.g., $CI(F)$), the probability $PR(f_i.e_j, F)$ equals to the existence probability of the clique instance (e.g., $\Pr[CI(F)]$). (2) If the event participates in more than one clique instance, then we can't simply add the probabilities of all involved clique instances. The reason is that the clique instances of a feature set are not independent. To address the issue, the following lemma gives the correct computation of $PR(f_i.e_j, F)$.

Lemma 1. *Let* $CIT'(F) = \{CI_1, CI_2, \ldots, CI_m\} \subseteq CIT(F)$ *be the set of clique instances of feature set* F *that an event* $f_i.e_j$ *participates in. The event participation ratio,* $PR(f_i.e_j, F)$, *can be given by:*

$$PR(f_i.e_j, F) = \sum_{CI_i \in CIT'} \Pr[CI_i] - \sum_{CI_i, CI_j \in CIT'} \Pr[CI_i, CI_j] + \ldots$$
$$+ (-1)^{m-1} \Pr[CI_1, \ldots, CI_m]. \quad (6)$$

The lemma can be proved based on the inclusion-exclusion principle in combinatorial mathematics [2]. Due to space constraints, we omit the details here. Although Lemma 1 provides a proper solution to calculate $PR(f_i.e_j, F)$, it suffers the computation efficiency problem, especially when an event participates in a large number of clique instances. That is, when m is large. This is because the number of terms in Eq. (6) is proportional to 2^m. Moreover, each $\Pr[CI_i, \ldots, CI_j]$ has to be obtained by sufficient number of samplings, which consumes considerable time.

In fact, since $PR(f_i.e_j, F)$ is the probability that $f_i.e_j$ participates in one of the clique instances in $CIT(F)$, we can skip calculating $\Pr[CI(F)]$ and deal with $PR(f_i.e_j, F)$ directly by using the Monte Carlo method.

Definition 11. *Let $CIT'(F) = \{CI_1, CI_2, \ldots, CI_m\} \subseteq CIT(F)$ be the set of clique instances of feature set F that event $f_i.e_j$ participates in. Let $\mathcal{W}$ denote the set of all samples and $CI_i^{(w)}$ represents a certain clique instance exists in the sample w. The event participation ratio, $PR(f_i.e_j, F)$, can be given by:*

$$PR(f_i.e_j, F) = \frac{\sum_{w \in \mathcal{W}} |CI_1^{(w)} \text{ or } CI_2^{(w)}, \ldots, \text{ or } CI_m^{(w)}|}{|\mathcal{W}|}. \tag{7}$$

The Monte Carlo method indicates that, by sampling the data set $\mathcal{W}$ times, the probability $PR(f_i.e_j, F)$ can be obtained as the fraction of data samples where any clique instance involving the event $f_i.e_j$ exists.

4 Probabilistic Co-location Mining Framework

In this section, before presenting the framework, we first addressed that the anti-monotonic property holds for the newly defined measure of probabilistic participation index. Due to limited space, the proof can be found in the full version of this paper [1].

*Property 1 (**Anti-monotonicity**).* The probabilistic participation index of a subset of features is monotonically non-increasing with respect to the number of features in the set.

According to the anti-monotonicity of probabilistic participation index, if a co-location pattern is prevalent, then all its sub-patterns must also be prevalent. Based on this property, an Apriori-like algorithm is developed to discover probabilistic co-location patterns. The main idea is illustrated in Algorithm 1. Given a set of Gaussian-based uncertain spatial data, we first construct a probabilistic neighborhood table (PNT) based on Definition 6 (line 1). Each entry of PNT is a pair of events that are probabilistic neighbors. Next, we generate size k candidate co-locations from those of size $k - 1$ using the Apriori-join method [5] (line 4). For each candidate co-location pattern, we derive its clique instance table (CIT) correspondingly (lines 5-8). Specifically, if the candidate co-location is of size 2, its CIT can be retrieved directly from PNT. Otherwise, we can construct

Algorithm 1. Probabilistic Co-location Mining

Input: A set of events of different features $\mathcal{F}$, a distance threshold τ, a probabilistic neighborhood threshold θ, a probabilistic participation index threshold min_{PPI}.

Variable: S_k: a set of CITs of size k, C_k: a set of size k candidate probabilistic co-locations, P_k: a set of size k probabilistic co-locations.

```
1:  PNT = gen_probabilistic_neighborhood_table(τ, θ)
2:  P₁ = F, k = 2
3:  while (P_{k-1} ≠ ∅) do
4:     C_k = gen_candidate_co-locations (P_{k-1})
5:     if k = 2 then
6:        CIT(C_k) = gen_clique_instance_table (PNT)
7:        add CIT(C_k) to S₂
8:     else
9:        CIT(C_k) = gen_clique_instance_table (S_{k-1}, PNT)
10:       add CIT(C_k) to S_k
11:    for all F ∈ C_k do
12:       PPI(F) = cal_ppi (F, CIT(F))
13:       if PPI(F) ≥ min_{PPI} then
14:          add F to P_k
15:    k = k + 1
16: return  P = P₂ ∪ P₃ ··· ∪ P_k
```

the CIT of size k from CITs of size $k - 1$ and PNT. For example, the clique instance A_1-B_1-C_1-D_1 may be obtained from clique instances A_1-B_1-C_1 and A_1-B_1-D_1, by verifying whether C_1 and D_1 are probabilistic neighbors in PNT. After CITs are generated, we examine whether a candidate is a valid probabilistic co-location pattern by computing the probabilistic participation ratio of each involved feature based on the generated CIT according to Definition 11 (line 12). An iterative loop is then carried out to generate co-locations of size $k + 1$ from those of size k.

We observe that one of the major costs of the mining process come from the generation of probabilistic neighborhood table (PNT), which verifies the probabilistic neighbor relationship between a great amount of event pairs. Therefore, in the next section, we address this issue by devising a filtering technique to improve the efficiency of PNT generation.

5   Finding Probabilistic Neighbors

When verifying probabilistic neighborhood relationship between events, the Monte Carlo method is adopted, which obtains an approximate probability by sufficient number of samplings. However, the sampling progress still engages high computation complexity. We are thus motivated to improve the efficiency by reducing the number of event pairs that need to be compared.

To this end, we propose an efficient filtering technique using Minimum Bounding Sphere (MBS), based on the ρ-region defined in Dong et al. [4].

Definition 12 [4]. *Consider a Gaussian-based event location variable e_i and the integration of its probability density function $P_{e_i}(x)$ over an ellipsoidal region $(x-\mu_i)^T \Sigma_i^{-1}(x-\mu_i) \leq r^2$. Let r_ρ be the value of r within which the result of the integration equals ρ, i.e., $\int_{(x-\mu_i)^T \Sigma_i^{-1}(x-\mu_i) \leq r_\rho^2} P_{e_i}(x)dx = \rho$. The ellipsoidal region $(x - \mu_i)^T \Sigma_i^{-1}(x - \mu_i) \leq r_\rho^2$ is called ρ-region of e_i.*

That is, the ρ-region represents an ellipsoidal region in which the probability that an event occurs is ρ. Given a specified probability ρ, the value of r_ρ can be obtained based on the following property:

Property 2 [6]. Given the normalized Gaussian distribution $P_{norm}(x) = \mathcal{N}(0, I)$, consider the integration of $P_{norm}(x)$ over $||x||^2 \leq \widetilde{r}_\rho^2$. For a given ρ ($0 < \rho < 1$), let $\widetilde{r}_\rho$ be the radius within which the integration becomes ρ, i.e., $\int_{||x||^2 \leq \widetilde{r}_\rho^2} P_{norm}(x)dx = \rho$, then $r_\rho = \widetilde{r}_\rho$ holds.

Although this property specifies that we may obtain r_ρ from $\widetilde{r}_\rho$, there is still no way to derive $\widetilde{r}_\rho$ from ρ analytically. Hence, we construct a $(\widetilde{r}_\rho, \rho)$ table in advance. Given a specified ρ, we return the matching $\widetilde{r}_\rho$, or if not matched, return the $\widetilde{r}_\rho$ corresponding to the smallest ρ' that is greater than ρ to guarantee correctness.

5.1   Minimum Bounding Sphere

It is difficult to examine the probabilistic neighborhood relationship between two events with locations represented by ellipsoidal ρ-regions. Hence, we adopt the Minimum Bounding Sphere (MBS) that tightly bounds the ρ-region. Examples of MBS of ρ-regions in 2-D space are shown in Fig. 1. In order to bound the ellipsoid region, the radius of the sphere should be the major axis, which is given by the following property.

Property 3. Given an ellipsoid $(x - \mu_i)^T \Sigma_i^{-1}(x - \mu_i) \leq r_\rho^2$, the radius r_{MBS} of its MBS can be calculated by $r_{\mathrm{MBS}} = \frac{\sqrt{r_\rho^2}}{\omega_{min}}$, where $\omega_{\min}$ is the minimum eigenvalue of the covariance matrix Σ_i^{-1}.

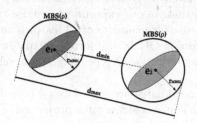

Fig. 1. Using minimum bounding spheres to bound ρ-regions.

5.2   The Filtering

We now explain how to quickly verify the probabilistic neighborhood relationship between two events based on MBS. Recall that, if two events e_1 and e_2 are probabilistic neighbors, $\Pr[\|e_1 - e_2\| \leq \tau] \geq \theta$ holds, where τ and θ are user specified distance threshold and probabilistic neighborhood threshold, respectively. Let d_min be the minimum distance between two MBS as illustrated in Fig. 1. In this case, we have $\Pr[\|e_1 - e_2\| > d\_min] > \rho^2$, since the probability of an event in the ρ-region is ρ and the probability that it occurs in the corresponding MBS is greater than ρ. That is, $\Pr[\|e_1 - e_2\| \leq d\_min] \leq 1 - \rho^2$. Let $1 - \rho^2 = \theta$ and consider the critical scenario $d\_min = \tau$, we have $\Pr[\|e_1 - e_2\| \leq \tau] \leq \theta$. In other words, if $d\_min > \tau$, then the two events are definitely not in a probabilistic neighborhood. We can then filter the pair of events without calculating the numerical integration.

Similarly, we consider the maximum distance between two MBSs. In this case, we have $\Pr[\|e_1 - e_2\| \leq d\_max] \geq \rho^2$. Let $\rho^2 = \theta$. If $d\_max \leq \tau$, $\Pr[\|e_1 - e_2\| \leq \tau] \geq \theta$ holds. We can conclude that the two events are probabilistic neighbors straightforwardly.

To sum up, given two events, we apply the following steps to filter event pairs before calculating the numerical integration:

1. Let $\rho = \sqrt{\theta}$ and derive the $d\_max = \|\mu_1 - \mu_2\| + r_{\text{MBS1}} + r_{\text{MBS2}}$. If $d\_max \leq \tau$, these two events are directly labeled as probabilistic neighbors.
2. Let $\rho = \sqrt{1 - \theta}$ and derive the $d\_min = \|\mu_1 - \mu_2\| - r_{\text{MBS1}} - r_{\text{MBS2}}$. If $d\_min > \tau$, the two events are not probabilistic neighbors.

6   Performance Study

6.1   Experiment Setup

Three real data sets are used in our experiments. (1) The US National Transportation Atlas Database with Intermodal Terminal Facilities (NTAD-ITF) of 2013 (http://www.rita.dot.gov/), in which every event is a facility with different types such as rail, airport, track, port and inter port. It consists of 5 features and 3087 events in total. (2) The EPA databases (http://www.epa.gov/), which contain environmental activities that affect air, land and water in United States. Different environmental interest types are used as spatial features and each facility represents a spatial event. In our experiment, we use the EPA data of Allen Counties in Indiana State, which consists of 23 features and 647 events in total. (3) The points of interest (POI) in California (http://www.usgs.gov/) which was used in [7]. There are 63 category types (e.g., dam, school, and bridge) and 104,770 data points.

All the geographic coordinates are transformed to 2-dimensional Cartesian coordinates by Universal Transverse Mercator projection. The uncertainty is generated synthetically by taking the coordinates as the mean values and assigning a covariance matrix to each event. By default, a covariance matrix $\left(\begin{smallmatrix} 100^2 & 0 \\ 0 & 100^2 \end{smallmatrix} \right)$ is assigned each event. The default sampling times employed by the Monte Carlo method is 1000.

6.2   Comparisons with Other Methods

We first compare our method with two approaches. One is the existing method [8], which considers several possible locations of an event. Given an uncertain spatial data set where locations of events are continuous variables, this method can be applied by randomly sampling several possible location points for each event from its location distribution. The other one simply *determinize* the data by considering the expected locations of events. Then, traditional co-location mining can be applied. We refer to the first method as the *discretization* method, and the second one as the *determinization* method.

Comparison with the Determinization Method. We conduct experiments to compare the two methods on the EPA data set, by varying the standard covariance σ of location distributions. More precisely, we generate the data uncertainty by using different standard covariances ($\sigma = 10, 100, 300$). We then compare the number of patterns discovered by the simple *determinization* approach and our method respectively. Other parameters are $\tau = 1050$, $\theta = 0.5$ and $min_{PPI} = 0.4$. Table 1 summarizes the experiment results. It can be seen that when $\sigma = 10$, the number of patterns found by both methods are same, so are the particular patterns (shown in the fourth column). As σ increases, the number of patterns output by our method decreases. For example, when σ varies from 10 to 100, the pattern *ASM-HWBR* (underlined in the table) will not be discovered as a valid co-location by our method. When σ varies from 100 to 300, even fewer number of patterns will be found by our method. This is because when σ increases, the location uncertainty of an event becomes more significant, resulting in the lower probability of two events being neighbors. However, the *determinization* method fails to recognize this and outputs the same 8 patterns under different uncertainty degrees.

Table 1. *Determinization* method vs. our method on EPA data

σ	$|\mathcal{P}'|$	$|\mathcal{P}|$	$\mathcal{P}$: Patterns found by our method
10	8	8	*CESQG-SQG, E-SQG, CA-Enforcement, CESQG-CA,* *ASM-HWBR, CESQG-E, HWBR-TRIR, ASM-CESQG*
100	8	7	*CA-E, CESQG-CA, CESQG-E, HWBR-TRIReporter,* *CESQG-SQG, ASM-CESQG, E-SQG*
300	8	2	*CESQG-CA, CESQG-E*

⋆ E=Enforcement, TRI=Toxics Release Inventory, SQG=Small Quantity Generators and CESQG=Conditionally Exempt Small Quantity Generators, TRIR=TRI Reporter, CA=Compliance Activity, ASM=Air Synthetic Minor, HWBR=Hazardous Waste Biennial Reporter.

Comparison with the Discretization Method. We further compare our method with the *discretization* approach proposed by Liu and Huang in [8]. In the *discretization* method, each event is associated with several location instances

Table 2. *Discretization* method vs. our method on ITF data

λ	$|\mathcal{P}'|$	$|\mathcal{P}|$	$\mathcal{P}$: Patterns found by discretization
30	3	3	*Airport-Rail, Airport-Truck, Rail-Truck*
10	2	3	*Airport-Rail, Airport-Truck*
3	1	3	*Rail-Truck*

within a region. In our experiment, we model the discrete location uncertainties as follows. For each event, we generate its location instances from its location distribution. The number of location instances per event is decided by a Poisson distribution with mean λ. We run the experiments on the ITF data set. The default parameters are: $\tau = 10000$, $\theta = 0.7$ and $min_{PPI} = 0.3$. By varying λ, we compare the numbers of patterns found by the *discretization* method and our method respectively. The results are shown in Table 2.

It can be observed that, when the density λ decreases, fewer number of patterns are discovered by the *discretization* method. This is because when λ gets smaller, fewer location instances will be sampled by the *discretization* method, resulting in the missing of some valid pairs of neighboring events. On the flip side, a large λ invokes more location instances being sampled, so that the result of the discretization method will be similar to that of our method. Our proposed method can thus be regarded as a generalization of the *discretization* method.

6.3 Effectiveness of Filtering

Next, we evaluate the effectiveness of the filtering method (proposed in Sect. 6) for finding probabilistic neighbors. We implement our pattern mining framework with and without the filtering technique, and record the running time respectively with respect to the variation of τ. The result is shown in Fig. 2. It can be seen that the filtering technique clearly contributes to the reduction of the running time. Moreover, as the data becomes dense (e.g., the POI data set is denser than the other two), the effect of filtering is more significant.

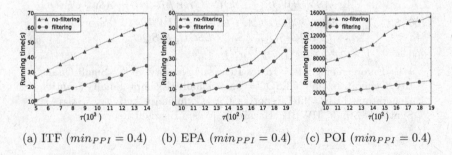

(a) ITF ($min_{PPI} = 0.4$)   (b) EPA ($min_{PPI} = 0.4$)   (c) POI ($min_{PPI} = 0.4$)

Fig. 2. Evaluation of filtering technique.

7   Conclusion

This paper addresses the problem of mining probabilistic co-locations from uncertain spatial data, where locations of events are modeled as Gaussian-based continuous variables. Prevalence measures are redefined to and computed with carefully designed strategy. A proper framework has been developed for probabilistic co-location mining, which integrates an effective filtering strategy to skip expensive verification of probabilistic neighborhood. Experimental results on multiple datasets demonstrate the effectiveness of the proposed algorithm.

References

1. Mining Co-locations from Continuously Distributed Uncertain Spatial Data (2016). https://goo.gl/Q3OYns
2. Allenby, R., Slomson, A.: How to Count: An Introduction to Combinatorics. Discrete Mathematics and Its Applications, 2nd edn. CRC Press (2010)
3. Bernecker, T., Kriegel, H.-P., Renz, M., Verhein, F., Züfle, A.: Probabilistic frequent itemset mining in uncertain databases. In: KDD, pp. 119–128 (2009)
4. Dong, T., Xiao, C., Guo, X., Ishikawa, Y.: Processing probabilistic range queries over Gaussian-based uncertain data. In: Nascimento, M.A., Sellis, T., Cheng, R., Sander, J., Zheng, Y., Kriegel, H.-P., Renz, M., Sengstock, C. (eds.) SSTD 2013. LNCS, vol. 8098, pp. 410–428. Springer, Heidelberg (2013)
5. Huang, Y., Shekhar, S., Xiong, H.: Discovering colocation patterns from spatial data sets: a general approach. IEEE Trans. Knowl. Data Eng. **16**(12), 1472–1485 (2004)
6. Ishikawa, Y., Iijima, Y., Yu, J.X.: Spatial range querying for gaussian-based imprecise query objects. In ICDE, pp. 676–687 (2009)
7. Li, F., Cheng, D., Hadjieleftheriou, M., Kollios, G., Teng, S.-H.: On trip planning queries in spatial databases. In: Medeiros, C.B., Egenhofer, M., Bertino, E. (eds.) SSTD 2005. LNCS, vol. 3633, pp. 273–290. Springer, Heidelberg (2005)
8. Liu, Z., Huang, Y.: Mining co-locations under uncertainty. In: Nascimento, M.A., Sellis, T., Cheng, R., Sander, J., Zheng, Y., Kriegel, H.-P., Renz, M., Sengstock, C. (eds.) SSTD 2013. LNCS, vol. 8098, pp. 429–446. Springer, Heidelberg (2013)
9. Niedermayer, J., Züfle, A., Emrich, T., Renz, M., Mamoulis, N., Chen, L., Kriegel, H.-P.: Probabilistic nearest neighbor queries on uncertain moving object trajectories. PVLDB **7**(3), 205–216 (2013)
10. Shekhar, S., Huang, Y.: Discovering spatial co-location patterns: a summary of results. In: Jensen, C.S., Schneider, M., Seeger, B., Tsotras, V.J. (eds.) SSTD 2001. LNCS, vol. 2121, pp. 236–256. Springer, Heidelberg (2001)
11. Thrun, S., Burgard, W., Fox, D.: Probabilistic Robotics (Intelligent Robotics and Autonomous Agents). The MIT Press, Cambridge (2005)
12. Wang, L., Wu, P., Chen, H.: Finding probabilistic prevalent colocations in spatially uncertain data sets. IEEE Trans. Knowl. Data Eng. **25**(4), 790–804 (2013)
13. Xia, Y., Yang, Y., Chi, Y.: Mining association rules with non-uniform privacy concerns. In: DMKD, pp. 27–34 (2004)
14. Xiong, H., Shekhar, S., Huang, Y., Kumar, V., Ma, X., Yoo, J.S.: A framework for discovering co-location patterns in data sets with extended spatial objects. In: SDM, pp. 78–89 (2004)

15. Yoo, J.S., Shekhar, S.: A joinless approach for mining spatial colocation patterns. IEEE Trans. Knowl. Data Eng. **18**(10), 1323–1337 (2006)
16. Yoo, J.S., Shekhar, S., Celik, M.: A join-less approach for co-location pattern mining: a summary of results. In: ICDM, pp. 813–816 (2005)
17. Zhang, X., Mamoulis, N., Cheung, D.W., Shou, Y.: Fast mining of spatial collocations. In: KDD, pp. 384–393 (2004)

An Online Approach for Direction-Based Trajectory Compression with Error Bound Guarantee

Bingqing Ke, Jie Shao[✉], Yi Zhang, Dongxiang Zhang, and Yang Yang

School of Computer Science and Engineering,
University of Electronic Science and Technology of China, Chengdu, China
{201521060331,yizhang}@std.uestc.edu.cn,
{shaojie,zhangdo,yang.yang}@uestc.edu.cn

Abstract. With the increasing usage of GPS-enabled devices which can record users' travel experiences, moving object trajectories are collected in many applications. Raw trajectory data can be of large volume but storage is limited, and direction-based compression to preserve the skeleton of a trajectory became popular recently. In addition, real-time applications and constrained resources often require online processing of incoming data instantaneously. To address this challenge, in this paper we first investigate two approaches extended from Douglas-Peucker and Greedy Deviation algorithms respectively, which are two most popular algorithms for trajectory compression. To further improve the online computational efficiency, we propose a faster approximate algorithm with error bound guarantee named *Angular-Deviation*. Experimental results demonstrate it can achieve low running time to suit the most constrained computation environments.

Keywords: GPS trajectory · Direction-based compression · Online algorithm

1 Introduction

With the popularity of GPS-enabled devices, trajectory data which records the traces of moving objects by sampling their positions according to a certain sampling rate is becoming ubiquitous. The raw trajectory data is massive, which leads to expensive storage and processing cost. Trajectory clustering employs various clustering algorithms (e.g., [13]), and representative trajectories can be extracted for dataset profiling [3]. Compressing each raw trajectory with a subset of important points is more commonly used, by eliminating points that contain little information. Moreover, the emergence of real-time applications and constrained resources demand intelligent online algorithms that can process the incoming points instantaneously.

There are many methods in the literature for trajectory compression [1,2,9,11,12] and most of them aim at preserving the position information. Existing online

© Springer International Publishing Switzerland 2016
F. Li et al. (Eds.): APWeb 2016, Part I, LNCS 9931, pp. 79–91, 2016.
DOI: 10.1007/978-3-319-45814-4_7

compression algorithms [5, 6, 10] also make the traditional assumption that the goal should be to compress trajectories such that the position information captured in the compressed trajectories is "similar" to the position information captured in the raw trajectories. We call them *position-based online trajectory compression* algorithms.

Recently, an alternative type of direction-based trajectory compression which aims at preserving direction information is proposed in [7], and demonstrated its superiority compared with position-based methods. In this paper, in order to solve this new type of trajectory compression in an online fashion, we first investigate two baseline solutions, namely *Direction-based Buffered Douglas-Peucker* (DBDP) and *Direction-based Buffered Greedy Deviation* (DBGD), by extending Douglas-Peucker [2] and Greedy Deviation (a variation of the generic sliding window algorithm [4]) respectively. The time complexities of DBDP and DBGD are both $\mathcal{O}(Bn)$, where B represents the size of buffer and n represents the number of trajectory points. To suit the more constrained computation environments, we further propose a faster approximate algorithm with error bound guarantee named *Angular-Deviation*, which can achieve $\mathcal{O}(n)$ time complexity.

In summary, we make the following contributions in this paper:

- We study online processing of direction-based trajectory compression, which is a novel problem and has important applications in practice.
- We introduce two baseline algorithms for online trajectory compression with $\mathcal{O}(Bn)$ time complexity, where B denotes the buffer size. To further improve the efficiency of online compression, a faster algorithm named *Angular-Deviation* which achieves $\mathcal{O}(n)$ time complexity is given.
- We conduct extensive experiments to verify that the proposed *Angular-Deviation* algorithm achieves lower running time than two baseline algorithms.

The rest of this paper is organized as follows. Section 2 discusses the related work. Section 3 introduce the property of direction-based trajectory compression. We present two baseline solutions in Sect. 4, and then design a faster version of online algorithm in Sect. 5. We show the experimental results in Sect. 6. Finally, we conclude in Sect. 7.

2   Related Work

2.1   Error Measurements for Trajectory Compression

According to different error measurements adopted, we can divide existing work on trajectory compression into two categories: (1) position-based [1, 2, 5, 6, 9–12]; and (2) direction-based [7, 8].

Position-Based Trajectory Compression. A position-based error of a compressed trajectory T' denoted by $\varepsilon_p(T')$ is usually defined to be the maximum Euclidean distance between a position on the original trajectory and its "mapped" position p' on the compressed trajectory. There are mainly two distance metrics to measure the position-based error of a compression: *perpendicular*

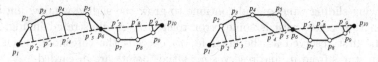

(a) Perpendicular Euclidean Distance   (b) Time Synchronized Euclidean Distance

Fig. 1. Distance metrics to measure the position-based compression error.

Euclidean distance and *time synchronized Euclidean distance* [12]. As illustrated in Fig. 1, suppose we compress a trajectory with 10 points into a representation T' with 3 points. The former metric defines the "mapped" position to be the closest position from p on the compressed trajectory and $\varepsilon_p(T') = dis(p_3, p_3')$. The latter metric assumes that the "mapped" position is the position with the same timestamp on the compressed trajectory as p, and we get $\varepsilon_p(T') = dis(p_4, p_4')$.

Direction-Based Trajectory Compression. Direction-preserving compression [7] is the state-of-the-art method for trajectory compression. The error of the compressed trajectory is the maximum angular difference between a segment and its corresponding compressed segment. In this work, we focus on direction-based trajectory compression for the reason of its superiority (Lemma 1). The details will be elaborated in Sect. 3.

2.2   Existing Trajectory Compression Methods

In this part, we review two representative trajectory compression methods.

Douglas-Peucker. As demonstrated in Fig. 2(a), the idea of Douglas-Peucker algorithm [1] is to replace the original trajectory by an approximate line segment (e.g., $\overline{p_1 p_8}$). If the replacement does not meet the specified error requirement (perpendicular Euclidean distance is used here), it recursively partitions the original trajectory into two sub-trajectories (e.g., $\{p_1,p_2,p_3\}$ and $\{p_3,p_4,p_5,p_6,p_7,p_8\}$) by selecting the point contributing the largest error as the splitting point (e.g., p_3).

Sliding Window. The major limitation of the above Douglas-Peucker algorithm is that it runs offline. This is not sufficient for location-aware applications which often require real-time data processing. A generic sliding window algorithm can run online. The idea of the sliding window algorithm [4] is to append the points

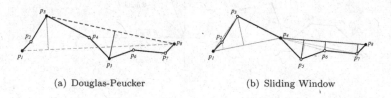

(a) Douglas-Peucker   (b) Sliding Window

Fig. 2. Existing trajectory compression methods.

to a growing sliding window and continue to grow the sliding window until the approximation error exceeds some error tolerance. As illustrated in Fig. 2(b), p_4 will be first reserved as the error for p_3 exceeds the threshold. Then, the algorithm starts from p_5 and reserves p_8. Other points are discarded.

Both of the two methods aim at preserving position information, and cannot be directly used for direction-based trajectory compression. We develop two baseline solutions extended from these two methods in Sect. 4.

3 Preliminaries

First, we give some definitions about trajectory compression.

Definition 1 (Trajectory). *A trajectory is represented by a sequence of n points in the form of $\{p_1(x_1, y_1, t_1), p_2(x_2, y_2, t_2), ..., p_n(x_n, y_n, t_n)\}$, where (x_i, y_i) denotes the longitude and latitude information of point p_i at timestamp t_i.*

Definition 2 (Compressed Trajectory). *Given a trajectory that contains n points $T = \{p_1, p_2, ..., p_n\}$, its compressed trajectory is $T' = \{p_{s_1}, p_{s_2}, ..., p_{s_m}\}$ where $s_1 = 1$, $s_m = n$ and $T' \subseteq T$.*

The notion of direction-based error for compression [7] is used in our work.

Definition 3 (Direction-Based Error of Compression). *Let the maximum value of angular difference between segment $\overline{p_i p_{i+1}}$ and corresponding simplified segment $\overline{p_{s_k} p_{s_{k+1}}}$ be $\varepsilon(\overline{p_{s_k} p_{s_{k+1}}})$. As shown in Fig. 3, $\overline{p_{s_k} p_{s_{k+1}}}$ is a simplified segment, and each segment $\overline{p_i p_{i+1}}$ between p_{s_k} and $p_{s_{k+1}}$ has an angular difference $\theta(\overline{p_i p_{i+1}}, \overline{p_{s_k} p_{s_{k+1}}})$. We define $\varepsilon(\overline{p_{s_k} p_{s_{k+1}}})$ as:*

$$\varepsilon(\overline{p_{s_k} p_{s_{k+1}}}) = \max_{p_{s_k} \leq i < p_{s_{k+1}}} |\theta(\overline{p_i p_{i+1}}) - \theta(\overline{p_{s_k} p_{s_{k+1}}})|.$$

Now, we can define the error of the whole compression as:

$$\varepsilon(T') = \max_{1 \leq k < m} \{\varepsilon(\overline{p_{s_k} p_{s_{k+1}}})\},$$

where T' is the result of the compression, and m is the number of points that T' contains. We regard $(n - m)/n$ as the compression rate when the raw trajectory is composed of n points.

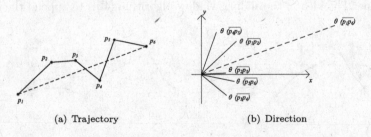

(a) Trajectory   (b) Direction

Fig. 3. Direction-based error measurement.

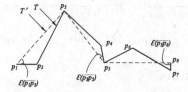

Fig. 4. An example illustrating direction-based error measurement.

Example 1. As shown in Fig. 4, a trajectory T is in the form of $\{p_1, p_2, ..., p_8\}$, and the compressed trajectory T' is $\{p_1, p_3, p_5, p_8\}$. Consider our example, $\varepsilon(\overline{p_1p_3}) = \pi/4$, $\varepsilon(\overline{p_3p_5}) = \pi/6$ and $\varepsilon(\overline{p_5p_8}) = \pi/2$. Thus, $\varepsilon(T') = \max\{\varepsilon(\overline{p_1p_3}), \varepsilon(\overline{p_3p_5}), \varepsilon(\overline{p_5p_8})\} = \varepsilon(\overline{p_5p_8}) = \pi/2$.

Direction-based trajectory compression not only preserves the direction information but also bounds position information loss. In this following, we give a brief proof.

Lemma 1 (Bounded Error of Distance). *The distance error of direction-based trajectory compression d_e meets the inequation*

$$d_e \leq \frac{1}{2}L_{max}\tan\varepsilon$$

where L_{max} is the maximum length of all segments in the compressed trajectory, and ε is the error tolerance of the compression.

Proof. An illustration is shown in Fig. 5, where $\overline{p_{s_k}p_{s_{k+1}}}$ is a compressed segment. We first prove that any p_i between p_{s_k} and $p_{s_{k+1}}$ in original trajectory is inside the rhomboid $\diamond_{ap_{s_k}bp_{s_{k+1}}}$. Assume that p_i is outside the rhomboid $\diamond_{ap_{s_k}bp_{s_{k+1}}}$ and it is over the segment $\overline{p_{s_k}a}$ as shown in Fig. 5, there must exist such a segment $\overline{p_{j-1}p_j}$ that intersects the segment $\overline{p_{s_k}a}$. Then the angular difference between $\overline{p_{j-1}p_j}$ and $\overline{p_{s_k}p_{s_{k+1}}}$ must exceed ε. This contradicts with the definition of the direction-based error of compression (Definition 4), and thus p_i must be inside the rhomboid $\diamond_{ap_{s_k}bp_{s_{k+1}}}$. Next, if p is the furthest point from $\overline{p_{s_k}p_{s_{k+1}}}$, we have

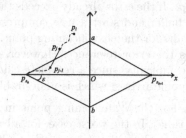

Fig. 5. Illustration of Lemma 1.

$$dis(p, \overline{p_{s_k} p_{s_{k+1}}}) \leq dis(a, \overline{p_{s_k} p_{s_{k+1}}}) = dis(a, O)$$
$$= \frac{1}{2} dis(p_{s_k}, p_{s_{k+1}}) \tan \varepsilon \leq \frac{1}{2} L_{max} \tan \varepsilon. \tag{1}$$

4 Baseline Solutions

In this section, we introduce two baseline solutions for *direction-based online tra-jectory compression*, namely *Direction-based Buffered Douglas-Peucker* (DBDP) and *Direction-based Buffered Greedy Deviation* (DBGD). They are extended from Douglas-Peucker and sliding window methods, respectively.

4.1 Direction-Based Buffered Douglas-Peucker

The major idea of Direction-based Douglas-Peucker (DDP) is to recursively split the trajectory into sub-trajectories, at the point which is the end of the segment that has the largest angular difference from the approximated segment (linking the start point and the end point of this trajectory), and finally we approximate each sub-trajectory until there are $W-1$ sub-trajectories where W is the number of points the resulting trajectory contains.

The DDP algorithm still runs offline, and its time complexity is $\mathcal{O}(n^2)$, which is inapplicable for data streams. To solve the problem, Direction-based Buffered Douglas-Peucker (DBDP) can be used. In this strategy, the incoming points are accumulated in a buffer, and then the points are processed by applying DDP algorithm when the buffer is full. Such a solution has an inferior compression rate, because at least two points should be kept whenever the buffer is full.

Complexity Analysis. Suppose a trajectory contains n points and the buffer size is B. Since there are $\mathcal{O}(n/B)$ times of calling DDP algorithm which causes $\mathcal{O}(B^2)$ time complexity in the worst case, DBDP has a time complexity of $\mathcal{O}(B^2 \cdot (n/B)) = \mathcal{O}(Bn)$.

4.2 Direction-Based Buffered Greedy Deviation

Direction-based Buffered Greedy Deviation (DBGD) is a variation of the generic sliding window algorithm. In this strategy, if the buffer is not full, whenever a point p_e arrives we append the point to the end of the buffer, and calculate the error of the compressed trajectory segment defined by the start point in the buffer and the end point p_e. If the error already exceeds the tolerance, we reverse the last point, clear the buffer and start a new compression at the last point. Otherwise, the process waits for the next incoming point. It is easy to find that the algorithm guarantees the error tolerance. However, its major weakness is the relatively high time complexity, since a complete calculation of the error is needed as long as a new point arrives, which is undesirable in online processing.

Complexity Analysis. For the ith incoming point in the buffer, we should calculate errors for $\mathcal{O}(i)$ times. In the worst case, for all B points in the buffer, calculations have to be made $\mathcal{O}(1 + 2 + 3 + ... + B) = \mathcal{O}(B^2)$ times. Therefore, the time complexity of DBGD is $\mathcal{O}(B^2 \cdot (n/B)) = \mathcal{O}(Bn)$.

5 Approximate Solution with Error Bound Guarantee

As discussed in Sect. 4, the complexities of two baseline solutions may not satisfy stringent performance requirements in resource-constrained environments. To improve the computational efficiency, we propose an approximate Angular-Deviation algorithm with guaranteed error bounds.

5.1 Definitions

Definition 4 (Angular Deviation). *The angular deviation $p_i.\varepsilon_d$ is the difference between two consecutive segments' directions,*

$$\Delta\theta = \theta(\overline{p_i p_{i+1}}) - \theta(\overline{p_{i-1} p_i}) \tag{2}$$

$$p_i.\varepsilon_d = \begin{cases} \Delta\theta + 2\pi & \Delta\theta \leq -\pi \\ \Delta\theta & -\pi < \Delta\theta \leq \pi \\ \Delta\theta - 2\pi & \Delta\theta > \pi \end{cases} \tag{3}$$

where $\theta(\overline{p_i p_{i+1}})$ represents the direction of the segment $\overline{p_i p_{i+1}}$, namely the angle between vector $\overrightarrow{p_i p_{i+1}}$ and the positive direction of x-axis. We set the range of $\theta(\overline{p_i p_{i+1}})$ as $(-\pi, \pi]$. In Fig. 6, $p_2.\varepsilon_d$ and $p_3.\varepsilon_d$ represent the change of direction of p_2 and p_3 respectively. Note that, the value of the angular deviation has a positive or negative sign.

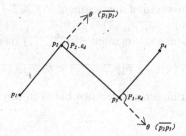

Fig. 6. Angular deviation ε_d of p_i.

Definition 5 (Accumulated Angular Deviation). *The accumulated angular deviation ε_a of point p_i is defined as:*

$$p_i.\varepsilon_a = \begin{cases} 0 & i = s_k \\ \sum_{m=s_k+1}^{i} p_m.\varepsilon_d & s_k < i < s_{k+1} \end{cases} \tag{4}$$

where p_{s_k} is the start point of the compressed segment $\overline{p_{s_k} p_{s_{k+1}}}$.

Combining with the Definition 4, we have another form of the expression of $p_i.\varepsilon_a$:

$$\Delta\phi = \theta(\overline{p_i p_{i+1}}) - \theta(\overline{p_{s_k} p_{s_{k+1}}}) \tag{5}$$

$$p_i.\varepsilon_a = \begin{cases} \Delta\phi + 2\pi & \Delta\phi \leq -\pi \\ \Delta\phi & -\pi < \Delta\phi \leq \pi \\ \Delta\phi - 2\pi & \Delta\phi > \pi \end{cases} \tag{6}$$

where $s_k \leq i < s_{k+1}$.

5.2 Theoretical Property

Before introducing the main idea of the approximate algorithm, we first introduce a fundamental lemma.

Lemma 2 (Bounded Error of Direction). *The error of the compression* $\varepsilon(T')$ *meets the inequation*

$$\varepsilon(T') < 2\varepsilon_t$$

where ε_t *is the threshold of accumulated angular deviation.*

Proof. As we defined before, the direction of any segment $\overline{p_i p_{i+1}}$ between p_{s_k} and $p_{s_{k+1}}$ can be calculated by Eqs. 4, 5 and 6:

$$\Psi = \theta(\overline{p_{s_k} p_{s_k+1}}) + \sum_{m=s_k+1}^{i} p_m.\varepsilon_d$$

$$\theta(\overline{p_i p_{i+1}}) = \begin{cases} \psi + 2\pi & \psi \leq -\pi \\ \psi & -\pi < \psi \leq \pi \\ \psi - 2\pi & \psi > \pi \end{cases}$$

As shown in Fig. 7, the direction of any segment $\theta(\overline{p_i p_{i+1}})$ between p_{s_k} and $p_{s_{k+1}}$ are limited within interval $[\theta(\overline{p_{s_k} p_{s_k+1}}) - \varepsilon_t, \theta(\overline{p_{s_k} p_{s_k+1}}) + \varepsilon_t]$. It is easy to find that the direction of compressed segment $\theta(\overline{p_{s_k} p_{s_{k+1}}})$ meets the inequation:

$$\theta(\overline{p_{s_k} p_{s_k+1}}) - \varepsilon_t < \theta(\overline{p_{s_k} p_{s_{k+1}}}) < \theta(\overline{p_{s_k} p_{s_k+1}}) + \varepsilon_t.$$

Therefore, the maximum angular difference between any segment and the simplified segment is less than $2\varepsilon_t$.

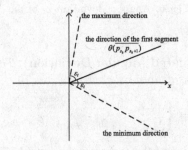

Fig. 7. Illustration of Lemma 2.

5.3   Angular-Deviation Algorithm

The *Angular-Deviation* algorithm is formally described in Algorithm 1. Besides, we present an example to show the details of the process.

Algorithm 1. Angular-Deviation for Direction-based Online Compression

---

Initialization:

Set the accumulated change of direction $\varepsilon_a = 0$, the resulting set $Traj = \emptyset$;

Input:

Start point s, the point currently being processed p_i, the previous point p_{i-1}, the new incoming point p_{i+1}, an error tolerance $error$, the threshold of accumulated angular deviation $\varepsilon_t = error/2$;

1: **if** p_{i+1}=null **then**
2:    $Traj \leftarrow Traj \bigcup \{p_i\}$;
3:    **return** $Traj$; $//p_i$ is the end of the whole trajectory
4: **else**
5:    $p_i.\varepsilon_d \leftarrow AngDiff(p_{i-1}, p_i, p_{i+1})$;
6:    $\varepsilon_a = \varepsilon_a + p.\varepsilon_d$;
7:    **if** $|\varepsilon_a| > \varepsilon_t$ **then**
8:        $Traj \leftarrow Traj \bigcup \{p_i\}$;
9:        $\varepsilon_a \leftarrow 0$;
10:   **end if**
11: **end if**

As shown in Algorithm 1, we set two variables: ε_a is the accumulated change of direction of the incoming point p_i, and $Traj$ is the resulting set which contains the compressed trajectory we expect (initialized as empty). Then we set two tolerance values: $error$ denotes the error tolerance of the compression and ε_t is equal to half of $error$ according to Lemma 2. The algorithm firstly checks if the current point p_i is the end point of the trajectory (line 1), because the end point must be reserved (line 2). If p_i is an intermediate point, the accumulated angular deviation ε_a is calculated by Eq. 4 (lines 5–6). If $|\varepsilon_a| > \varepsilon_t$, we cannot guarantee that the maximum angular difference between a segment on original trajectory and corresponding compressed segment less than $error$ by Lemma 2, so we should reverse p_i and start a new segment compression (lines 8–9), otherwise p_i should be discarded since it carries little direction information.

Example 2. Consider an example shown in Fig. 8, $T = \{p_1, p_2, p_3, p_4, p_5, p_6, p_7\}$, and we set $error = \pi/2 = 1.57$ radian and $\varepsilon_t = \pi/4 = 0.785$ radian. Firstly, $p_i = p_2$ and $|\varepsilon_a| = |p_2.\varepsilon_d = +0.785| \leq \varepsilon_t$, hence p_2 should be deleted. Then, $p_3.\varepsilon_d = -2.356$ and we can get $|\varepsilon_a| = |0.785 - 2.356| = 1.571 > \varepsilon_t$, hence $Traj = \{p_1, p_3\}$ and $\varepsilon_a = 0$. Then, the algorithm starts from p_4, $|\varepsilon_a| = p_4.\varepsilon_d = +0.524 < \varepsilon_t$. Next, $p_5.\varepsilon_d = +1.571$, the accumulated angular deviation $|\varepsilon_a| = 0.524 + 1.571 = 2.095 > \varepsilon_t$, hence $Traj = \{p_1, p_3, p_5\}$. After that, p_6 is deleted and p_7 is reversed in the same way. Finally, we get the compressed trajectory $Traj = \{p_1, p_3, p_5, p_7\}$.

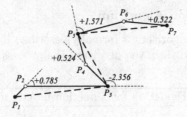

Fig. 8. An example illustrating Angular-Deviation algorithm.

6   Experiments

We conducted extensive experiments to evaluate two baseline algorithms DBDP and DBGD, and approximate algorithm *Angular-Deviation*. All algorithms are implemented in C++ and run on a Linux platform (Intel Xeon E5-2620 2.1GHz CPU with 6 cores and 200GB RAM). Two real datasets are used: (1) Chengdu taxi dataset, which contains taxi trajectories in Chengdu, China in August 2014 and each trajectory contains 3038 points on average; (2) T-Drive dataset which contains taxi trajectories generated by over 10000 taxis in a period of one week in Beijing [14].

Running Time with Buffer Size: In this experiment, as *Angular-Deviation* algorithm does not require a buffer, we study the effect of buffer size (i.e., B) for two baseline algorithms DBDP and DBGD. The values used for buffer size B are 6, 12, 25, 50, 100, 200 (the error tolerance ε is fixed to be 1). The running time results are shown in Fig. 9. We can see that the running time of both DBDP and DBGD will rise linearly with the increase of B. This is reasonable as the time complexities of both algorithms are $\mathcal{O}(Bn)$. Besides, it is worth noting that the running time of DBGD becomes stable when the buffer size is larger than some value. This is because that if buffer is large enough, the error of the compressed trajectory segment will always exceed ε before the buffer is full, which leads to the situation that buffer size has little impact on the compression. Therefore, the running time no longer depends on buffer size.

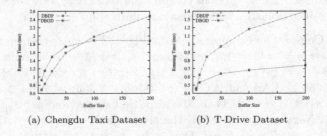

(a) Chengdu Taxi Dataset   (b) T-Drive Dataset

Fig. 9. Effect of buffer size B.

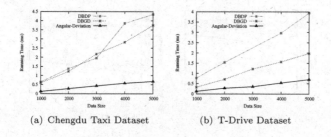

Fig. 10. Effect of data size $|T|$.

Running Time with Data Size: In this experiment, we study the effect of data size (i.e., $|T|$) on the performance of three algorithms (DBDP, DBGD and *Angular-Deviation*). The values used for $|T|$ are around 1 k, 2 k, 3 k, 4 k and 5 k (error tolerance ε is fixed to be 1 and buffer size B is fixed to be 100). For each data size, we select a set of 10 trajectories each of which has its size close to this value and run three algorithms on each of these trajectories. Then, we average the experimental results on these trajectories. The result is shown in Fig. 10. As we can see, *Angular-Deviation* runs much faster than the other two algorithms due to its low time complexity. Besides, we observe that the running time of all three algorithms increases approximatively linearly with data size. This is because the time complexity of each algorithm is proportional to data size (DBDP and DBGD are $\mathcal{O}(Bn)$, and *Angular-Deviation* is $\mathcal{O}(n)$).

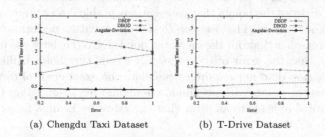

Fig. 11. Effect of error tolerance ε.

Running Time with Error Tolerance: We increase the error tolerance ε from 0.2 to 1 and report the results in Fig. 11 (buffer size B is set to be 100). As shown in Fig. 11, *Angular-Deviation* achieves the best performance, and the running time is not sensitive to ε. This is because whatever ε is, *Angular-Deviation* just needs to calculate the direction of each segment and the angular difference. The running time of DBDP decreases when ε becomes larger. The reason is that the number of segments which leads to a split (Sect. 4.1) is reduced. For DBGD, when ε increases, it needs more time to compress which is contrast to DBDP. Consider that we give each point a number in ascending order when it is added

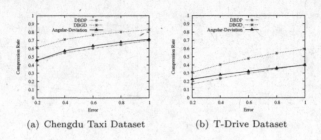

(a) Chengdu Taxi Dataset   (b) T-Drive Dataset

Fig. 12. Compression rate.

in the buffer. For i^{th} point in the buffer, it leads to $i + 1$ times of calculation (2 times for direction and $i - 1$ times for angular difference). That is, the larger number the point has, the more calculations caused. Hence, when ε becomes larger, the number of points that have relatively large number decreases, and the running time is reduced consequently.

Compression Rate with Error Tolerance: In this experiment, we compare the compression rate with increasing error tolerance ε and the experimental results are shown in Fig. 12. The compression rate β is defined as $\beta = (|T| - |T'|)/|T|$, where T denotes the number of raw trajectory and T' represents the result of compression. For β, higher is better. When error tolerance ε becomes larger, compression rates of three algorithm all increase. That is because the number of segments which cause the error of compression exceeds ε decreases. The DBDP algorithm achieves the worst performance. Since for every B points in the buffer, besides the points that cause unacceptable errors there are at least two points to be reversed. The *Angular-Deviation* algorithm performs worse than DBGD. The reason is that, for the incoming point, DBGD deletes the points that exactly cause that the error of compression exceeds threshold. Unlike DBGD, *Angular-Deviation* does not require to calculate the exact error of compression, and it discards the points with accumulated angular deviation larger than $1/2\varepsilon$, which just probably leads to an error that exceeds ε by Lemma 2.

7   Conclusion and Future Work

This paper studies a novel problem of online algorithms for direction-based trajectory compression. We first introduce two baseline solutions extended from Douglas-Peucker and Greedy Deviation algorithms respectively. To further improve the computational efficiency, we propose a faster *Angular-Deviation* algorithm with error bound guarantee. Experimental results show that *Angular-Deviation* runs fastest. All methods on trajectory compression mentioned in this paper aim at preserving skeleton information of trajectory in Euclidean space. In the future, we plan to study how to compress trajectories on road networks.

Acknowledgments. This work is support in part by the Fundamental Research Funds for the Central Universities (No. ZYGX2015J058, No. ZYGX2015J055 and No. ZYGX2014Z007), and the National Nature Science Foundation of China (No. 61572108).

References

1. Douglas, D.H., Peucker, T.K.: Algorithms for the reduction of the number of points required to represent a digitized line or its caricature. Can. Cartographer **10**(2), 112–122 (1973)
2. Heckbert, P.S., Garland, M.: Survey of polygonal surface simplification algorithms. Technical report, Carnegie Mellon University (1997)
3. Jiang, W., Zhu, J., Xu, J., Li, Z., Zhao, P., Zhao, L.: HV: a feature based method for trajectory dataset profiling. In: Cellary, W., Wang, D., Wang, H., Chen, S.-C., Li, T., Zhang, Y. (eds.) WISE 2015. LNCS, vol. 9418, pp. 46–60. Springer, Heidelberg (2015). doi:10.1007/978-3-319-26190-4_4
4. Keogh, E.J., Chu, S., Hart, D.M., Pazzani, M.J.: An online algorithm for segmenting time series. In: ICDM, pp. 289–296 (2001)
5. Kolesnikov, A.: Efficient online algorithms for the polygonal approximation of trajectory data. In: MDM, pp. 49–57 (2011)
6. Liu, J., Zhao, K., Sommer, P., Shang, S., Kusy, B., Jurdak, R.: Bounded quadrant system: error-bounded trajectory compression on the go. In: ICDE, pp. 987–998 (2015)
7. Long, C., Wong, R.C., Jagadish, H.V.: Direction-preserving trajectory simplification. PVLDB **6**(10), 949–960 (2013)
8. Long, C., Wong, R.C., Jagadish, H.V.: Trajectory simplification: on minimizing the direction-based error. PVLDB **8**(1), 49–60 (2014)
9. Meratnia, N., Park, Y.-Y.: Spatiotemporal compression techniques for moving point objects. In: Bertino, E., Christodoulakis, S., Plexousakis, D., Christophides, V., Koubarakis, M., Böhm, K. (eds.) EDBT 2004. LNCS, vol. 2992, pp. 765–782. Springer, Heidelberg (2004)
10. Muckell, J., Hwang, J., Patil, V., Lawson, C.T., Ping, F., Ravi, S.S.: SQUISH: an online approach for GPS trajectory compression. In: COM.Geo, pp. 13:1–13:8 (2011)
11. Muckell, J., Olsen, P.W., Hwang, J., Lawson, C.T., Ravi, S.S.: Compression of trajectory data: a comprehensive evaluation and new approach. GeoInformatica **18**(3), 435–460 (2014)
12. Potamias, M., Patroumpas, K., Sellis, T.K.: Sampling trajectory streams with spatiotemporal criteria. In: SSDBM, pp. 275–284 (2006)
13. Yang, Y., Ma, Z., Yang, Y., Nie, F., Shen, H.T.: Multitask spectral clustering by exploring intertask correlation. IEEE Trans. Cybern. **45**(5), 1069–1080 (2015)
14. Yuan, J., Zheng, Y., Zhang, C., Xie, W., Xie, X., Sun, G., Huang, Y.: T-drive: driving directions based on taxi trajectories. In: ACM-GIS, pp. 99–108 (2010)

A Data Grouping CNN Algorithm
for Short-Term Traffic Flow Forecasting

Donghai Yu[1], Yang Liu[1(✉)], and Xiaohui Yu[1,2]

[1] School of Computer Science and Technology,
Shandong University, Jinan 250101, China
{yliu,xyu}@sdu.edu.cn
[2] School of Information Technology, York University,
Toronto, ON M3J 1P3, Canada

Abstract. In this paper, a data grouping approach based on convolutional neural network (DGCNN) is proposed for forecasting urban short-term traffic flow. This approach includes the consideration of spatial relations between traffic locations, and utilizes such information to train a convolutional neural network for forecasting. There are three advantages of our approach: (1) the spatial relations of traffic flow are adopted; (2) high-quality features are extracted by CNN; and (3) the accuracy of forecasting short-term traffic flow is improved. To verify our model, extensive experiments are performed on a real data set, and the result shows that the model is more effective than other existing methods.

Keywords: Convolution Neural Network · Traffic flow forecasting · CBOW · Deep learning

1 Introduction

Traffic problems are crucial issues in the rapidly developing society. Traffic flow forecasting can be an important problem in Intelligent Transportation System (ITS) for urban construction. To build the Wisdom City, forecasting traffic flow in real time is a realistic demand. In a modern city, the change of traffic flow has a profound impact on people's daily life, such as the route selection for drivers. Long-term traffic flow forecasting usually refers to predicting the traffic by month or year in advance. Due to unpredictable disruptions, long-term forecasting may not be accurate enough for practical use [1]. Therefore, it might be more helpful to make a shot-term prediction, e.g., we may just want to know the traffic flow volume within 10 min to decide our route while driving.

A variety of techniques have been applied in the context of short-term traffic flow forecasting, including moving average methods [19], k-nearest-neighbor methods [6], auto regressive MA (ARIMA) or seasonal ARIMA (SARIMA) model [20,21], and neural networks (NNs) [7,15]. Forecasting traffic flow heavily depends on historical and real-time traffic data collected from various sensor sources, such as inductive loops, radars, cameras, mobile Global Positioning System, social media, etc.

© Springer International Publishing Switzerland 2016
F. Li et al. (Eds.): APWeb 2016, Part I, LNCS 9931, pp. 92–103, 2016.
DOI: 10.1007/978-3-319-45814-4_8

Despite the popularity of this research, existing work tends to suffer from the following problems. Firstly, in the downtown area of a city where traffic jam occurs, the traffic flows at different locations will influence each other. In a complicated traffic network, the influences between different locations are decided by traffic network structure, signal light, regionalism, commercial layout, etc. Previous methods just built the model based on time-series data regardless of spatial influences. Secondly, most existing studies only adopt few useful features, which can not provide sufficient information for forecasting.

In recent years, deep learning has drawn a lot of academic and industrial attention [2], and it has been applied with success in the tasks of classification, natural language processing, dimensionality reduction, object detection, motion modeling, etc. There are also early attempts to apply deep learning to traffic flow forecasting. For example, using SAE [15] and using deep belief networks [10]. Both methods first use the deep network to rebuild input features, and then train regression network for forecasting the traffic flow. However, neither has considered the traffic impact among different locations.

In this paper, We propose a Data Grouping Convolutional Neural Network (DGCNN) approach. In this method, to solve above problems, we apply Convolutional Neural Network (CNN) to forecast traffic flow and use CBOW model to find the spatial influences that exist in locations. In DGCNN, the inputs are the historical traffic flow data from the target location and historical data of other locations. We group these data to build feature matrix for forecasting. To discover the traffic flow relations at different locations, we calculate the location vectors by CBOW model [16], and we expect that the distances between location vectors can reflect each relation.

The contributions of this paper can be summarized as follows.

(1) We propose a Data Grouping Convolutional Neural Networks to forecast short-term traffic flow. To the best of our knowledge, it is the first time that the CNN has been applied to traffic flow forecasting.
(2) We identify a spatial relationship between different locations in the city by calculating the distance between location vectors, and utilize the relationship to increase the accuracy of traffic flow forecasting.
(3) We perform extensive experiments using a real data set and the results demonstrate the effectiveness of DGCNN.

The rest of this paper is organized as follows. Section 2 reviews related work. Section 3 provides a brief description of preliminary preparations. Section 4 designs and describes the DGCNN Modeling for the short-term traffic flow forecasting. In Sect. 5, the experiment details are discussed. Finally, the conclusions and future directions, regarding short-term traffic flow predictor design using the DGCNN approach, are presented in Sect. 6.

2   Related Work

Researchers have been trying to make traffic flow forecasting more accurate since 1970s [15]. Based on the length of time interval, there are long-term forecasting and short-term forecasting. Based on the parameter mode, there are parametric methods, non-parametric models, and simulation methods. In the following, we will introduce some literatures which are mostly related to our work.

Long-term forecasting indicates making a prediction by month or by year. The volume of traffic flow is large and relatively stable, and it is slightly affected by daily accident. For example, Papagiannaki et al. [12] proposed a method based on ARIMA to forecast network traffic flow in a month after they revealed that their time series data had the long term trend and the fluctuations at the 12 hour time scale.

For short-term forecasting, auto-regressive integrated moving average (ARIMA) [5] models and artificial neural network (ANN) [13] models are widely exploited. Lv et al. [14] proposed a plan moving average algorithm for utilizing previous days' historical data. In addition, there are some integration models for this task. For example, Chang et al. [4] proposed a 3-stage model which integrates ARIMA and ANN, which uses the ARIMA forecasting data as a part of input of ANN. Moretti et al. [17] studied an ensemble model which consists of a statistical method and a neural network bagging model. Besides, researchers also consider the use of integrated data. As another example, Oh et al. [18] proposed a Gaussian mixture model clustering (GMM) method to partition the data set for training ANN. Deep learning methods have also gained a lot of attention recently. Lv et al. [15] used a stacked autoencoder (SAE) to learn generic traffic flow features. Huang, Song and Huang et al. [10] applied a deep belief networks model in traffic flow prediction, which adopts multitask learning to reduce the error.

There are also some simulation methods in the same vein. For example, Fusco et al. [9] proposed a quasi-dynamic traffic assignment model to simulate the traffic flow. Fusco et al. [8] applied network-based machine learning models and dynamic traffic assignment models on a large urban traffic network. Abadi [1] used a simulation system to solve the problem of data sparseness where the simulation system produces simulative traffic data to supplement the traffic flow data set.

3   Preliminaries

In this section, we will define a few terms that are required for the subsequent discussion, and describe our modeling strategy for the vehicle traffic flow forecasting.

Definition 1 (Traffic Flow). *For each location, we record the total number of vehicles that have past during a certain period of time. Traffic flow is defined as $f_{l_i}^d(n)$ where l_i is the location ID, d is the date and n stands for n-th period (e.g., when the period is set to 10 min, $f_0^{0201}(0)$ is the traffic flow at location whose id is 0 in February 1st from 00:00 to 00:10).*

Definition 2 (Trajectory). *For each vehicle, trajectories are collected by recording the locations the vehicle has passed sequentially. A trajectory T_p is defined as a series of location IDs as follows.*

$$T_p : \{l_0, l_1, l_2, ..., l_{s-1}, l_s\}$$

Here, p is the ID of a vehicle, $l_0, l_1, ..., l_s$ are the IDs of the locations. $\{l_0, l_1, ..., l_s\}$ is ordered by passing time, and s is the length of the trajectory (e.g., if a vehicle of which the ID is 0001 has orderly passed location l_a, l_b, l_c and l_d, here is a Trajectory $T_{0001} : \{l_a, l_b, l_c, l_d\}$).

4   DGCNN Modeling

In this paper, we propose a novel DGCNN model to forecast the short-term traffic flow. In order to develop a robust model for the traffic flow prediction, we first need to accumulate a sufficient amount of training data that would well explain the traffic patterns. However, it is clear that for each single spot, such traffic flow data fade out rapidly, leaving low impact on its future traffic flow prediction. Meanwhile, we notice that in a dense traffic network, the traffic flows at different locations may have strong relationship between each other. For example, if there are no exits on the road between location l_a and l_b, the traffic flows of l_a and l_b would be similar. So, the relationships in the traffic network are also considered in our model.

Multi-layers Convolutional Neural Network (CNN) contains convolution layer, pooling layer, loss-function layer, etc. Based on different types of loss-function layers, we can perform different tasks such as classification and regression. With the assist of convolution layer and pooling layer, CNN can help understanding the relations between the features. Properly handled, CNN can be a powerful tool for modeling the future traffic flow.

CBOW model [16] has been applied in natural language processing, with its capacity to represent the word in vector with more comprehensive information. In the CBOW model, the more co-occurences two words have, the more similar patterns they share, and the larger similarity values the two vectors have. Motivated by this idea, we aim to find the spatial relations of traffic flow from trajectory data. When the traffic flow at different locations has strong relation among each other, therefore can be grouped together. Further, to address the problem that arises from data sparsity, we can borrow features of similar locations to train a more robust CNN for flow prediction.

There are two stages in training DGCNN: grouping and training. In the first part, we use the vectors to represent the locations so that we can find every location's k-nearest locations by evaluating their similarities, e.g., using a standard Euclidean Distance. The output of this first step is the grouping result of these locations. In the second part, we will train a deep convolutional neural networks to forecast traffic flow. The input features for the network comes from the locations which are in the same group.

4.1 Grouping

In order to discover locations that are similar to the target, we first need to gain a comprehensive understanding of the location, and represent it in a proper way. Word2vec can be used for learning high-quality word vectors from huge data sets with billions of words when Word2vec has implemented the CBOW model based on Negative Sampling [16]. In the task of prediction, we regard the locations in trajectories as words while trajectories represent documents.

CBOW is a 3-layer neural network which includes an input layer, a projection layer, and an output layer.

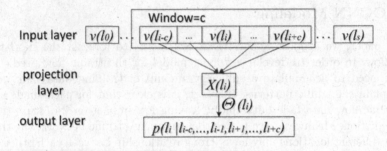

Fig. 1. CBOW model

Figure 1 shows the structure of the CBOW model, where $v(l_i)$ is the location vector for l_i, $\theta(l_i)$ is the parameter vector, and Eqs. (1) and (2) indicate the outputs of projection layer and output layer respectively.

$$X(l_i) = \sum_{x=i-c}^{i-1} v(l_x) + \sum_{x=i+1}^{i+c} v(l_x) \tag{1}$$

$$p(l_i|l_{i-c}, ..., l_{i-1}, l_{i+1}, ..., l_{i+c}) = \frac{1}{1+e^{-X^{\mathrm{T}}(l_i)\theta(l_i)}} \tag{2}$$

In order to reduce the cost of training, the strategy of negative sampling is adopted. When we train the vector of l_i, we can get a corresponding negative sample set $NEG(l_i)$. If $l_x \in NEG(l_i)$, then $l_x \neq l_i$. Mikolov et al. [16] introduces the detailed method to find negative samples. The detailed process of updating system parameters are shown in Eqs. (3) and (4). Here, η is the learning rate.

$$L^{l_i}(l_x) = \begin{cases} 0, l_x = l_i; \\ 1, l_x \neq l_i; \end{cases}$$

$$\theta(l_x) := \theta(l_x) + \eta[L^{l_i}(l_x) - \frac{1}{1+e^{-X^{\mathrm{T}}(l_i)\theta(l_x)}}]X(l_i), l_x \in NEG(l_i) \tag{3}$$

$$v(l_x) := v(l_x) + \eta \sum_{l_x \in \{l_i\} \bigcup NEG(l_i)} [L^{l_i}(l_x) - \frac{1}{1 + e^{-X^{\mathrm{T}}(l_i)\theta(l_x)}}]\theta(l_x),$$

$$l_x \in \{l_{i-c}, ..., l_{i-1}, l_{i+1}, ..., l_{i+c}\} \tag{4}$$

The input for word2vec are all the trajectories $\{T_p, \ p \in V\}$ where V is the set of vehicle IDs. The output are the key-value tuples. For each tuple, the key is the location ID and the value is a w-dimension vector. The structure of a tuple can be represented as follows:

$$< l_i \ : \ v(l_i) > \ (v(l_i) \in \mathbb{R}^w)$$

In this way, each location can be represented as a vector. We use the Euclidean Distance between the vectors to represent closeness between locations.

$$distance(l_i, l_j) = ||v(l_i) - v(l_j)|| \tag{5}$$

For each location l_a, we calculate the distances to other locations and sort out the result. Further, we select the top-k locations $\{l_b, l_c, l_d, \cdots\}$, where k is usually less than 5. Then, location l_a and the top-k locations can be grouped together in group $G_a : \{l_a, l_b, l_c, l_d, \cdots\}$. Finally, we obtain $k + 1$ locations in group G_a, and the locations after l_a are ordered by their distances to l_a.

4.2   Feature Matrix

To integrate the temporal-spatial traffic flow information into the input, we build a feature matrix. Note that to provide a more comprehensive understanding of traffic flow, not only the traffic data, but also its mean value and the mode are considered as the elements of the input matrix.

Average Traffic Flow. For every location, as the traffic flow is relatively stable at the same time of a day, we calculate the mean value for historical traffic flow data. For a historical volume $f_{l_i}^d(n)$, the corresponding average traffic flow $m_{l_i}^d(n)$ is calculated as follows:

$$m_{l_i}^d(n) = \frac{1}{7} \sum_{a=d-1}^{d-7} f_{l_i}^a(n) \tag{6}$$

Here, we choose a previous week's mean value to represent the average traffic flow.

Mode. During the process of prediction, outliers can make the average traffic flow volume abnormal. To eliminate such effect, we include the mode $r_{l_i}^d(n)$ for $f_{l_i}^d(n)$. However, as the traffic flow volumes are continuous, they need to be discretized before use as input. Accordingly, it can be evaluated as follows:

$$r_{l_i}^d(n) = mode(\lfloor \frac{f_{l_i}^{d-1}(n)}{\beta} \rfloor \times \beta, \lfloor \frac{f_{l_i}^{d-2}(n)}{\beta} \rfloor \times \beta, ..., \lfloor \frac{f_{l_i}^{d-7}(n)}{\beta} \rfloor \times \beta) \tag{7}$$

Here, β is an integer which is smaller than the maximum value of traffic flow, and the mode is calculated according to the result of the previous week.

In practice, the best β can be data dependent. We vary the value of β from 1 to 10, and notice that the best result can be obtained at 4 when the average traffic flow is smaller than 100, otherwise β is 10. We apply such setting in the rest of the paper.

We build a traffic flow feature matrix using all the three types of traffic flow data: the real data, the average data, and the mode data. Besides the three types data, we combine the traffic flow features that comes from the locations in a same group. For target $f_{l_0}^d(\mathrm{x})$, the following matrix shows the configuration of the feature matrix.

$$
\begin{bmatrix}
f_{l_0}^d(x-p) & \dots & f_{l_0}^d(x-2) & f_{l_0}^d(x-1) \\
m_{l_0}^d(x-p) & \dots & m_{l_0}^d(x-2) & m_{l_0}^d(x-1) \\
r_{l_0}^d(x-p) & \dots & r_{l_0}^d(x-2) & r_{l_0}^d(x-1) \\
\\
f_{l_1}^d(x-p) & \dots & f_{l_1}^d(x-2) & f_{l_1}^d(x-1) \\
m_{l_1}^d(x-p) & \dots & m_{l_1}^d(x-2) & m_{l_1}^d(x-1) \\
r_{l_1}^d(x-p) & \dots & r_{l_1}^d(x-2) & r_{l_1}^d(x-1) \\
\cdot & \dots & \cdot & \cdot \\
\cdot & \dots & \cdot & \cdot \\
f_{l_k}^d(x-p) & \dots & f_{l_k}^d(x-2) & f_{l_k}^d(x-1) \\
m_{l_k}^d(x-p) & \dots & m_{l_k}^d(x-2) & m_{l_k}^d(x-1) \\
r_{l_k}^d(x-p) & \dots & r_{l_k}^d(x-2) & r_{l_k}^d(x-1)
\end{bmatrix}
$$

Here, p is number of previous period, and it should be smaller than x, and $l_0, l_1, \,,\,, l_k$ are in the same group of G_0. The size of the matrix is decided by p and the group size $k+1$. In horizontal and vertical directions, the elements are sorted by time and location relation. Therefore, in the matrix, the spatial impacts can be reflected in the vertical direction and the temporal influences are extracted in the horizontal direction.

4.3 Training

In the second part of DGCNN, to use the 2-dimensional temporal-spatial traffic flow features for forecasting, we adopt the Convolutional Neural Network (CNN) with its capacity to process the temporal-spatial features at the same time.

CNN is built by convolution layers, sub-sampling layers, and full connected layers, etc. It stands with an array of alternating convolution and sub-sampling operations, and then continues with generic multi-layer network. The last few layers (closest to the outputs) are fully connected with 1-dimensional layers [3] for regression analysis. Figure 2 shows the structures of the convolution layer, and Fig. 3 is the sub-sampling layer.

The input and output of convolution layers are both feature maps. The active function that used in the convolution layer is shown in Eq. (8). M_j indicates the kernel j's projection positions on the feature map, k_{ij}^l is the kernel weights, b_j is a bias for kernel j, x_j^l and x_i^{l-1} are the volumes of output and input feature maps' elements, and $f(\sim)$ is a *sigmoid* function.

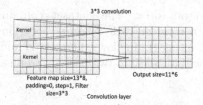
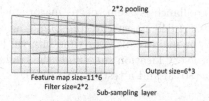

Fig. 2. The structures of convolution layer

Fig. 3. The structures of sub-sampling layer

$$x_j^l = f(\sum_{i \in M_j} x_i^{l-1} \times k_{ij}^l + b_j^l) \tag{8}$$

Equation (9) shows the sub-sampling layer function. Different from convolution layers, the $pool(\sim)$ is a sampling function, usually using the maximum or average value as output. Note the matrices on the feature map are mutually disjoint, and Fig. 3 shows this character.

$$x_j^l = f(pool(x_j^{l-1}) + b_j^l) \tag{9}$$

Apparently, we can see that the sub-sampling layer would reduce the feature dimensions, and the convolution layer can extend the feature dimension by increasing the number of kernels. In this study, we choose 2 convolution layers with 15 kernels and 7 kernels, and 1 sub-sampling layer. Table 1 shows an example whose input matrix size is set respectively to be $k = 1, p = 20$. Note that in the last layer, there is only one node, and its task is to apply regression for prediction. Hence, the output of CNN is the forecasting value.

Table 1. CNN layers' size

No	type	input size	output size
1	data	1 * 6 * 20(120)	1 * 6 * 20(120)
2	convolution	1 * 6 * 20(120)	15 * 4 * 8(480)
3	pool(sub-sampling)	15 * 4 * 8(480)	15 * 2 * 9(270)
4	convolution	15 * 2 * 9(270)	7 * 1 * 8(56)
5	full-connection	7 * 1 * 8(56)	300
6	full-connection	300	1
7	EuclideanLoss	1	1

5 Experiment

5.1 Data

The dataset used in the experiments consists of real vehicle passage records from December 1, 2015 to February 29, 2016 that were collected from the traffic surveillance system of a major metropolitan area. The dataset contains 479,451,848 passage records, and involves a total of 1323 detecting locations on the main roads. The time interval for counting traffic flow volume is 12 min. For every location, we divide the traffic flow data set into two parts, a training set (80 %) and a testing set (20 %).

5.2   Evaluation Matrics

We use two widely employed evaluation measures to assess the forecast performance:

(1) The Root Mean Squared Error (RMSE) is a way to measure the average error of the forecasting results and is calculated by

$$RMSE = \sqrt{\frac{1}{N} \sum_{n=1}^{N} (y_n - \hat{y}_n)^2}$$

Here, y_n and $\hat{y}_n$ are the observed and the forecast values of observation n. N is the total number of locations.

(2) The Mean Relative Error (MRE) is a way to measure the proportional error of the forecasting results and is calculated by

$$MRE = \frac{1}{N} \sum_{n=1}^{N} \frac{|y_n - \hat{y}_n|}{y_n} \times 100\%$$

5.3   Evaluation of DGCNN

There are two steps in the experiment. Firstly, trajectory data are used to train the word2vec, and we find every location's k nearest other locations. Secondly, the feature matrix is fed into CNN for prediction.

In the first step, we get 1,592,562 trajectories for training the word2vec. To generate a robust model, we ignore locations which appear less than 5 times. For one location in a trajectory, we consider that its previous 3 and next 3 locations as the impact factors ($window = 3$). One iteration is not enough for training high-quality location vectors, so $iteration$ can be more than one, Here we set it as 5. To make the vectors more effective, we set the vector dimension to be 350.

In the second step, We employ Caffe [11] to build the CNN. Different values of p(the height of feature matrix) and k(the width of feature matrix) are tested in the experiment. We want to minimize the MRE while training the CNN. Figure 4 shows the MREs of CNN with different values of group size $k + 1$ and Fig. 5 shows MREs with different previous interval number p. Two group of G_1 and G_2 have been tested.

From Figs. 4 and 5, we find that $k = 1, p = 20$ are the best values for DGCNN.

5.4   Comparison of Results

We compare SingleCNN($k = 0, p = 20$) with DGCNN($k = 1, p = 20$), Fig. 6. shows the RMSEs with different iterations and Fig. 7 shows the MREs. The result demonstrates that the grouped data can effectively decrease the MRE and RMSE and improve the performance.

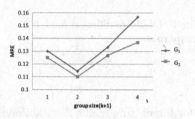

Fig. 4. MREs with different values of $k+1$

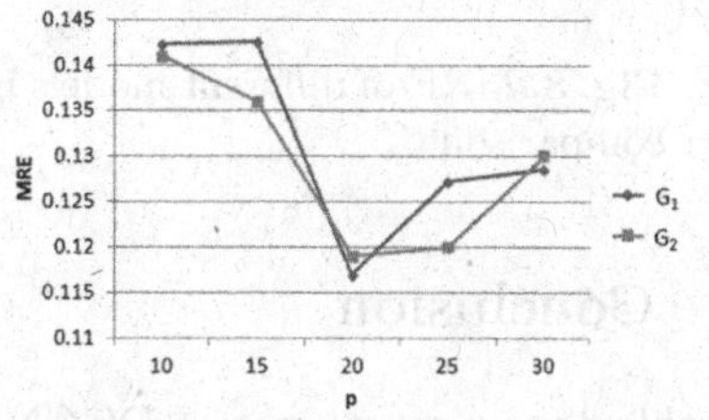

Fig. 5. MREs with different values of p

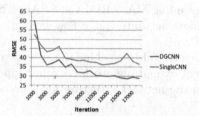

Fig. 6. RMSEs for SingleCNN and DGCNN

Fig. 7. MREs for SingleCNN and DGCNN

We also compare other models'.forecasting results with DGCNN. These models include MA, ARIMA and SAE.

(1) MA Model: An MA of order is computed by

$$\hat{y}_{t+1} = \frac{y_t + y_{t-1} + y_{t-2} + \dots + +y_{t-k+1}}{k},$$

where k is the number of terms in the MA. The MA technique deals only with the latest k periods of known data, and the number of data points in each average dose not change as time continues.

(2) ARIMA Model: We adopt a general ARIMA model of order (r, d, s), where d is the order of differencing, and orders r and s are the AR and MA operators.

(3) SAE Model: SAE model uses stacked autoencoder to extract features, and trains a regression network for forecasting.

Figure 8 shows their average MREs, and Fig. 9 reports their average RMSEs. All these results demonstrate that DGCNN model can make a more accurate short-term traffic flow forecasting.

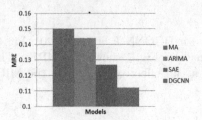 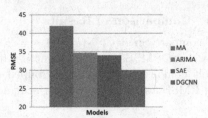

Fig. 8. MRE of different models in comparison

Fig. 9. RMSE of different models in comparison

6   Conclusion

In this paper, we propose a DGCNN model for forecasting traffic flow. Different to previous models, which have paid most of the attention on time-series data, this model uses a spatio-temporal dataset to build 2 networks. A CBOW model is used to find the spatio-relations, and the CNN is used to forecast traffic flow when the input is a spatio-temporal data matrix. We have compared DGCNN model with some state-of-the-art methods including MA, ARIMA and SAE. The result shows that the DGCNN can make more accurate traffic flow prediction than other models.

For further work, it is extraordinarily meaningful to combine the two stages of DGCNN for the forecasting task. We will extend our work on other types of input features, and we want to optimize the network structure for a better result.

Acknowledgment. This work was supported in part by the National Basic Research 973 Program of China under Grant No. 2015CB352502, the National Natural Science Foundation of China under Grant Nos. 61272092 and 61572289, the Natural Science Foundation of Shandong Province of China under Grant Nos. ZR2012FZ004 and ZR2015FM002, the Science and Technology Development Program of Shandong Province of China under Grant No. 2014GGE27178, and the NSERC Discovery Grants.

References

1. Abadi, A., Rajabioun, T., Ioannou, P.A.: Traffic flow prediction for road transportation networks with limited traffic data. IEEE Trans. Intell. Transp. Syst. **16**(2), 653–662 (2015)
2. Bengio, Y.: Learning deep architectures for AI. Found. Trends Mach. Learn. **2**(1), 1–127 (2009)
3. Bouvrie, J.: Notes on convolutional neural networks. Neural Nets (2006)
4. Chang, S.C., Kim, R.S., Kim, S.J., Ahn, M.H.: Traffic-flow forecasting using a 3-stage model. In: Proceedings of the IEEE Intelligent Vehicles Symposium, IV 2000, pp. 451–456 (2000)
5. Chen, C., Jianming, H., Meng, Q., Zhang, Y.: Short-time traffic flow prediction with arima-garch model. In: 2011 IEEE Intelligent Vehicles Symposium (IV), pp. 607–612 (2011)

6. Davis, G.A., Nihan, N.L.: Nonparametric regression and short-term freeway traffic forecasting. J. Transp. Eng. **117**(2), 178–188 (1991)
7. Dia, H.: An object-oriented neural network approach to short-term traffic forecasting. Eur. J. Oper. Res. **131**(2), 253–261 (2001)
8. Fusco, G., Colombaroni, C., Comelli, L., Isaenko, N.: Short-term traffic predictions on large urban traffic networks: Applications of network-based machine learning models and dynamic traffic assignment models. In: International Conference on MODELS and Technologies for Intelligent Transportation Systems (2015)
9. Fusco, G., Colombaroni, C., Gemma, A., Sardo, S.L.: A quasi-dynamic traffic assignment model for large congested urban road networks. Int. J. Math. Models Methods Appl. Sci. **7**(1), 63–74 (2013)
10. Huang, W., Song, G., Hong, H., Xie, K.: Deep architecture for traffic flow prediction: deep belief networks with multitask learning. IEEE Trans. Intell. Transp. Syst. **15**(5), 2191–2201 (2014)
11. Jia, Y., Shelhamer, E., Donahue, J., Karayev, S., Long, J., Girshick, R., Guadarrama, S., Darrell, T.: Caffe: Convolutional architecture for fast feature embedding. Eprint Arxiv, pp. 675–678 (2014)
12. Konstantina, P., Nina, T., Zhi-Li, Z., Christophe, D.: Long-term forecasting of internet backbone traffic. IEEE Trans. Neural Netw. **16**(5), 1110–1124 (2005)
13. Ledoux, C.: An urban traffic flow model integrating neural networks. Transp. Res. Part C Emerg. Technol. **5**(5), 287–300 (1997)
14. Lv, L., Chen, M., Liu, Y., Yu, X.: A plane moving average algorithm for short-term traffic flow prediction. In: Cao, T., Lim, E.-P., Zhou, Z.-H., Ho, T.-B., Cheung, D., Motoda, H. (eds.) PAKDD 2015. LNCS, vol. 9078, pp. 357–369. Springer, Heidelberg (2015)
15. Lv, Y., Duan, Y., Kang, W., Li, Z., Wang, F.-Y.: Traffic flow prediction with big data: a deep learning approach. IEEE Trans. Intell. Transp. Syst. **16**(2), 865–873 (2015)
16. Mikolov, T., Chen, K., Corrado, G., Dean, J.: Efficient estimation of word representations in vector space. arXiv preprint arXiv:1301.3781 (2013)
17. Moretti, F., Pizzuti, S., Panzieri, S., Annunziato, M.: Urban traffic flow forecasting through statistical and neural network bagging ensemble hybrid modeling. Neurocomputing **167**, 3–7 (2015)
18. Oh, S.D., Kim, Y.J., Hong, J.S.: Urban traffic flow prediction system using a multifactor pattern recognition model. IEEE Trans. Intell. Transp. Syst. **16**(5), 1–12 (2015)
19. Smith, B.L., Williams, B.M., Oswald, R.K.: Comparison of parametric and nonparametric models for traffic flow forecasting. Transp. Res. Part C: Emerg. Technol. **10**(4), 303–321 (2002)
20. Williams, B.: Multivariate vehicular traffic flow prediction: evaluation of ARIMAX modeling. Transp. Res. Rec. J. Transp. Res. Board **1776**(1), 194–200 (2001)
21. Williams, B.M., Hoel, L.A.: Modeling and forecasting vehicular traffic flow as a seasonal arima process: theoretical basis and empirical results. J. Transp. Eng. **129**(6), 664–672 (2003)

Efficient Evaluation of Shortest Travel-Time Path Queries in Road Networks by Optimizing Waypoints in Route Requests Through Spatial Mashups

Detian Zhang[1(⊠)], Chi-Yin Chow[2], Qing Li[2,3], and An Liu[4]

[1] School of Digital Media, Jiangnan University, Wuxi, China
`detian.cs@gmail.com`
[2] Department of Computer Science, City University of Hong Kong,
Kowloon Tong, Hong Kong
`{chiychow,itqli}@cityu.edu.hk`
[3] Multimedia Software Engineering Research Centre, City University of Hong Kong
Shenzhen Research Institute, Shenzhen, Guangdong, China
[4] School of Computer Science, Soochow University, Suzhou, China
`anliu@suda.edu.cn`

Abstract. In the real world, the route with the shortest travel time in a road network is more meaningful than that with the shortest network distance for location-based services (LBS). However, not every LBS provider has adequate resources to compute/estimate travel time for routes by themselves. A cost-effective way for LBS providers to estimate travel time for routes is to issue external requests to Web mapping services (e.g., Google Maps, Bing Maps, and MapQuest Maps). Due to the high cost of processing such external requests and the usage limits of Web mapping services, we take the advantage of direction sharing and waypoints supported by Web mapping services to reduce the number of external requests and the query response time for shortest travel-time route queries in this paper. We model the problem of selecting the optimal waypoints for an external route request as finding the longest simple path in a weighted bipartite digraph. As it is a NP-complete problem, we propose a greedy algorithm to find the best set of waypoints in an external route request. We evaluate the performance of our approach using real Web mapping services, a real road network, real and synthetic data sets. Experimental results show the efficiency, scalability, and applicability of our approach.

Keywords: Spatial mashups · Location-based services · Direction sharing · Waypoints · Web mapping services

1 Introduction

A spatial (or GIS/mapping) mashup [1–4] provides a cost-effective way for a Web or mobile application that combines data, representation, and/or functionality

© Springer International Publishing Switzerland 2016
F. Li et al. (Eds.): APWeb 2016, Part I, LNCS 9931, pp. 104–115, 2016.
DOI: 10.1007/978-3-319-45814-4_9

from at least one Web mapping service and other local/external services to create a new application. It becomes more and more popular along with the development of Web mapping services and the ubiquity of Internet access and GPS-enabled mobile devices. Typical Web mapping services are Google Maps, MapQuest Maps, Microsoft Bing Maps, Yahoo! Maps, and Baidu Maps. Based on the latest statistics of Programmable Web [5], the spatial mashup is the most popular one among all types of mashups including search mashups, social mashups, etc.

In the real world, the route with the shortest travel time (e.g., driving, walking and cycling time) in a road network is more meaningful than the route with the shortest network distance for location-based services (LBS). Since travel time is highly dynamic due to many realistic factors, e.g., heterogeneous traffic conditions and traffic accidents, it is difficult for an LBS provider to perform travel route estimation/computation because of huge deployment cost. However, it is not a big problem for Web mapping service providers, as they have adequate resources to collect data from historical traffic statistics and/or continuously monitor the real-time traffic in road networks; besides, most of existing Web mapping services provide user-friendly APIs for applications to access travel route information, e.g., the Google Maps Directions API [6] and the MapQuest Directions Web Service [7]. Therefore, a typical and popular application scenario using spatial mashups is that an LBS provider subscribes the travel route information based on live traffic conditions from Web mapping services through their APIs to answer location-based queries for its own users [1–4].

However, retrieving the travel route information through spatial mashups suffers from the following two critical limitations [1,2]: (1) It is costly to access travel route information from a Web mapping service, e.g., retrieving travel time from the Microsoft MapPoint Web service to a database engine takes 502 ms while the time needed to read a cold and hot 8 KB buffer page from disk is 27 ms and 0.0047 ms, respectively [8]. (2) There is a charge on the number of requests issued to a Web mapping service, e.g., the Google Maps Directions API [6] allows only 2,500 requests per day for evaluation users and 100,000 requests per day for business license users [9]. An LBS provider needs to pay to have higher usage limits. Therefore, when an LBS provider endures high workload, e.g., a large number of concurrent user queries, it needs to issue a large number of external Web mapping requests, which not only result in high business operation cost, but also induce long response time to querying users.

In this paper, we aim to explore a direction (or route/path) sharing optimization and select an appropriate set of waypoints in *external route requests* sent to a Web mapping service to reduce the number of external request and the response time to querying users for LBS providers. Given user u_i's *shortest travel-time path query* $R(o_i \rightarrow d_i)$, the basic idea of the direction sharing optimization is that the route information from query origin o_i to query destination d_i can be shared with another user u_j's query $q_j = (o_j, d_j)$ if both of its origin o_j and destination d_j are located in $R(o_i \rightarrow d_i)$. To reflect the opportunity of sharing the direction information of an external route request, we also formally define the *sharing ability* of a shortest travel-time path query. As most of existing Web mapping services support adding waypoints into a route request, e.g., Google Maps allows up to eight intermediate waypoints in a route request for evaluation users and 23 intermediate waypoints for premier users [9], we attempt to select optimal waypoints in a route request to further reduce the number of external requests and the query response time to users. We first model the problem of selecting optimal waypoints in a road network, i.e., the route through those waypoints can be shared by the largest number of queries, as the problem of finding the longest simple path in a weighted bipartite digraph. Since it is a NP-complete problem, we propose a greedy algorithm to find the best set of waypoints in an external route request, the time complexity of which is only $O(mn)$, where m is the maximum allowed number of waypoints in a route request and n is the number of vertices in a graph (m is usually much smaller than n).

To evaluate the performance of our proposed algorithm, we compare it with two basic algorithms using real Web mapping services, a real road network, real and synthetic data sets. Experimental results show that our proposed algorithm is efficient and scalable with various numbers of queries and waypoints.

2   Related Work

Existing shortest path query processing algorithms in road networks can be categorized into two main models. (1) **Time-independent road networks.** In this model, shortest path query processing algorithms assume that the cost or weight of a road segment, which can be in terms of distance or travel time, is constant (e.g., [10–12]). These algorithms mainly rely on pre-computed distance or travel time information of road segments in road networks. However, the actual travel time of a road segment may vary significantly during different time of a day due to dynamic traffic on road segments [13]. (2) **Time-dependent road networks.** The shortest path query processing algorithms designed for this model have the ability to support the dynamic weight of a road segment and topology of a road network, which can change with time. This model is more realistic but more challenging. George et al. [14] proposed a time-aggregated graph, which uses time series to represent time-varying attributes. The time-aggregated graph can be used to compute the shortest path at a given start time or to find the best start time for a path that leads to the shortest travel time. In time-dependent road networks where the weight of each road segment is a function of time, problems of the time-dependent shortest-path [13,15] also have been extensively studied. The solutions for those problems may capture the effects of periodic events (e.g., rush hours, and weekdays). However, they still cannot reflect live traffic information, which can be affected by sudden events, e.g., congestions, accidents, and road maintenance.

This paper focuses on the shortest travel-time route (or path) queries in time-dependent road networks. Our work distinguishes itself from previous work in that it does not model the underlying road network based on different criteria [13–15]. Instead, it employs external Web mapping services, e.g., Google Maps, to provide the real-time route information in a road network.

There are some work about query processing with expensive attributes that are accessed from external Web services (e.g., [1–4,8,16,17]). To minimize the number of external requests, [8,16,17] mainly focus on using either some *cheap* attributes that can be retrieved from local data sources [8,16] or sampling methods [17] to prune a whole set of objects into a smaller set of candidate objects; then, they only issue necessary external requests for retrieving candidate objects. The closest work to our paper are [1–4]. In [1–3], the authors proposed k-NN query processing algorithms that utilize grouping, direction sharing, shared query execution, pruning techniques and parallel requesting to reduce the number of external requests to Web mapping services and provide highly accurate query answers. In [4], route logs are employed to derive tight lower/upper bounding travel times to reduce the number of external Web mapping requests for answering range and k-NN queries. However, none of these existing techniques focuses on how to utilize waypoints in an external route request to reduce the number of such expensive requests, which is the problem that we address in this paper.

3   System Model

In this section, we describe our system architecture, road network model, problem definition, and the objectives of the paper.

System Architecture. Figure 1 depicts the system architecture that consists of three entities: users, an LBS provider, and a Web mapping service provider (e.g., Google Maps). Users send shortest travel-time path queries with query origins and destinations to the LBS provider, where the origin could be a user's current location or any other location and the destination could be any POI or location that the user wants to go or is recommended by the LBS provider. Besides turning in its own business information to a querying user, the LBS provider also returns the detailed direction information of the shortest route based on the user selected travel model (e.g., driving, walking, or cycling), which is accessed from the Web mapping service provider.

Road Network Model. In this paper, a road map is modeled as a graph $G = (V, E)$ comprising a set V of vertices with a set E of edges, where each road segment is an edge and each intersection of road segments is a vertex. As the

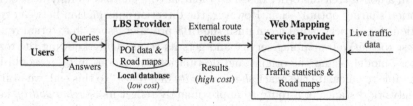

Fig. 1. System architecture.

weight of each edge is static (i.e., the network distance), G cannot provide real-time travel-time-based route information.

Problem Definition. Given a set of user queries $Q = \{q_1, q_2, \ldots, q_n\}$ arrived concurrently or within a short time period at the LBS provider, and each user query q_i is in a form of (o_i, d_i), where o_i and d_i are query origin and destination, respectively, the LBS provider returns the shortest travel-time route with the detailed direction information from o_i to d_i for each user query q_i based on live traffic conditions, which are provided by the Web mapping service provider.

Our Objectives. Since there are two critical limitations to access the travel route information from a Web mapping service as stated above, i.e., high cost and usage limits, our objectives are to reduce the number of external Web mapping requests issued by the LBS provider and the query response time to its querying users.

4   Direction Sharing Optimization

For an external route request with an origin and a destination, the Web mapping service will return the shortest travel-time route from the origin to the destination with detailed turn-by-turn direction information. Shortest routes exhibit the optimal sub-route property [2,18,19], i.e., every sub-route of the shortest route is also the shortest sub-route. In other words, the shortest route $R(o_i \rightarrow d_i)$ returned from the Web mapping service for a query $q_i = (o_i, d_i)$ can be shared used by another query $q_j = (o_j, d_j)$, if both o_j and d_j are located in the route in order[1], i.e., $o_j \in R(o_i \rightarrow d_i)$ and $d_j \in R(o_i \rightarrow d_i)$; thus, there is no need for the LBS provider to issue an external Web mapping request for $q_j = (o_j, d_j)$ any more. As a result, the number of external requests can be reduced.

To reflect the possibility of sharing the route information of a route with others, we formally define its *sharing ability* as follows:

Definition 1. (Sharing Ability). *Let Q be a set of user queries, and $R(o_i \rightarrow d_i)$ be the shortest travel-time route for query $q_i = (o_i, d_i)$ $(q_i \in Q)$. The sharing ability of $R(o_i \rightarrow d_i)$ with respect to Q (denoted as $SA(R(o_i \rightarrow d_i), Q)$) is the number of queries in Q, where the direction information of $R(o_i \rightarrow d_i)$ can be shared with these queries.*

Since the *sharing ability* can reflect the possibility of sharing the direction information of a route with others, to minimize the number of external route requests to the Web mapping service, the LBS provider should process queries in Q in a non-increasing order based on their *sharing abilities* to fully utilize the direction sharing optimization. However, the direction information between two locations for a query is unknown until the LBS provider issues an external route request to the Web mapping service and gets its result. Only knowing the road network model G and the origins and destinations of queries in Q, it would be impossible to calculate the exact *sharing ability* of a route. To this end, we utilize the network distance of a route to approximately reflect its *sharing ability*, i.e., a route with a longer network distance has a higher *sharing ability*.

---

[1] In this paper, we assume that a route is directional.

5   Route Waypoint Optimization

Most of existing Web mapping services allow users to specify waypoints in a route request. Waypoints alter a route by routing it through the specified locations. For example, given a route request with a set of ordered waypoints $W = \{w_1, w_2, \cdots, w_m\}$, where w_1 and w_m are the origin and destination locations, respectively, and the others are intermediate locations, Web mapping services will return the detailed direction information from w_1 to w_2 to $\cdots$ to w_m as a result for the route request (i.e., $R(w_1 \rightarrow w_2 \rightarrow \cdots \rightarrow w_m)$), and the sub-route $R(w_i \rightarrow w_{i+1})$ (where $1 \leq i \leq m - 1$) between any two successive waypoints in the result is with the shortest travel time.

However, most of queries issued by users only seek for the shortest travel-time route information from their own origins to corresponding destinations, i.e., only two waypoints, which is less than the maximum allowed number of waypoints provided by Web mapping service providers. For example, Google Maps allows up to 8 intermediate waypoints in a route request for evaluation users and 23 intermediate waypoints for premier users [9]. On the other hand, there is usually a limit or charge on the number of requests to a Web mapping service.

It is unwise for the LBS provider to simply issue external route requests to the Web mapping service provider for each query in $\mathcal{Q}$ separately without utilizing waypoints. Thus, a more efficient way is composing multiple user queries in $\mathcal{Q}$ as one external route request by taking the origins and destinations of those queries as waypoints.

5.1   Problem Modeling

To minimize the number of external route requests, the LBS provider should make full utilization of the direction sharing and waypoints provided by Web mapping services. The optimal ordered waypoint set $W = \{w_1, \cdots, w_m\}$ should be determined for an external route request to maximize the possibility of sharing its direction information with other queries in $\mathcal{Q}$. In other words, we should maximize the *sharing ability* of $R(w_1 \rightarrow \cdots \rightarrow w_m)$ with respect to the queries in $\mathcal{Q}$ by maximizing the *sharing ability* of the sub-routes in W with respect to the queries in $\mathcal{Q}$; hence,

$$SA(R(w_1 \rightarrow \cdots \rightarrow w_m), \mathcal{Q}) = \sum_{i=1}^{m-1} SA(R(w_i \rightarrow w_{i+1}), \mathcal{Q}). \tag{1}$$

To find the optimal $R(w_1 \rightarrow \cdots \rightarrow w_m)$, we first model the origins and destinations of the queries in $\mathcal{Q}$ and possible routes between them as a weighted semicomplete bipartite digraph $G_w = (O_w, D_w, E_w)$ comprising a set O_w of origin vertices, a set D_w of destination vertices, and a set E_w of weighted directed edges. The vertex set is composed by two disjoint sets O_w and D_w, where O_w and D_w consist of the origins, i.e., o_i ($i = 1, 2, \cdots, n$), and destinations, i.e., d_i ($i = 1, 2, \cdots, n$), of the queries in $\mathcal{Q}$, respectively. The set E_w consists of two parts of edges: (1) the edges starting from the query origins and ending at their corresponding destinations, i.e., $o_i d_i$ ($i = 1, 2, \cdots, n$), and (2) the edges starting from the query destinations and ending at all other origins except for their corresponding origins, i.e., $d_i o_j$ ($i, j = 1, 2, \cdots, n$ and $i \neq j$). The weight of each edge in E_w is represented by the *sharing ability* of its corresponding route, e.g., $weight(d_i o_j) = SA(R(d_i \rightarrow o_j), \mathcal{Q})$.

5.2 Waypoint Selection

After modeling the queries in $\mathcal{Q}$ as a weighted bipartite digraph $G_w = (O_w, D_w, E_w)$, we should find a waypoint set $\mathcal{W} = \{w_1, \cdots, w_m\}$ for $\mathcal{Q}$ such that the route $R(w_1 \rightarrow \cdots \rightarrow w_m)$ will result in the best *sharing ability*. We can transform this problem into how to find a simple path in $G_w = (O_w, D_w, E_w)$ with the largest weight via at most m vertices.

Since finding the longest simple path in a weighted graph is a NP-complete problem [18,20], in this work, we aim to find the best set of waypoints in an external route request for all the queries in $\mathcal{Q}$ by designing a greedy algorithm that attempts to find the best path in $G_w = (O_w, D_w, E_w)$. The time complexity of the proposed greedy algorithm is only $O(mn)$, where m is the maximum allowed number of waypoints in a route request supported by Web mapping services, n is the number of vertices in G_w, and m is usually much smaller than n.

The main idea of the greedy algorithm is to select the edge with the largest weight (i.e., the best *sharing ability*) from E_w, and then insert the vertices of the selected edge to waypoint set $\mathcal{W}$. The vertices that are connected to the selected edges are considered as waypoints in a route request. The details of the greedy algorithm are as follows.

The Greedy Algorithm. Given a query set $\mathcal{Q}$ and the maximum number of waypoints m, the proposed greedy algorithm consists of two main steps:

1. **Step 1: Initial waypoint selection step.** For each query $q_i = (o_i, d_i)$ in $\mathcal{Q}$, the *sharing ability* of its query route with respect to $\mathcal{Q}$ (i.e., $SA(R(o_i \rightarrow d_i), \mathcal{Q})$ is calculated, and then the origin o_l and destination d_l of the query q_l with the highest *sharing ability* are selected. o_l is inserted into $\mathcal{W}$ as the first waypoint, and it is removed from the query origin set O_w. If the number of waypoint in $\mathcal{W}$ (i.e., $|\mathcal{W}|$) is less than m, d_l is also inserted into $\mathcal{W}$ as the second waypoint and removed from the query destination set D_w. The algorithm proceeds to the next step if $|\mathcal{W}| < m$ and O_w is not empty.

2. **Step 2: Waypoint selection step.** In this step, d_l is the *last* waypoint inserted into $\mathcal{W}$. The next waypoint o_t is selected from all the origins in O_w and inserted into $\mathcal{W}$ such that $R(d_l \rightarrow o_t)$ has the highest *sharing ability*. o_t is then removed from O_w. If $|\mathcal{W}| < m$, the corresponding destination d_t of o_t is also added to $\mathcal{W}$, and d_t is removed from D_w. This step is repeated until $|\mathcal{W}| = m$ or both the origin and destination sets (i.e., O_w and D_w) become empty.

Finally, a route request with the waypoints in $\mathcal{W}$ is sent to the Web mapping service to retrieve the required direction information for the queries in $\mathcal{Q}$.

Algorithm 1. The algorithm with the direction sharing and route waypoint optimizations.

1: **input:** query set Q and the maximum number of waypoints m
2: **while** Q is not empty **do**
3:     Find the waypoint set $W = \{w_1, \cdots, w_m\}$ by the greedy algorithm;
4:     Retrieve $R(w_1 \rightarrow \cdots \rightarrow w_m)$ by issuing a route request to the Web mapping service;
5:     **for** each query $q_i = (o_i, d_i)$ in Q **do**
6:         **if** o_i and d_i is located in any $R(w_i \rightarrow w_{i+1})$ $(i = 1, \cdots, m-1)$ **then**
7:             Use the retrieved direction information to answer q_i;
8:             Remove q_i from Q;
9:         **end if**
10:     **end for**
11: **end while**

5.3 Algorithm

Given a query set $Q = \{q_1, q_2, \ldots, q_n\}$ and the maximum number of waypoints m, Algorithm 1 shows the pseudo code of the algorithm with the direction sharing and route waypoint optimizations. After we find the waypoint set $W = \{w_1, \cdots, w_m\}$ by the greedy algorithm, the LBS provider issues an external route request with the ordered waypoints in W to retrieve the direction information of $R(w_1 \rightarrow \cdots \rightarrow w_m)$ (Lines 3 to 4). For each query $q_i = (o_i, d_i)$ in Q, if o_i and d_i is located in any $R(w_i \rightarrow w_{i+1})$ $(i = 1, \cdots, m-1)$, the algorithm can use the retrieved direction information to answer q_i and q_i is removed from Q (Lines 7 to 8). The algorithm repeatedly deals with the remaining queries in Q until Q becomes empty.

6 Performance Evaluation

In this section, we first give our evaluation model in Sect. 6.1, including basic approaches, performance metrics, and experiment settings. Then, we analyze the experimental results in Sect. 6.2.

6.1 Evaluation Model

Basic Approaches. To the best of our knowledge, there is no other existing work about taking advantage of waypoints in route requests to reduce the number of external Web mapping requests for spatial query processing. To this end, we design two baseline algorithms to show the effectiveness of our proposed approach.

The direction sharing optimization has been proposed in our previous work [2] that is considered as a baseline algorithm (denoted as DS). The basic idea of DS is that: (1) the LBS provider processes the queries in Q based on the non-ascending order of their network distance in G by issuing external requests to the

Web mapping service provider. (2) When a travel route is returned, its direction information is used to answer the queries with origins and destinations located in the route and these queries are removed from Q. DS repeatedly processes the remaining queries in Q based on these two steps until Q becomes empty.

To reflect the effectiveness of our proposed greedy algorithm that aims to select the best set of waypoints in an external route request, (denoted as Greedy), we design another baseline algorithm by randomly selecting waypoints from the sets of query origins and destinations (denoted as Random). The basic idea of Random is that: (1) the LBS provider randomly selects waypoints from O_w and D_w in G_w into W. (2) The LBS provider next issues an external route request with the waypoints in W to retrieve the direction information of $R(w_1 \rightarrow \cdots \rightarrow w_m)$. (3) When the direction information of the route is returned, it is used to answer the the queries with their origins and destinations located in any $R(w_i \rightarrow w_{i+1})$ $(i = 1, \cdots, m - 1)$ and these queries are removed from Q. These three steps are repeatedly executed until Q becomes empty.

Performance Metrics. We evaluate the performance of the basic approaches (i.e., DS and Random) and our proposed algorithm (i.e., Greedy) in terms of two metrics: (1) the average number of external route requests submitted to the Web mapping service per user query and (2) the average query response time per user query. The query response time of a query is the time from the time when the query is received by the LBS provider to the time when the answer is returned to the querying user.

Experiment Settings. We evaluated the performance of DS, Random, and Greedy using C++ with a real road network of Hennepin County, Minnesota, USA. We selected an area of 8×8 km^2 that contains 6,109 road segments and 3,593 intersections, and the latitude and longitude of its left-bottom and right-top corners are (44.898441, -93.302791) and (44.970094, -93.204015), respectively. For all queries in our experiment, the query origins are randomly generated and the query destinations are randomly selected from a real POI dataset (including 126 bars, 320 restaurants, and 491 food places), which is collected from Google Places API [21] within the selected area. MapQuest Maps is taken as the Web mapping service in our experiments.

6.2 Experimental Results

We evaluate the scalability, efficiency, and applicability of the proposed algorithm (i.e., Greedy) and the basic approaches (i.e., DS and Random) with respect to various numbers of queries and waypoints.

Effect of the Number of Queries. In this section, the number of waypoints provided by the Web mapping service is set to six (i.e., $m = 6$), and the number of queries varies from 2,000 to 10,000 to evaluate the performance of DS, Random, and Greedy with respect to various numbers of queries as depicted in Fig. 2. Without any optimization, the LBS provider needs to issue one external request for each query. With the direction sharing optimization, the average number of external requests of DS decreases to less than one request, i.e., nearly 20 % reduction (Fig. 2a). With the utilization of waypoints, the average number of external route requests further reduces, i.e., nearly 40 % and 75 % are reduced by Random and Greedy, respectively. These experimental results show the effectiveness of Greedy for the waypoint section as it performs much better than Random. When the number of queries gets larger, there are more outstanding queries in the road network; hence, the direction information of a route has a higher chance to be shared with more queries (i.e., the power of the direction sharing optimization increases); thus, the number of external route request gets lower for all evaluated algorithms, as show in Fig. 2a. The results also show the scalability of Greedy with the increase of the number of queries.

In all the experiments, we do not consider parallel requesting to the Web mapping service provider, i.e., the LBS provider issues external route requests to the Web mapping service in a sequential manner. When there are more queries arrived, a query encounters a longer waiting time. Therefore, the average query response time gets longer with the increase of the number of queries, as depicted in Fig. 2b. Greedy yields the least average query response time compared with the other two basic algorithms, which benefits from the direction sharing and route waypoint optimizations.

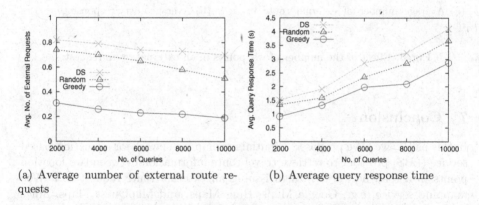

(a) Average number of external route requests

(b) Average query response time

Fig. 2. Effect of the number of queries.

Effect of the Number of Waypoints in an External Route Request. In this section, the number of queries is set to 6,000, and the number of waypoints provided by the Web mapping service provider varies from 2 to 10, to study the performance of DS, Random, and Greedy with respect to various numbers of waypoints, as depicted in Fig. 3. When the number of waypoints gets larger, the average number of external route requests for DS remains the same, while that for Random and Greedy drops significantly (Fig. 3a). This is because both Random and Greedy employ the route waypoint optimization while DS does not. With more waypoints utilized in an external route request, its direction information can be used to answer more queries. Figure 3a also shows that Greedy is the most effective algorithm with different numbers of waypoints as it consistently yields less external requests than Random. Similarly, the average query response time of DS is not affected by the number of waypoints in an external route request, while that of Random and Greedy drops gradually along with the increase of the number of waypoints (Fig. 3b), as Random and Greedy can utilize more waypoints in one route request. As a result, our Greedy can achieve not only the best number of external route requests but also the shortest query response time compared with the other two basic algorithms.

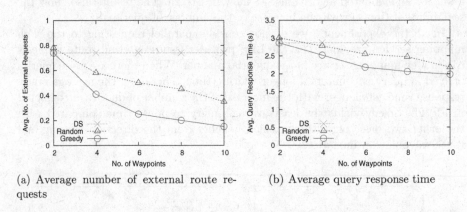

(a) Average number of external route requests

(b) Average query response time

Fig. 3. Effect of the number of waypoints in an external route request.

7   Conclusion

In this paper, we have proposed a spatial mashup framework for a location-based service (LBS) provider to retrieve travel route information between two location points in a road network through issuing an external route request to a Web mapping service, e.g., Google Maps, Bing Maps, and MapQuest Maps. Since retrieving external data is much more expensive than accessing local data, we employ the direction sharing and route waypoint optimizations to reduce the number of such external route requests. To reflect the possibility of sharing the direction information of a route, we first formally define a method to measure its *sharing ability*. Then, we model the problem of selecting the optimal set of waypoints in a road network for an external route request, i.e., the *sharing ability* of the sub-routes in a route request should be maximized, as the problem of finding the longest simple path in a weighted bipartite digraph. We design a greedy algorithm that aims to find the best waypoint set for an external route request. We evaluate the performance of our algorithm using real Web mapping services, a real road network, real and synthetic data sets. Our experimental results show that our proposed algorithm is efficient and scalable with various numbers of queries and waypoints in a route request.

Acknowledgments. This work was supported in part by the Fundamental Research Funds for the Central Universities in China (Project No. JUSRP11557), the National Natural Science Foundation of China (Project No. 61572336 and 61472337).

References

1. Zhang, D., Chow, C.Y., Li, Q., Zhang, X., Xu, Y.: Efficient evaluation of k-NN queries using spatial mashups. In: SSTD (2011)
2. Zhang, D., Chow, C.Y., Li, Q., Zhang, X., Xu, Y.: SMashQ: spatial mashup framework for k-NN queries in time-dependent road networks. Distrib. Parallel Databases DAPD **31**(2), 259–287 (2013)
3. Zhang, D., Chow, C.Y., Li, Q., Zhang, X., Xu, Y.: A spatial mashup service for efficient evaluation of concurrent k-nn queries. IEEE Trans. Comput. (accepted to appear)
4. Li, Y., Yiu, M.L.: Route-saver: leveraging route apis for accurate and efficient query processing at location-based services. IEEE TKDE **27**(1), 235–249 (2015)
5. ProgrammableWeb. http://www.programmableweb.com/category-api
6. The Google Directions API. https://developers.google.com/maps/documentation/directions
7. MapQuest Directions Web Service. http://www.mapquestapi.com/directions
8. Levandoski, J.J., Mokbel, M.F., Khalefa, M.E.: Preference query evaluation over expensive attributes. In: CIKM (2010)
9. Google Maps/Google Earth APIs Terms of Service. http://code.google.com/apis/maps/terms.html
10. Wu, L., Xiao, X., Deng, D., Cong, G., Zhu, A.D., Zhou, S.: Shortest path and distance queries on road networks: an experimental evaluation. In: VLDB (2012)
11. Zhu, A.D., Ma, H., Xiao, X., Luo, S., Tang, Y., Zhou, S.: Shortest path and distance queries on road networks: towards bridging theory and practice. In: ACM SIGMOD (2013)
12. Sommer, C.: Shortest-path queries in static networks. ACM Comput. Surv. (CSUR) **46**(4), 45:1–45:31 (2014)
13. Demiryurek, U., Banaei-Kashani, F., Shahabi, C., Ranganathan, A.: Online computation of fastest path in time-dependent spatial networks. In: Pfoser, D., Tao, Y., Mouratidis, K., Nascimento, M.A., Mokbel, M., Shekhar, S., Huang, Y. (eds.) SSTD 2011. LNCS, vol. 6849, pp. 92–111. Springer, Heidelberg (2011)
14. George, B., Kim, S., Shekhar, S.: Spatio-temporal network databases and routing algorithms: a summary of results. In: Papadias, D., Zhang, D., Kollios, G. (eds.) SSTD 2007. LNCS, vol. 4605, pp. 460–477. Springer, Heidelberg (2007)
15. Ding, B., Yu, J.X., Qin, L.: Finding time-dependent shortest paths over large graphs. In: EDBT (2008)
16. Bruno, N., Gravano, L., Marian, A.: Evaluating top-k queries over web-accessible databases. In: IEEE ICDE (2002)
17. Chang, K.C.C., Hwang, S.W.: Minimal probing: supporting expensive predicates for top-k queries. In: ACM SIGMOD (2002)
18. Cormen, T.H., Leiserson, C.E., Rivest, R.L., Stein, C.: Introduction to Algorithms, 3rd edn. MIT Press, Cambridge (2009)
19. Thomsen, J.R., Yiu, M.L., Jensen, C.S.: Effective caching of shortest paths for location-based services. In: ACM SIGMOD (2012)
20. Karger, D., Motwani, R., Ramkumar, G.: On approximating the longest path in a graph. Algorithmica **18**(1), 82–98 (1997)
21. The Google Places API. https://developers.google.com/places/

Discovering Companion Vehicles from Live Streaming Traffic Data

Chen Liu[1,2(✉)], Xiongbin Wang[1,2], Meiling Zhu[1,2,3],
and Yanbo Han[1,2]

[1] Beijing Key Laboratory on Integration and Analysis of Large-Scale
Stream Data, North China University of Technology, Beijing, China
{liuchen,hanyanbo}@ncut.edu.cn, {suifengzhewxb,
meilingzhu2006}@126.com
[2] Cloud Computing Research Center, North China University of Technology,
Beijing, China
[3] School of Computer Science and Technology, Tianjin University,
Tianjin, China

Abstract. Companions of moving objects are object groups that move together in a period of time. To quickly identify companion vehicles from a special kind of streaming traffic data, called Automatic Number Plate Recognition (ANPR) data, this paper proposes an approach to discover companion vehicles. Compared to related approaches, we transform the companion discovery into a frequent sequence-mining problem. We make several improvements on top of a recent frequent sequence-mining algorithm, called SeqStream, to handle customized time constraints among sequence elements when discovering traveling companions. We also use pseudo projection technique to improve the performance of our algorithm. Finally, extensive experiments are done using a real dataset to show efficiency and effectiveness of our approach.

Keywords: Companion vehicles · ANPR data · Moment companion ·
Traveling companions · Frequent sequence-mining

1 Introduction

Discovering companions of moving objects has become a hot topic and attracted many research studies [1–7]. Its goal is to discover object groups that move together in a lasting period of time. Lots of companion patterns have been proposed, like flock [1], convoy [2], swarm [3], traveling companion [4], gathering [5, 6], platoon [7] and so on. These techniques have wide applications in various areas like transportation management, protecting special kind of vehicles and military surveillance [4].

Vehicles are typical moving objects. Lots of researches have been done to discover companion vehicles based on GPS data [1–7]. However, to obtain GPS data, vehicles are required to install a GPS device and turn it on. Compared with GPS data, ANPR (Automatic Number Plate Recognition) data are very suitable for the round-the-clock and wide-range monitoring of vehicles without requiring pre-installing or turning on any extra devices. ANPR data come from Traffic Enforcement Cameras that have been

© Springer International Publishing Switzerland 2016
F. Li et al. (Eds.): APWeb 2016, Part I, LNCS 9931, pp. 116–128, 2016.
DOI: 10.1007/978-3-319-45814-4_10

installed at most road crossings in China. These cameras continuously take pictures of passing vehicles with approximate one second interval at rush hour. Vehicle information, e.g., plate number and passing time, is automatically recognized and transmitted to a data center of traffic management department in form of data stream.

In our previous work, we have designed several algorithms to effectively discover companion vehicles on ANPR data. In [8], we borrow ideas from frequent itemset mining algorithm, like Apriori algorithm, to mine companion vehicles from historical ANPR dataset and apply it into a scenario of carpooling recommendations. Then, in [9], we try to discover companion vehicles from live ANPR data stream. We propose a concept called *Moment Companion*, which records a snapshot of companions for a vehicle when it passes through a camera. However, this algorithm doesn't take historical companions into account. It cannot assure the discovered companions are occasional or will last for a certain period. In [10], we borrow ideas from SeqStream algorithm and explore a new way to encapsulate the ability of companion discovery into a service. Compared to it, further improvements are made in this paper. We redesign the core tree data structure and bind its nodes to a variable time range by clustering the timestamps of data records. More operations and corresponding optimization techniques for manipulating tree nodes are also proposed.

Following our previous work, in this paper, we try to instantly discover companion vehicles that keep passing through several cameras in a lasting period. To reach this goal, we transform the discovery of companion vehicles into a frequent sequence-mining problem. SeqStream algorithm was developed for mining frequent sequence with a sliding window [11]. However, traditional frequent sequence-mining algorithms, like SeqStream, cannot directly solve our problem. It is because traditional algorithms only consider the order of elements in sequences. To discover companion vehicles, underlying time constraints among sequence elements need to be considered. Besides, how to instantly discover companion vehicles is also a challenge when new ANPR data records are received.

Hence, on top of SeqStream algorithm, this paper proposes an approach to instantly discover companion vehicles from live ANPR data. Our main contributions include: (1) We make several improvements on SeqStream algorithm to take arriving time of each ANPR data record into account and mine the frequent sequences under a customized time constraint where the elements in a frequent sequence should be close enough in time. (2) We integrate pseudo projection technique into our tree node-manipulating algorithm to improve its performance. Finally, several experiments are done to show efficiency and effectiveness of our algorithm.

2 Problem Analysis

Figure 1 shows the effects of our algorithm, which tries to instantly compute companions of a given vehicle passing through a camera. As this figure shows, traffic enforcement cameras are spread over road crossings. Each camera is a stream data source, which continuously generates ANPR data records.

An ANPR data record $r = (c, v, t)$ means a vehicle v passes through a camera c at time t. For example, (CAM04612111, JingCN8R21, 2013-1-1 11:27:00), (CAM04612111,

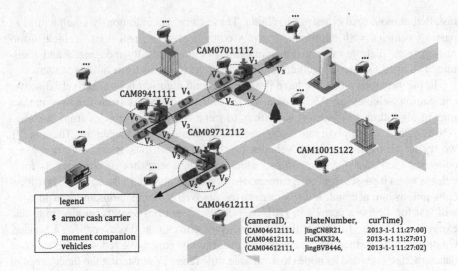

Fig. 1. The illustration of discovering traveling companions

HuCMX324, 2013-1-1 11:27:01) and (CAM04612111, JingBVB446, 2013-1-1 11:27:02)
are all ANPR data records.

We try to discover companion vehicles in accordance with traveling companion
pattern proposed by Tang et al. [4]. However, the definition of traveling companion is
given based on GPS data. ANPR data is very different. It is essentially discrete. Hence,
referring to [4], we first propose the concept of Moment Companion [9]. Then, based
on it, the traveling companion on ANPR data is defined as follows.

Definition 1 (Moment Companion). Let Δt be the time threshold, a moment companion for camera c at time t can be defined as:

$$MC(c, t, \Delta t) = \{r | t - \Delta t \leq r.t \leq t\}, \text{ where } r = (c, v, t) \text{ is an ANPR data record.}$$

Example. In Fig. 1, given vehicle v_1, $\{v_2, v_4, v_5\}$, $\{v_2, v_5, v_6\}$ and $\{v_2, v_5, v_7\}$ are all
moment companions when v_1 passing through camera CAM07011112, CAM89411111
and CAM09712112 successively.

Definition 2 (Traveling Companion): Given a time threshold Δt, a vehicle number
threshold δ_{veh}, a camera number threshold δ_{cam}, an ANPR data record set R, a vehicle
set $V = \{v_1, v_2, \ldots, v_m\}$, a camera set in sequence $C = <c_1, c_2, \ldots, c_n>$, and a
timestamp set in sequence $T = <t_1, t_2, \ldots, t_n>$, a traveling companion is a three
tuple: $TC = (V, C, T)$, which should satisfy the following conditions: (1) $|V| \geq \delta_{veh}$,
$|C| \geq \delta_{cam}$; (2) $i = 1, 2, \ldots, n$, $\exists r \in R$, s.t. $r.v \in V \bigwedge r \in MC(c_i, t_i, \Delta t)$.

Example: When ANPR data record (CAM09712112, v_1, t) in Fig. 1 arrives, let
$\delta_{veh} = 2$, $\delta_{cam} = 2$, $\{\{v_1, v_2, v_5\}, \{\text{CAM07011112, CAM89411111}\}, \{t_1, t_2\}\}$ and
$\{\{v_1, v_2, v_5\}, \{\text{CAM07011112, CAM89411111, CAM09712112}\}, \{t_1, t_2, t_3\}\}$ are both
Traveling Companions.

3   Discovery of Traveling Companions

A. Preliminaries

Frequent sequences are sequences that appear frequently in a data set [12]. Let $I = \{i_1, i_2, \ldots, i_m\}$ be a set of items. A subset of I is an **itemset**. A **sequence** is an ordered item list $s = <t_1, t_2, \ldots, t_m>$ $(t_i \subseteq I, i = 1, 2, \ldots, m)$. A **sequence database** is a set of sequences, and each sequence is associated with an identifier, called a *SID*. A **support** of a sequence is the number of it is contained in the sequences of a sequence database D. A sequence becomes **frequent** if its support exceeds a pre-specified minimum support threshold in database D. A frequent sequence becomes **closed** if there is no super-sequence of it with the same support can be found in database D. An **s-projection database** in D is defined as $D_s = \{\alpha | \gamma \in D, \gamma = \beta \cdot \alpha$ such that β is the minimum prefix (of γ) containing $s\}$.

The **inverse sequence** of a sequence s, denoted by s', can be obtained by inverting the item order of s. A sequence $\alpha = \langle \alpha_1 \alpha_2 \ldots \alpha_n \rangle$ is called a **subsequence** of another sequence $\beta = \langle \beta_1 \beta_2 \ldots \beta_m \rangle$, if there exist integers $1 \leq j_1 < j_2 < \ldots < j_n \leq m$, such that $\alpha_1 = \beta_{j1} \wedge \alpha_2 = \beta_{j2} \wedge \ldots \wedge \alpha_n = \beta_{jn}$, denoted as $\alpha \sqsubseteq \beta$; and β is a **supersequence** of α, denoted as $\beta \sqsupseteq \alpha$.

Traditional frequent sequence mining algorithms can be categorized into two classes: Apriori-based algorithms and projection-based pattern growth algorithms. In recent years, some researchers put focuses on how to mine frequent sequence from live data stream. SeqStream is a typical work [11], which mines closed frequent sequence in a sliding window for arriving data records. To facilitate the removal of expired data, it first transforms original data sequence database into inverse sequence database (inverse sequence of < abc > is < cba >). Then, it proposes a core data structure, called IST (Inverse Closed Sequence Tree), which keeps closed sequential patterns in the inverse sequence database of current sliding window. Next, the algorithm continuously updates IST by inserting or removing nodes when new data arrives.

B. Running Example

Table 1 shows a simple example of an inverse sequence database based on ANPR dataset. We divide this table into two parts. The first part is from column t_{10} to t_1. It contains historical data records and stores them in a window. The column t_{12} to t_{11} are data records newly arrived. We establish a sequence for each vehicle by its passing cameras and corresponding timestamps. Table 2 shows a sample sequence database based on the ANPR dataset from column t_{10} to t_1 in Table 1.

To illustrate rationales of our algorithm, we first give explanations of some symbols. D: the set of inverse sequence database in a sliding window; $\hat{D}$: intermediate database, the sequence database in current sliding window after inserting a new set of elements (but before removing expired ones); D^+: insertion database, the sequence database of ANPR records arriving at $[t - \Delta t, t]$, t is the arriving time of newest records; D^-: removal ANPR record database, the set of expired ANPR records.

Table 1. A sample of an inverse sequence database on ANPR dataset

	$t_{12}=$ 9:43:39	$t_{11}=$ 9:43:38	$t_{10}=$ 9:43:31	$t_9=$ 9:43:29	$t_8=$ 9:43:28	$t_7=$ 9:42:26	$t_6=$ 9:42:20	$t_5=$ 9:40:23	$t_4=$ 9:39:20	$t_3=$ 9:39:16	$t_2=$ 9:36:03	$t_1=$ 9:36:00
v_1			c_5		c_4	c_3			c_2		c_1	
v_2				c_4			c_3			c_2		c_1
v_3	c_4										c_2	
v_4	c_5		c_4					c_2				
v_5		c_4						c_3				
v_6					c_4					c_3		
v_7								c_4				

Table 2. A sample sequence database

SID	Sequence
s_1	$<c_5,09{:}43{:}31> \rightarrow <c_4,09{:}43{:}28> \rightarrow <c_3, 09{:}42{:}26> \rightarrow <c_2, 09{:}39{:}20> \rightarrow <c_1, 09{:}36{:}03>$
s_2	$<c_4,09{:}43{:}29> \rightarrow <c_3,09{:}42{:}20> \rightarrow <c_2,09{:}39{:}16> \rightarrow <c_1,09{:}36{:}00>$
s_3	$<c_2, 09{:}36{:}03>$
s_4	$<c_4,09{:}43{:}31> \rightarrow <c_2,09{:}40{:}23>$
s_5	$<c_3, 09{:}40{:}23>$
s_6	$<c_4,09{:}42{:}26> \rightarrow <c_3, 09{:}39{:}16>$
s_7	$<c_4,09{:}40{:}23>$

C. IST* Tree

We propose a new data structure, called IST*, to record discovered traveling companions. It is a variation of the IST data structure. An IST is used to keep closed sequential patterns in the inverse sequence database of current sliding window. Figure 2(a) shows a sample of IST with the running example in Table 1. As Fig. 2(a) shows, a node n (contains camera ID and its support) of an IST corresponds to a sequence that starts from the root node to that node, and the sequence is denoted by s_n. The root node of an IST is a NULL node, which represents an empty sequence ϕ. Except for the root node, nodes of an IST can be divided into following three types.

- c-node (closed node): If s_n is a closed sequential pattern in D^+, n is a c-node.
- t-node (termination node): let D' be the inverse database of D, n is a t-node if (1) there exists a frequent sequence β in D' such that $\beta \sqsupset s_n$ and $D'_\beta = D'_{s_n}$; (2) it does not have any t-node ancestor.
- i-node (intermediate node): If s_n is frequent, n is neither a c-node nor a t-node, and n has no t-node ancestor, then n is an i-node.

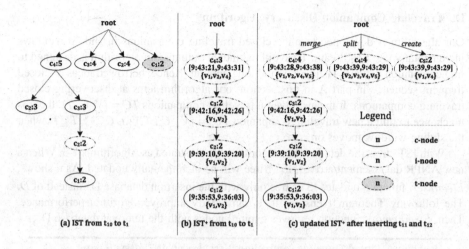

Fig. 2. Comparisons between IST and IST* tree

We can use IST tree to represent frequent camera sequences just like Table 2 shows. For example, as Fig. 2(a) shows, let $\delta_{cam} = 2$, we can find 6 closed frequent camera sequences $< c_4 >$:5, $< c_4, c_2 >$:3, $< c_4, c_3 >$:3, $< c_4, c_3, c_2, c_1 >$:2, $< c_3 >$:4 and $< c_2 >$:4. Although IST tree can represent the camera sequence, it cannot be directly used to represent the traveling companions. It is because IST and corresponding SeqStream algorithm do not take time threshold into account. Essentially, IST and SeqStream algorithm only can mine sequences that do not have any constraints among sequence elements. However, mining traveling companion is very different. Each element in a camera sequence has a timestamp, which is very important to judge a traveling companion. For example, let $\Delta t = 10$ s, the closed frequent camera sequence $< c_4, c_2 >$:3 and its corresponding vehicle set $\{v_1,v_2,v_4\}$ in Fig. 2(a) is not a traveling companion. It is because the time interval between v_2 and v_4 passing through c_2 is greater than 10 s.

Hence, we design IST* to take the time attribute of each element of a camera sequence into account. Like IST, IST* is a rooted prefix tree as well. Each node n in an IST* stores a traveling companion, including its type ($n.type$), a cameraID ($n.cameraID$), a vehicle set ($n.VS$) and a time interval [$n.t_1, n.t_2$]. The time interval of an IST* node is formed by clustering the timestamps of involved data records according to time threshold Δt in Definition 2. It records allowable time scope about which elements can join this sequence.

The node type ($n.type$) of an IST* node is in accordance with IST tree. As Definition 2 shows, based on the camera number threshold δ_{cam}, the types of nodes in an IST* can be defined as follows: Given an IST* node n_i, the corresponding sequence from root to it is $s_i = \{root, n_1, ..., n_i\}$, $|s_i| \geq \delta_{cam}$, then n_i is a c-node, else n_i is an i-node. A difference between IST* and IST tree is IST* tree doesn't keep any live t-nodes. If a c-node becomes a t-node, then it will be abandoned.

D. Traveling Companion Discovery Algorithm

Our algorithm is developed to mine closed traveling companions in sliding windows incrementally like SeqStream. To avoid redundant outputs, our algorithm is designed to find companion vehicles lasting a period long enough. Hence, borrowed ideas of closed frequent sequence in part A in this section, our algorithm aims at discovering closed traveling companions. It means, given a traveling companions $TC = (V, C, T)$, there is no chance to find a new traveling companion $TC' = (V', C', T')$ s.t. $C \subseteq C', T \subseteq T'$ when the sliding window moves on.

With IST* tree, the details of our algorithm are illustrated as Algorithms 1-3. When a new ANPR data element arrives, IST* tree will be incrementally updated. As it shows, *ElemInsert* function updates IST* tree only with the insertion database D^+ instead of $\hat{D}$. The following Theorem 1 ensures it. It can greatly improve algorithm performance. Then, *ElemRemove* function removes expired nodes with the removal database D^-.

Algorithm 1. Closed Traveling Companion Discovery on ANPR Data Stream

Input: *root* : root node of current IST*;
      elem: a new arrived ANPR data element;
      Δt; δ_{veh}; δ_{cam}; //defined in Definition 1 and 2;
Output: the updated IST*;

1.   get D^+;
2.   call ElemInsert(*root, elem, D^+*);
3.   get D^-;
4.   call ElemRemove(*root, D^-*);

Theorem 1 (Intermediate Sequence Database Shrink). Intermediate sequence database $\hat{D}$ can shrink to D^+ without losing any traveling companions.

Proof: If a new arrived record leads to the generation of a new closed traveling companion $(V, C, T) = (\{v_1, v_2, \ldots, v_m\}, <c_1, c_2, \ldots, c_n>, <t_1, t_2, \ldots, t_n>)$, it should satisfy that $V \subseteq MC(c_n, t_n, \Delta t)$. Let t_i^n be the time of vehicle v_i in V passing through camera c_n, we have $t_n - t_i^n \leq \Delta t$. It means the sequence corresponds to vehicle v_i is included in D^+.

Algorithm 2 shows the algorithm about inserting a new element. Line 1 is responsible for merging, splitting or creating depth-1 nodes in an IST* when a new data element arrives. According to Theorem 2 in SeqStream algorithm [11], only depth-1 nodes newly updated will be extended. Hence, when a new ANPR data record r arrives, it will first be compared with existing depth-1 nodes in an IST* tree. Given a depth-1 node n, there are four comparison results: (1) if $(n.cameraID == r.c \wedge (r.t - n.t_1) \leq \Delta t)$, then r can be merged into node n; (2) if $(n.cameraID == r.c \wedge (r.t - n.t_1) > \Delta t)$, then n is split into two nodes; (3) if $\nexists n(n.cameraID == r.c)$, and $|MC(r.c, r.t, \Delta t)| \geq \delta_{cam}$, then a new node is created for $MC(r.c, r.t, \Delta t)$; (4) $\nexists n(n.cameraID == r.c)$ and $|MC(r.c, r.t, \Delta t)| < \delta_{cam}$, then none of the depth-1 nodes will include r.

As Table 1 and Fig. 2(c) show, the new arriving element (c_4, v_5, 09:43:38) is merged into the left depth-1 node in Fig. 2(c). The new arriving element (c_4, v_3, 09:43:39) splits the above node to the middle depth-1 node. And the new arriving element (c_5, v_4, 09:43:39) leads to the creation of the right depth-1 node in Fig. 2(c).

Line 2 in Algorithm 1 establishes sequence database $D^+_{n_{elem}}$ according to Theorem 2. This theorem ensures that the sequence database can be further shrunk when extending each newly updated depth-1node in the following.

Theorem 2 (Sequence Database Shrink). To extend a newly changed or created depth-1 node n in an IST*, mining in sequence database $D^+_n = \{s|s \in D^+ \land s.SID \in n.VS\}$ is the same to mining in D^+.

Proof: Given a descendant node n_{des} under node n, the node n_{des} corresponds to a newly generated three tuple (V, C, T), where $V = n_{des}.VS$, $C = <n_{des}.cameraID, ...,$ $n.cameraID >$ and $T = <n_{des}.t_2, ..., n.t_2 >$. According to the construction of IST*, the (V, C, T) satisfies with definition of traveling companion except for $|C| \geq \delta_{cam}$. Hence, $V \subseteq MC(n.cameraID, n.t_2, \Delta t)$. Since $V = n_{des}.VS \subseteq n.VS$, sequences corresponding to vehicles in V are included in D^+_n.

Line 3 in Algorithm 1 generates the timestamp sequence T according to $D^+_{n_{elem}}$. Line 4 defines two pointers p_vs and p_ts. They point to vehicle set of a node and generated time sequence respectively. They are used in the construction of pseudo projection database [12].

The pseudo projection [12] is an optimization technique, which can efficiently improve performance of our algorithm. When a sequence database in stored in memory, we can use pointers to realize pseudo projection instead of establishing a real projection database. For instance, given a sequence $s_1 = <abcd>$, the projection of 'a' in s_1 is $< bcd >$. However, without establishing a real database to record $< bcd >$, we can calculate the projection database using a pointer pointing to s_1 and an index $i = 1$. This technique is called as *pseudo projection*. The projection database generated by this technique, composed of list of pointers as well as corresponding indexes, is called as the pseudo projection database.

According to the time constraints of traveling companions, the pseudo projection database can be generated based on pointers to *SID* of a camera sequence and timestamps of corresponding elements. Given a sequence database D^+, each element in each sequence can be identified by a pointer to its *SID* and a pointer to its timestamp. Thus, we construct pseudo projection database from D^+ as a list of pointers to *SID* of sequences and a sequence of pointers to corresponding timestamps.

Algorithm 2. Element Insert Algorithm

Input: *root* : root node of current IST*;
      elem: a new arrived ANPR data element;
      D^+: the insertion database;
Output: the updated IST*;

---

1.   merging, splitting or creating depth-1 nodes according to elem;
2.   get $D^+_{n_{elem}} = \{s | s \in D^+ \wedge s. SID \in n_{elem}. VS\}$ // n_{elem} is the IST* node contains elem
3.   get time sequence *TS* from $D^+_{n_{elem}}$;
4.   Pointer $p\_vs \to n. VS$, $p\_ts \to TS$;
5.   call ExtendNode(n_{elem} , *descendants, p_vs, p_ts*);

Function ExtendNode(n_{elem} :Node ,descendants:List<Node>, p_vs:Pointer, p_ts:Pointer)
6.   int *sq_size* = size of sequence pointed by *p_ts*;
7.   Array *pArray* = new Array [*sq_size*];
8.   for int *i* from 0 to *sq_size*
9.   Let *pArray[i]* be the sub sequence of p_ts starting from *i*;
10.   construct pseudo projection database *PDB* by *p_vs* and *pArray* ;
11.   compute moment companion set *MC* from *PDB* on timestamps pointed by *pArray*;
12.   for each *mc* in *MC*
13.   add a new descendant node according to mc to descendants;
14.   ExtendNode(n_{elem}, descendants, p_ts, pArray[i+1]);
15.   remove newly added node from descendants;
16.   endfor
17.   endfor
18.   if $|descendants| \geq \delta_{cam}$ extend node n_{elem} on IST* by descendants;
19.   delete t-node;
20.   return;

Lines 6−20 in Algorithm 2 implement a procedure called as *ExtendNode* to extending newly updated depth-1 nodes. Lines 9 define a temporary pointer array *pArray* to point subsequence of timestamps from *p_ts*. Line 10 constructs a pseudo projection database *PDB* by pointers *p_vs* and *pArray*. Line 11 computes all moment companions *MC* according to *pArray*. New descendant node corresponding to *mc* in *MC* is added to *descendants* in Line 11. Then the method enters into a recursion in Line 12. When a recursion is finished, line 13 removes the newly added node in *descendants*. If *descendants* has no less than δ_{cam} nodes, line 14 extends the newly updated depth-1 node n_{elem} according to *descendants*. Line 19 deletes t-nodes.

ElemRemove() is presented in Algorithm 3, which removes expired data without extending any node. First, it abandons invalid depth-1 nodes in the IST*. Given a new arriving ANPR data element *e* and an existing depth-1 node *n*, if the time span between the upper bound t_2 of the time interval in node *n* and *elem.t₁* is more than Δt, node *n* is abandoned. Secondly, for each expired data record *e* in D^-, find all closed nodes including element *e*. For each closed node *m*, remove *e* from the bottom up.

Algorithm 3. Element Removal Algorithm
Input: *root* : root node of current IST*;
elem: a new arriving ANPR data element;
D^-: the removal database;
Output: the updated IST*;
1.   for each depth-1 node n
2.   if $n.t_2 < elem.t_1$ and n has no descendant delete node n;
3.   for each element e in D^-
4.   for each closed node m including element e
5.   remove e from m;
6.   if $|m.VS| < \delta_{veh}$ remove m;
7.   if the depth of the parent of m is less than $\delta_{cam} - 1$
8.   delete all ancestor nodes of m, whose depth is more than 2;
9.   endfor
10. endfor
11. return;

4 Experiment

A. Experiment Setup

We do experiments to measure the effectiveness and efficiency of our algorithm. Our algorithm involves three key parameters $\Delta t, \delta_{veh}$ and δ_{cam}, defined in Definitions 1 and 2 respectively. They should be pre-assigned before running the algorithm. Generally, δ_{veh} is set to 2 as a companion should contain at least two vehicles. However, the values of the rest two parameters depend on user experiences. Their values will greatly impact the results of discovered traveling companions. Hence, we verify the effectiveness and efficiency of our algorithm using different parameter values.

The following experiments use a real ANPR dataset in Beijing, China. The dataset contains vehicle information from 2012-10-17 06:00:00 to 2013-01-04 20:00:00. Totally 1040 cameras and 336,268,812 ANPR data records are involved. We simulate the data set as a stream. The time interval between two adjacent data records is in accordance with real intervals when they were shot by cameras. Our algorithm takes the data stream as inputs and output discovered companions instantly when a new data element arrives. The experiments are done on a PC with four Intel Core i5-2400 CPUs 3.10 GHz and 4.00 GB RAM. The operating system is Windows 7 Ultimate. All the algorithms are implemented in Java with JDK 1.8.0.

B. Effectiveness

Values of parameters Δt and δ_{cam} will greatly impact the results of discovered traveling companions. To evaluate the effectiveness of our algorithm, we did some experiments to verify the numbers of discovered traveling companions with different parameter values. The experiment results can provide some references about how to choose parameter values when the algorithm is applied into various scenarios.

To be objective, we run our algorithm 20 times for each parameter configuration and compute average number of discovered traveling companions for each time.

variation of average discovered traveling companions number under different Δt and δ_{cam} (x axis:Δt)

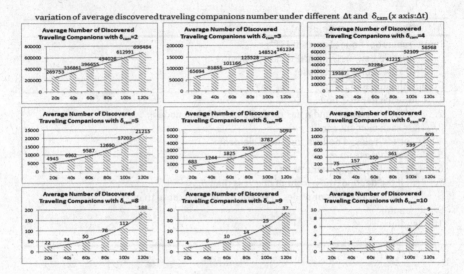

Fig. 3. Variation of average discovered traveling companion number with different Δt and δ_{cam}

Figure 3 shows the final results. It shows that, when Δt increases, the average number of discovered companions grows significantly under same δ_{cam}. Furthermore, when $\delta_{cam} \leq 4$, the average number of discovered companions rises linearly with the growth of Δt. But when $\delta_{cam} > 4$, it increases exponentially. In this case, the bigger δ_{cam} is, the greater curve it has. From another perspective, by comparing the average numbers under same Δt with different δ_{cam}, we can find that it decreases exponentially when δ_{cam} rises.

C. Efficiency

We further verify our algorithm's performance with different values of Δt through experiments. Actually, we also do experiments to verify the performance under different δ_{cam} values. However, the experiment results show that δ_{cam} doesn't have obvious impacts on performance. Again, we run our algorithm 20 times for each parameter configuration. Each execution lasts for 20 min to continuously to receive arriving ANPR data records and instantly output companion results under different values of Δt. For each execution, we will compute the latency value based on Definition 3. Finally, the average latency values are shown in Fig. 4.

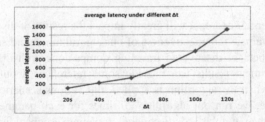

Fig. 4. Values of average latency under different Δt($\delta_{cam} = 2$)

Definition 3 (Latency): Given an ANPR data set $R = \{r_1, r_2, ..., r_m\}$, the corresponding vehicle set V, let t_{in} be the arriving time of a new ANPR data record, t_{out} be the corresponding companion result, then the average latency of our algorithm can be defined as follows: $t_{lat} = \frac{\sum_{i=1}^{m}(t_{out}^i - t_{in}^i)}{m}$.

Figure 4 shows the average latency increases exponentially from $\Delta t = 20s$ to $\Delta t = 120s$. When $\Delta t \leq 100s$, the average latency is less than 1000 ms. The minimum value of time interval between two ANPR data records is 1 s. It means our algorithm can instantly discover companions with $\Delta t \leq 100s$ when vehicles passing through camera. However, when $\Delta t > 100s$, our algorithm shows longer latencies. In our future work, we try to parallelize our algorithm to improve performance.

5   Related Work

Many researchers have put their interests on traveling companion study. Most of the studies are designed to work on static datasets on 2D Euclidean space. Laube et al. proposed *flock* as a group of moving objects moving in a disc of a fixed size for k consecutive timestamps [1]. Jeung et al. proposed *convoy*, an extension of *flock*, where spatial clustering is based on density [2]. Li et al. proposed *swarm* which enables the discovery of interesting moving object clusters with relaxed temporal constraint [3]. Li et al. proposed a new type of patterns, platoon patterns, that describes object clusters that stay together for time segments, each with some minimum consecutive duration of time [7]. Recently, researchers begin to pay attention to large scale trajectory. Zheng et al. developed a set of techniques to improve the performance of discovering gathering patterns over static large-scale trajectory databases [5]. Zhang et al. proposed a gathering retrieving algorithm to retrieve gathering pattern by searching a spatio-temporal graph composed of the moving object clusters [6].

The above studies cannot effectively handle streaming data. In recent years, more and more studies began to process traffic data stream. Tang et al. proposed a framework to discover travelling companion among streaming trajectories and output results in an incremental manner [13]. Yu et al. studied on a density-based clustering algorithm for trajectory data stream and tried to discover trajectory clusters in real time [14]. All these related work aimed at processing GPS data stream and provide some foundations for our study.

6   Conclusion

In this paper, we propose an approach to instantly discover companion vehicles from a special kind of streaming traffic data, called ANPR. Compared to related approaches, we transform the companion discovery into a frequent sequence-mining problem. On top of SeqStream algorithm, we make several improvements to handle customized time constraints among sequence elements when discovering traveling companions. We also use pseudo projection technique to improve the performance. Our experiments show efficiency and effectiveness of our algorithms. In the future, we plan to parallelize our algorithms to further improve their performance.

Acknowledgment. The research work is supported by the projects: Key Program of Beijing Municipal Natural Science Foundation (No. 4131001); Training Plan of Top Young Talent in North China University of Technology, "An Incremental Approach to Instant Discovery of Data Correlations among Multi-Source and Large-scale Sensor Data".

References

1. Laube, P., Imfeld, S.: Analyzing relative motion within groups of trackable moving point objects. In: Egenhofer, M., Mark, D.M. (eds.) GIScience 2002. LNCS, vol. 2478, pp. 132–144. Springer, Heidelberg (2002)
2. Jeung, H., Shen, H., Zhou, X.: Convoy queries in spatio-temporal databases. In: International Conference on Data Engineering, pp. 1457–1459. IEEE Computer Society, Washington, DC (2008)
3. Li, Z., Ding, B., Han, J., et al.: Swarm: mining relaxed temporal moving object clusters. VLDB Endowment 3(1), 723–734 (2010)
4. Tang, L.A., Zheng, Y., Yuan, J., et al.: A framework of traveling companion discovery on trajectory data streams. ACM Trans. Intell. Syst. Technol. 5(1), 992–999 (2013)
5. Zheng, K., Zheng, Y., Yuan, N.J., et al.: On discovery of gathering patterns from trajectories. In: IEEE 29th International Conference on Data Engineering, pp. 242–253. IEEE Computer Society, Washington, DC (2013)
6. Zhang, J., Li, J., Wang, S., et al.: On retrieving moving objects gathering patterns from trajectory data via spatio-temporal graph. In: IEEE Int. Congr. Big Data, pp. 390–397. IEEE Computer Society, Washington, DC (2014)
7. Li, Y., Bailey, J., Kulik, L.: Efficient mining of platoon patterns in trajectory databases. Data Knowl. Eng. 100(PA), 167–187 (2015)
8. Han, Y., Wang, G., Yu, J., et al.: A service-based approach to traffic sensor data integration and analysis to support community-wide green commute in China. IEEE Trans. Intell. Transp. Syst. PP(99), 1–10 (2015)
9. Zhu, M., Liu, C., Wang, J., et al.: Instant discovery of moment companion vehicles from big streaming traffic data. In: International Conference on Cloud Computing and Big Data, pp. 4–6. IEEE, Taipei (2015)
10. Zhu, M., Liu, C., Wang, J., et al.: A service-friendly approach to discover traveling companions based on ANPR data stream. In: IEEE International Conference on Services Computing, San Francisco USA (2016)
11. Chang, L., Wang, T., Yang, D., et al.: SeqStream: mining closed sequential patterns over stream sliding windows. In: 8th IEEE International Conference on Data Mining, pp. 83–92. IEEE Computer Society, Washington, DC (2008)
12. Mooney, C.H., Roddick, J.F.: Sequential pattern mining: approaches and algorithms. ACM Comput. Surv. 45(2), 94–111 (2013)
13. Tang, L.A., Zheng, Y., Yuan, J., et al.: On discovery of traveling companions from streaming trajectories. In: IEEE 28th International Conference on Data Engineering, pp. 186–197. IEEE Computer Society, Washington, DC (2012)
14. Yu, Y., Wang, Q., Wang, X., et al.: Online clustering for trajectory data stream of moving objects. Comput. Sci. Inf. Syst. 10(3), 1293–1317 (2013)

Time-Constrained Sequenced Route Query
in Indoor Spaces

Wenyi Luo[1], Peiquan Jin[1,2(✉)], and Lihua Yue[1,2]

[1] School of Computer Science and Technology,
University of Science and Technology of China, Hefei 230027, China
jpq@ustc.edu.cn
[2] Key Laboratory of Electromagnetic Space Information,
Chinese Academy of Sciences, Hefei 230027, China

Abstract. Location-based services (LBSs) in indoor spaces have emerged as a
new research direction. In this paper, we study a new kind of indoor LBSs that is
called *Time-Constrained Sequenced Route (TCSR) query*. A TCSR query
returns a route consisting of a sequence of indoor locations before a given
deadline such that each location matches a given location type as well as a given
stay-time period. Such queries are popular in indoor spaces, e.g., in a business
center, people may want to first stay at a toy shop for 30 min, and then go to a
coffee room for one-hour rest, and finally arrive at a cinema before 18:00 PM.
Classic route-search algorithms like Dijkstra have to search a large set of possible routes and thus are inefficient for TCSR queries. In addition, they do not
consider the multi-floor feature of indoor spaces. In this paper we present a
two-stage approach to evaluate a TCSR query. First, we find the optimal floor
sequence for a TCSR query. Next, we propose a multi-source Dijkstra algorithm
to get the time-constrained sequenced locations in a single floor. We conduct
experiments on a synthetic indoor space and the results suggest that our proposal
is efficient and scalable.

Keywords: Indoor space · Route query · Time constraint

1 Introduction

People spend much time in indoor spaces such as office building, apartments, shopping
centers, and airports. The availability of indoor positioning techniques is increasing
with the rapid development of wireless telecommunication technologies such as Wi-Fi,
Bluetooth, and RFID [1, 2], which enables a variety of indoor location-based services
(LBSs).

In this paper, we study a new kind of indoor LBSs that is called *Time-Constrained
Sequenced Route* (TCSR) query. A TCSR query returns a route consisting of a
sequence of indoor locations before a given deadline such that each location matches a
given location type as well as a given stay-time period. Such queries are popular in
indoor spaces, e.g., in a business center, people may want to first stay at a toy shop for
30 min, and then go to a coffee room for one-hour rest, and finally arrive at a cinema
before 18:00 PM.

© Springer International Publishing Switzerland 2016
F. Li et al. (Eds.): APWeb 2016, Part I, LNCS 9931, pp. 129–140, 2016.
DOI: 10.1007/978-3-319-45814-4_11

Evaluating TCSR queries in indoor spaces is challenging because of two reasons. First, indoor spaces are usually three-dimensional. Existing route search algorithms like Dijkstra are not suitable for three-dimensional indoor spaces because they are mainly designed for 2D spaces such as road networks. Second, the distance measurement in indoor spaces is different from that in outdoor spaces [8]. The latter usually employ the Euclidean distance. However, this is not applicable in indoor spaces, due to the existence of doors and rooms.

In this paper, we present a two-stage approach to answer a TCSR query. At the first stage, we find the optimal floor sequence so as to decrease the search area and the cost of route search. At the second stage, we propose a multi-source Dijkstra algorithm to get the time-constrained sequenced locations in a single floor. Briefly, we make the following contributions in this paper:

1. We explore a new kind of indoor LBSs that is called *Time-Constrained Sequenced Route* (TCSR) query.
2. We propose a two-stage approach to evaluate TCSR queries. We first present a cost-based algorithm to find the optimal floor sequence so as to limit the search area. Then, we propose a multi-source Dijkstra algorithm to get the time-constrained sequenced locations in a single floor.
3. We conduct experiments on a synthetic indoor space and the results suggest that our proposal is efficient and scalable.

2 Related Work

Sequenced route queries have been studied in outdoor spaces. In the literature [4], an A* based algorithm was proposed, which used a remove set to save all the routes from the start point to current extended points. Optimal solutions to a sequenced route query in vector and metric spaces were first proposed in [5] and later extended in [6]. For vector spaces, a light threshold-based iterative algorithm named LORD was proposed [5], which utilizes various thresholds to filter out the locations that are not possible in the optimal route. The authors also proposed R-LORD, an extension of LORD that uses R-trees to examine the threshold value more efficiently. In [6], the authors exploited the geometric properties of the solution space and theoretically proved its relation to additively weighted Voronoi diagrams. In [7], a new method for the partial sequenced route query with travel rules in road network was proposed to get all the possible access sequences. In addition, trajectory queries towards imprecise locations were studied in [9, 16]. However, these previous approaches are not suitable for indoor spaces because indoor spaces have to consider different distance measures and location models.

In recent years, there have been some studies focusing one valuating indoor-space-based queries. For example, P. Jin et al. [1, 3] proposed algorithms for similarity search on indoor moving-object trajectories and for extracting hotspots from trajectories in indoor spaces. They also developed tools to generate indoor trajectories [12]. Yang et al. [13] proposed a complete set of techniques for computing probabilistic threshold kNN queries in indoor environments. They proposed the minimum indoor walking distance as the distance metric for indoor spaces and designed a hash-based indexing scheme for indoor moving objects. Other works have been done on context-aware queries in indoor spaces [10, 11] and distance-aware queries in indoor spaces [8, 14, 15]. However, these existing works mainly concentrated on indoor location queries or range queries. To the best of our knowledge, there are no direct works focusing on the time-constrained route queries in indoor spaces.

3   Problem Statement

3.1   Indoor Space

In this paper, we define indoor space by the following symbolic model.

Definition 1 (Indoor Space). An indoor space can be modeled as a graph which is represented as follows:

$$IndoorSpace = (Sensor, Connectivity)$$

Here, *Sensor* is a set of positioning sensors deployed in the indoor space to identify rooms. Typical sensors are RFID readers, Bluetooth detectors, and Wi-Fi signal receivers. *Connectivity* records the connectivity between rooms. That means, if two rooms are connected by a common door, we record a tuple of $(s_i, s_j)|s_i, s_j \in Sensor$ in *Connectivity*.   □

In real applications, the *Connectivity* information of sensors can be pre-determined and maintained in a database. In order to introduce semantics into the model, we assign semantic labels to each sensor. As a result, the set *Sensor* is defined as follows.

$$Sensor = \{s|s = (sensorID, location, label, floor)\}$$

The *label* of a sensor provides descriptions on the semantic attributes of the location identified by *sensorID*. For instance, we can use labels like "*elevator*" and "*stair*" to indicate the functions of the cell identified by a sensor. We can also use other labels like "*Starbucks*" and "*Burger King*" to annotate semantic features of the cell. The *location* of a sensor is a three-dimensional coordinate (x, y, z), which reflects the relative position of the sensor inside the indoor space where the sensor is deployed. The floor of a sensor is recorded in the element *floor*.

3.2   Time-Constrained Sequenced Route(TCSR) Query

Definition 2 (TCSR Query). Given an *IndoorSpace* = (*Sensor, Connectivity*), a list of time-constrained labels $(<l_1, t_1>, <l_2, t_2>, \ldots, <l_n, t_n>)$, a start location with a start time (s, t_s), a destination location with an arrival time (e, t_e), and a time threshold d, a TCSR query returns a $Route = \{(s, s_1, s_2, \ldots, s_n, e)|s, e, s_i \in Sensor\}$, such that:

(1) s is the start location and e is the destination location
(2) $s_i.label = l_i$

(3) $t_i \leq$ The time period that the query object spends in s_i

(4) $d \geq t_e$ □

Figure 1 shows an example of a TCSR query. Here, the start and end location of the query are identified by a star symbol in the figure. We suppose that the list of time-constrained labels is (<"*fashion shop*", 30>, <"*theatre*", 60>, <"*restaurant*", 60>, <"*hair salon*", 60>), where the numbers represent minutes. We can get 16 routes from the start location to the end location, which all satisfy the constraints given by the label list. Given that the departure time is 9:00 AM and the time threshold is 2:00 PM, the best answer to the TCSR query is <*start*, L_1, L_2, L_3, L_8, *end*>, as shown in Table 1, because its cost of floor crossing is minimum.

It is not efficient to use traditional methods such as exhaustive search, Dijkstra, SPFA, and Floyd to evaluate TCSR queries, because they need to search the whole indoor space to get all possible routes. In this paper, we aim to optimize the query

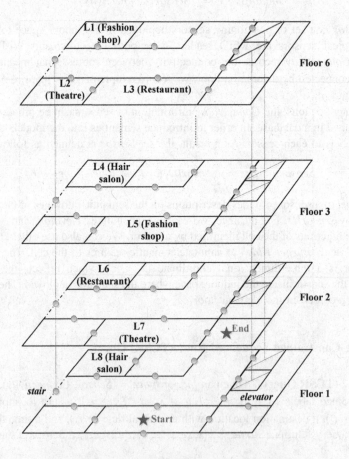

Fig. 1. An example of TCSR queries in indoor spaces

Table 1. A possible route for a TCSR query on the indoor space shown in Fig. 1

Location	Label	Stay time	Departure/Arrival time
Start			9:00 AM
L1	Fashion shop	30 min	
L2	Theatre	60 min	
L3	Restaurant	60 min	
L8	Hair salon	60 min	
End			2:00 PM

processing algorithm for TCSR queries. Particularly, we propose a two-phase approach to decrease the search space and therefore reduce the search time.

4 Evaluating TCSR Queries

We propose a two-phase approach to evaluate TCSR queries. At the first stage, we aim to get the best floor sequence for a TCSR query. Then, we compute the shortest route within each floor. Differing from the exhaustive search scheme and the Dijkstra algorithm, our design does not require searching on the whole set of possible routes, but only need to search the shortest route in one floor. This is owing to the best floor sequence that is computed at the first stage.

Algorithm 1. *TCSR (G,T, Label, start, end)*

Input: *G*: graph of indoor space; *T*: Threshold of travel time; *Label*: a sequenced list of categories; *start*: start point; *end*: destination

Output: A route that has the minimum cost.

Preliminary: Initially, *Pathfrom* and a set of sources *S* are empty.

---

1. *Floors = getFloorList (dist, end, 0);* // *get the best floor sequence*
2. *Pathfrom[start] ← <start>;* //*initialize the path of start point*
3. Add *start* into S_0; //*decide sources in a single floor*
4. *i = 0;*
5. **While** *i < Floors.size* **do**
6.   *Multi_Source_Dijkstra(G, S_i, Pathfrom, T, Label[i]);*
7.   **If** *Floors(i) ≠Floors(i+1)* //*find the sources in the next single floor*
8. S_{i+1} ← a collection of stairs, elevators, escalators in *Floors(i+1);*
9.   **Else** S_{i+1} ← *Label*[i + 1] //*next category points to be visited.*
10.   *i++;* //*visit the next floor*
11. **End While**
12. **Return** *Pathfrom[end]*

The general algorithm for evaluating a TCSR query is shown in Algorithm 1. We first get the best floor sequence using the algorithm Get FloorList, which will be discussed in Sect. 4.1. Then, we visit the floor sequence in order and we use the algorithm *Multi_Source_Dijkstra* to search the shortest route within a single floor, which will be discussed in Sect. 4.2.

4.1   Determining the Best Floor Sequence

In order to prune a large of points in a coarse granularity, we first discuss the floor cost for TCSR queries. Firstly, if we use different ways to across a floor, it will cost different time. Secondly, we assume that the average travel time between two floors is influenced by the probability of using which ways to cross the floors. i.e., people will probably choose an elevator to cross two floors if the current position is near an elevator. Then, we utilize the average time as the edge weight and build a floor graph to show the relation between categories and floors. Finally, we traverse the *Floor Graph* to get the best access sequences of floors. As a result, we can simply remove the floors that are not in the access sequences.

Definition 3 (*Floor Graph*). A *Floor Graph* is defined by $G_f = \{start, end, C, V, E\}$. *start* and *end* are indoor locations. Each element in C is a set of floors that have the same label, e.g., "*fashion shop*" or "*restaurant*". V is the set of floors, where each floor is an vertex in V. Each edge in $E = \{(v_i, v_j)|v_i \in V\}$ represents the travel cost between floor v_i and v_j.   □

Figure 2 shows an example of floor graph. The floor graph is defined for determining the best floor sequence for a TCSR query.

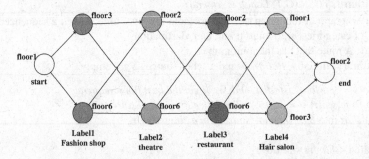

Fig. 2. Floor graph for a TCSR query

In order to find the best floor sequence, we first propose different ways to calculate the time cost of crossing floors. Then, we traverse *Floor Graph* to get the best floor sequence. We assume that the travel time of elevator can be omitted comparing with the time for waiting for an elevator. Thus, if the average time for waiting an elevator is denoted as T_{ele}, the time cost of taking an elevator is defined as follows.

$$Cost_{ele}(floor_i, floor_j) = T_{ele} \tag{4.1}$$

The travel time through an escalator or a stair is proportional to the number of crossed floors. If we denote the travel time of adjacent floors through a stair and an escalator as T_{sta}, T_{esc} respectively, we can get the travel time by escalators, stairs respectively as follows.

$$Cost_{esc}(floor_i, floor_j) = T_{esc} \times |floor_j - floor_i| \tag{4.2}$$

$$Cost_{sta}(floor_i, floor_j) = T_{sta} \times |floor_j - floor_i|^{\alpha} \text{ if } j > i, \alpha > 1, \text{ if } j < i, \alpha = 1 \tag{4.3}$$

As the travel speed will decrease with the increasing of the floors to be traveled, we introduce a parameter α and represent the travel time using an exponential function. Particularly, we assume that $\alpha = 1$ when we go down the stair and $\alpha > 1$ when we up the stairs.

The whole time spent in the adjacent floors is not only influenced by the number of stairs, escalators and elevators, but also influenced by the position of stairs, escalators and elevators. Consequently, we use the following equation to simulate the cost between different floors.

$$Cost(floor_i, floor_j) = (1 - w_1 - w_2) \times Cost_{ele} + w_1 \times Cost_{esc} + w_2 \times Cost_{sta} \tag{4.4}$$

Algorithm 2. *getFloorList(dist, i, j)*

Input: *dist*: an array recording the edge weights between adjacent points; *i*: category index in the visit list; *j*: index of the points with the same category.
Output: *ans*: the minimal cost of floor sequence
Preliminary: *f[i,j]* denotes the minimum cost from start to the *j* point with category *i.ahead[i]* is an array storing all the points with category *i-1*.

---
1. **If** $(i == 0)$ **Return** 0;
2. **If** $(f[i,j] < \text{INF})$ **Return** $f[i,j]$;
3. **For** each x in $ahead[i]$
4.   **If** $dist(start, x) + dist(x, j) < ans$
5. $ans = min\{getFloorList(dist, i - 1, x) + dist(x, j)\}$
6. $f[i,j] = ans$;
7.   **End If**
7. **End For**
8. **Return** ans;

We use a dynamic programming algorithm to traverse the *Floor Graph* and get the best cross floors sequence, as shown in Algorithm 2. The time cost is also $O(N^2), N = \sum_{i=1}^{n} m_i, m_i << k$, k is the number of floor. m_i shows the number of floor which have category i. n is the number of categories we needed to visit. Consider the worst situation, each floor have all the visited categories, which means each category distribute at k floors, and N equals to $k*n$.

4.2   Finding the Shortest Route within One Floor

Based on the access floor sequence we get from the *Floor Graph*, we propose a multi-source Dijkstra algorithm (Algorithm 3) to evaluate a TCSR query in a single floor. We need to initial a collection Paths to save the minimum paths from one of the sources to other points in the same floor and insert the vertex of sources into queue Q. When the queue is not null, we pop a vertex and calculate its travel cost and route. And when a neighbor vertex is in queue Q already, we consider whether its original path is minor than the path from source to it crossing current point or not. And update the path from source to it.

Algorithm 3. *Multi_Source_Dijkstra(G, S, Pathfrom, T, COI)*

Input: G: graph of indoor space; S: source set; *Pathfrom*: minimum cost paths from start to S; T: time threshold; *COI*: a set of current visited category points.

Output: a set of paths *Pathfrom*[COI]

Preliminary: Q initialize as $\emptyset$.p: point in the same floor with S; *extendNode* initialize as $\emptyset$; Paths: minimum paths from S to other vertices in the same floor;

---

1. **For** each vertex s in S
2. Put $<s>$ in $Paths[s]$; Insert (Q, s);
3. **While** $Q \neq \emptyset$
4. $min \leftarrow$ EXTRACT_MIN(Q)
5.   add min into *extendNode*;
6.   **For** each $(min, v) \in Edge$
7.   **If** $v \notin entendNode$ and not in Q //initial the path from S to v
8. $Paths[v] \leftarrow Paths[min] +<v>$;
9.   Insert (Q,v);
10.   **ElseIf** $v \notin entendNode$ and
                $cost(Pathfrom[Paths[min].from])+cost(Paths[min])+cost(min,v) <$
                    $cost(Paths[v].from) +cost(Paths[v])$
11. $Paths[v] \leftarrow Paths[min] +<v>$;//update the path from S to v
12.   Update(Q, v);
13.   **End If**
14.   **End For**
15. **End While**
16. **For** each v in COI
17.   Update $Pathfrom[v] = Pathfrom[Paths[v].from]+Paths[v]$;
18. **Return** *Pathfrom*;

4.3 Time Complexity

Traditional methods including the exhaustive search, Floyd-Warshall, and Dijkstra may also be used for TCSR queries. Using the exhaustive search must traverse all the possible routes. It may have $\prod_{i=1}^{n} |S_i|$ routes, $S_i = \{s | s \in Sensor, s.label = label_i, label_i \in Labels\}$. In order to get the route cost, we need to traverse all the routes by Dijkstra, each Dijkstra need to traverse all the indoor space and the time complexity is $O\left(|Sensor|^2\right)$. The time complexity of the exhaustive search is $O\left(\prod_{i=1}^{n} |S_i| \, |Sensor|^2\right)$, where n is the number of visited categories. The time complexity is exponential.

If using the Floyd-Warshall algorithm, an intuitional algorithm to calculate the cost between any two points. We need to spend $O\left(|Sensor|^3\right)$ time and $O\left(|Sensor|^2\right)$ space.

Compare with the traditional algorithms, we first consider the floor information with different categories. We can aggregate points with the same label in the same floor into a new vertex of *Floor Graph*. It can also eliminate the influence about searching in the same floor and only care about the travel cost crossing floors. Building the *Floor Graph* will cost $O(N^2)$ time. After that, using the recurrence method shown as Algorithm 2 to traverse the *Floor Graph*, we can get the best floor sequence. Finally, we need to use the *Multi_Source_Dijkstra* algorithm to traverse each floor in the best floor sequence. The time complexity of this step is $O\left(N^*(|Sensor|/k)^2\right)$. Consequently, the two-phase approach has time complexity of $O\left(N^*(|Sensor|/k)^2\right) + O(N^2)$. As $N << |Sensor|$, the $O(N^2)$ can be regarded as a constant in real applications. Thus, the time complexity of our design can be simplified to $O\left(N^*(|Sensor|/k)^2\right)$. This time complexity is much better than $O\left(|Sensor^3|\right)$ and $O\left(\prod_{i=1}^{n} |S_i| \, |Sensor|^2\right)$.

5 Performance Evaluation

We conduct experiments using different synthetic datasets. We compare our proposal to the Dijkstra algorithm according to the number of expended vertices in the search and the processing time of the query.

We generate a synthetic indoor space where each room, corridor, stair, escalator or elevator is denoted as a sensor in *Indoor Space*. Points with categories are uniformly distributed over the *Indoor Space* and each category has, in average, the same number of points. The edge cost is the travel time which calculates by using the distance to divide average walking speed.

We evaluate how each of the solution perform with respect to the patameters shown in Table 2. Regarding the network, we varied its size (number of rooms), the sequence size (the number of the category we should visit) and the floor number in the network.

Table 2. Experimental Settings

Network size (room number)	1000, 2000, 3000, 4000
Sequence size	5, 10, 15
Floor number	3, 5, 10, 15, 20, 30

For each experiment, we varied a parameter and set the others parameters to their default values. For each combination of different parameters, we generate one graph to express the indoor space, one *Floor Graph* to get the best sequence of floors, and execute 15 randomly generated queries.

Analysis on the Network Size. Figure 3 shows both the number of vertices expanded and the processing time increase with the network size in all solutions. The TCSR query algorithm is more steady for different variable than the original Dijkstra. This happens because each time using the original Dijkstra, it must traverse all the vertices, our two-phase method only search the floor in the optimal floor list, which can prune a lot of vertices when searching. Also, the original Dijkstra always repeat visit vertex when expanded different categories.

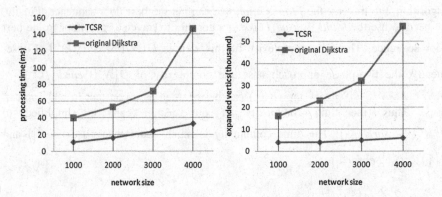

Fig. 3. Processing time of the queries and number of expanded vertices w.r.t the network size

Analysis on the Sequence Size. Figure 4 shows that the processing time and the number of expanded vertices increase with the number of labels we visited. Since the point of interest are uniformly distributed among the categories, increasing the sequence size means increase the number of labels we should expand. So, we should check more vertices to find a sequence and the processing time will increase. Also, the more categories we visit the more floors we may visit and the more vertices we need to expand.

Analysis on the Number of Floors. With the number of floor increasing, the number of vertices in each floor decrease. In addition, we may prune more vertices with a larger number of floors after the phase one. So, the processing time and the number of expanded vertices will decrease with the number of floor increasing (Fig. 5).

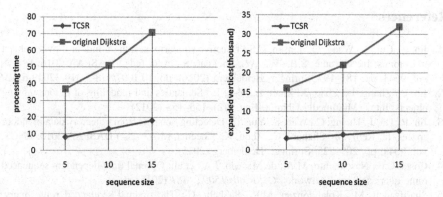

Fig. 4. Processing time and number of expanded vertices w.r.t the sequence size

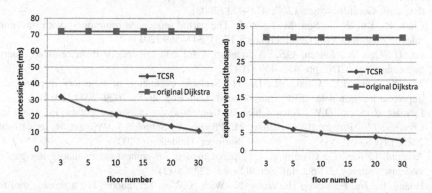

Fig. 5. Processing time and number of expanded vertices w.r.t. the number of floors

6 Conclusions

In this paper, we study a new kind of indoor LBSs that is called *Time-Constrained Sequenced Route* (TCSR) Query. We present a two-phase approach to answer a TCSR query. At the first stage, we find the optimal floor sequence so as to reduce the cost of route search. At the second stage, we propose a multi-source Dijkstra algorithm to get the time-constrained sequenced locations in a single floor.

In future work, we will investigate how to deal with the situation that the whole travel time cannot satisfy the travel time constraints. One possible solution is to replace some unimportant categories with other similar categories in order to decrease the travel time. We can also reduce the stay time of some unimportant categories.

Acknowledgements. This work is supported by the National Science Foundation of China (61379037, 61472376, & 61272317).

References

1. Jin, P., Cui, T., Wang, Q., Jensen, C.S.: Effective similarity search on indoor moving-object trajectories. In: Navathe, S.B., Wu, W., Shekhar, S., et al. (eds.) DASFAA 2016. LNCS, vol. 9643, pp. 181–197. Springer, Heidelberg (2016). doi:10.1007/978-3-319-32049-6_12
2. Jin, P., Zhang, L., Zhao, J., Zhao, L., Yue, L.: Semantics and modeling of indoor moving objects. Int. J. Multimedia Ubiquit. Eng. 7(2), 153–158 (2012)
3. Jin, P., Du, J., Huang, C., Wan, S., Yue, L.: Detecting hotspots from trajectory data in indoor spaces. In: Renz, M., Shahabi, C., Zhou, X., Cheema, M.A. (eds.) DASFAA 2015. LNCS, vol. 9049, pp. 209–225. Springer, Heidelberg (2015)
4. Costa, C.F., Nascimento, M.A., de Macêdo, J.A., et al.: Optimal time-dependent sequenced route queries in road networks. *CoRR abs/1509.01881* (2015)
5. Sharifzadeh, M., Kolahdouzan, M.R., Shahabi, C.: The optimal sequenced route query. VLDB J. 17(4), 765–787 (2008)
6. Sharifzadeh, M., Shahabi, C.: Processing optimal sequenced route queries using voronoi diagrams. GeoInformatica 12(4), 411–433 (2008)
7. Chen, H., Ku, W.-S., Sun, M.-T., et al.: The partial sequenced route query with traveling rules in road networks. GeoInformatica 15(3), 541–569 (2011)
8. Lu, H., Cao, X., Jensen, C.S.: A foundation for efficient indoor distance-aware query processing. In: ICDE, pp. 438–449 (2012)
9. Xie, X., Yiu, M.L., Cheng, R., Lu, H.: Scalable evaluation of trajectory queries over imprecise location data. IEEE Trans. Knowl. Data Eng. 26(8), 2029–2044 (2014)
10. Lyardet, F., Szeto, D.W., Aitenbichler, E.: Context-aware indoor navigation. In: Aarts, E., Crowley, J.L., Ruyter, B., Gerhäuser, H., Pflaum, A., Schmidt, J., Wichert, R. (eds.) AmI 2008. LNCS, vol. 5355, pp. 290–307. Springer, Heidelberg (2008)
11. Afyouni, I., Ray, C., Claramunt, C.: Spatial models for context-aware indoor navigation systems: a survey. J. Spat. Inf. Sci. 4(1), 85–123 (2012)
12. Huang, C., Jin, P., Wang, H., Wang, N., Wan, S., Yue, L.: IndoorSTG: a flexible tool to generate trajectory data for indoor moving objects. In: MDM, pp. 341–343 (2013)
13. Yang, B., Lu, H., Jensen, C.S.: Probabilistic threshold k nearest neighbor queries over moving objects in symbolic indoor space. In: EDBT, pp. 335–346 (2010)
14. Xie, X., Lu, H., Pedersen, T.B.: Distance-aware join for indoor moving objects. IEEE Trans. Knowl. Data Eng. 27(2), 428–442 (2015)
15. Xie, X., Lu, H., Pedersen, T.B.: Efficient distance-aware query evaluation on indoor moving objects. In: ICDE, pp. 434–445 (2013)
16. Xie, X., Jin, P., Yiu, M.L., Du, J., Yuan, M., Jensen, C.S.: Enabling scalable geographic service sharing with weighted imprecise voronoi cells. IEEE Trans. Knowl. Data Eng. 28(2), 439–453 (2016)

Near-Duplicate Web Video Retrieval
and Localization Using Improved Edit Distance

Hao Liu[✉], Qingjie Zhao, Hao Wang, and Cong Zhang

Beijing Key Lab of Intelligent Information Technology, School of Computer Science
and Technology, Beijing Institute of Technology, Beijing, China
{liuhao,zhaoqj,2120141053,zhangcong}@bit.edu.cn

Abstract. With the development of network, there exists many near-duplicate videos online shared by individuals. These ones cause problems such as copyright infringement and search result redundancy. To solve the issues, this paper proposes a filter-and-refine framework for near-duplicate video retrieval and localization. By regarding video sequences as strings, Edit distance is used and improved in the approach. Firstly, bag-of-words (BOW) model is utilized to measure the similarities between frames. Then, non-near-duplicate videos are filtered out by computing the proposed relative Edit distance similarity (REDS). Next, a dynamic programming strategy is proposed to rank the remained videos and localize the similar segments. Experiments demonstrate the effectiveness and robustness of the method in retrieval and localization.

Keywords: Near-duplicate video retrieval · Near-duplicate video localization · Edit distance

1 Introduction

With the popularity of the social network services, web video sharing is of great convenience. Users can easily create and share interesting contents through sharing websites. Among these, there is a significant part of near-duplicates. Near-duplicate videos refer to videos containing the same contents but with certain variations [9]. As depicted in [15], nearly 27 % of web videos are near-duplicates. With so many similar multimedia contents, two significant problems arose: copyright protection and search result refinement [9,13].

Therefore, numerous approaches, called near-duplicate video retrieval (NDVR), are proposed to solve above issues. What's more, consumers may also want to know the positions of similar segments of the retrieved results. Thus, the technique of near-duplicate video localization (NDVL) also has attracted much attention.

For NDVR and NDVL, video signature and sequence matching are two major concerns. Some methods represent a whole video with a global signature [6,10], which can be efficiently managed. But this kind of features are not suitable for dealing with NDVL. Chou *et al.* [4] encodes each frame with a signature, but the

© Springer International Publishing Switzerland 2016
F. Li et al. (Eds.): APWeb 2016, Part I, LNCS 9931, pp. 141–152, 2016.
DOI: 10.1007/978-3-319-45814-4_12

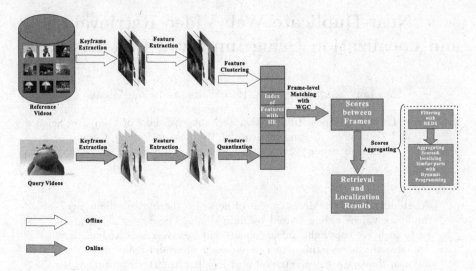

Fig. 1. Framework of the proposed approach.

process of compressing so many details into a compact signature usually leads to much information loss. What's more, this kind of frame-level global signature is not suitable for detecting the partial duplicate images, e.g., picture-in-picture.

Existing work on sequence matching can be generally categorized into two main classes: video-level and frame-level. The video-level methods summarize or index the videos or clips with compact signatures for matching [5,12]. However, these methods ignore relations between frames and are unsuitable for partial NDVR and NDVL. Meanwhile, frame-level matching technique [4,16], which analyzes the spatial and temporal information between frames, is a reasonable choice to overcome the disadvantages. But the precision needs to be enhanced.

To overcome the above drawbacks, we propose a novel framework, which is effective and robust for retrieval and localization (Fig. 1). Firstly, extracted keypoints are quantized into bag-of-words (BOW) model with Hamming Embedding (HE) and weak geometry consistency (WGC) [7]. For videos to be compared, we compute the scores between the frame-pairs and formulate a matching score matrix. With the matrix, the Edit distance can be calculated. Moreover, a metric, termed REDS, is proposed to filter out non-near-duplicates. Then, we use the matching score matrix to generate the path matrix and video similarity matrix alternately, which can be used for localization and retrieval, respectively.

To summarize, we make the following contributions:

(1) Taking videos as strings, we propose a Edit-distance-based method, under the filter-and-refine framework, which could deal with full and partial NDVs and localize the similar fragments. Both methods used in the two steps are based on Edit distance, which is consistent in logic. Firstly, the adhoc REDS, which is sequence length independent, is designed for fast filter. Then, a detect-and-refine algorithm by employing dynamic programming method,

which exploits the temporal relations between video, is proposed in the refine stage for video results ranking and localizing.

(2) Experiments are conducted on two standard datasets, CC_WEB_VIDEO [14] and TRECVID CBCD 2011 [8], and three state-of-the-art methods are implemented for comparison. The experimental results demonstrate the effectiveness of our method.

The rest of this paper is organised as follows. We survey the related work in Sect. 2. In Sect. 3, we present our approach. Comprehensive experiments are conducted in Sect. 4. Finally, we conclude our work in Sect. 5.

2 Related Work

Video Signature. Video signature representations can be broadly grouped into two categories: video-level and frame-level. video-level signature, such as bounded coordinate system [6] and reference-based histogram [10]. These methods can be efficiently managed, however, are easy to get false conclusions, due to large information loss. Frame-level methods, such as [4,14,17] can cope with heavy spatial edit, but the trade-off between precision and recall should be paid attention to.

Sequence Matching. There are also two kinds of matching methods: video-level and frame-level. Methods like fingerprinting [5], feature hashing [12], regard videos or clips as the basic units, are proposed to match similar videos. The video-level methods are efficient in online retrieval, but unsuitable in detecting partial near-duplicates and localizing similar fragments. Frame-level methods take frames as the basic units, are reasonable choices for coping with partial NDVR and NDVL tasks. In [3,11], they employed dynamic programming for video comparisons. But these methods, using exhaustive search, are far from web-scale applications.

3 The Proposed Method

In this section, we first present the frame-level similarity measurement method. Then we describe the sequence matching algorithm proposed in details.

Symbol Definition. Let $Q = \{q_1, q_2, ..., q_m\}$ be the query video of m frames, $R = \{R^1, R^2, ..., R^K\}$ be the reference videos in dataset of total K videos. Particularly, $R^k = \{r_1^k, r_2^k, ..., r_n^k\}$ represents the kth video of n frames. Moreover, q_i represents the ith frame of the query video Q and r_j^k is the jth frame of R^k.

3.1 Frame-Level Similarity Measurement

In order to reduce data amount while maintaining the distinctness of frames, we adopt the BOW model for precise frame matching. Firstly, we choose RootSIFT as the local keypoint descriptor. Compared with SIFT, RootSIFT can yield more matches during the searching because it reduces the weight of the larger bin values [1]. Moreover, the transformation from SIFT to RootSIFT neither requires much computational time nor any additional storage space, and it can be done online using a few lines of code.

Then K-Means algorithm is used to generate visual word codebook and each keypoint is quantized into nearest visual word using Euclidean distance. To improve matching speed, inverted file index is introduced into the constructing of the retrieval index. Since keypoints with the same visual word are always matched without considering their real distance, false matching is unavoidable. Thus, Hamming embedding (HE) [7] is adopted for handling this problem. Then, the similarity of query keyframe q_i and reference keyframe r_j^k is calculated through following equation,

$$fs(q_i, r_j^k) = \frac{\sum_{x \in q_i, y \in r_j^k} f_{HE}(x, y).idf(q(x))^2}{\| q_i \|_2 \| r_j^k \|_2}, \tag{1}$$

where $f_{HE}(x, y)$ is a Hamming distance matching function, $idf(q(x))$ represents the conventional inverse document frequency (IDF) weight function.

To further improve accuracy, Weak Geometric Consistency (WGC) [17] is employed for filtering visual words that are not consistent in terms of angle and scale. And the keyframe similarity is calculated as,

$$fs(q_i, r_j^k)_{wgc} = min(max(h^\theta), max(h^s)) \times fs(q_i, r_j^k), \tag{2}$$

where θ, s and h represent the angle information, scale information and Hamming distance respectively. If the score between frame pairs is over the threshold β, we regard these two are similar. For the compared videos, the scores of each frame pairs will be stored in fs table (details will be given in the next section). Both HE and WGC can further enhance the matching precision and are easy to be embedded into the BOW model.

3.2   Sequence Matching

In this process, we use four tables (illustrated in Fig. 2) to record related information. The fs table, $dist$ table, loc table, vs table are used to record the keyframe similarity information, Edit distance information, backtracking path and video similarity score, respectively.

Using a query video $Q\{q_1, q_2, \cdots, q_m\}$ and a reference video $R_k\{r_1^k, r_2^k, \cdots, r_n^k\}$ as the example, we first calculate the keyframe similarity using Eq. (2) for each query keyframe and record the score in the fs table. Then, the Edit distance of the two sequences is calculated. Mathematically, the Edit distance between two sequences Q and R^k is given by $dist(m, n)$,

$$dist(i, j) = \begin{cases} max(i, j) & \text{if } min(i, j) = 0 \\ min \begin{cases} dist(i-1, j) + 1 \\ dist(i, j-1) + 1 \\ dist(i-1, j-1) + f(i, j) \end{cases} & \text{otherwise} \end{cases} \tag{3}$$

where $dist(i, j)$ is the distance between the first i frames of Q_i and the first j frames of R^k. Particularly, $f(i, j)$ is the value in i^{th} row and j^{th} column of fs table. Then, based on Edit distance, we will calculate the REDS for fast filter.

Previous approaches based on Edit distance, such as [16], usually use the Edit distance directly as the criteria for measuring video dissimilarities. However, this method has one critical limitation: Because the number of keyframes in reference videos varies from one to hundreds, directly using Edit distance value might not be reliable enough. Thus, we propose relative Edit distance similarity (REDS) to solve the above problem. From Eq. (3), we can know that the Edit distance of the two is $dist(m, n)$. Then the REDS is given as,

$$REDS(Q, R^k) = \frac{max(m, n) - dist(m, n)}{min(m, n)}. \tag{4}$$

where Q and R^k are video sequences of frame length m and n, respectively. The REDS of two sequences ranges from 0 to 1.

As we know, Edit distance measures the dissimilarity of the sequences, i.e., the higher $dist(m, n)$ is, the more dissimilar they are. And REDS reflects the relative similarity between the two videos, i.e., higher REDS means more similar. Given Q and R^k, if $REDS(Q, R^k)$ is less than the predefined threshold T_{REDS}, R^k will be filtered out. Otherwise, R^k is regarded as near-duplicate candidates of Q and the scores between them will be aggregated in further steps, details will be discussed in the following section.

From Eq. (4), we can see that REDS is sequence-length-independent, which is suitable for partial near-duplicate retrieval. Specially, suppose that a short video segment is inserted into a long video sequence. Although the length difference between them is large, but Edit distance approaches to $(m - n)$, and the REDS score between them is relative high. On the other hand, suppose that two sequences share the same length but non-related content, the REDS score will be low and the reference video will be filtered out. The key factor which affects REDS is the amount of matched frames in temporal order.

3.3 Results Ranking and Near-Duplicate Segments Localizing

After REDS checking, if the video is a candidate near-duplicate of the query video, we combine the $dist$ table with the fs table to calculate the video similarity score (vs table) and path information (loc table).

As Fig. 2(c) illustrates, the loc table uses 4 states to record the path for backtracking, and each state reflects where $dist(i, j)$ is from. According to Fig. 2(b), $dist(i, j)$ may come from one or more candidate directions or states, i.e., $dist(i-1, j)+1$, $dist(i, j-1)+1$, $dist(i-1, j-1)+0$ and $dist(i-1, j-1)+1$, or we can say, the value can be from three orientations: up, left, up-left. In this condition, $loc(i, j)$ is uncertain, we thus propose a algorithm using detect-and-refine strategy. Firstly, we detect the candidate states, then we choose the most optimal direction from the candidate directions that ensuring the sub video sequence has the largest score. We complete this process by employing dynamic

(a) fs

	r_1'	r_2'	r_3'	r_4'	r_5'	r_6'	r_7'
	0	0	0	0	0	0	0	0
q_1	0	1.21	0.01	0.02	0.01	0.03	0.02	0.00
q_2	0	0.02	1.15	0.01	0.02	0.02	0.02	0.03
q_3	0	0.02	0.03	0.00	0.97	0.02	0.01	0.04
q_4	0	0.03	0.02	0.01	0.02	0.00	1.83	0.03
q_5	0	0.01	0.01	0.01	0.02	0.01	0.03	1.02

(b) dist

	r_1'	r_2'	r_3'	r_4'	r_5'	r_6'	r_7'
	0	1	2	3	4	5	6	7
q_1	1	0	1	2	3	4	5	6
q_2	2	1	0	1	2	3	4	5
q_3	3	2	1	1	1	2	3	4
q_4	4	3	2	2	2	2	2	3
q_5	5	4	3	3	3	3	3	2

(c) loc

	r_1'	r_2'	r_3'	r_4'	r_5'	r_6'	r_7'
	0	2	2	2	2	2	2	2
q_1	3	0	2	2	2	2	2	2
q_2	3	3	0	2	2	2	2	2
q_3	3	3	3	1	0	2	2	2
q_4	3	3	3	3	3	1	0	2
q_5	3	3	3	3	3	3	3	0

(d) vs

	r_1'	r_2'	r_3'	r_4'	r_5'	r_6'	r_7'
	0	0	0	0	0	0	0	0
q_1	0	1.21	1.21	1.21	1.21	1.21	1.21	1.21
q_2	0	1.21	2.36	2.36	2.36	2.36	2.36	2.36
q_3	0	1.21	2.36	2.36	3.33	3.33	3.33	3.33
q_4	0	1.21	2.36	2.36	3.33	3.33	5.16	5.16
q_5	0	1.21	2.36	2.36	3.33	3.33	5.16	6.18

Fig. 2. A toy example of the matrices used in the procedure of subsequence matching. (a) fs matrix: the similarity of keyframes; (b) $dist$ matrix: the process of Edit distance computing; (c) loc matrix: the spatial and temporal relations between keyframes for backtracking; (d) vs matrix: the similarity information of video segments.

programming and forming a vs table. The $vs(i,j)$ value is calculated through following equations:

$$
vs(i,j) = \begin{cases} vs(i-1,j) + \dfrac{1}{2} \times \alpha \times fs(i,j) & up \\[2mm] vs(i,j-1) + \dfrac{1}{2} \times \alpha \times fs(i,j) & left \\[2mm] vs(i-1,j-1) + \alpha \times fs(i,j) & up-left \end{cases} \tag{5}
$$

where α is set as 1 when $fs(i,j)$ exceeds the threshold β and 0, otherwise. Moreover, $vs(i,j)$ calculated with direction of up and left is incremented by half for penalty to reduce the effect of fast forwarding and slow motion. Finally, $vs(m,n)$ is the score of reference video and query video, which reflects the similarity. Any videos under the threshold η of video similarities will be discarded. We then rerank the remained results through this metric.

Note that the process of filter by REDS and the process of aggregating scores are not unnecessary or conflictive. REDS focuses on the correlation between videos, i.e., whether the reference is a duplicate or near-duplicate of the query. Score aggregating emphasizes the similarity of the two videos, i.e., how similar they are. Take the example listed in the former section again. Given a short video segment and insert it into a long video as the reference. From the perspective of content, they are high related, so the value of REDS should be high and we should retain the video. But related as they are, the similarity between them is low due to the huge amount of unrelated content in the other segment. So when reranking the retrieved results, this kind of videos will get a low ranking point (low similarity score). So both REDS and video aggregating are necessary.

After video similarity calculation, *loc* table is used to localize near-duplicate segments, as depicted in Fig. 2(c). Generally, diagonal property is exploited in the *loc* table because a near-duplicate video usually temporally related to the origin video. That's to say, a block with high video segment similarity can be regarded as a near-duplicate segment. In order to obtain the diagonal block, we use the *loc* table for backtracking by following the direction of each state represented. In order to reduce the effect of fast forwarding, slow motion and error judgement [16], no more than two consecutive break is allowed. Finally, the remaining blocks can be regarded as near-duplicate segments of the original video.

4   Experiments and Results

To evaluate the effectiveness of the proposed method, we conduct experiments on two standard datasets. We evaluate the retrieval accuracy on CC_WEB_VIDEO and TRECVID-CBCD 2011, localization accuracy on TRECVID-CBCD 2011.

4.1   Datasets and Evaluation Metrics

CC_WEB_VIDEO is a famous NDVR benchmark cooperated by VIREO group from City University of Hong Kong, and Informedia group from Carnegie Mellon University. This dataset consists of 24 queries and 12,790 videos which were collected from YouTube, Google Video, and Yahoo! Video.

TRECVID-CBCD 2011 is one of the largest benchmarks for video copy detection and localization, consists of 11,503 reference videos. In the collection of videos, there are three types of original queries: reference video only, reference video embedded into non-reference video, and non-reference video only. What's more, for each type of original videos, there are also eight transformations applied, as listed in Table 1 [2].

On CC_WEB_VIDEO, the effectiveness in terms of NDVR is measured with mean average precision (MAP). Moreover, precision-recall curves are adopted to intuitively show our experiment results.

On TRECVID-CBCD 2011, we follow the benchmarks provided by [2]. For NDVR accuracy, normalized detection cost rate (NDCR) is adopted and smaller NDCR value indicates the better performance. Meanwhile, Mean F1 score is used to evaluate the NDVL performance. It is defined as the harmonic mean

Table 1. 8 types of video transformations.

Type	Descriptions
T1	Camcording
T2	Picture-in-picture
T3	Insertions of pattern
T4	Compression
T5	Change of gamma
T6	Decrease in quality - 3 random selected transformations including blur, gamma, frame dropping, contrast, compression, ratio, white noise
T8	Post production - 3 random selected transformations including crop, shift, contrast, caption, flip, insertion of pattern, picture-in-picture
T10	3 random selected transformations from each of (T1 to T5), T6, T8

of precision and recall of relevant correctly retrieved results and larger F1 score reflects better performance of the approach.

4.2 Experiment Settings

The size of codebook used in this paper is 20 K. The thresholds β, T_{REDS}, η used in frame similarity measurement, filter stage and refine stage are set as 0.07, 0.1, 0.1, respectively. All the experiments are run on a PC with a 3.4 GHz Intel Core processor and 8 GB memory.

4.3 Results on NDVR

As illustrated in Table 2, Fig. 3 and Table 3, our proposed method outperforms the other three approaches on both datasets. Among these three algorithms, the method proposed by Yeh et al. [16] utilized Edit distance for exhaustive matching, while Song et al. [12] employed hash methods and Chou et al. [4] adopted pattern based method under hierarchical framework.

Table 2. Comparison in terms of MAP on CC_WEB_VIDEO

Method	Yeh et al. [16]	Song et al. [12]	Chou et al. [4]	Ours
MAP	0.8852	0.9582	0.9582	**0.9828**

On CC_WEB_VIDEO, we reached 0.9828 in terms of MAP, which is the best result among the four (Table 2). Compared with the second best methods proposed by Chou et al. [4] and Song et al. [12], our method has 2.46 % improvement, owing to better signature representations and sequence matching strategy. Specifically, in [4], frame-level global signature has insufficient discriminative to

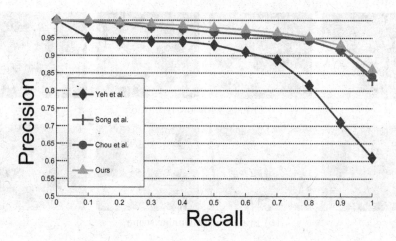

Fig. 3. Precision-recall curves on CC_WEB_VIDEO.

Table 3. Comparison of NDCR on TRECVID-CBCD 2011 (better = lower).

	T1	T2	T3	T4	T5	T6	T8	T10	Avg.
Yeh *et al.* [16]	0.847	0.437	0.233	0.147	0.166	0.331	0.563	0.412	0.3920
Song *et al.* [12]	0.813	0.884	0.562	0.681	0.637	0.734	0.738	0.803	0.7315
Chou *et al.* [4]	0.132	0.208	**0.033**	0.061	0.046	0.162	**0.082**	0.226	0.1188
Ours	**0.128**	**0.193**	0.036	**0.059**	**0.037**	**0.157**	**0.082**	**0.221**	**0.1141**

cope with local object matching. Meanwhile, the method in [12] ignores the effect of the temporal transformations during aggregating the similarity scores.

From Fig. 3, we can see that the curve of proposed method decreased slower and the method by Yeh *et al.* [16] has a quick drop with the recall rate increases. That is mainly because exhaustive searching may cause more false positives when the recall rate increases.

Besides, the frames with low resolution and similar scenes will affect the local keypoint matching and cause false positives. Figure 4 shows an example of poor queries on this dataset.

As depicted in Table 3, our method remains the best, but the improvement to the second best one decreases. This is because the considerable proportion of frames with more complicated situations, e.g., low resolution and with constant color background, which is not good for feature extracting, see examples in Fig. 5. The large amount of low score videos will cause false positives. Besides, the NDCR score of Yeh *et al.* [16] on T1 is only 0.847 while ours is 0.128, this mainly because that semi-global feature can't handle simulated camcording transformations properly. Our method can save more content information, making the video similarity measurement more precise and robust. Also, we can see hashing method proposed by Song *et al.* [12] did worse than that on the first

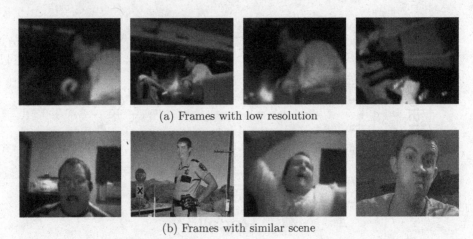

(a) Frames with low resolution

(b) Frames with similar scene

Fig. 4. Poor queries on CC_WEB_VIDEO. Low resolution (a) and similar scene (b) will cause false matching of local descriptors.

Fig. 5. Poor queries On TRECVID-CBCD 2011 dataset. The constant color background and letter foreground is not good for descriptors extracting

dataset, for the video level matching strategy is weak to cope with videos under temporal transformations.

4.4 Results on NDVL

We show the performance results on NDVL, in terms of mean F1, in Table 4. The symbol '-' means this method can not cope with the near-duplicate video localization problems.

As depicted in Table 4, our method outperforms other methods on the overall performance, due to the use of suitable signature and the Edit-distance-based aggregating method. In the table, For nearly every transformations, the method achieved the best or the second best result. Although both use Edit distance, our method shows much better results than that of the method in [16], which shows the superiority of the sequence matching strategy we proposed.

Table 4. Comparison of mean F1 on TRECVID-CBCD 2011 (better = higher).

	T1	T2	T3	T4	T5	T6	T8	T10	Avg.
Yeh *et al.* [16]	0.762	0.833	0.865	0.902	0.931	0.877	0.836	0.841	0.8559
Song *et al.* [12]	–	–	–	–	–	–	–	–	–
Chou *et al.* [4]	0.928	**0.940**	0.943	0.948	**0.948**	0.932	**0.936**	**0.930**	0.9381
Ours	**0.938**	**0.940**	**0.948**	**0.950**	**0.948**	**0.934**	0.932	**0.930**	**0.9400**

5   Conclusion

In this paper, we proposed a novel approach for near-duplicate video retrieval and localization task, which combines BOW-based frame matching and improved Edit-distance-based sequence aggregating techniques. BOW-based frame matching was introduced to achieve an accurate frame-wise similarity score, which is the foundation of precise sequence matching. For sequence matching, we proposed the REDS to firstly filter out non-related videos and a dynamic programming strategy to rerank the videos and localize the similar segments. The effectiveness and robustness in dealing NDVR and NDVL of the proposed method was validated though experiments on CC_WEB_VIDEO and TRECVID-CBCD 2011 datasets compared with state-of-the-art methods.

Acknowledgement. This work is supported by the National Science Foundation of China (No. 61175096) and Specialized Fund for Joint Building Program of Beijing Municipal Education Commission.

References

1. Arandjelović, R., Zisserman, A.: Three things everyone should know to improve object retrieval. In: IEEE Conference on Computer Vision and Pattern Recognition (CVPR), pp. 2911–2918 (2012)
2. Awad, G., Over, P., Kraaij, W.: Content-based video copy detection benchmarking at TRECVID. ACM Trans. Inf. Syst. (TOIS) **32**(3), 14 (2014)
3. Chiu, C.Y., Chen, C.S., Chien, L.F.: A framework for handling spatiotemporal variations in video copy detection. IEEE Trans. Circ. Syst. Video Technol. **18**(3), 412–417 (2008)
4. Chou, C.L., Chen, H.T., Lee, S.Y.: Pattern-based near-duplicate video retrieval and localization on web-scale videos. IEEE Trans. Multimedia **17**(3), 382–395 (2015)
5. Esmaeili, M.M., Fatourechi, M., Ward, R.K.: A robust and fast video copy detection system using content-based fingerprinting. IEEE Trans. Inf. Forensics Secur. **6**(1), 213–226 (2011)
6. Huang, Z., Shen, H.T., Shao, J., Zhou, X., Cui, B.: Bounded coordinate system indexing for real-time video clip search. ACM Trans. Inf. Syst. (TOIS) **27**(3), 17 (2009)
7. Jegou, H., Douze, M., Schmid, C.: Hamming embedding and weak geometric consistency for large scale image search. In: Forsyth, D., Torr, P., Zisserman, A. (eds.) ECCV 2008, Part I. LNCS, vol. 5302, pp. 304–317. Springer, Heidelberg (2008)

8. Kraaij, W., Awad, G.: TRECVID 2011 content-based copy detection: task overview. In: Online Proceedings of TRECVID (2011)
9. Liu, J., Huang, Z., Cai, H., Shen, H.T., Ngo, C.W., Wang, W.: Near-duplicate video retrieval: current research and future trends. ACM Compu. Surv. (CSUR) **45**(4), 44 (2013)
10. Liu, L., Lai, W., Hua, X.-S., Yang, S.-Q.: Video histogram: a novel video signature for efficient web video duplicate detection. In: Cham, T.-J., Cai, J., Dorai, C., Rajan, D., Chua, T.-S., Chia, L.-T. (eds.) MMM 2007. LNCS, vol. 4352, pp. 94–103. Springer, Heidelberg (2006)
11. Roopalakshmi, R., Reddy, G.R.M.: A novel spatio-temporal registration framework for video copy localization based on multimodal features. Signal Process. **93**(8), 2339–2351 (2013)
12. Song, J., Yang, Y., Huang, Z., Shen, H.T., Luo, J.: Effective multiple feature hashing for large-scale near-duplicate video retrieval. IEEE Trans. Multimedia **15**(8), 1997–2008 (2013)
13. Wang, J., Shen, H.T., Song, J., Ji, J.: Hashing for similarity search: a survey. CoRR abs/1408.2927 (2014). http://arxiv.org/abs/1408.2927
14. Wu, X., Hauptmann, A.G., Ngo, C.W.: Practical elimination of near-duplicates from web video search. In: Proceedings of the 15th International Conference on Multimedia, pp. 218–227. ACM (2007)
15. Wu, X., Ngo, C.W., Hauptmann, A.G., Tan, H.K.: Real-time near-duplicate elimination for web video search with content and context. IEEE Trans. Multimedia **11**(2), 196–207 (2009)
16. Yeh, M.C., Cheng, K.T.: Video copy detection by fast sequence matching. In: Proceedings of the ACM International Conference on Image and Video Retrieval, p. 45. ACM (2009)
17. Zhao, W.L., Wu, X., Ngo, C.W.: On the annotation of web videos by efficient near-duplicate search. IEEE Trans. Multimedia **12**(5), 448–461 (2010)

Efficient Group Top-k Spatial Keyword Query Processing

Kai Yao[1], Jianjun Li[1(✉)], Guohui Li[1], and Changyin Luo[2]

[1] School of Computer Science and Technology, Huazhong University
of Science and Technology, Wuhan, China
{kaiyao,jianjunli,guohuili}@hust.edu.cn
[2] School of Computer, Central China Normal University, Wuhan, China
luochangyin@hust.edu.cn

Abstract. With the proliferation of geo-positioning and geo-tagging, spatial web objects that possess both a geographical location and textual description are gaining in prevalence. Given a spatial location and a set of keywords, a top-k spatial keyword query returns the k best spatio-textual objects ranked according to their proximity to the query location and relevance to the query keywords. To our knowledge, existing study on spatial keyword query processing only focuses on single query point scenario. In this paper, we take the first step to study the problem of multiple query points (or group queries) top-k spatial keyword query processing. We first propose a threshold-based algorithm, which first performs incremental top-k spatial keyword query for each query point and then combines their results. Next, we propose another more efficient algorithm by treating the whole query set as a query unit, which can significantly reduce the objects to be examined, and thus achieve higher performance. Extensive experiments using real datasets demonstrate that our approaches are efficient in terms of runtime and I/O cost, as compared to the baseline algorithm.

Keywords: Spatial keyword query · Top-k query · Spatial databases

1 Introduction

In modern applications such as Google Maps, Points of Interest (PoI) are typically augmented with textual descriptions. This development gives prominence to spatial keyword queries [1–4]. A typical such query takes a location and a set of keywords as arguments and returns relevant content that matches the arguments. To our knowledge, most of the existing proposals for answering spatial keyword queries are only focused on single query point scenario. However, in some cases,

The work is partially supported by the State Key Program of National Natural Science of China under Grant No. 61332001, National Natural Science Foundation of China under Grants Nos. 61572215, 61300045, China Postdoctoral Science Special Foundation under Grant No. 2015T80802, and the Fundamental Research Funds for the Central Universities, HUST-2016YXMS076.

© Springer International Publishing Switzerland 2016
F. Li et al. (Eds.): APWeb 2016, Part I, LNCS 9931, pp. 153–165, 2016.
DOI: 10.1007/978-3-319-45814-4_13

there may be multiple queries. For example, two friends at different locations want to find a restaurant to have dinner together, and they may have different preferences (e.g., one might like restaurant that offers "pizza" and "dessert" in a "cozy" environment, the other might prefer restaurant that offers "pizza" and "coffee" and has "friendly" service). As another example, consider a group of tourists who visit a city and stay in different hotels. These people may want to visit an attraction of the city, which is not far from their locations and at the same time is relevant to their different preferences (e.g., some tourists would like to visit "museum" and "library", while others want to visit "art gallery" and "theater").

In this paper, we take the first step to study the group location-aware top-k text retrieval (GLkT) query, which takes into account multiple query points. This query enables multiple mobile users (or drivers) to get the k spatio-textual objects that best match their queries with respect to location and text relevancy.

Note that each query point may have different locations and keywords. It can be widely utilized in various decision support systems and multiple domains like service recommendation, investment planning, etc. The most relevant study on this problem is IR-tree [1] which embeds an inverted index in each node of R-tree [5]. The inverted index at each node refers to a pseudo-document that represents all the objects under the node. Therefore, in order to verify if a node is relevant for a set of query keywords, the current approaches access the inverted index at each node to evaluate the similarity between the query keywords and the pseudo-document associated with the node. However, they only consider single query point and their methods are not suitable for GLkT query. Most traditional spatial queries on spatial database such as group nearest neighbor queries [6] and aggregate nearest neighbor queries [7], which find the k objects with the minimum total distance to a group of query points. However, both group NN and aggregate NN queries do not concern non-spatial information (e.g., name, description, and type etc.). To address this limitations and improve the performance, we first propose a threshold-based algorithm, which first performs incremental top-k spatial keyword query for each query point and then combines their results. To achieve higher performance, we propose another more efficient algorithm by treating the whole query set as a query unit, which can significantly reduce the objects to be examined.

The main contributions of this paper can be summarized as follows:

- We propose a threshold algorithm, which first performs incremental top-k spatial keyword query for each query point and then combines their results, to address the GLkT query processing problem.
- By treating the whole query set as a query unit, we propose a more efficient algorithm namely UA. We also propose a heuristic to prune the search space and reduce the number of node accesses significantly.
- Extensive experiments using real datasets demonstrate that our approaches are efficient in terms of both runtime and I/O cost, as compared to the baseline algorithm.

The remainder of the paper is organized as follows. Section 2 gives a formal problem definition. Section 3 presents a threshold algorithm and a Unit-based algorithm for GLkT query. Experimental results are provided in Sect. 4. We review related work in Sect. 5 and make a conclusion in Sect. 6.

2   Preliminaries

Problem Statement. Let $D = \{p_1, p_2, ..., p_n\}$ be a dataset of spatio-textual objects. Each object $p \in D$ includes a spatial location $p.l$ and textual description $p.d$, denoted by $p = \{p.l, p.d\}$. Let $Q = \{q_1, q_2, ..., q_m\}$ be a collection of query points. Each $q \in Q$ is represented by $q = \{q.l, q.d, k, \alpha\}$, where $q.l$ is the query location, $q.d$ is the set of query keywords and parameter $\alpha \in (0, 1)$ is used to balance between the spatial and textual components. Given a query set Q, our goal is to return k spatio-textual objects $\{r_1, r_2, ..., r_k\}$ from D with the highest relevance score $T(r_1, Q) \geq T(r_2, Q) \geq ... \geq T(r_k, Q)$. We use a linear interpolation function to compute the spatio-textual relevance score. This paper's proposals are applicable to a wide range of ranking functions, namely all functions that are monotone with respect to spatial proximity and text relevancy.

Definition 1 (Spatial Proximity). *Given an object p and a query q, the spatial proximity is defined in the following equation:*

$$\delta(p, q) = 1 - \frac{D_\varepsilon(p.l, q.l)}{D_{\max}} \tag{1}$$

where $D_\varepsilon(p.l, q.l)$ is the Euclidean distance between $p.l$ and $q.l$, and $D_{\max}$ is the maximum distance in the location space. The maximum distance may be obtained by getting the largest diagonal of the Euclidean space of the application.

Definition 2 (Textual Relevance). *Given an object p and a query q, the textual relevance id defined in the following equation:*

$$\theta(p, q) = \frac{\sum_{t \in q.d} \omega_{t,p.d} \cdot \omega_{t,q.d}}{\sqrt{\sum_{t \in p.d} (\omega_{t,p.d})^2 \cdot \sum_{t \in q.d} (\omega_{t,q.d})^2}} \tag{2}$$

There are several similarity measures that can be used to evaluate the textual relevance between the query keywords $q.d$ and the text description $p.d$ [8]. In order to compute the cosine, we adopt the approach employed by Zobel and Moffat [9]. Therefore, the weight $\omega_{t,p.d}$ is computed as $\omega_{t,p.d} = 1 + \ln(f_{t,p.d})$, where $f_{t,p.d}$ is the number of occurrences (frequency) of t in $p.d$; and the weight $\omega_{t,q.d}$ is obtained from the following formula $\omega_{t,q.d} = \ln(1 + \frac{|P|}{df_t})$, where $|P|$ is the total number of documents in the collection. The document frequency df_t of a term t gives the number of documents in P that contains t. In this paper, we adopt the well-known cosine similarity between the vectors composed by the weights of the terms in $q.d$ and $p.d$. The textual relevance is a value within the range $[0, 1]$ (property of cosine).

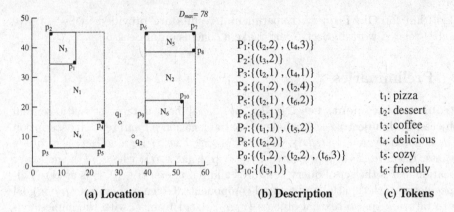

(a) Location   **(b) Description**   **(c) Tokens**

P_1:{$(t_2,2)$, $(t_4,3)$}
P_2:{$(t_3,2)$}
P_3:{$(t_2,1)$, $(t_4,1)$}
P_4:{$(t_1,2)$, $(t_2,4)$}
P_5:{$(t_2,1)$, $(t_6,2)$}
P_6:{$(t_3,1)$}
P_7:{$(t_1,1)$, $(t_5,2)$}
P_8:{$(t_2,2)$}
P_9:{$(t_1,2)$, $(t_2,2)$, $(t_6,3)$}
P_{10}:{$(t_3,1)$}

t_1: pizza
t_2: dessert
t_3: coffee
t_4: delicious
t_5: cozy
t_6: friendly

Fig. 1. Group Top-k spatial keyword query

Definition 3 (*Spatio-textual Relevance Score*). *Given an object p and a query q, the spatio-textual relevance score between p and q is defined as:*

$$\tau(p,q) = \alpha \cdot \delta(p,q) + (1-\alpha) \cdot \theta(p,q) \tag{3}$$

where $\alpha \in (0,1)$ is a parameter used to balance between the spatial and textual components. For example, $\alpha = 0.5$ means that spatial proximity and textual relevance are equally important.

Definition 4 (*Group Spatio-textual Relevance Score*). *Given an object p and a query set $Q = \{q_1, q_2, ..., q_m\}$, the group spatio-textual relevance score between p and Q is defined as:*

$$T(p,Q) = \sum_{q \in Q} \tau(p,q) \tag{4}$$

Example 1. Suppose in a geographic search engine there is a set $D = \{p_1, ..., p_{10}\}$, each of which is associated with one spatial location and a set of keywords. Figure 1(a) and (b) plot the spatial location of the objects and the frequencies of keyword, respectively.

Assume that two friends q_1 and q_2 want to find a restaurant to have dinner together, and q_1 prefer restaurant that offers "pizza" and "dessert" in a "cozy" environment, and q_2 prefer restaurant that offers "pizza" and "coffee" and has "friendly" service. Therefore, the query set Q is consist of $q_1 = \{[30,15], t_1, t_2, t_5, 1, 0.5\}$ and $q_2 = \{[35,10], t_1, t_3, t_6, 1, 0.5\}$. For object p_4 where $p_4.l = (25, 15)$, its spatial proximity between q_1 and q_2 is $\delta(p_4, q_1) = 0.94$ and $\delta(p_4, q_2) = 0.86$ respectively, its textual relevance between q_1 and q_2 is $\theta(p_4, q_1) = 0.55$ and $\theta(p_4, q_2) = 0.31$ respectively. Thus, the spatio-textual relevance score $\tau(p_4, q_1) = 0.5 * 0.94 + (1 - 0.5) * 0.55 = 0.75$ and $\tau(p_4, q_2) = 0.5 * 0.86 + (1 - 0.5) * 0.31 = 0.58$. According to Definition 4, the group spatio-textual relevance score $T(p_4, Q) = 0.75 + 0.58 = 1.33$. Similarly, for object p_9,

$T(p_9, Q) = 0.71 + 0.81 = 1.52$. Thus, p_9 is better than p_4 since $1.52 > 1.33$. By calculating the relevance score of all the objects, we take the one object with highest score $\{p_9/1.52\}$ as the result.

3   GLkT Query Algorithm

In this section, we present efficient algorithms for processing GLkT query with the assumption that spatio-textual objects are organized by IR-Tree. Section 3.1 gives a brief description of the baseline method. Section 3.2 gives the threshold algorithm and the Unit-based algorithm will be introduced in Sect. 3.3.

3.1   Baseline Algorithm

The idea of the baseline algorithm is as follows. For each query $q_i \in Q$, we compute the ranked list of all objects according to the ranking function τ respectively. Then, we combine them to gain the comprehensive relevance score, thus we can use classical selecting algorithm to get the k most relevant spatio-textual objects from dataset D.

Though this method can solve our problem, it is rather inefficient and may generate large amounts of candidates. For each $q_i \in Q$, it requires retrieving all the spatio-textual objects and computing relevance score (high I/O cost), without any pruning.

3.2   Threshold Algorithm

In this section, we propose a threshold algorithm (TA) to efficiently find top-k most relevant objects from dataset D. The basic idea is that, it first performs incremental top-k spatial keyword query for each $q_i \in Q$ and then combines their results. As shown in Fig. 1, the query set Q is consist of $q_1 = \{[30, 20], t_1, t_2, t_5, 1, 0.5\}$ and $q_2 = \{[35, 10], t_1, t_3, t_6, 1, 0.5\}$, TA retrieves the most relevant object of q_1 (object p_4 with $\tau(p_4, q_1) = 0.75$) and computes the spatio-textual relevance about q_2 and p_4 ($\tau(p_4, q_2) = 0.58$). Similarly, TA finds the most relevant object of q_2 (object p_9 with $\tau(p_9, q_2) = 0.81$) and computes the spatio-textual relevance about q_1 and p_9 ($\tau(p_9, q_1) = 0.71$). Thus, object p_9 with the larger group spatio-textual relevance score than p_4 ($1.52 > 1.33$) becomes the temporary result.

For each query q_i, TA stores a threshold λ_i, which is the score of the currently most relevant object, i.e., $\lambda_1 = 0.75$ and $\lambda_2 = 0.81$. The total threshold Λ is defined as the sum of each λ_i, i.e., $\Lambda = \lambda_1 + \lambda_2 = 1.56$. Continuing our algorithm, since $\Lambda > \tau(p_9, Q)$, it is possible that there exists an object in D whose relevance score is larger than p_9. So we continue to retrieve the second most relevant object of q_1 and q_2. As shown in Fig. 2, the second most relevant object of q_1 and q_2 is p_9 and p_5, respectively. Then we update $\lambda_1 = 0.71$, $\lambda_2 = 0.72$ and $\Lambda = 1.43$. Continuing TA, since $\Lambda < \tau(p_9, Q)$, it indicates that there exists no object in D

Object	$\tau(p, q_1)$
p_4	0.75
p_9	0.71
p_5	0.51
...	...

Object	$\tau(p, q_2)$
p_9	0.81
p_5	0.72
p_4	0.58
...	...

Object	$T(p, Q)$
p_9	1.52
p_4	1.33
p_5	1.23
...	...

(a) Rank list of q_1 (b) Rank list of q_2 (c) Rank list of Q

Fig. 2. Example of threshold algorithm

whose relevance score is larger than p_9, so we terminate algorithm and return p_9 as the ultimate result.

Algorithm 1 shows the pseudo code for TA, where C is the candidate list of the k most relevant objects found so far, and $best\_k$ is the relevance score of the k-th most relevant object in C. Note that if C contains fewer than k members, $best\_k$ is set to zero. When an object is inserted into C, an existing member is replaced if the list already contains k members.

Algorithm 1. TA(Query Q, Index IR-$tree$)

```
1  begin
2      Λ = ∞; best_k = 0; C = ∅;
3      while Λ > best_k do
4          foreach qᵢ in Q do
5              get the next relevant object pⱼ of qᵢ;
6              λᵢ = τ(pⱼ, qᵢ);
7              update Λ;
8              compute the relevance score between pⱼ and Q : T(pⱼ, Q);
9              if T(pⱼ, Q) > best_k then
10                 Add pⱼ to C;
11                 if |C| ≥ k then
12                     update best_k;
13     return C;
```

3.3   Unit-Based Alogorithm

TA may cause multiple accesses to the same node of the index structure(i.e., IR-Tree) due to multiple query points. To avoid this problem, we propose Unit-based algorithm (UA) which could process GLkT query by a single traversal. The main idea is to treat the whole query set as a query unit, then we could use it to prune the search space. Specifically, starting from the root of the IR-tree for dataset D, UA visits only nodes that may contain candidate objects. In the following, we will use symbol $\mathcal{U}$ to represent the query unit. Firstly, we give the

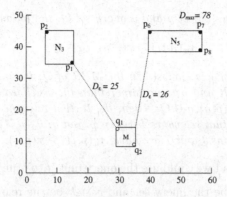

Fig. 3. Example of Definition 5

definitions of spatial proximity and textual relevance between object p and $\mathcal{U}$ (i.e., $\delta_u(p,\mathcal{U})$ and $\theta_u(p,\mathcal{U})$).

Definition 5 $(\delta_u(p,\mathcal{U}))$. *Given an object p and a query set Q. The spatial proximity between p and $\mathcal{U}$ is defined in the following equation:*

$$\delta_u(p,\mathcal{U}) = 1 - \frac{D_\varepsilon(p.l, MBR.l)}{D_{\max}} \tag{5}$$

where MBR is the minimum bounding rectangle of Q, and D_{max} is the maximum distance in the location space.

According to the characters of minimum bounding rectangle, we have $D_\varepsilon(p.l, MBR.l) \leq D_\varepsilon(p.l, q.l), q \in Q$. Therefore, we have $\delta_u(p,\mathcal{U}) \geq \delta(p,q), q \in Q$. Note that p could be an intermediary-node N of IR-tree. As shown in Fig. 3, Q is consist of q_1 and q_2, N_3 and N_5 are intermediary-nodes of IR-tree, we have $D_\varepsilon(N_3.l, MBR.l) = 25$ and $D_\varepsilon(N_5.l, MBR.l) = 26$.

Definition 6 $(\theta_u(p,\mathcal{U}))$. *Given an object p and a query set Q. The textual relevance between p and $\mathcal{U}$ is defined in the following equation:*

$$\theta_u(p,\mathcal{U}) = MAX\{\theta(p,q)\}, q \in Q \tag{6}$$

That is, $\theta_u(p,\mathcal{U})$ is the largest textual relevance score of p with q that belongs to Q. Therefore, it is easy to see that $\theta_u(p,\mathcal{U}) \geq \theta(p,q), q \in Q$.

Definition 7 $(\tau_u(p,\mathcal{U}))$. *Given an object p and a query set Q. The spatio-textual relevance score between p and $\mathcal{U}$ is defined as:*

$$\tau_u(p,\mathcal{U}) = \alpha \cdot \delta_u(p,\mathcal{U}) + (1-\alpha) \cdot \theta_u(p,\mathcal{U}) \tag{7}$$

Theorem 1. *Given an object p and a query set Q, the following is true:*

$$\tau_u(p,\mathcal{U}) \geq \tau(p,q), \forall q \in Q$$

Proof. Since the query q is enclosed in the MBR of Q, the minimum Euclidean distance between MBR and p is no larger than the Euclidean distance between q and p: $\delta_u(p,\mathcal{U}) \geq \delta(p,q), q \in Q$; Then, the textual relevance between p and $\mathcal{U}$ is larger than the textual relevance between p and q: $\theta_u(p,\mathcal{U}) \geq \theta(p,q), q \in Q$; According to Definitions 3 and 7, we have: $\tau_u(p,\mathcal{U}) \geq \tau(p,q), \forall q \in Q$.

Based on the Theorem 1 we could use the query unit $\mathcal{U}$ to prune the search space.

Heuristic 1: Let Q be the query set and $best\_k$ be the relevance score of the k-th most relevant object found so far. A node N will be pruned if:

$$\tau_u(N,\mathcal{U}) \leq \frac{best\_k}{|Q|}$$

where $\tau_u(N,\mathcal{U})$ is the spatio-textual relevance score between N and $\mathcal{U}$, and $|Q|$ is the size of Q. The concept of heuristic 1 also applies to the leaf entries of IR-tree. When an object p is encountered, we first compute relevance score between p and $\mathcal{U}$. If $\tau_u(p,\mathcal{U}) \leq \frac{best\_k}{|Q|}$, p is discarded since its relevance score cannot be larger than $best\_k$. In this way we avoid performing the computations between p and the members of Q.

Based on example 1, assuming we try to find top-1 spatio-textual object and $best\_k = 1.52$. When N_3 is encountered, as shown in Fig. 3, according to Definitions 5 and 6 we could have $\delta_u(N_3,\mathcal{U}) = 1 - \frac{D_e(N_3,\mathcal{U})}{D_{\max}} = 1 - \frac{25}{78} = 0.68$ and $\theta_u(N_3,\mathcal{U}) = MAX\{\theta(N_3,q_1), \theta(N_3,q_2)\} = MAX\{0.25, 0.28\} = 0.28$. Since $\tau_u(N_3,\mathcal{U}) = 0.5 \cdot 0.68 + 0.5 \cdot 0.28 = 0.48 < \frac{best\_k}{|Q|} = 0.76$, N_3 can be pruned without being visited.

Algorithm 2 shows the pseudo code for UA, where $TopkList$ denotes the candidate list of the k most relevant objects found so far, and $best\_k$ is the relevance score of the k-th most relevant object in $TopkList$. Note that if $TopkList$ contains fewer than k members, $best\_k$ is set to zero. When an object is inserted into $TopkList$, an existing member is replaced if the list already contains k members.

In UA, if $Node$ is a non-leaf node, its entries are sorted in a $List$ according to $\tau_u(entry,\mathcal{U})$ (lines 2–4). Next, we iterate through the $List$ (in the sorted order) and recursively invoke UA on the entries (lines 5–9). Once the first entry with $\tau_u(entry,\mathcal{U}) \leq \frac{best\_k}{|Q|}$ has been found, we ignore this entry and the rest entries in $List$ (line 7). We do this because no object in the subtree of this entry will be inserted into $TopkList$. If $Node$ is a leaf node, its entries are sorted in a $List$ according to $\tau_u(object,\mathcal{U})$ and are visited in this order (lines 10–12). When an object passes heuristic 1 and $T(object,Q) > best\_k$, we insert it into $TopkList$ and update $best\_k$ (lines 13–19). Note that if $TopkList$ contains fewer than k members, $best\_k$ is kept at zero.

Algorithm 2. UA($Node$: IR-tree node, Q: collection of query)

```
 1  begin
 2  │  if Node is an intermediate node then
 3  │  │   List ← entries in Node;
 4  │  │   sort entries in List according to τ_u(entry, U);
 5  │  │   foreach entry in List do
 6  │  │   │   if τ_u(entry, U) ≤ (best_k)/|Q| then
 7  │  │   │   │   break;  /* pruned by heuristic 1 */
 8  │  │   │   else
 9  │  │   │   │   UA(entry, Q);
10  │  else
11  │  │   List ← objects in Node;
12  │  │   sort objects in List according to τ_u(object, U);
13  │  │   foreach object in List do
14  │  │   │   if τ_u(object, U) ≤ (best_k)/|Q| then
15  │  │   │   │   break;  /* pruned by heuristic 1 */
16  │  │   │   else
17  │  │   │   │   if T(object, Q) > best_k then
18  │  │   │   │   │   Add object to TopkList;
19  │  │   │   │   │   update best_k;
20  │  return TopkList;
```

---

4 Experiments

In this section we evaluate the efficiency of our algorithms, using two real datasets: EURO[1] and TWITTER[2]. Table 1 summarizes these two datasets. EURO is a real dataset that contains points of interest (e.g., restaurant, hotel, park) in Europe. Each point of interest, which can be regarded as a spatial web object, contains a geographical location and a short description (name, features, etc.). TWITTER is a dataset generated from 3 million real tweets with location and textual information. All algorithms were implemented in Java, and an Intel(R) Core(TM) i5-4210M CPU @2.60 GHz with 4 GB RAM was used for the experiments. The index structure is disk resident, and the page size is 4 KB and the maximum number entries in internal nodes is set to 100.

In order to make the query Q resemble what users would like use, each $q_i \in Q$ is distributed uniformly in the MBR of Q and has 3 tokens which are randomly generated from datasets. Unless stated explicitly, parameters are set as follows by default: $\alpha = 0.5$, $k = 10$. In all experiments, we use workloads of 100 queries and report average costs of the queries.

---

[1] http://www.pocketgpsworld.com.
[2] http://twitter.com.

Table 1. Dataset properties

Property	EURO	TWITTER
Total number of objects	193,263	3,000,000
Total number of keywords	105,877	957,716
Average number of keywords per object	11.52	8.74
Data size	43.8 M	264 M

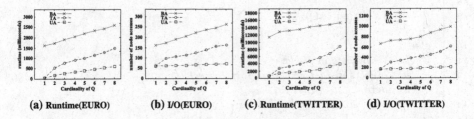

(a) Runtime(EURO)   (b) I/O(EURO)   (c) Runtime(TWITTER)   (d) I/O(TWITTER)

Fig. 4. Results for varying cardinality of Q

The first experiment shows the effect of the cardinality of Q. We fix MBR of Q to 5 % of the workspace of dataset and vary $|Q|$ from 1 to 8. The results are shown in Fig. 4. The CPU cost increases in all algorithms when the cardinality of Q increases from 1 to 8. This is because the distance and textual relevance computations for qualifying objects increase with the number of query points. When $|Q|$ equals 1, the problem is converted to a LkT query. In this case TA has similar performance as UA. However, the performance of UA better than TA when $|Q|$ is larger than 1, due to the high pruning power of heuristic 1. On the other hand, the cardinality of Q has little effect on the node accesses of UA because it does not influence the pruning power of heuristic 1.

To evaluate the effect of the MBR size of Q, we set $|Q|$ to 2 and vary MBR from 2 % to 30 % of the datasets. The results are shown in Fig. 5. We can see that the MBR size of Q does not influence the CPU cost and node accesses of BA significantly, because it does not play an important role in the process of BA. However, the cost of the other two algorithms increase with the MBR size. For TA, the termination condition is that the total threshold Λ should be smaller than $best\_k$, which increases with the MBR size. Therefore, TA retrieves more spatio-textual objects for each query $q_i \in Q$. For UA, the reason is the degradation of pruning power of heuristic 1 with the MBR size of Q, but it still has remarkable performance than BA and TA.

To evaluate the performance under different number of query keywords, we vary the number of keywords of each $q_i \in Q$ from 1 to 6 by setting $|Q|$ to 2 and MBR to 5 %. The results are shown in Fig. 6. We can see that UA outperforms BA and TA in terms of both runtime and I/O cost. When the number of query keywords increase, all three algorithms need more time and node accesses to finish the search query.

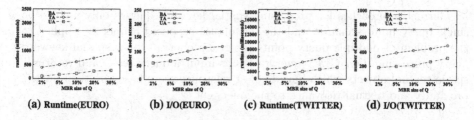

Fig. 5. Results for varying MBR size of Q

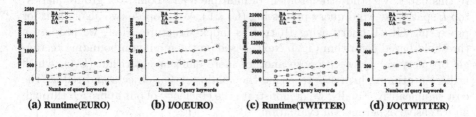

Fig. 6. Results for varying number of keywords

5 Related Work

Spatial keyword queries [3] extends classic keyword search to retrieve spatio-textual objects considering relevance to a set of input keywords as well as proximity to the location of the query. A comprehensive comparison of indexing techniques for spatial keyword search appears in [4]. Several proposals use loose combinations of text indexing (e.g., inverted lists) and spatial indexing (e.g., the R*-tree or the grid index) [10,11]. They usually employ the two structures in separate stages. Later, Ian de Felipe et al. [2] proposed a data structure that integrates signature files and R-trees. The main idea was indexing the spatial objects in an R-tree employing a signature on the nodes to indicate the presence of a given keyword in the sub-tree of the node. Cong et al. [1] and Li et al. [12] propose augmenting the nodes of an R-tree with textual indexes such as inverted files. The inverted files are used to prune nodes that cannot contribute with relevant objects. The mCK query [13] takes a set of m keywords as an argument, and retrieves groups of spatial keyword objects. It returns m objects of minimum diameter that match the m keywords. Rocha-Junior et al. [14] propose an indexing structure that associates each term to a different data structure (block or aggregated R-tree) and can process top-k spatial keyword queries more efficiently. Wu et al. [15] proposed an index structure called WIR-tree, which partitions objects into multiple groups such that each group shares as few keywords as possible. Their work aims to find the object closet to the query location that covers all the keywords specified in the query. However, our problem definition is differ significantly from joint spatial keyword query processing.

Current approaches for processing spatial keyword queries only focuses on single query point scenario, and the techniques proposed cannot be applied to the problem of multiple query points (or group queries) top-k spatial keyword query processing. Our GLkT query takes as input a query set Q and returns k spatio-textual objects which have highest relevance score based on their spatial proximity and textual relevance to the query set Q.

6   Conclusion

In this paper, we study the problem of multiple query points (or group queries) top-k spatial keyword query processing (GLkT). We propose two algorithms for processing GLkT query using IR-tree as index structure. First, we propose a threshold-based algorithm (TA). Next, based on the minimum bounding rectangle of query set Q, we propose another more efficient algorithm (UA), which utilizes a pruning heuristic to prune the spatio-textual objects and reduce the cost significantly. Finally, we demonstrate the efficiency of our approach through an extensive experimental evaluation.

References

1. Cong, G., Jensen, C.S., Wu, D.: Efficient retrieval of the top-k most relevant spatial web objects. Proc. VLDB Endow. **2**(1), 337–348 (2009)
2. De Felipe, I., Hristidis, V., Rishe, N.: Keyword search on spatial databases. In: IEEE 24th International Conference on Data Engineering, ICDE 2008, pp. 656–665. IEEE (2008)
3. Cao, X., Chen, L., Cong, G., Jensen, C.S., Qu, Q., Skovsgaard, A., Wu, D., Yiu, M.L.: Spatial keyword querying. In: Atzeni, P., Cheung, D., Ram, S. (eds.) ER 2012 Main Conference 2012. LNCS, vol. 7532, pp. 16–29. Springer, Heidelberg (2012)
4. Chen, L., Cong, G., Jensen, C.S., Dingming, W.: Spatial keyword query processing: an experimental evaluation. In: Proceedings of the VLDB Endowment, vol. 6, pp. 217–228. VLDB Endowment (2013)
5. Guttman, A.: R-trees: a dynamic index structure for spatial searching, vol. 14. ACM (1984)
6. Papadias, D., Shen, Q., Tao, Y., Mouratidis, K.: Group nearest neighbor queries. In: Proceedings of 20th International Conference on Data Engineering, pp. 301–312. IEEE (2004)
7. Yiu, M.L., Mamoulis, N., Papadias, D.: Aggregate nearest neighbor queries in road networks. IEEE Trans. Knowl. Data Eng. **17**(6), 820–833 (2005)
8. Manning, C.D., Raghavan, P., Schütze, H., et al.: Introduction to Information Retrieval, vol. 1. Cambridge University Press, Cambridge (2008)
9. Zobel, J., Moffat, A.: Inverted files for text search engines. ACM Comput. Surv. (CSUR) **38**(2), 6 (2006)
10. Zhou, Y., Xie, X., Wang, C., Gong, Y., Ma, W.-Y.: Hybrid index structures for location-based web search. In: Proceedings of the 14th ACM International Conference on Information and Knowledge Management, pp. 155–162. ACM (2005)
11. Chen, Y.-Y., Suel, T., Markowetz, A.: Efficient query processing in geographic web search engines. In: Proceedings of the 2006 ACM SIGMOD International Conference on Management of Data, pp. 277–288. ACM (2006)

12. Li, Z., Lee, K.C., Zheng, B., Lee, W.-C., Lee, D., Wang, X.: Ir-tree: an efficient index for geographic document search. IEEE Trans. Knowl. Data Eng. **23**(4), 585–599 (2011)
13. Zhang, D., Chee, Y.M., Mondal, A., Tung, A.K., Kitsuregawa, M.: Keyword search in spatial databases: towards searching by document. In: IEEE 25th International Conference on Data Engineering, ICDE 2009, pp. 688–699. IEEE (2009)
14. Rocha-Junior, J.B., Gkorgkas, O., Jonassen, S., Nørvåg, K.: Efficient processing of Top-k spatial keyword queries. In: Pfoser, D., Tao, Y., Mouratidis, K., Nascimento, M.A., Mokbel, M., Shekhar, S., Huang, Y. (eds.) SSTD 2011. LNCS, vol. 6849, pp. 205–222. Springer, Heidelberg (2011)
15. Dingming, W., Yiu, M.L., Cong, G., Jensen, C.S.: Joint top-k spatial keyword query processing. IEEE Trans. Knowl. Data Eng. **24**(10), 1889–1903 (2012)

Research Full Paper:
Social Media Data Analysis

Learn to Recommend Local Event Using Heterogeneous Social Networks

Shaoqing Wang[1,2], Zheng Wang[1], Cuiping Li[1(✉)], Kankan Zhao[1], and Hong Chen[1]

[1] Key Lab of Data Engineering and Knowledge Engineering of MOE, Renmin University of China, Beijing, China
{wsq,zheng.wang,licuiping,zhaokankan,chong}@ruc.edu.cn
[2] School of Computer Science and Technology, Shandong University of Technology, Zibo, China

Abstract. Event-based social networks (EBSNs), which link the online and offline social networks, are increasing popular online services. Along with dramatic rise of the users and events in EBSNs, it is necessary to recommend event to users. Taking full advantage of social networks information can significantly improve predictive accuracy in recommender systems. The intuition here is that the user's response to events are determined by his/her instinct and behaviours of friends. We propose a Heterogeneous Social Poisson Factorization(HSPF) model which combines online and offline social networks into one framework, and integrates the tie strength of online and offline friend relationships to the model. We test HSPF on Meetup dataset. Experimental results demonstrate that HSPF outperforms state-of-the-art recommendation methods.

Keywords: Event recommendations · Social recommendation · Poisson factorization · Event-based social networks

1 Introduction

With the rapid development of event-based social network services, it becomes increasing popular to participate local events through the online services such as Meetup (www.meetup.com) and Douban Events (beijing.douban.com/events). Event-based social networks (EBSNs) not only contain online social networks by joining the same groups where users can organize, participate, comment,

This work is supported by National Basic Research Program of China(973)(No. 2014CB340403, No.2012CB316205), National High Technology Research and Development Program of China (863) (No.2014AA015204) and NSFC under the grant No.61272137, 61033010, 61202114, 61532021, 61502421 and NSSFC (No.12&ZD220), and the Fundamental Research Funds for the Central Universities, and the Research Funds of Renmin University of China(15XNLQ06). It was partially done when the authors worked in SA Center for Big Data Research in RUC. This Center is funded by a Chinese National 111 Project Attracting.

© Springer International Publishing Switzerland 2016
F. Li et al. (Eds.): APWeb 2016, Part I, LNCS 9931, pp. 169–182, 2016.
DOI: 10.1007/978-3-319-45814-4_14

share and advertise events, but also include offline social networks where users make friends through attending the same face-to-face activities. The core goal of EBSNs is to make easy for neighbors (users located in the same city) together to do what they are commonly interested in [20]. To better organize events, these services allow users to join online groups, in which a user can publish and announce events to other group members [14]. For example, the online social networks on Meetup are the groups that organizers create and other numbers join, while the offline social networks on Meetup are captured in offline activities. Consequently, the users in EBSNs have two kinds of heterogeneous friend relationships that one is the online friend relationships and another is offline friend relationships.

Along with dramatic rise of the users and events in EBSNs, how to choose the interesting events for users is become very important and difficult. Meetup, the world's largest network of local groups, currently has 25.72 million users with more than 240,000 Meetup groups and 580,000 monthly events[1]. Therefore, it becomes essential to recommend from so many dazzling events to solve information overload problem. In contrast to traditional social network services (SNS), user behavior in EBSNs is predominantly driven by offline activities and highly influenced by a set of unique factors, such as spatio-temporal constraints and special social relationships [3]. As a result, event recommendation in EBSNs becomes very difficult. In addition, the event recommendation problem is arguably more challenging than classic recommendation scenarios (e.g. movies, books), since events have time limited efficacy, which means that event recommender systems have to recommend the event after it created and before it terminate.

Generally, users can only participate local events because of the limitation of distance. Different from existing recommendation problems, the characteristic of users' friend relationships plays very important role in event recommendation. Particularly worth mentioning is that online and offline friend relationship are not identical in EBSNs. On Meetup, the online friends mean the users who are in the same online group, and the offline friends mean the users who participate the same offline activities.

In this paper, we introduce online and offline friend relationships to recommend events to users and also introduce the tie strength of online and offline friend relationships. Based on these factors and the multi-factor model, we propose a novel method, named Heterogeneous Social Poisson Factorization (HSPF). In summary, the main contributions of this work lie in the following three aspects:

1. We propose the Heterogeneous Social Poisson Factorization(HSPF) model which combines online and offline social networks into one framework.
2. We integrate the tie strength of online and offline friend relationships to proposed HSPF model and use the coordinate ascent alogrithm for inference of the HSPF model parameters.
3. Our experiments on the Meetup dataset show that for the task of event recommendation, our HSPF model outperforms other models.

---

[1] http://www.meetup.com/about.

2   Related Works

2.1   Social Recommendation

Recent studies have proposed various methods to include social information in matrix factorization process [19]. For instance, [8,11,12] include trust in recommendation process; zhou et al. [21] exploits users interactions to improve recommendation qualities; duan et al. [4] uses locations to build a personalized recommendation system; and chaney et al. [2] proposes SPF (Social Poisson Factorization) model using trusted friend relationships. All the above methods for recommender systems only include a single factor. In our work, we not only divide trusted friends into online and offline friends, but also introduce the tie strength of trusted friends.

2.2   Event Recommendation

Currently, there are a small amount research works studying on event recommendation in EBSNs. EBSNs are first analyzed in data mining field in [10]. Du et al. [3] explores the modeling of EBSNs users by utilizing content preference, spatio-temporal context, and social influence (the event organizer) features. Pham et al. [14] transforms the recommendation problems into node proximity calculation problem and proposes a general graph-based model. Macedo et al. [13] takes several context-aware recommenders as input features such as content-based, social, locational and temporal signals. Zhang et al. [20] proposes CBPF (collective Bayesian Poisson factorization) model utilizing users' relationship information and events organizer, location, and textual content information to recommend local events. However, all the above methods do not introduce the heterogenous online+offline social relationships or cannot reflect the difference between online and offline social relationships. Qiao et al. [15] presents a Bayesian latent factor mode that unify the data, i.e., the geographical features, heterogenous online+offline social relationships and user implicit rating, for event recommendation. But it ignores the tie strength of the social relationships.

3   Preliminaries

Bayesian Poisson Factorization (BPF), proposed by Gopalan [5], is a probabilistic model of users and items for recommendation. BPF assumes that an observed rating matrix y_{ui} comes from a Poisson distribution:

$$y_{ui} \sim Poisson(\theta_u^T \beta_i)$$

where θ_u is a non-negative K-vector of user preference and β_i is a non-negative K-vector of item attributes. θ_u and β_i are hidden variables with Gamma priors:

$$\theta_{u,k} \sim Gamma(\lambda_{ua}, \lambda_{ub})$$
$$\beta_{i,k} \sim Gamma(\lambda_{ia}, \lambda_{ib})$$

where λ_{ua} and λ_{ia} are the shape parameters of the Gamma distribution and λ_{ub} and λ_{ib} are the rate parameters of the Gamma distribution.

BPF can handle sparse data well and is more robust to the issue of overfitting [20]. We build on BPF to develop a model of data where users attend face-to-face offline events and the same users are organized in a online network.

4   Proposed HSPF Model

In this section, we describe the Heterogeneous Social Poisson Factorization(HSPF). We are given data about users, events and groups, where each user who belongs to some groups has attend some events. The groups mean the online social networks while the events represent the offline social networks.

HSPF is a latent variable model of user-event interaction and user-group interaction. HSPF uses Poisson factorization to model both of the interactions that are typically sparse. The user's responses to events are determined by his/her instinct and behaviours of friends. That is the users would attend the events that are not consistent with his/her preference, just because that his/her friends attend it. The intuition is that the closer the relationship between of them, the greater the impact of the choices of RSVPs(Reply, if you please). EBSNs link the online and offline social worlds [10]. For different users, the impact of online and offline social worlds is different. As the graphical model of HSPF shown in Fig. 1, HSPF captures this intuition because there are three parts in the model: user's preference θ_u, online social network influence δ_{uv}, and offline social network influence τ_{uf}. Then, we provide modeling details for the Heterogeneous Social Poisson Factorization(HSPF).

Heterogeneous Social Poisson Factorization(HSPF). We assume that the rating y_{ui} of a (user u, event i) pair equals one if the choice of RSVP is yes, i.e., the user u attends event i, and is zero otherwise. The relation r_{uv} of a (user u, online friend v) pair is regarded as the degree of online closeness between them, which can be computed by the number of groups that both of them belong to. The relation r_{uf} of a (user u, offline friend f) pair is considered as the degree of offline closeness between them, which is in proportion to the number of events that both of them attend. Each user u is represented by a vector of K latent preference θ_u and each event i by a vector of K latent attributes β_i. $N^{on}(u)$ and $N^{off}(u)$ are the set of online and offline social friends of user u respectively. δ_{uv} and τ_{uf} are the influences of online friend v and offline friend f respectively.

The distribution of the observation y_{ui} is denoted by

$$y_{ui}|y_{\neg ui} \sim Poisson(\theta_u^T\beta_i + \sum_{f \in N^{off}(u)} \tau_{uf}r_{uf}y_{fi} + \sum_{v \in N^{on}(u)} \delta_{uv}r_{uv}y_{vi}), \quad (1)$$

where $y_{\neg ui}$ denotes the responses of other users. To complete the specification of the variables, we place Gamma priors with shape and rate parameters on the user's preference θ_{uk}, item's attribute β_{ik}, online social network influence δ_{uv}, and offline social network influence τ_{uf}. This is because that Gamma distribution

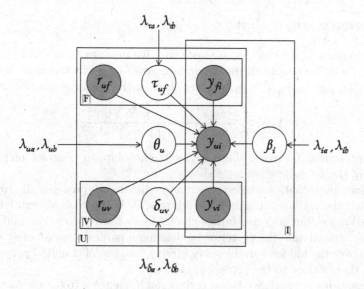

Fig. 1. Graphical model of the Heterogeneous Social Poisson Factorization(HSPF)

is the conjugate prior for Poisson distribution. It is very facilitate to Bayesian learning of model parameters. Furthermore, the Gamma prior encourages sparse representations of users, events and influences. More specifically, by setting the shape parameters, i.e., $\lambda_{ua}, \lambda_{ia}, \lambda_{\tau a}$ and $\lambda_{\delta a}$, to be small (e.g., 0.3), most of the generated values will be close to zero. Note that $\lambda_{ub}, \lambda_{ib}, \lambda_{\tau b}$ and $\lambda_{\delta b}$ are rate parameters.

Based on the above description, the generative process of the Heterogeneous Social Poisson Factorization(HSPF) is as follows:

(1) For each user $u = 1, ..., U$ and each component $k = 1, ..., K$, draw latent factor
$$\theta_{uk} \sim Gamma(\lambda_{ua}, \lambda_{ub}).$$

(2) For each event $i = 1, ..., I$ and each component $k = 1, ..., K$, draw latent factor
$$\beta_{ik} \sim Gamma(\lambda_{ia}, \lambda_{ib}).$$

(3) For each influence δ_{uv} of (user u, online-friend v) pair, draw latent factor
$$\delta_{uv} \sim Gamma(\lambda_{\delta a}, \lambda_{\delta b}).$$

(4) For each influence τ_{uf} of (user u, offline-friend f) pair, draw latent factor
$$\tau_{uf} \sim Gamma(\lambda_{\tau a}, \lambda_{\tau b}).$$

(5) For each (user u, event i) pair, draw response y_{ui} through Eq. (1).

5   Inference Algorithm

The key inferential problem that we need to solve in order to use HSPF is posterior inference. We denote all the Gamma priors of latent factors with λ. Given $\Theta = \{\theta_{uk}, \beta_{ik}, \delta_{uv}, \tau_{uf}\}$, $Y = \{y_{ui}, r_{uv}, r_{uf}\}$, then the posterior distribution,

$$p(\Theta|Y, \lambda) = \frac{p(\Theta, Y|\lambda)}{p(Y|\lambda)} = \frac{p(\Theta, Y|\lambda)}{\int p(\Theta, Y|\lambda)d\Theta} \qquad (2)$$

which is intractable for exact inference due to the coupling between integration variables of the normalization term shown in Eq. (2).

We adopt the variational inference to approximate the posterior distribution because that the variational inference tends to scale better than alternative sampling based algorithm such as Markov chain Monte Carlo. Variational inference algorithms approximate the posterior by defining a parameterized family of distributions over the hidden variables, i.e., $q(\Theta|\gamma)$, and then fitting the parameters of $q(\Theta|\gamma)$ that is close to the posterior, i.e., $p(Y|\lambda)$.

The basic idea of convexity-based variational inference is to take advantage of Jensen's inequality to obtain an evidence lower bound on the log likelihood [9].

$$\begin{aligned} logp(Y|\lambda) \\ = log \int p(\Theta, Y|\lambda)d\Theta \\ = log \int \frac{q(\Theta|\gamma)}{q(\Theta|\gamma)}p(\Theta, Y|\lambda)d\Theta \\ \geq \int q(\Theta|\gamma)logp(\Theta, Y|\lambda)d\Theta - \int q(\Theta|\gamma)logq(\Theta|\gamma)d\Theta \\ = E_q[logp(\Theta, Y|\lambda)] - E_q[logq(\Theta|\gamma)] \\ = \mathcal{L}(\gamma; \lambda) \end{aligned} \qquad (3)$$

The difference between $logp(Y|\lambda)$ and $\mathcal{L}(\gamma; \lambda)$ of Eq. (3) is the Kullback-Leibler(KL) divergence between $q(\Theta|\gamma)$ and $p(Y|\lambda)$, that is

$$logp(Y|\lambda) = \mathcal{L}(\gamma; \lambda) + D(q(\Theta|\gamma)||p(Y|\lambda)). \qquad (4)$$

This shows that maximizing the evidence lower bound $\mathcal{L}(\gamma; \lambda)$ equivalent to minimizing the KL divergence, thus the problem of posterior inference becomes an optimization problem.

5.1   Auxiliary Variables

Our inference algorithm for HSPF makes use of general results about the class of conditionally conjugate models [5,7]. We first give an alternative formulation of HSPF in which we add some auxiliary variables to facilitate derivation and description of the algorithm. Without the auxiliary variables, HSPF is not conditionally conjugate model.

Note that a sum of independent Poisson random variables is itself a Poisson with rate equal to the sum of the rates. We introduce the auxiliary latent variables z_{uik}^M, z_{uif}^{off} and z_{uiv}^{on} for each (user u, event i) pair, each (user u, offline friend f) pair and each (user u, online friend v) pair respectively such that

$$y_{ui}|y_{\neg ui} = \sum_{k=1}^{K} z_{uik}^M + \sum_{f=1}^{F} z_{uif}^{off} + \sum_{v=1}^{V} z_{uiv}^{on}, \tag{5}$$

where

$$z_{uik}^M \sim Poisson(\theta_{uk}\beta_{ik}),$$
$$z_{uif}^{off} \sim Poisson(\tau_{uf}r_{uf}y_{fi}),$$
$$z_{uiv}^{on} \sim Poisson(\delta_{uv}r_{uv}y_{vi}),$$

and $F = |N^{off}(u)|$, $V = |N^{on}(u)|$.

After adding the auxiliary variables, the variational distribution, i.e., $q(\Theta|\gamma)$, turns to $q(\Theta, Z|\gamma, \phi)$ where Z and ϕ denote all the auxiliary variables and added parameters respectively.

5.2 Mean-Field Variational Family

We resort to the mean-field variational family, where each latent variable is independent and governed by its own variational parameter. Omitting the parameters γ and ϕ for simplicity, the mean-field variational family is

$$q(\Theta, Z) = \prod_{u,k} q(\theta_{uk}) \prod_{i,k} q(\beta_{ik}) \prod_{u,f} q(\tau_{uf}) \prod_{u,v} q(\delta_{uv}) \prod_{u,i,k} q(z_{uik}),$$

Each factor in the mean-field family usually is set to the same type of distribution as its *complete conditional* [2,5,6,20]. A *complete conditional* which is the conditional distribution of a latent variable given the observations and other latent variables in the model.

Complete Conditional. Firstly, we compute the *complete conditionals* of all the latent variables in the model. For the user preferences θ_{uk}, the *complete conditional* is a Gamma shown as follows,

$$p(\theta_{uk}|\lambda, \beta, \tau, \delta, z, y)$$
$$\propto p(\theta_{uk}|\lambda_{ua}, \lambda_{ub}) \prod_i p(z_{uik}^M|\theta_{uk}, \beta_{ik})$$
$$= Gamma(\theta_{uk}; \lambda_{ua}, \lambda_{ub}) \prod_i Poisson(z_{uik}^M; \theta_{uk}, \beta_{ik})$$
$$= Gamma(\lambda_{ua} + \sum_i z_{uik}^M, \lambda_{ub} + \sum_i \beta_{ik}). \tag{6}$$

We can similarly derive the *complete conditionals* for β_{ik}, τ_{uf}, and δ_{uv}.

$$\beta_{ik}|\lambda, \theta, \tau, \delta, z, y \sim Gamma(\lambda_{ia} + \sum_u z_{uik}^M, \lambda_{ib} + \sum_u \theta_{uk}), \tag{7}$$

$$\tau_{uf}|\lambda, \theta, \beta, \delta, z, y \sim Gamma(\lambda_{fa} + \sum_i z_{uif}^{off}, \lambda_{fb} + \sum_i r_{uf} y_{fi}); \tag{8}$$

$$\delta_{uv}|\lambda, \theta, \beta, \tau, z, y \sim Gamma(\lambda_{va} + \sum_i z_{uiv}^{on}; \lambda_{vb} + \sum_i r_{uv} y_{vi}) \tag{9}$$

The conditional distribution of a set of Poisson variables, given their sum, is a multinomial for which the parameter is their normalized set of rates [1]. So the *complete conditionals* for $z_{ui} = (z_{ui}^M, z_{ui}^{off}, z_{ui}^{on})$ is multinomial, i.e., $z_{ui} \sim Mult(y_{ui}, \psi_{ui})$, where $\psi_{ui} = (\psi_{ui}^M, \psi_{ui}^{off}, \psi_{ui}^{on})$ is a point in the $(K + F + V)$-simplex, i.e.,

$$\psi_{ui}^M \propto \langle \theta_{u1}\beta_{i1}, ..., \theta_{uK}\beta_{iK}\rangle,$$
$$\psi_{ui}^{off} \propto \langle \tau_{u1}r_{u1}y_{1i}, ..., \tau_{uF}r_{uF}y_{Fi}\rangle,$$
$$\psi_{ui}^{on} \propto \langle \delta_{u1}r_{u1}y_{1i}, ..., \delta_{uV}r_{uV}y_{Vi}\rangle,$$

Note that the parameters of $\psi_{ui}^M, \psi_{ui}^{off}$ and ψ_{ui}^{on} should be normalized together to ensure their sum to be one.

Variational Parameter. Then, we set each factor in the mean-field family to be the same type of distribution as its complete conditional.

The complete conditionals of $\theta_{uk}, \beta_{ik}, \delta_{uv}$, and τ_{uf} are Gamma distributions, so their variational parameters are Gamma parameters, i.e., variational distributions for $\theta_{uk}, \beta_{ik}, \tau_{uf}$, and δ_{uv} are $Gamma(\gamma_{uk}^{shp}, \gamma_{uk}^{rte})$, $Gamma(\gamma_{ik}^{shp}, \gamma_{ik}^{rte})$, $Gamma(\gamma_{uf}^{shp}, \gamma_{uf}^{rte})$, and $Gamma(\gamma_{uv}^{shp}, \gamma_{uv}^{rte})$ respectively. We can similarly deduce the variational distribution for $z_{ui} = (z_{ui}^M, z_{ui}^{off}, z_{ui}^{on})$ is $Mult(y_{ui}, \phi_{ui})$, where $\phi_{ui} = (\phi_{ui}^M, \phi_{ui}^{off}, \phi_{ui}^{on})$.

We set each variational parameter equal to the expected parameter (under q) of the complete conditional because of the conditionally conjugate models [7].

For variational Gamma distributions, we take θ_{uk} as an example to derive the close-form update solution of parameters, i.e., shape parameter γ_{uk}^{shp} and rate parameter γ_{uk}^{rte}.

$$\gamma_{uk}^{shp} = E_q[\lambda_{ua} + \sum_i z_{uik}^M] = \lambda_{ua} + \sum_i y_{ui}\phi_{uik}^M, \tag{10}$$

$$\gamma_{uk}^{rte} = E_q[\lambda_{ub} + \sum_i \beta_{ik}] = \lambda_{ub} + \sum_i \frac{\gamma_{ik}^{shp}}{\gamma_{ik}^{rte}}, \tag{11}$$

The update solutions for the parameters of β_{ik}, τ_{uf}, and δ_{uv} are similarly derived, we omit the details and provide the final results.

$$\gamma_{ik}^{shp} = \lambda_{ia} + \sum_u y_{ui}\phi_{uik}^M, \tag{12}$$

$$\gamma_{ik}^{rte} = \lambda_{ib} + \sum_u \frac{\gamma_{uk}^{shp}}{\gamma_{uk}^{rte}}, \tag{13}$$

$$\gamma_{uf}^{shp} = \lambda_{fa} + \sum_i y_{uf}\phi_{uif}^{off}, \tag{14}$$

$$\gamma_{uf}^{rte} = \lambda_{fb} + \sum_i r_{uf}y_{fi}, \tag{15}$$

$$\gamma_{uv}^{shp} = \lambda_{va} + \sum_i y_{uv}\phi_{uiv}^{on}, \tag{16}$$

$$\gamma_{uv}^{rte} = \lambda_{vb} + \sum_i r_{uv}y_{vi}. \tag{17}$$

For variational multinomial distribution, we take z_{ui}^M as an example to derive update solution of parameter, i.e., ϕ_{ui}^M.

$$
\begin{aligned}
\phi_{uik}^M &\propto G_q[\theta_{uk}\beta_{ik}] \\
&= exp\{E_q[log(\theta_{uk}) + log(\beta_{ik})]\} \\
&= exp\{E_q[log(\theta_{uk})] + E_q[log(\beta_{ik})]\} \\
&= exp\{\Psi(\gamma_{uk}^{shp}) - log\gamma_{uk}^{rte} + \Psi(\gamma_{ik}^{shp}) - log\gamma_{ik}^{rte})\},
\end{aligned} \tag{18}
$$

where $k = 1, ..., K$; $\Psi(\cdot)$ is the digamma function and $G_q[\cdot] = exp(E_q[log(\cdot)])$ denotes the geometric expectation [17]. Similarly,

$$\phi_{uif}^{off} \propto exp\{\Psi(\gamma_{uf}^{shp}) - log\gamma_{uf}^{rte}\} + r_{uf}y_{fi}, \tag{19}$$

$$\phi_{uiv}^{on} \propto exp\{\Psi(\gamma_{uv}^{shp}) - log\gamma_{uv}^{rte}\} + r_{uv}y_{vi}, \tag{20}$$

where $f = 1, ..., F$ and $v = 1, ..., V$. Note that the parameters of $\phi_{ui}^M, \phi_{ui}^{off}$ and ϕ_{ui}^{on} should be normalized together to ensure their sum to be one.

5.3   Coordinate Ascent Algorithm

The coordinate ascent algorithm is illustrated in Fig. 2, which iteratively optimize each variational parameter while holding the others fixed.

5.4   Event Recommendation Using HSPF

Once the posterior is fit, we use the HSPF to recommend events, which are unattend and ongoing events, to users who will like to attend. Firstly, we compute the predicting scores $\hat{y}_{ui}$ (i.e., Eq. 21) for each (user u, unattend and ongoing event i) pair by their posterior expected Poisson parameters. Then, we rank the events for each user using the scores. Lastly, Top-N events will be recommended to each user.

$$
\begin{aligned}
\hat{y}_{ui} &= E_q[\theta_u^T\beta_i + \sum_{f\in N^{off}(u)} \tau_{uf}r_{uf}y_{fi} + \sum_{v\in N^{on}(u)} \delta_{uv}r_{uv}y_{vi}] \\
&= \sum_{k=1}^K E_q[\theta_{uk}\beta_{ik}] + \sum_{f\in N^{off}(u)} E_q[\tau_{uf}]r_{uf}y_{fi} + \sum_{v\in N^{on}(u)} E_q[\delta_{uv}]r_{uv}y_{vi} \\
&= \sum_{k=1}^K \frac{\gamma_{uk}^{shp}}{\gamma_{uk}^{rte}}\frac{\gamma_{ik}^{shp}}{\gamma_{ik}^{rte}} + \sum_{f\in N^{off}(u)} \frac{\gamma_{uf}^{shp}}{\gamma_{uf}^{rte}}r_{uf}y_{fi} + \sum_{v\in N^{on}(u)} \frac{\gamma_{uv}^{shp}}{\gamma_{uv}^{rte}}r_{uv}y_{vi}
\end{aligned} \tag{21}
$$

Initialize parameters of variational Gamma distributions.
Repeat until convergence:

1. For each observation $y_{ui} > 0$ of (user u, event i) pair, update the multinomial using Equation (18), (19) and (20).
2. For each user u, update the parameter using Equation (10) and (11).
3. For each event i, update the parameter using Equation (12) and (13).
4. For influence of each offline friend f, update the parameter using Equation (14) and (15).
5. For influence of each online friend v, update the parameter using Equation (16) and (17).

Fig. 2. The coordinate ascent algorithm for HSPF

6   Experiments

6.1   Dataset

We conduct experiments on the Meetup dataset [13]. The dataset contains all public activities on Meetup from January, 2010 to April, 2014, which we use to evaluate the HSPF model proposed in this work. We select two cities located in the USA, namely Chicago and Phoenix Table 1.

Table 1. Statistics of the Meetup dataset.

City	$|G|$	$|U|$	$|E|$	RSVPs
Chicago	2,138	133,357	100,701	810,213
Phoenix	842	43,112	64,255	326,913

6.2   Setup and Metrics

We select data during each of last 10 months as the test set for each city. The data during preceding 6 months of each test set is regarded as the corresponding training set. We firstly compute the scores for each (user u, event i) pair in the test set, then sort the triplets by these scores according to user, lastly recommend Top-N events to each user. We report the average value of the predicted performance over the 10 test set. We choose NDCG@N and Precision@N as the metrics.

6.3   Comparison Methods

To demonstrate the effectiveness of our HSPF model, we compare it to the following state-of-the-art recommenders:

- **CMF** [18] solves the recommendation task by reconstructing multiple relation matrices which are associated with some shared elements. In this work, the users are the shared elements.
- **FM** [16] a general predictor and is easily applicable to a wide variety of contexts, which include user, event, online and offline social networks in Meetup dataset, by specifying only the input data.
- **BPF**, proposed by Gopalan [5], is a probabilistic model of users and items for recommendation. BPF can handle sparse data well and models the long-tail of users and items.
- **SPF-on** and **SPF-off** stem from SPF [2] which aims to bridge the gap between preference- and social-based recommendations. However, SPF only incorporates one kind of social networks information into a Poisson factorization method and can't capture the tie strength of friend relationships. **SPF-on** and **SPF-off** means only incorporates online and offline social networks respectively.
- **HSPF-on** and **HSPF-off** are two variants of our HSPF model and can be considered as two special cases of our model. The difference of them is that they only consider the influence of online and offline social networks respectively.

6.4 Performance Comparison

For HSPF, the latent dimensionality, i.e., K, is set to 50 for both Chicago and Phoenix dataset. The Gamma priors parameters, i.e., shape and rate parameters, of all the latent factors are fixed to be 0.3.

Effectiveness Comparison. The detailed comparison results are shown in Fig. 3. From the Fig. 3, we can observe that:

- The performance of **HSPF,CMF** and **FM** are better than **SPF-off** and **SPF-on**. The reason is that either **SPF-off** or **SPF-on** can only integrate one kind of social networks. The performance of event recommendation can be improved by integrating the heterogeneous social networks.

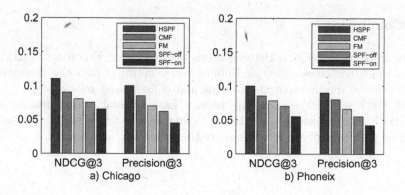

Fig. 3. Comparisons of different methods on NDCG@N and Precision@N

- The proposed **HSPF** model achieves the best performance comparing to the other models. It can be explained that the user responses to events are implicit feedback [5] while HSPF is more suitable for modeling implicit user feedback.

Factor Contribution. From the detailed comparison results which are shown in Fig. 4, we can draw the conclusions that:

- The performance of **HSPF,HSPF-off** and **HSPF-on** outperforms that of **BPF** which can be considered as a special case of HSPF without social network information. It shows that the performance of recommender systems can be improved by integrating the social network information into the model.
- **HSPF-off** achieves better results than **HSPF-on**, which means that the influence of offline social networks is greater than the online social network. This explains that face to face communication helps to become a true friend.

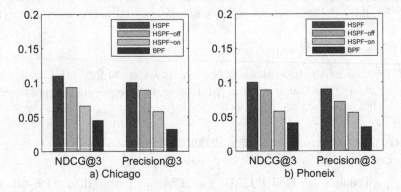

Fig. 4. Factor contribution of HSPF model

7   Conclusion

In this work, we propose a Heterogeneous Social Poisson Factorization(HSPF) model which combines online and offline social networks into one framework, and integrates the tie strength of online and offline friend relationships to the model. We test HSPF on Meetup dataset. Experimental results demonstrate that HSPF outperforms state-of-the-art recommendation methods. In the future work, we plan to exploit the influence of time, location, and content.

References

1. Cemgil, A.T.: Bayesian inference for nonnegative matrix factorisation models. Comput. Intell. Neurosci. **2009** (2009)
2. Chaney, A.J., Blei, D.M., Eliassi-Rad, T.: A probabilistic model for using social networks in personalized item recommendation. In: Proceedings of the 9th ACM Conference on Recommender Systems, pp. 43–50. ACM (2015)
3. Du, R., Yu, Z., Mei, T., Wang, Z., Wang, Z., Guo, B.: Predicting activity attendance in event-based social networks: content, context and social influence. In: Proceedings of the 2014 ACM International Joint Conference on Pervasive and Ubiquitous Computing, pp. 425–434. ACM (2014)
4. Duan, R., Goh, R.S.M., Yang, F., Tan, Y.K., Valenzuela, J.F.: Towards building and evaluating a personalized location-based recommender system. In: 2014 IEEE International Conference on Big Data (Big Data), pp. 43–48. IEEE (2014)
5. Gopalan, P., Hofman, J.M., Blei, D.M.: Scalable recommendation with poisson factorization (2013). arXiv preprint arXiv:1311.1704
6. Gopalan, P.K., Charlin, L., Blei, D.: Content-based recommendations with poisson factorization. In: Advances in Neural Information Processing Systems, pp. 3176–3184 (2014)
7. Hoffman, M.D., Blei, D.M., Wang, C., Paisley, J.: Stochastic variational inference. J. Mach. Learn. Res. **14**(1), 1303–1347 (2013)
8. Jamali, M., Ester, M.: A matrix factorization technique with trust propagation for recommendation in social networks. In: Proceedings of the Fourth ACM Conference on Recommender systems, pp. 135–142. ACM (2010)
9. Jordan, M.I., Ghahramani, Z., Jaakkola, T.S., Saul, L.K.: An introduction to variational methods for graphical models. Mach. Learn. **37**(2), 183–233 (1999)
10. Liu, X., He, Q., Tian, Y., Lee, W.C., McPherson, J., Han, J.: Event-based social networks: linking the online and offline social worlds. In: Proceedings of the 18th ACM SIGKDD International Conference on Knowledge Discovery and Data Mining, pp. 1032–1040. ACM (2012)
11. Ma, H., Yang, H., Lyu, M.R., King, I.: Sorec: social recommendation using probabilistic matrix factorization. In: Proceedings of the 17th ACM Conference on Information and Knowledge Management, pp. 931–940. ACM (2008)
12. Ma, H., Zhou, D., Liu, C., Lyu, M.R., King, I.: Recommender systems with social regularization. In: Proceedings of the Fourth ACM International Conference on Web Search and Data Mining, pp. 287–296. ACM (2011)
13. Macedo, A.Q., Marinho, L.B., Santos, R.L.: Context-aware event recommendation in event-based social networks. In: Proceedings of the 9th ACM Conference on Recommender Systems, pp. 123–130. ACM (2015)
14. Pham, T.A.N., Li, X., Cong, G., Zhang, Z.: A general graph-based model for recommendation in event-based social networks. In: 2015 IEEE 31st International Conference on Data Engineering (ICDE), pp. 567–578. IEEE (2015)
15. Qiao, Z., Zhang, P., Cao, Y., Zhou, C., Guo, L., Fang, B.: Combining heterogenous social and geographical information for event recommendation. In: Twenty-Eighth AAAI Conference on Artificial Intelligence (2014)
16. Rendle, S., Gantner, Z., Freudenthaler, C., Schmidt-Thieme, L.: Fast context-aware recommendations with factorization machines. In: Proceedings of the 34th International ACM SIGIR Conference on Research and Development in Information Retrieval, pp. 635–644. ACM (2011)

17. Schein, A., Paisley, J., Blei, D.M., Wallach, H.: Bayesian poisson tensor factorization for inferring multilateral relations from sparse dyadic event counts. In: Proceedings of the 21th ACM SIGKDD International Conference on Knowledge Discovery and Data Mining, pp. 1045–1054. ACM (2015)

18. Singh, A.P., Gordon, G.J.: Relational learning via collective matrix factorization. In: Proceedings of the 14th ACM SIGKDD International Conference on Knowledge Discovery and Data Mining, pp. 650–658. ACM (2008)

19. Yang, X., Guo, Y., Liu, Y., Steck, H.: A survey of collaborative filtering based social recommender systems. Comput. Commun. **41**, 1–10 (2014)

20. Zhang, W., Wang, J.: A collective bayesian poisson factorization model for cold-start local event recommendation. In: Proceedings of the 21th ACM SIGKDD International Conference on Knowledge Discovery and Data Mining, pp. 1455–1464. ACM (2015)

21. Zhou, T., Shan, H., Banerjee, A., Sapiro, G.: Kernelized probabilistic matrix factorization: exploiting graphs and side information. In: SDM, vol. 12, pp. 403–414. SIAM (2012)

Top-k Temporal Keyword Query
over Social Media Data

Fan Xia, Chengcheng Yu, Weining Qian$^{(\boxtimes)}$, and Aoying Zhou

ECNU-RMU-Infosys Data Science Joint Lab, Institute for Data Science
and Engineering, East China Normal University, Shanghai, China
{52101500012,52111500011}@ecnu.cn,
{wnqian,ayzhou}@sei.ecnu.edu.cn

Abstract. Analytic jobs over social media data typically need to explore data of different periods. However, most existing keyword search work merely use creation time of items as the measurement of their recency. In this paper we propose top-k temporal keyword query that ranks data by their aggregate sum of shared times during the given time window. A query algorithm that can be executed over a general temporal inverted index is provided. The complexity analysis based on the power law distribution reveals the upper bound of accessed items. Furthermore, two-tiers structure and piecewise maximum approximation sketch are proposed as refinements. Extensive empirical studies on a reallife dataset show the combination of two refinements achieves remarkable performance improvement under different query settings.

Keywords: Social media · Temporal keyword query · Top-k query

1 Introduction

Recent years we have witnessed social media service to become one of the main sources of accessing information on the web. Apart from default publish-subscribe mechanism, social media services also provide keyword search as an easy-to-use tool to retrieve data. Yet most of recent work on keyword search infrastructure of social media data utilize temporal information to evaluate recency of results. Those work are based on the assumption that people concern more about recent popular items.

However, a kind of more complex analytic query that promotes the temporal dimension as the first class attribute is desired. Marketers need to analyze opinion of consumers during the campaign of some product. Journalists need to find out hot topics of different periods to portray the development of some event. In those scenarios, the query that filters items by keywords and ranks results by their importance during the time window is more handy. In December 2014, Twitter announced in a blog article [14] that they have built a full tweet index to support searching tweets posted during the past 8 years. However, it is not clear how to account for retweets in their system. Furthermore, it seems that

© Springer International Publishing Switzerland 2016
F. Li et al. (Eds.): APWeb 2016, Part I, LNCS 9931, pp. 183–195, 2016.
DOI: 10.1007/978-3-319-45814-4_15

only tweets published within the query window áre returned. We argue that an item should be visible throughout its lifespan.

In this paper we propose top-k temporal keyword query over social media data that returns top k social items, but ranks them by the number of times they are shared during the query window. The main contributions of this paper are listed as following:

1. We propose the query algorithm based on a general temporal inverted index, which is the baseline index. Based on the long tail distribution, a complexity analysis is given to bound the maximum number of social items that must be accessed. The statistic of our real-life dataset shows our solution can scale quite well to the value of k for most queries.
2. Two techniques are proposed to improve the performance of the baseline solution. First, the two-tiers posting list with two different disk structures is designed to deter trivial items from polluting tier-one. Second, a piecewise maximum approximation sketch is proposed to prune unpromising candidates at early stage.
3. Extensive experiments have been conducted on a real-life microblog dataset. The empirical study shows that the combination of proposed techniques achieves significant performance improvement under different query setups.

The rest part of the paper is organized as follows. The related work is given in Sect. 2. Then we present the definition of our problem in Sect. 3. Section 4 describes details of index structure. Query algorithm are proposed in Sect. 5 and a complexity analysis is also provided. The PMA sketch is also defined. Empirical studies with a reallife dataset are given in Sect. 6. We conclude the paper in Sect. 7.

2   Related Work

At the first glance, the problem we study is quite similar to time-travel text search [1,3,6,7] in Internet archives. Each document in the archive contains multiple versions and the importance of each term in the document, such as the frequency, keeps constant during the timespan of each version. [3,6,7] only emphasize on efficiency of retrieving document versions intersecting with a time point or time window. Among those work, [1,7] are the most related to ours. They apply aggregation functions, e.g. min, max and avg, to scores of selected versions for each document. However, those version-ed data can be viewed as a special case of social items associated with time series in some sense.

The other closely related area is the top-k query on temporal data proposed in [8,11]. Temporal objects are tuples that contain some fields whose value change over time, which is formally named as time series. This problem of ranking temporal data orders the tuple in terms of the value computed by some aggregated function, e.g. sum, on a temporal field and returns top k tuples. [8] studies an approximate algorithm that selects time points called as breakpoint and pre-computes top-k tuples for time windows composing of those breakpoints.

Due to the ad-hoc combination of keywords and time window, using their approximation method to materialize top k items for each keyword may not produce enough results for the combination of keywords.

A bunch of work [4,13] have been devoted to the problem of keyword search in social media data. Both [4,13] use log-structured merge tree to the leverage the high performance of memory access and sequential write to disk. On the other hand, [4] postpones indexing trivial tweets by serving popular tweets as first citizen to keep the current index collection filled with hot tweets related to hot queries. The recency is not the most urgent problem in our case and those techniques can be combined to query recent data.

3 Problem Definition

In this paper we formalize social item p_i as a tuple $< i, t, c, o >$, where i is the item's identifier and t is the timestamp of creation time. Field c represents the content of the item, which is simply modeled by a subset of the overall vocabulary set Ω. o equals the identifier of its parent or -1 if it is an original item. It captures the linkage relationship among social items. Let symbol $p_i \rightarrow p_j$ denote that social item p_j directly shares social item p_i, i.e. $p_j.o = i$. Furthermore, social item p_j originates from social item p_i, denoted by $p_i \rightsquigarrow p_j$, if there exists a linkage path from p_i to p_j. At last we define the linkage tree rooted at item p_i with the node set $\{I_m | m = i \vee p_i \rightsquigarrow I_m\}$ and the edge set $\{I_m \rightarrow I_n | m = i \vee p_i \rightsquigarrow I_m\}$.

The function $TSize$ in Eq. 1 calculates the number of p_i's descendants that are created during the time window. Given the minimal time unit, e.g., seconds, minutes and hours, $TSize$ is used to obtain a time series data $\{v_{i,1}, v_{i,2}, ..., v_{i,n}\}$. The time series reflects the evolution of an item's importance.

$$TSize(p_i, [t_s, t_e]) = |\{m | p_i \rightsquigarrow p_j \wedge t_s <= p_j.t <= t_e\}| \tag{1}$$

The **top-k temporal keyword query over social media data** consists of three parts: a keyword set K, a temporal window $W[t_s, t_e]$ and an integer value k. The content of k items returned by the query should contain all keywords in K. Given the time window $W[t_s, t_e]$, the aggregate sum over time series of items can be computed. This aggregation score, which is given by Eq. 2, can be used to measure the temporal social influence of an item. Our proposed query aims to find out k social items with the largest aggregate score.

$$TInf(p_i, [t_s, t_e]) = \sum_{t_s <= t_j <= t_e} v_{i,j} \tag{2}$$

As time series data still occupy too much storage space, they are further approximated by segmentation techniques. In terms of segmentation, the time series will be segmented into consecutive non-intersect sub-sequence $\{I_1, I_2, ...\}$, where each sub-sequence I_i is approximated by some chosen function. There already exist plenty number of time series segmentation methods, e.g. piecewise linear function [9], piecewise polynomial function [5] and their combinations [10]. It has also been studied in [12] that the piecewise sigmoid function is quite suitable to fit the retweet time series of popular tweets. The problem of designing an approximation approach is beyond the scope of this paper. Any piecewise approximation approaches could be plugged into our solution.

4   Index Structure

Our query processing framework is illustrated by Fig. 1(a). It consists of three main components: the temporal inverted index, the segment store and the query processor. In this section we introduce first two components in details while leaving the last one to the next section.

4.1   Two-Tiers Structure

Partitioning Social Items. Each entry stored in the posting list is represented by a tuple $< i, t_s, t_e, ts >$, where i represents the identifier of a social item, t_s and t_e together constitute the lifespan of an item at the index construction time. The field ts stores the size of the linkage tree rooted at p_i, which is the maximum score the social item can get. The decreasing chronological order of $< ts, i >$ is used to sort social items by default. Currently we rely on the offline process to calculate the histogram for ts of items for each posting list. Given a configured partition ratio γ, the partition boundary β is selected such that the percentage of items with ts greater than β is about γ. Such an offline process could be implemented straightforwardly using the MapReduce framework. We maintain β in the meta data of a posting list. Both update and query rely on β to decide which tier an item will be assigned to.

Tier-One Posting List. The tier-one posting list manages items with ts greater than β. It is implemented by a variant of external interval tree [2], which is a disk based index structure managing intervals. Weighted B+tree is used as the base tree to index the endpoints of intervals. An inserting interval is assigned to the lowest node in B+tree whose subtree contains its both endpoints. The intervals themselves are stored in separate data structure which we call atomic

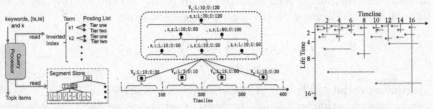

(a) Query Processing Framework (b) Structure of Tier-one Posting List (c) Illustration of Tier-two Posting List

Fig. 1. Temporal inverted index

structure for ease of discussion. Intervals stored in an internal node are further partitioned and organized as multislabs. A multislab $s_{i,j}$ of a node stores intervals whose start point and end point are contained in the subtrees rooted at ith and jth children respectively. Atomic structures are implemented by B+tree and stored in a single file. Leaf nodes and multslabs only store meta data pointing to corresponding B+tree.

Figure 1(b) gives an example of our interval tree variant for the tier-one posting list. Four leaf nodes of the base tree, i.e. v2, v3, v4 and v5, lie at the bottom while an internal node v1 represented by the dotted square lies at the upper part. Each black circle represents a reference to an atomic structure. We extend the external interval tree by maintaining two fields representing minimum and maximum bounds of ts in meta data of various components. For internal nodes of base tree, those values equal correspond value of ts of items stored in the subtree rooted at them. For leaf nodes and multislabs, those bounds are only determined by items assigned to them. We use fields L and U to represent those bounds in the figure.

Tier-Two Posting List. It has been studied the lifetime distribution of social items complies with power law distribution. Figure 2(c) shows the long tail phenomenon also exists in lifetime distribution of items in a popular posting list. As most items in a posting list are trivial, we assume the lifetime distribution of less significant items is still a power law. Based on such an assumption, those items are organized in sublists of multiple levels, which is illustrated by Fig. 1(c). Firstly, entries are partitioned into multiple levels according to their lifetime. The range of lifetime for items in the ith level is $(2^{i-1}, 2^i]$. Then we divide the time axis of the ith level into consecutive intervals of width 2^i. Thus the jth interval in ith level is $[(j-1) * 2^i, j * 2^i)$ and items with start time falling in that interval are stored in it. Each sublist can be implemented as a sequence file or B+tree sorted in descending order of $< ts, i >$.

We also index meta data of sublists using B+tree. The index key is the composition of their level and the interval ordinal in the corresponding level. Given the time window $[t_s, t_e]$, we can use B+tree to obtain meta data of sublists whose ordinal belongs to $[t_s/2^i - 1, t_e/2^i]$ for each level i. As the lifetime of all items in level i belongs to $(2^{i-1}, 2^i]$, it is possible that an item published later than $t_s - 2^i$ would also satisfy the query window. $(t_s - 2^i)/2^i$ equals the start ordinal $t_s/2^i - 1$. Basically, this multi-level structure can be viewed a coarse and static partition mechanisms compare to the external interval tree.

4.2   Managing Segmented Items

The time series of items are approximated by segments using segmentation techniques mentioned previously. Those segments are managed by the segment store to efficiently compute the accurate aggregation score for a given item. Let $s_{i,j}$ denote the jth segment of item p_i. It can be represented by a tuple $< i, t_{i,j}, v_{i,j}, t_{i,j+1}, v_{i,j+1} >$, where $t_{i,j}, v_{i,j}, t_{i,j+1}$ and $v_{i,j+1}$ represents coordinators of the segment's two endpoints. Besides, we also use prefix sum approach

[8] to extend each segment with an additional field $agg_{i,j}$. It equals aggregation sum of value at points before the time $t_{i,j}$. At last, B+tree is used to index those segments with $< i, t_{i,j} >$ as the key. Given an item p_i and a query window $W[t_s, t_e]$, we search the B+tree to find out the first segment $s_{i,s}$ $(s_{i,e})$ whose start time is not greater than t_s $(t_e$ respectively). Let $s_{i,j}(t)$ stand for the function that computes the aggregated sum of value for item i up to time point t. Equation 3 gives the detail of function $s_{i,j}(t)$ when linear function is used. Then the temporal importance of an item could be computed by Eq. 4. In case of null returned by B+tree, its correspond part in Eq. 4 is simply replaced by zero. It is easy to derive that the I/O cost to compute the aggregation for an item is $O(log_{B_p}(N_s/B_s))$, where B_p (B_s) is the maximum number of points (segments) stored in a disk block and N_s is the total number of segments.

$$s_{i,j}(t) = \frac{(v_{i,j+1} - v_{i,j}) * (t - t_{i,j})}{(t_{i,j+1} - t_{i,j})} + v_{i,j} + agg_{i,j} \tag{3}$$

$$score(id, i) = s_{i,e}(t_e) - s_{i,s}(t_s) \tag{4}$$

5   Query Processing

In this section we describe how to execute queries over the index structure and how to traverse in the two-tiers posting lists to prune the data space.

5.1   Querying the Index

Query Algorithm. Given a query time window, we define the virtual posting list as a view of the two-tiers posting list. It supports returning social items that satisfy the time window and are sorted in the items' default sort order. In addition, the virtual posting list provides $skipTo$ function to position itself to the first entry with value greater than the field ts of the passed item p_i. The function returns a true value if p_i appears in the posting list. Our query algorithm behaves similar to sort-merge-join algorithm. It polls the entry p_s with smallest ts from posting lists. Then it invokes $skipTo$ of remaining posting lists with p_s passed as the argument. if p_s appears in all posting lists, function $tweetHit$ is invoked to compute the aggregate score of p_s and update the current $topk$ answer. The algorithm terminates when the minimum aggregate score of items in $topk$ is not smaller than the maximum ts value of entries in all posting lists.

Complexity Analysis. Now we analyze the upper bound for the number of items needed to be accessed by the query algorithm. We assume the linkage size distribution of items in posting list and virtual posting list complies with power law distribution. Figure 2(b) plots the #retweet distribution of a posting list. The model $f(x) = \frac{N}{x^\alpha}$ is used to represent the power law distribution. Let the model parameters of both posting lists are presented by N_o, α_o and N_q, α_q respectively. Then the upper bound for the number of accessed items is given by Theorem 1.

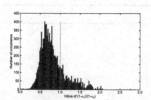 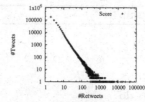 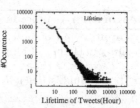

(a) The distribution of the exponent $(1 - \alpha_o)/(1 - \alpha_q)$ (b) The #retweet Distribution of tweets containing a popular word   (c) The lifetime Distribution of tweets containing a popular word

Fig. 2. Statistics of the Sina Weibo dataset

Theorem 1. *Given k of the top-k query, the number of visited items is upper bounded by $\frac{N_o}{\alpha_o - 1}(\frac{k(1-\alpha_q)}{N_q})^{\frac{1-\alpha_o}{1-\alpha_q}}$.*

Proof. Let s denote the score of the kth item in the answer set. Equation 5 computes s by integrating the distribution function of the virtual posting list.

$$k = \int_s^\infty \frac{N_q}{x^{\alpha_q}} dx = \frac{N_q}{(1 - \alpha_q)} x^{1-\alpha_q}|_s^\infty = \frac{N_q}{(\alpha_q - 1)} s^{1-\alpha_q} \tag{5}$$

By solving the integration, we obtain the value of s represented by Eq. 6.

$$s = (\frac{k(\alpha_q - 1)}{N_q})^{1/(1-\alpha_q)} \tag{6}$$

According to our algorithm, those temporally intersected items with ts greater than s are visited. Equation 7 counts all items with ts greater than s, hence derives an upper bound for the number of accessed items.

$$\#items = \int_s^\infty \frac{N_o}{x^{\alpha_o}} dx = \frac{N_o}{(\alpha_o - 1)} s^{1-\alpha_o} = \frac{N_o}{(\alpha_o - 1)} (\frac{k(\alpha_q - 1)}{N_q})^{\frac{1-\alpha_o}{1-\alpha_q}} \tag{7}$$

$\square$

N_o, α_o, N_q and α_q are fixed once keyword set and the time window are provided. We study the distribution of exponent $(1 - \alpha_o)/(1 - \alpha_q)$ for queries used in our experiment to study the complexity of the algorithm in terms of k. First, we choose top ten thousands most popular keyword pairs to compose keyword part of queries. The time window are generated in a manner similar to the experiment studying influence of k. Then the aggregate score distribution of items satisfying the constraints of keywords and time window is fitted by the power law distribution. We choose maximum value of αs corresponding to two keyword pairs as α_o. Figure 2(a) shows the distribution of exponent, which is akin to a Poisson distribution. It is noteworthy that a large fraction of those exponents are smaller than one and most of them are smaller than 2. It means the algorithm scales quite well in terms of k in most cases.

Algorithm 1. procedure skipTo

   Input: Item t of form $< i, t_s, t_e, ts >$

   Output: Item r of form $< i, t_s, t_e, ts >$

1 **if** $t.score < \beta$ **then**

2 | loadSubListRef();

3 **while** $qrs.first().U >= t.ts$ **do**

4 | ref = qrs.poll();

5 | **if** $ref.L <= t.ts$ **then**

6 | | loadRef(ref);

7 **while** $qc.first().U >= t.ts$ **do**

8 | c = qc.poll();

9 | **if** $c.skipTo(t)$ **then**

10 | | r = c.next();

11 | **if** $c.hasNext()$ **then**

12 | | qc.insert(c);

13 **return** r

5.2 Traversing in the Two-Tiers Posting List

The posting list cursor exposes procedure $skipTo$ defined in Algorithm 1 to guide traverse in the two-tiers structure. According to previous introduction, there exists various meta data referring to nodes of base tree, atomic structures and sublists. Those data are stored in priority queue qrs in decreasing order of upper bound U. Another queue qc is used to store cursors for atomic structure and sublists. Each cursor also exposes upper bound field U that equals ts of next un-accessed items.

First the procedure checks whether item t may be stored in tier-two by comparing its score with β in line 1. The subroutine $LoadSubListRef$ will load meta data of all satisfying sublists into qrs according to the logic introduced in the previous section. Then we find out all meta data whose bounds contain ts of t. $loadRef$ will insert all meta data of temporally intersected multislabs and children into qrs if ref points to an internal nodes. Otherwise, a cursor for atomic structure or sublist is opened and inserted into qc. Finally, we visit all cursors in qc which may contain t. The condition is given in line 7. The procedure $skipTo$ of atomic structure cursor locates its position to the first item that is not smaller than t. It can be implemented by the search algorithm of B+tree. A true value returned by the function indicates t is contained in the current atomic structure.

Let T' be the number of tweets that have been passed to the $tweetHit$ procedure before query algorithm terminates. In the worst case, all nodes in the base B+tree that intersects with query window W are visited. Let $|W|$ represent the width of W. We assume each time points exist and the number of blocks visited is $O(|W|/B_p)$. It also takes at most $O(T_i/B_s)$ I/Os to visit items in atomic structure, where T_i represents the number items visited in posting list i. For each candidate passed to $tweetHit$, it takes $O(log_{B_p}(N_s/B_s))$ I/Os to compute the aggregation. Hence the overall I/O cost is $O(\sum_{k_i \in K}(|W|/B_p + T_i/B) + T'log_{B_p}(N_s/B_s))$. The value of T_is can be bounded by Theorem 1 and T' can be bounded by $min(T_i)$.

5.3 The Piecewise Maximum Approximation Sketch

The I/O cost may be dominated by $T'log_{B_p}N_s/B_s$ due to a large value of T'. It is caused by the fact that value ts is an over-estimation of the actual aggregate score. Thus we propose the piecewise maximum approximation(PMA) sketch as the additional payload data. Suppose the lifetime of an item is $[l, u]$. The time series is partitioned into B sub-intervals $I_1[b_0, b_1],...,I_B[b_{B-1}, b_B]$, where b_0 and b_B equal l and u respectively. A value h_i, which equals the maximum value of points in sub-interval I_i, is used to approximate points in that interval. To determine the optimal boundaries of sub-intervals, the objective function $\sum_{0 \leq i < B} \sum_{v_j \in [b_i, b_{i+1}]}(h_i - v_j)$ is minimized. Its goal is to minimize estimation error when the sketch is used to compute the upper bound for an item.

Currently we heuristically use $log_{10}(u - l)$ as B. The exact optimal solution could be find using dynamic programming. Suppose $f[a, b]$ records optimal error when the interval $[l, a]$ is divided into b sub-intervals. Let $error(a - m + 1, u)$ equal the estimation error when interval $[a-m+1, u]$ is chosen as one sub-interval. Equation $f[a, b] = max_{m \in (l+b-2, a)}\{f[m, b-1] + error(m + 1, a))\}$ helps us derive a dynamic program. As a and m could take at most $l - u$ possible values and b could take at most B possible values, the total time complexity is $O((u - l)^2 log_{10}(u - l))$. Using PMA sketch, we can use equation $\sum_{|I_i \cap [t_s, t_e]| > 0} h_i * |I_i \cap [t_s, t_e]|$ to compute a tighter upper bound of an item's aggregation score. Subroutine $tweetHit$ is modified such that only items with estimation value greater than aggregation score of the current kth answer are passed to segment store for evaluation.

6 Experimental Results

We implement four approaches in Java language to study the performance of two-tiers structure and PMA sketch under different query setups. Implementations with or without two-tiers structure are annotated with prefix "hpl" and "ipl" respectively. Similarly, approaches with or without sketch are annotated with suffix "s" or "ns" correspondingly. According to our knowledge, temporal keywords search is the most related area. Our implementation with single tier and without sketch can be viewed as a variant of temporal inverted index. Thus "ipl_ns" is used as the baseline index in our experiments.

6.1 Read-Life Weibo DataSet and Query Workload

The dataset used in the empirical study is an event repository that is extracted from a large dataset collected from Sina Weibo, the most popular microblogging service in China. For details of the large dataset, please refer to [12]. We select 190 popular events and define rules based on regular expressions to extract tweets for each event. Those tweets constitute the event repository used in our experiment. Tweets that are never retweeted are discarded and about 13 millions of tweets are retained.

As there are no publicly available temporal keyword search query logs for Sina Weibo or our event repository, we generate query workloads in the following steps. First, we sort pairs of keywords in descending order of their occurrence in tweets. Then, we retrieve the top one tweet for each of the top 10 keyword pairs using the unbounded time window. We assume those top ranked tweets correspond to some popular events, and thus choose them as start time of the query window. Finally, the top 10 keyword pairs together with correspond selected start time are used to compose the query. The average query latency of ten queries are reported for each query configuration.

6.2 Performance Under Different Query Setups

The Influence of k. First we study how the query performance is influenced by the value of k. Figure 3(a) shows the query latency of four approaches with different value of k. The horizontal axis is the value of k, while the vertical axis is the average query latency of ten queries. The start time of all queries is set to the generated start time and the width is fixed to be 12 h. It is noteworthy that those tendencies coincide with our previous complexity analysis when $1 - \alpha_o$ is smaller than $1 - \alpha_q$.

We also note that the combination of two-tiers structure and adoption of PMA sketch significantly improve the query performance. Figure 4(a) presents the number of disk blocks that are incurred by the base B+tree, atomic structure, segment store, and sublist. On one hand, the adoption of PMA sketch reduces the number of candidate items, which leads to the remarkable decrease in I/O incurred by segment store. The additional space needed by the sketch only causes a minor increase in I/O of atomic structure. On the other hand, utilization of the two-tiers structure further brings down the disk cost of atomic structure. Curiously there exists a slight inconsistency between the I/O cost and performance improvement indicated by the query latency. Thus, we analyze the

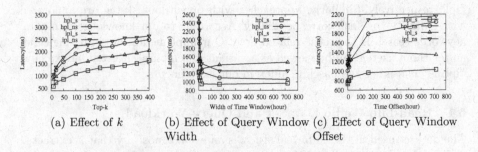

(a) Effect of k (b) Effect of Query Window Width   (c) Effect of Query Window Offset

Fig. 3. Performance under different query setups

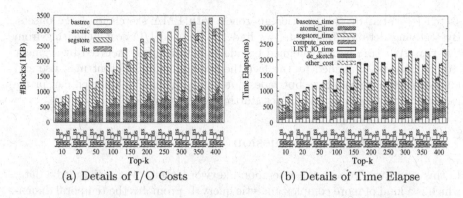

(a) Details of I/O Costs   (b) Details of Time Elapse

Fig. 4. Detail costs for effect of k

detail time cost taken by different parts, which is given in Fig. 4(b). It can be explained by the fact that disk access is dominated by the random access to the segment store. Thus the benefit of reduced I/O cost of segment store may be amplified, which is different from case of sequence access.

Width of Query Window. Secondly we study how the query performance changes with the width of query window. The start time of all queries is set to the generated start time and k is fixed to be 100. The result is reported in Fig. 3(b). The detailed costs are omitted due to the limitation of space. The performance of all methods decreases exponentially when $width$ is relative small, but tends to increase slightly with large $widths$. We also notice the aberrant phenomena that the performance of single tier with sketch behaves worse than that without sketch. By examining I/O cost we find the I/O cost of segment store and atomic structure decreases significantly with larger $widths$ when $width$ is relatively small. It is because the wider the query window is, the larger the aggregate score of kth item is. Hence more items can be excluded from candidates of top-k answers, which can be verified by the decrease in I/O cost of atomic structure and segment store. However, the gap between the aggregate score and the final linkage tree size of an item also shrinks, which lessens the pruning ability of the sketch. At last the negative effect of additional storage incurred by the sketch overwhelms its benefit.

Deviation from Occurrence Time of Events. As analytic jobs typically explore topics of an event at different periods, we also study how the performance of proposed methods reacts to different choices of start time of query window. Figure 3(c) presents the average latency of queries starting at different time offsets from their generated start time. The width is fixed to be 12 h and k is set to be 50. The horizontal axis is hours that the start time of queries differs from that of corresponding events. The query performance of all methods fluctuates during the initial stages of events, and then degrades dramatically at further offsets. Finally it trends to keep steady. The combination of both refinements is still the best among four implementations. We also observe two-tiers approach without sketch is beaten by the one-tier approach with PMA sketch for large offsets. By studying detailed I/O costs, we find the reduced I/O cost benefiting from two-tiers structure is overwhelmed by additional disk cost incurred by segment store. It can be explained by the fact the popularity of an event may rise and fall during the initial stage but fades out finally. Thus PMA sketch can prune items more aggressively for queries starting at longer offsets.

7   Conclusion and Discussion

In this paper we propose top-k temporal keyword query over social media data, which is a kind of more complex analytic query. It promotes the temporal dimension as an important query constraint, rather than treating it as the measurement of recency. We develop an efficient solution combining the two-tiers posting list and PMA sketch. Given the long tail phenomenon existing in the data, the complexity analysis proves the efficiency of our approaches. The experiments on a real-life dataset verify our analysis and show that the combination of proposed techniques significantly reduces the I/O cost and improves the overall performance.

Acknowledgements. This work is partially supported by National High-tech R&D Program (863 Program) under grant number 2015AA015307 and National Science Foundation of China under grant number 61432006.

References

1. Anand, A., Bedathur, S.J., Berberich, K., Schenkel, R.: Efficient temporal keyword search over versioned text. In: CIKM, pp. 699–708 (2010)
2. Arge, L., Vitter, J.S.: Optimal external memory interval management. SIAM J. Comput. **32**(6), 1488–1508 (2003)
3. Berberich, K., Bedathur, S., Neumann, T., Weikum, G.: A time machine for text search. In: SIGIR, p. 519 (2007)
4. Chen, C., Li, F., Ooi, B.C., Wu, S.: Ti: an efficient indexing mechanism for real-time search on tweets. In: SIGMOD Conference, pp. 649–660 (2011)
5. Fuchs, E., Gruber, T., Nitschke, J., Sick, B.: Online segmentation of time series based on polynomial least-squares approximations. IEEE Trans. Pattern Anal. Mach. Intell. **32**(12), 2232–2245 (2010)
6. He, J., Suel, T.: Faster temporal range queries over versioned text. In: SIGIR, p. 565 (2011)
7. Huo, W., Tsotras, V.J.: A comparison of Top-k temporal keyword querying over versioned text collections. In: Liddle, S.W., Schewe, K.-D., Tjoa, A.M., Zhou, X. (eds.) DEXA 2012, Part II. LNCS, vol. 7447, pp. 360–374. Springer, Heidelberg (2012)
8. Jestes, J., Phillips, J.M., Li, F., Tang, M.: Ranking large temporal data. PVLDB **5**(11), 1412–1423 (2012)
9. Keogh, E.J., Chu, S., Hart, D.M., Pazzani, M.J.: An online algorithm for segmenting time series. In: ICDM, pp. 289–296 (2001)

10. Lemire, D.: A better alternative to piecewise linear time series segmentation. In: SDM, pp. 545–550 (2007)
11. Li, F., Yi, K., Le, W.: Top- k queries on temporal data. VLDB J. **19**(5), 715–733 (2010)
12. Ma, H., Qian, W., Xia, F., He, X., Xu, J., Zhou, A.: Towards modeling popularity of microblogs. Front. Comput. Sci. **7**(2), 171–184 (2013)
13. Wu, L., Lin, W., Xiao, X., Xu, Y.: LSII: an indexing structure for exact real-time search on microblogs. In: ICDE, pp. 482–493 (2013)
14. Zhuang, Y.: Building a complete Tweet index. Tuesday, 18 November 2014 (2014). https://blog.twitter.com/2014/building-a-complete-tweet-index. Accessed 21 Nov 2014

When a Friend Online is More Than a Friend in Life: Intimate Relationship Prediction in Microblogs

Yunshi Lan[(⊠)], Mengqi Zhang, Feida Zhu, Jing Jiang, and Ee-Peng Lim

Singapore Management University, Singapore, Singapore
yslan.2015@phdis.smu.edu.sg, mqzhang.2015@mais.smu.edu.sg,
{fdzhu,jingjiang,eplim}@smu.edu.sg

Abstract. Microblogging services such as Twitter and Sina Weibo have been an important, if not indespensible, platform for people around the world to connect to one another. The rich content and user interactions on these platforms reveal insightful information about each user that are valuable for various real-life applications. In particular, user offline relationships, especially those intimate ones such as family members and couples, offer distinctive value for many business and social settings. In this study, we focus on using Sina Weibo to discover intimate offline relationships among users. The problem is uniquely interesting and challenging due to the difficulty in mining such sensitive and implicit knowledge across the online-offline boundary. We introduce deep learning approaches to this relationship identity problem and adopt an integrated model to capture features from both user profile and mention message. Our experiments on real data demonstrate the effectiveness of our approach. In addition, we present interesting findings from behavior between intimate users in terms of user features and interaction patterns.

Keywords: Intimate relationship · Relationship identification · Deep learning · Microblogging platform

1 Introduction

Online microblogging services provide popular public platforms where users can establish their own social networks online. For example, Sina Weibo[1], which is widely used in China, now has 76 million active users every day and 1.67 hundred million users every month, and still counting. With the common usage of such social networks, some interesting questions arise. For example, profiling a user's topic interests can benefit product recommendation and advertisements to serve targeted customers [14,15]; profiling a user's location helps push localized news or weather information [1]. Such profiling can not only offer personalized services to users but also give potential commercial opportunities to businesses.

---

[1] http://www.weibo.com.

© Springer International Publishing Switzerland 2016
F. Li et al. (Eds.): APWeb 2016, Part I, LNCS 9931, pp. 196–207, 2016.
DOI: 10.1007/978-3-319-45814-4_16

Fig. 1. Examples of intimate relationship

A closely related task to user profiling is the task of identifying offline friends from online microblogging platform. This task has also attracted researchers' attention and been applied to applications such as friend recommendation and product recommendation, since offline friendship is arguably a more reliable connection between users.

In this paper, we are particularly interested in identifying intimate relationships in Weibo. We define an intimate relationship as the relationship between romantic partners, parents and children, and other close relatives. Some examples are shown in Fig. 1. An intimate relationship is a much stronger relationship than general offline relationship which represents an elevated level of mutual trust and association in many aspects such as financial ones. As such, it is of great value to leverage these relationships among users for a great number of business and financial applications.

For example, Word-of-mouth marketing has been observed to be more effective among users with intimate relationships as a result of deeper trust and shared lifestyle. On the other hand, in banking industry, it is crucial to be able to identify customers of potential credit issues and one useful approach is to propagate credit-related labels from identified ill-credit users to those unknown ones. As people with intimate relationships are usually financially connected, these user attributes and labels such as personal credit and financial status are only legitimate to be propagated among users of intimate relationships [16].

Intuitively, a user's profile and Weibo mention messages provide valuable signals for identifying her intimate relationships. In this work, we have the following assumptions: (1) Intimate user pairs share certain common properties such as their geolocations and interests. (2) Intimate user pairs would more likely mention each other in specific messages. The challenges of the problem lie in the difficulty in inferring these implicit sensitive user information from noisy social data across the online-offline boundary.

In this paper, we employ an integrated model to combine user profile information as well as mention interaction. The main contributions of our work are as follows:

- We identify the important and interesting problem of mining user intimate offline relationship from microblogging platforms, which offers unique value for a wide range of real-life applications.

- We introduce effective TransE and CNNs models to the problem and propose an integrated model which captures both user profile and interaction features.
- We conduct experiments on real data sets to demonstrate the effectiveness of our model. We also summarize specific features and behavior patterns between pairs of intimate users to offer a better understanding of the characteristics of intimate relationships on microblogging platform.

In the rest of this work, we introduce related work in Sect. 2 and then formulate our research approaches in detail in Sect. 3. The dataset is introduced in Sect. 4. Experiments and result analysis are described in Sect. 5. We conclude this work in Sect. 6.

2 Related Work

2.1 Strong Tie

Strong tie plays an important role in social network systems. Researchers are interested in identifying strong ties since they give us more reliable information. A strong tie is typically "embedded" in a social network, with more mutual friends [11] or homogeneity. It benefits us in identifying the most important individuals in a social network. Recently, some work developed methods analyzing social network structures and measuring tie strengths [12,13]. Most of their methods take advantage of the number of mutual friends as embeddedness between two users.

2.2 Intimate Relationships

In the work of [2], the author proposed a new network-based characterization for intimate relationships, those involving spouses or romantic partners. For our research, we broaden the definition of intimate relationships. The intimate relationships can be between romantic partners, parents and children, and relatives. Such relationships are important to study for several reasons. From a substantive point of view, romantic relationships are of course singular types of social ties that play powerful roles in social processes over a person's whole life [6], from adolescence [5] to older age [7]. They also form an important aspect of the everyday practices and uses of social media [8]. There are important challenges from a methodological point of view: they are evidently among the very strongest ties, but it has not been clear whether standard structural theories based on embeddedness are sufficient to characterize them, or whether they possess singular structural properties of their own [9]. In comparison, our work utilizes additional rich context features to identify intimate relationship.

3 Model

3.1 Profile-Based Model

As we introduced above, people usually use friendship to measure similarity of users online, however, as we know, it is not enough to predict intimate relationships. More useful information like profiles of users may also contribute to intimate relationship prediction. Based on this motivation, we extend a social relation network into a knowledge network and characterize users with rich information.

To better explain our model, we demonstrate in a precise way. In network graph G, triplets can be displayed as $< h, r, t > \in G$, where $h, t \in E$ represents a head entity and a tail entity, E is a set of all entities, and $r \in R$ is a set containing all relationships between entities.

Given the above definition, we construct our knowledge network graph, which is a collection of relational fact triplets but with many missing parts in need of completion. We leverage TransE [4], which is an advanced deep learning technique for knowledge graph completion problems. This model learns embeddings of entities and relationships by mapping entities and relations into the same vector space where entity embeddings occupy positions in the knowledge graph space and relation embeddings connect them. Such low-dimensional continuous representation encodes inherent relationships between entities and relations [18]. Its high performance and low complexity triggered many modifications and applications [17,19]. The basic idea behind the TransE framework is the following. For a true triplet, $h+r$ is approximatedly equal to t, that is, $h+r \approx t$, while $h+r$ should be far away from t for a false triplet. For some dissimilarity measurement d, if $< h, r, t >$ holds, $d(h+r, t)$ should be relatively small compared with a false triplet $d(\acute{h}+r, t)$ or $d(h+r, \acute{t})$. In order to ensure $d(h+r, t)$ is smaller than a false triplet, we minimize the following margin-based penalty function:

$$L_1 = \sum_{(h,r,t) \in S} \sum_{(\acute{h}, r, \acute{t}) \in \acute{S}} [\gamma + d(h+r, t) - d(\acute{h}+r, \acute{t})]_+ \qquad (1)$$

where all $h, r, t \in R^{k*1}$, $[x]_+$ denotes positive part of x, γ is a margin hyper parameter, and

$$\acute{S} = \{(\acute{h}, r, t) | \acute{h} \in E\} \cup \{(h, r, \acute{t}) | \acute{t} \in E\} \qquad (2)$$

Which means false triplet is obtained from replacing one entity in triplet with another one in E randomly. When there are many same relationships attached to two entities, their embeddings will be much similar.

3.2   Mention-Based Model

Since Weibo is a microblogging platform where users can interact with each other by repost or @ or reply, and interactions between intimate users and regular friends may be distinct, we would like to investigate Weibo mention messages between two users to predict whether they have an intimate relationship.

In the mention space, definition m is a sequence of many tokens $\{e_1, e_2, ..., e_n\}$, which is Weibo message in practice. $m \in M_{h,t}$ represents existing mentions, which include entities h and t, and M means the set of all mentions. We would like to build a convolutional neural network (CNN) [10] to classify it into target classes, i.e., intimate relationship or not. CNNs are an effective type of sentence-level neural networks extensively applied in social media tasks, like Tweet sentiment

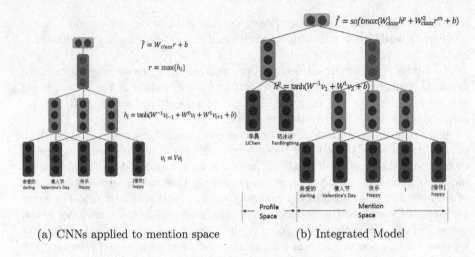

(a) CNNs applied to mention space   (b) Integrated Model

Fig. 2. Neural network model architecture

analysis [20]. Figure 2(a) shows the neural network architecture. In the first layer, input is a parsed mention message. Then every token is mapped into a vector v_i using embedding matrix V. Then convolution layer convolutes input by windows, and a position-specific parameter W implies weight of feature at different positions. After that, a max pooling operation is applied to select the most significant feature for every dimension among the various windows. Finally, the output of the convolution layer is sent to the classifier layer, and classified into two classes. The objective function is as follows:

$$L_2 = argmin \sum_{m \in M_{h,t}} \| f_m - \hat{f}_m \|_2, \tag{3}$$

where $\hat{f}_m$ is the predicted label. The reason why we choose CNNs is that this neural network can capture relatively global features inside a sequence instead of local features. For example, two sequences containing *Valentine's Day*, convolutional neural network tells difference between *Happy Valentine's Day* and *Valentine's Day is a holiday*.

In our model, the embedding of tokens are all pre-trained by Word2Vec, and we keep embedding matrix static and feed CNNs model Weibo mention messages to learn parameters.

3.3   Integrated Model

Even though mention message can be a powerful evidence to decide whether there is an intimate relationship between two users, we still cannot say as long as two users communicate with an intimate expression then they have an intimate relationship. So we attempt to integrate these two models. In the prediction part, we combine profile information and mention information together and make predictions jointly.

Figure 2(b) shows the architecture of the integrated model. The left side is in profile space, where we leverage pre-trained embedding from profile-based model. The right side is in mention space, where we still use pre-trained Word2Vec embeddings. But we finally connect these two sides together by applying a linear combination of these two sides and finally send into a softmax layer to obtain output.

In the profile space, the output of the convolution layer is obtained in a similar way as in mention-based model, that is:

$$h^p = tanh(W^{-1}v_1 + W^1 v_2 + b), \tag{4}$$

Where v_1 and v_2 are entity embeddings learned from Profile-based model. The output of the mention space is the output of the max pooling layer denoted as r^m. So the classifier layer tries to merge this two linearly and obtain the final label with a softmax function:

$$\hat{f}_m = softmax(W^1_{class}h^p + W^2_{class}r^m + b) \tag{5}$$

Thus the parameter W^1_{class} and W^2_{class} will act as projectors projecting two spaces into a single space. And this model carries both types of information from the profile space as well as the mention space. The objective function is the same as the one in the mention-based model.

4 Dataset

The Weibo dataset we collected contains details of their online activities on Weibo of 1622 user pairs between October 2012 and June 2013. These users are randomly picked from Weibo.

The ground truth is manually labeled based on their Weibo messages by major votes. From the content of each Weibo, we distinguish whether a user and his mentioned user may have any intimate relationship by specific and distinct words. The intimate relationships include three types: romantic partners, parents and children, and other close relatives. Since many user pairs have fuzzy intimate relationships and certain relationship has rare samples, we narrow down all kinds of intimate classes into one, even though our model can handle multi-classes classification problem well theoretically.

Overall, we have 642 pairs who have certain intimate relationship; 980 pairs who are not shown any evidence that can prove that they have any intimate relationship. In our research, based on the collected user pairs, we extend their social relation network by collecting information of both themselves and their neighbors.

For baseline, we consider three forms of network. Forming with only followee relationship, only follower relationship and only mutual follow relationship, denoted as Follower&Followee (Table 1). Obviously, network forming with Follower&Followee contains much fewer edges than the other two.

Table 1. Network specification

Network	Avg degree	Nodes	Edges
Followee	4.2881	185241	397168
Follower	2.0212	209372	211594
FollowerFollowee	1.9042	57510	54756

Besides follower and followee relationship, details of users' gender, interests, education and Weibo messages .etc are collected, allowing for an analysis of their profile and mention.

4.1   Weibo Mention Message

For these 1622 user pairs, we crawled their Weibo messages from users' homepages. These Weibo messages are also randomly picked which can be employed as evidence to predict intimate relationship between two users in literal way.

For every mention message containing a mentioned user, we replace the mentioned username with specific symbol $account$ in the sentence, since it doesn't matter who is mentioned in our mention-based model. In terms of pre-processing, all the emotion is transformed into text and text are simplified and tokenized by the tool Jieba[2]. Thus we obtain clear mention message for every user pair.

4.2   User Profile

We extract 8 types of data from user profiles to define the relationships between entities. Table 2 contains a summary of this data.

Table 2. Relation summary

Relation	Description
StudyIn	Schools that users have studied in. A user can have this kind of relation with several schools
WorkIn	City and district where a user is working
LiveIn	City and district where a user is living
GenderIs	Gender of user, right entities are "male" or "female"
InterestedIn	Topics that users are interested in which is extracted from description of Weibo user
BelongTo	A user belongs to "celebrity" if he is verified by official Weibo, always celebrity is very influential user among other people
Follower	User's Follower, left and right entity of this triplet are users
Followee	User's Followee, left and right entity of this triplet are users

---

[2] https://github.com/fxsjy/jieba

We collected 966705 triplets about listed relations. Finally, we add "IntimateWith", "UnitimateWith" triplets inside together to comprise all triplets for profile-based model.

5   Experiment

5.1   Experiment Setup

In the experiment, given two users, we are going to identify whether they have intimate relationship between them, based on dataset we collected.

For baseline, we construct Weibo social relation network by connecting two users if there is online friendship between them, these online friendships variate as Follower, Followee and Follower&Followee. Some widely-used algorithms for link prediction are implemented to measure similarity between user pair. *Jaccard* and *AdamicAdar* coefficients [3] represent how many common neighbors two users are sharing, which may reveal how intimate they are. *Dispersion* [2] motivates from the idea that people can easily have common neighbor online in a cluster but intimate users would more likely to have common neighbors from different clusters and . And *MutualFollow* is binary feature applied in Followee&Fllower network, which values 1 if there is an edge holding between two users.

In the training part, we split dataset into two train and test part by 80 % and 20 % randomly. We obtain similarity scores for every user pairs, then we apply SVM^{light} to do classification.

For the profile-based model, in TransE, we apply L_2 distance to measure dissimilarity $(h+r)$ and t, and learning rate is set as 0.01 and iterate 1000 times, we learn 50 dimension embedding for every entity. All these parameters are tuned to achieve best performance for validation set. When predicting relationship of one user pair, we rank triplet containing "IntimateWith" and "UnitimateWith", relation with smaller dissimilarity will be our predicted label.

For mention-based model, all embedding of tokens are pre-train from a large corpus collected from Weibo[3]. Here, we still learn 50 dimension for every word, and set windows size as 3, learn rate as 0.05 for every step, iterate for 100 times. Same parameters are also used in integrated model.

For evaluation, we use recall, precision, F1 and accuracy index, which are widely used in classification problem.

5.2   Experiment Result

The result can be seen in Table 3. Notice traditional *Jaccard* gains a low F1 score, while *Adar* performs relatively better. Both of them use information of common neighbors, but *Jaccard* reduces common neighbor's effect by adding union of their neighbor inside. *Dispersion*, which was proved to have impressive performance, however obtains a low recall, this may because most of people would not

---

[3] We randomly picked 100 active Weibo users, and crawled all their Weibo messages and constituted corpus with 12 k Weibo messages.

Table 3. Experiment results

Method	Algorithm	Recall	Precision	F1	Accuracy
Followee	Jaccard	3.68	83.33	7.05	59.88
	Adar	41.18	90.32	56.57	73.86
	Dispersion	35.29	94.12	51.33	72.34
Follower	Jaccard	1.47	100	2.90	59.27
	Adar	26.47	92.31	41.14	68.69
	Dispersion	17.65	92.31	29.63	65.35
Follower&Followee	Mutual	47.06	72.73	57.14	70.82
	Jaccard	3.68	100	7.10	60.18
	Adar	13.64	100	24.01	64.13
	Dispersion	8.82	100	16.21	62.13
Profile-based model		99.26	67.17	71.42	67.17
Mention-based model		58.82	73.39	65.30	74.16
Integrated model		88.24	75.47	**81.35**	**83.28**

like to establish their neighbors circle in Weibo, so the feature is unavailable, which leads to performance depreciation. From the table, the followee relationship performs better than follower relationship which may indicates that followee relationship is more reliable and applicable compared with follower relationship in this dataset.

Profile-based method performs not very well, even if it captures not only friendship between users but also their interest or education or work information, its low precision indicates it tends to make false positive prediction.

Mention-based method doesn't meet our expectation, note that even though CNNs is a powerful tool to capture text features in Weibo mention message, during learning process, it gives high weight to some intimate tokens or phrases, like *I love you*, *miss you*. Moreover, some specific names entities like *IKEA*, *Resort hotel* are also given high weights implying intimate relationship. However, in Weibo, even though text maybe the most straightforward for us to recognize intimate users, the precision is relatively low, this may because intimate expression does not mean intimate relationship between users, most false positive prediction happens because of this. Many Weibo on Valentines' Day contains intimate expression, but it turns out to be advertisement from one company reposted by user. Meanwhile, fans would like to express intimate words to their idols but there is no intimate relationship between them. Thus it's reasonable for us to combine these two spaces together and make prediction.

When we combine these two features together, integrated model outperforms all other models, improving mention-based model F1 score by 16.05 % and profile-based model by 9.93 %. Integrated model consider not only distance of two users, but also their message interaction on Weibo, so that it can achieve highest performance.

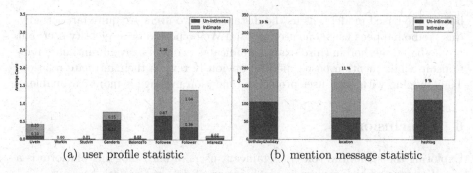

(a) user profile statistic   (b) mention message statistic

Fig. 3. Profile and interaction understanding

5.3   Result Analysis

More than prediction, it would be interesting to figure out significant feature and understand interaction behavior of intimate users online. We analyze average sharing relation statistics for both intimate and un-intimate users and plot Fig. 3(a). Obviously, intimate user pairs share more relations especially followee and follower. However, intimate user pairs in our dataset seems to have different genders, this implies majority of intimate user pairs are couple. Figure 3(b) records count of specific text appearing in Weibo mention message like, birthday or holiday name, location, or hashtag. 19 % Weibo messages mention birthday or specific holiday like *Valentines' Day*, *Mother's Day* etc., 11 % Weibo messages contain location name like hotel, restaurant. Intimate user pairs mention them more frequently by comparison. However, hashtag appear in un-intimate user pairs more often, because actually, Weibo message containing hashtag is likely to be advertisement.

Table 4 displays particular examples we extracted from dataset, the Weibo mention message in first example shows nothing about intimate literally, but our mention-based model can learn it as intimate expression because of *ResortHotel* is likely to be some romantic place, meanwhile, profile-based model implies they are intimate user pair for the sake of relations they share in their profile; in second example, Weibo content seems to happen between father and son,

Table 4. Case study

Golden Standard	Profile-based	Mention-based	user&user mention	Weibo mention message
+	+	+	Devil_住在云上的人&昊小磊(regular user&regular user)	我在千潮馨度假酒店。与 $account$ 在一起。朔景房～真心美哦～晚上要多拍照片 (I'm in Thousand - Islet Lake Cozy Island Resort Hotel with $account$. Lake house~really great~I'll take lots of photo tonight)
-	-	+	維尼小標題&微博搞笑排行榜(regular user&celebrity user)	Dad happy father's day.I love you//$account$: 親愛的爸爸，父親節快樂！愛你 (Dad happy father's day. I love you//$account$: Dear father happy father's day! love you)
+	-	+	Scarlet-Jason&Miss-小甲菌菌子(regular user&regular user)	$account$ 老婆，我没有這感覺你有嗎？[偷笑][偷笑] ($account$ wife, I don't have such feeling, do you?)

but profile-based model tells us the distance of two users are quite far, actually, this Weibo happens when one user repost a Weibo posted by a celebrity user; but profile-based method in third example indicates two users have un-intimate relationship while mention-based method obviously reveals their intimate relation. Thus judging with both user profile and mention message is more reasonable.

6   Conclusion

Exploring the structural roles of significant users in microblogging platform is a broad question which requires a combination of different approaches.

In this paper, we introduce TransE to profile space, which reveals similarity among users based on user profile, as well as CNNs to mention space which encodes mention messages, then an integrated model is proposed to adopt both rich context features and contribute to prediction.

Overall, predicting intimate relationship in microblogging platform depends heavily on various signature we extract. In the future, photo and URL link can be used as other useful signals for prediction; sounder baseline should be employed for comparison and more fine-grained intimate classes would like to be explored. To better investigate specific relationship within microblogging platform, more potential feature and methods should be considered in the right way.

Acknowledgements. This research is partially funded by the National Research Foundation, Prime Minister's Office, Singapore under its International Research Centres in Singapore Funding Initiative and Pinnacle Lab for Analytics at Singapore Management University.

References

1. Li, R., Wang, S., Deng, H., Wang, R., Chang, K.C.-C.: Towards social user profiling: unified and discriminative influence model for inferring home locations. In: KDD (2012)
2. Backstrom, L., Kleinberg, J.: Romantic partnerships and the dispersion of social ties: a network analysis of relationship status on Facebook. In: Proceedings of 17th ACM Conference on Computer Supported Cooperative Work and Social Computing (CSCW) (2014)
3. Liben-Nowell, D., Kleinberg, J.: The link-prediction problem for social networks. J. Am. Soc. Inf. Sci. **58**, 1019–1031 (2007)
4. Bordes, A., Usunier, N., Garcia-Duran, A.: Translating embeddings for modeling multi-relational data. In: NIPS (2013)
5. Bearman, P., Moody, J., Stovel, K.: Chains of affection: the structure of adolescent romantic and sexual networks. Am. J. Sociol. **110**, 44–91 (2004)
6. Bott, E.: Family and Social Network: Roles, Norms, and External Relationships in Ordinary Urban Families. Tavistock Press, London (1957)
7. Cornwell, B.: Spousal network overlap as a basis for spousal support. J. Marriage Fam. **74**, 229–238 (2012)

8. Zhao, X., Sosik, V.S., Cosley, D.: It's complicated: how romantic partners use Facebook. In: Proceedings of 30th ACM Conference on Human Factors in Computing Systems (2012)
9. Felmlee, D.H.: No couple is an island: a social network perspective on dyadic stability. Soc. Forces **79**, 1259–1287 (2001)
10. Kim, Y.: Convolutional neural networks for sentence classification. In: EMNLP (2014)
11. Coleman, J.S.: Social capital in the creation of human capital. Am. J. Sociol. **94**, S95–S120 (1988)
12. Marsden, P.V., Campbell, K.E.: Measuring tie strenth. Soc. Forces **63**, 482–501 (1984)
13. Jones, J.J., Settle, J.E., Bond, R.M., Fariss, C.J., Marlow, C., Fowler, J.H.: Inferring tie strength from online directed behaviour. PLoS ONE **8**, e52168 (2013)
14. Ahmed, A., Low, Y., Aly, M., Josifovski, V., Smola, A.J.: Scalable distributed inference of dynamic user interests for behavioral targeting. In: KDD 2011, pp. 114–122 (2011)
15. Provost, F., Dalessandro, B., Hook, R., Zhang, X., Murray, A.: Audience selection for on-line brand advertising: privacy-friendly social network targeting. In: KDD 2009, pp. 707–716 (2009)
16. Kawachi, I., Kennedy, B.P., Wilkinson, R.G.: Crime: social disorganization and relative deprivation. Soc. Sci. Med. **48**(6), 719–731 (1999)
17. Lin, Y., Liu, Z., Sun, M., liu, Y., Zhu, X.: Learning entity and relation embeddings for knowledge graph completion. In: Proceedings of AAAI, pp. 2181–2187 (2015)
18. Chang, K.-W., Yih, W.-t., Yang, B., Meek, C.: Typed tensor decomposition of knowledge bases for relation extraction. In: EMNLP (2014)
19. Toutanova, K., Chen, D., Pantel, P., Poon, H., Choudhury, P., Gamon, M.: Representing text for joint embedding of text and knowledge bases. In: Empirical Methods in Natural Language Processing (EMNLP). ACL Association for Computational Linguistics, September 2015
20. Severyn, A., Moschitti, A.: UNITN: training deep convolutional neural network for twitter sentiment classification. In: Proceedings of the 9th International Workshop on Semantic Evaluation, SemEval 2015, Denver, Colorado. Association for Computational Linguistics, June 2015

Community Inference with Bayesian Non-negative Matrix Factorization

Xiaohua Shi[1,2(✉)] and Hongtao Lu[1]

[1] MOE-Microsoft Laboratory for Intelligent Computing and Intelligent Systems, Shanghai, China
[2] Library, Shanghai Jiaotong University, Shanghai, China
{xhshi,htlu}@sjtu.edu.cn

Abstract. In terms of networks, the clustering is based on the topology structure of the network and the groups found are called Communities. We might expect a coherent group to be one which has more links between members of the group than it has to nodes outside the group in other clusters. Detection Communities in a large network can efficiently simplify network structure, help to understand the network topology and learn how the network works.

As a dimension reduction method, Non-negative Matrix Factorization (NMF) aims to find two non-negative matrices whose product approximates the original matrix well, and is widely used in graph clustering condition with good physical interpretability and universal applicability. Based on the consideration that there is no any physical meaning to reconstruct a network with negative adjacency matrix, using NMF to obtain new representations of network with non-negativity constraints can achieve much productive effect in community analysis.

Incorporating Bayesian methods with prior knowledge for NMF, we can gain further insights into the data and determinate the optimal parameters for detecting model. In this paper, we propose a Bayesian non-negative matrix factorization method with Symmetric assumption (BSNMF), which not only achieve better community detection results in undirected network, but also effectively predict most appropriate count of communities in a large network with Automatic Relevance Determination model. We compare our approaches with other NMF-based methods in Email social networks, and experimental results for community detection show that our approaches are effective to find the communities number and achieve better community detection results.

Keywords: Community detection · Non-negative matrix factorization · Symmetric matrix · Bayesian inference · Automatic relevance determination

1 Introduction

Community Detection is an important approach in analysis works of network science [29], such as social network, collaborative network [19] and biological

© Springer International Publishing Switzerland 2016
F. Li et al. (Eds.): APWeb 2016, Part I, LNCS 9931, pp. 208–219, 2016.
DOI: 10.1007/978-3-319-45814-4_17

network [17]. Community Detection is an important approach in complex networks to understand and analysis large network character, and it can find most correlated sub-communities (we also call sub-modules, sub-groups, sub-clusters etc.) to simplify global structure to understand the network topology, and learn how the network works with existing information. It is a recognition with community detection that nodes in same community are densely connected, and nodes in different communities are sparsely connected [24]. Nodes in a social network can be people and edges can be friendship relations. Communities can be detected based on the graph partitioning approach, which tries to minimize the number of edges between communities [6,22,34]. We can also obtain communities information with methods as objective function optimization (modularity or other function) [15,20,38] or statistical inference [1,23,30].

In many clustering applications, object data is nonnegative due to their physical nature, e.g., images are described by pixel intensities and texts are represented by vectors of word counts. As to a graph-based network, the adjacency matrix (or weighted adjacency matrix) $\mathbf{A}$ as well as the Laplacian matrix completely represents the structure of network, and $\mathbf{A}$ is non-negative naturally. Meanwhile, Nonnegative Matrix Factorization (NMF) was originally proposed as a method for dimension reduction and finding matrix factors with parts-of-whole interpretations [9,14]. Based on the consideration that there is no any physical meaning to reconstruct a network with negative adjacency matrix, using NMF to obtain new representations of network with non-negativity constraints can achieve much productive effect in community analysis [12,27,37]. It is likely an efficient network partition tool to find the communities because of its powerful interpretability and close relationship with other clustering methods. Community detection with NMF can capture the underlying structure of network in the low dimensional data space with its community-based representations.

NMF aims at decomposing a given nonnegative data matrix $\mathbf{X}$ as $\mathbf{X} \approx \mathbf{U}\mathbf{V}^{\mathbf{T}}$ where $\mathbf{U} \geq \mathbf{0}$ and $\mathbf{V} \geq \mathbf{0}$ (meaning that U and V are *component-wise nonnegative*). Tan *et al.* [28] addresses the estimation of the latent dimensionality in nonnegative matrix factorization (NMF) with the β-divergence, and proposes for maximum a posteriori (MAP) estimation with majorization-minimization (MM) algorithms. Psorakis *et al.* [23] presents a novel approach to community detection that utilizes the Bayesian non-negative matrix factorization (NMF) model to extract overlapping modules from a network.

The symmetric NMF (SNMF) decomposition is a special case of NMF, in which both factors are identical. Kuang *et al.* [11] proposes a SNMF as a general framework for graph clustering, which inherits the advantages of NMF by enforcing nonnegativity on the clustering assignment matrix. Wang *et al.* [31] deploy a symmetric NMF algorithm for multi-document summarization and sentence-level semantic analysis.

In this paper, we propose a novel Bayesian non-negative symmetric matrix factorization (BSNMF) method for community detection. In a Bayesian framwork, BSNMF assumpts that original matrix $\mathbf{X}$ with object symmetric matrix $\mathbf{S}$ follow a certain probability distribution. Specifically, we define $\mathbf{U} = \mathbf{V}$, that is

$\mathbf{X} \approx \mathbf{U}\mathbf{U}^{\mathbf{T}}$ [32], and this assumption can effectively reduce arithmetic time and simplify the factorizing process with good performance. In this way, we expect that BSNMF can alse obtain a relavant count of communities for the network data. To achieve this, we design a new non-negative matrix factorization objective function by incorporating Bayesian Inference and symmetric expression. Our experimental evaluations show that the proposed approach can validly estimate relevant dimension in lower space and achieves much better performance than the state-of-arts methods.

2   Related Works

Given a set of data points $\mathbf{x}_1, \mathbf{x}_2, \cdots, \mathbf{x}_n$, they form the data matrix $\mathbf{X} = [\mathbf{x}_1, \mathbf{x}_2, \cdots, \mathbf{x}_n] \in \mathbf{R}^{m \times n}$, where $\mathbf{x}_j$, $j = 1, \cdots, n$, is an m-dimensional non-negative vector, denoting the j-th data point. NMF aims to factorize $\mathbf{X}$ into the product of two non-negative matrices $\mathbf{U}$ and $\mathbf{V}$. The product of $\mathbf{U}$ and $\mathbf{V}^T$ is expected to be a good approximation to the original matrix, i.e.,

$$\mathbf{X} \approx \mathbf{U}\mathbf{V}^T \tag{1}$$

The sizes of the factorized matrices $\mathbf{U}$ and $\mathbf{V}$ are $m \times k$ and $n \times k$, respectively. The dimensionality of $\mathbf{U}$ and $\mathbf{V}$ is k. In most NMF algrothmns, the dimensionality k of the factorized matrices is given, and we should set k to be the same as known count of communities in community detection. Each column of decomposed matrix $\mathbf{U}$ can be regarded as the center of one community of the network, and each node can be represented by an additive combination of all column vectors of the decomposed matrix $\mathbf{U}$. Each entry in the j-th row of the factorized matrix $\mathbf{V}$ is the projection of the j-th node $\mathbf{x}_j$ of the matrix $\mathbf{X}$ onto corresponding column vector of matrix $\mathbf{U}$. Hence, the community membership of each node can be determined by finding the basis (one column of $\mathbf{U}$) with which the node has the largest projection value. More specifically, we examine each row of $\mathbf{V}$, and assign node $\mathbf{x}_j$ to community c if $c = \arg\max_c v_{jc}$ [35].

In order to obtain the two non-negative matrices, we can quantify the quality of the approximation by using a cost function with some distance metric. Generally we use β-Divergence $\mathbf{D}_\beta(\mathbf{X}; \mathbf{U}\mathbf{V}^{\mathbf{T}})$ [7]. When $\beta=0,1,2$, $\mathbf{D}_\beta(\mathbf{X}; \mathbf{U}\mathbf{V}^{\mathbf{T}})$ is proportional to the (negative) log-likelihood of the Itakara-Saito (IS), KL and Euclidean noise models up to a constant.

Recently, Bayesian inference has been introduced into NMF. With a noise E between $\mathbf{X}$ and $\mathbf{U}\mathbf{V}^{\mathbf{T}}$.

$$\mathbf{X} = \mathbf{U}\mathbf{V}^{\mathbf{T}} + \mathbf{E} \tag{2}$$

Schmidt *et al.* [25] present a Bayesian treatment of NMF based on a normal likelihood and exponential priors, and approximate the posterior density of the NMF factors. This model equals to minimize the squares Euclidean distance $\mathbf{D}_2(\mathbf{X}; \mathbf{U}\mathbf{V}^{\mathbf{T}})$ for NMF. We assume the noise $\mathbf{E}$ is i.i.d Gaussian with variance σ_n^2, and the likelihood can be written as

$$P(X|U,V) = \left(-\frac{1}{\sqrt{2\pi}\sigma_r}\right)^{MN} \prod_i \prod_j \frac{1}{\sqrt{2\pi}\sigma_r} e^{\left\{\frac{1}{2}\left(\frac{X_{ij}-[UV^T]_{ij}}{\sigma_r}\right)^2\right\}} \qquad (3)$$

So the cost function for log-likelihood is

$$ln(P(X|U,V)) = -MNln(\sqrt{2\pi}\sigma_r) - \underbrace{\frac{1}{2\sigma_r^2}\sum_i\sum_j(X_{ij} - [UV^T]_{ij})^2}_{D_2(X;UV^T)} \qquad (4)$$

Cemgil [5] proposes NMF models with a KL-divergence error measure in a statistical framework with a hierarchical generative model consisting of an observation and a prior component. We can see that this models of $\mathbf{D_1(X; UV^T)}$ is equals to NMF model with poisson noise likelihood.

$$P(n;\lambda) = \frac{\lambda^n}{n!}exp(-\lambda) \qquad (5)$$

$$P(X|U,V) = \prod_i \prod_j \frac{[UV^T]_{ij}^{X_{ij}}exp(-[UV^T]_{ij})}{X_{ij}!} \qquad (6)$$

We further assume that all entries of X are independent of each other (the dependency structure is later induced by the matrix product), we can write:

$$ln(P(X|U,V)) = \sum_i\sum_j X_{ij}ln[UV^T]_{ij} - [UV^T]_{ij} - ln(X_{ij}!) \qquad (7)$$

We use Stirling's formula $ln(n!) \approx nln(n)-n$ for $n >> 1$ to get approximated expression:

$$ln(P(X|U,V)) \approx \sum_i\sum_j X_{ij}ln\frac{[UV^T]_{ij}}{X_{ij}} - [UV^T]_{ij} + X_{ij} \qquad (8)$$

NMF can derive the latent characteristic structure space $\mathbf{U}$ using the matrix factorization in the clustering process [16,33]. When NMF is used to deal with community detection tasks, as reference as adjacency matrix $\mathbf{A}$, $\mathbf{x}_j$ represents the relationship of j-th nodes with other nodes in the network, and we define $\mathbf{x}_{ij}=1$ if node j associates with node i, otherwise $\mathbf{x}_{ij}=0$. In undirected network, $\mathbf{X}$ is symmetric. We can keep symmetric community form in low dimension [11], for symmetric NMF(SNMF) can naturally capture the cluster structure, and the objective function of SNMF is defined as:

$$\mathbf{X} \approx \mathbf{SS}^T \quad where \quad \mathbf{X}^T = \mathbf{X} \qquad (9)$$

He et $al.$ [10] focuses on symmetric NMF (SNMF), which is a special case of NMF decomposition, and three parallel multiplicative update algorithms using level 3 basic linear algebra subprograms directly are developed. Shi *et al.* [26] discusses SNMF decomposition with beta divergences. The multiplicative update algorithm can iteratively find a factorization for SNMF problem by minimizing beta divergences between an input nonnegative semidefinite matrix and its SNMF approximation.

3 Community Inference with Bayesian Symmetric NMF

3.1 Bayesian Symmetric NMF Model

In this section, we introduce our Bayesian symmetric NMF(BSNMF) model. Given an undirected and unweighed network $\mathbf{G}$ consisting of n nodes $\mathbf{a}_1, \mathbf{a}_2, \cdots, \mathbf{a}_n$, and we can represent the network as matrix $\mathbf{X}$ transformed from adjacency matrix. In our NMF processing, the diagonal elements are defined to be 1 rather than 0, as in the usual clustering cases. $\mathbf{X}$ is symmetric and nonnegative, and we alse consider no overlapping communities in network $\mathbf{G}$ i.e. each node is only assigned into a unique community.

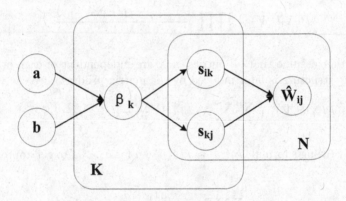

Fig. 1. Graphical model of Bayesian symmetric NMF.

In Fig. 1, X is combined with symmetric latent structure S which have scale hyperparameters β. The hyper-hyperparameters a,b are fixed in this model.

3.2 Model Selection of BSNMF

We consider there lies a relation between original community matrix $\mathbf{X}$ and combination of factorized matrix $\mathbf{SS^T}$. The distribution of this relation can be Gaussian [25] or Poisson model [13]. In maximum-likelihood solution to find $\mathbf{S}$, $\mathbf{P(X|S)}$ is maximized, or its energy function $-\mathbf{logP(X|S)}$ is minimized.

Poisson Likelihood. As in Poisson model, the log-likelihood of $\mathbf{X}$ and $\mathbf{S}$ is:

$$- ln(P(X|S)) = - \sum_i \sum_j \left\{ X_{ij} ln \frac{[SS^T]_{ij}}{X_{ij}} - [SS^T]_{ij} + X_{ij} \right\} \tag{10}$$
$$= -X ln(SS^T) + 1SS^T 1^T + const(X)$$

where $\mathbf{1}$ is an n×n matrix with every elements equal to 1.

Prior of Community Number β_k. We use independent half-normal prior over every column of $\mathbf{S}$, where the mean is zero and precisian is β_k:

$$p(s_{ik}|\beta_k) = \mathcal{HN}(x|0, \beta_k^{-1}) \tag{11}$$

when

$$\mathcal{HN}(x|0, \beta^{-1}) = \sqrt{\frac{2}{\pi}} \beta^{\frac{1}{2}} exp(-\frac{1}{2}\beta x^2) \tag{12}$$

We define the diagonal matrix $\mathbf{B}$ with $[\beta_1,...,\beta_K]$ and zeros elsewhere, and the negative log prior of $\mathbf{S}$ is:

$$- ln(p(S|\beta)) = \sum_i \sum_k \frac{1}{2}\beta_k s_{ik}^2 - \sum_k \frac{N}{2} log\beta_k + const \tag{13}$$
$$= tr(BS^T S) - \frac{N}{2}tr(logB) + const$$

At last, we set the independent prior distribution of β_k as a Gamma distribution with parameters a_k and b_k:

$$p(\beta_k|a_k, b_k) = \frac{b_k^{a_k}}{\Gamma(a_k)}\beta_k^{a_k-1}exp(-\beta_k b_k) \tag{14}$$

So the negative log of β_k is:

$$- ln(p(\beta)) = \sum_k [\beta_k b_k - (a_k - 1)ln\beta_k] + const \tag{15}$$

The MAP(Maximum a Posteriori) of BSNMF is:

$$\mathcal{U} = -lnP(X|S)) - lnP(S|\beta)) - lnP(\beta)) \tag{16}$$

3.3   Iteration Rules of BSNMF

From Eq. (16), we can derive the iteration rules of BSNMF with Poisson likelihood. Let ϕ_{ij} be the Lagrange multiplier for constraint $s_{ij} \geq 0$, respectively, and $\mathbf{\Phi} = [\phi_{ij}]$. The Lagrange function $\mathcal{L}$ is

$$\mathcal{L} = \mathcal{U} + tr(\mathbf{\Phi S}^T) \tag{17}$$

Let the derivatives of $\mathcal{L}$ with respect to $\mathbf{S}$ vanish, we have:

$$\frac{\partial \mathcal{L}}{\partial S} = -2 * \frac{X}{SS^T}S + 2 * 1S + 2 * BS + \mathbf{\Phi} = 0 \tag{18}$$

Using the KKT conditions $\phi_{ik}s_{ik} = 0$, we get the following equation for s_{ik}:

$$s_{ik} \longleftarrow s_{ik}\left(\frac{X}{SS^T}\right)_{ik}\left(\frac{S}{1S + SB}\right)_{ik} \tag{19}$$

and the β_k will be updated below:

$$\beta_k \longleftarrow \frac{n + a_k - 1}{\frac{1}{2}\sum_n s_{ik}^2 + b_k} \tag{20}$$

As for Gaussian likelihood the update rule will be:

$$s_{ik}^g \longleftarrow s_{ik}^g\left(\frac{2XS}{2SS^TS + SB}\right)_{ik} \tag{21}$$

and the β_k^g will be updated below:

$$\beta_k^g \longleftarrow \frac{n + a_k - 1}{\frac{1}{2}\sum_n s_{ik}^2 + b_k} \tag{22}$$

We can get an approximate fixed value in convergence for both iteration with two likelihood. And suppose the multiplicative updates stop after t iterations with parameters from Table 1, the overall computational complexity for BSNMF will be $O(tn^2k + n^2)$.

Table 1. Parameters used in complexity analysis

Parameters	Description
n	number of network nodes
k	number of initial communities count
β_k	paraments of communities number
a_k	hyper-hyperparaments a
b_k	hyper-hyperparaments b

3.4 Inference for Community Number K_C

In regular NMF methods for clustering, the object factorized dimension K should be given. But in community detection situation, we just know the relation of nodes without prior information of community number K, and it's hard to count out the suitable number of communities. If K is too small, some communities will be very large and the model can not be fitted well. On contrary, If K is too large, we can not catch the group character effectively from an entire network and occur into overfitting. We try to find K with a suitable solution between network fineness and overfitting.

To solve this problem, we propose a statistical *shrinkage* method in a Bayesian framework to find the number of communities and build a model selection based on Automatic Relevance Determination (ARD) [18, 28]. In Symmetric Bayesian NMF, we principally iterated out s_{ik} with gradual change, and the prior will try to promote a *shrinkage* to zero of s_{ik} with a rate constant proportional to β_k. A large β_k represents a belief that the half-normal distribution over s_{ik} has small variance, and hence s_{ik} is expected to get close to zero. We can see the priors and the likelihood function (quantifying how well we explain the data) are combined with the effect that columns of S which have little effect in changing how well we explain the observed data will shrink close to zero. We can effectively estimate the number of communities by computer the non-zero column number from S.

3.5 Performance Comparisons in Email Network

We compare our algorithm with other five state-of-the-art NMF methods. Seven algorithms are listed below, where algorithm 1,3,4 are asymmetric methods, and algorithm 2,5,6,7 are symmetric methods. Some of them have been applied in community detection tasks before.

1. Non-negative Matrix Factorization based clustering to be used in community detection (NMF) [35].
2. Symmetric Non-negative Matrix Factorization (SNMF) for undirected network [31, 32].
3. Graph Regularized Non-negative Matrix Factorization for Data Representation(GNMF) [4].
4. Bayesian Non-negative Matrix Factorization (BNMF) which takes likelihood as Poisson distribution [23].
5. Graph Regularized Non-negative Matrix Factorization for community detection in Symmetric form(GSNMF).
6. Symmetric Bayesian Non-negative Matrix Factorization with Gaussian likelihood (BSNMF-G).
7. Symmetric Bayesian Non-negative Matrix Factorization with Poisson likelihood (BSNMF-P).

We conduct the performance evaluations using social network datasets of different sizes with ground-truth to test semi-supervised feature. The definition of ground-truth network datasets is based on practical significance [36]. In supervised condition, we randomly select some nodes from each ground-truth network community. The details of experiments are stated below:

(1) In our methods, we apply 10 independent experiments and every experiment iterates for 500 times.

(2) We use Email[1] network [8] with decided communities numbers to compare BSNMF with other methods that should be given K manually. We calculate 3 metrics to evaluate the test results of six methods. They are Accuracy (AC), Normalized Mutual Information (NMI) [2] and Modularity(Q) [21].

$$Q = \frac{1}{4n}\mathbf{s}^T\mathbf{B}\mathbf{s} \quad where \quad B_{ij} = A_{ij} - \frac{k_ik_j}{2n} \tag{23}$$

Modularity is a measure of networks structure, and it is designed to measure the strength of communities detection. In Eq. 23, s is the column vector whose elements are s_i and B is defined as a symmetric matrix to represent the difference between the actual number of edges between node i and j and the expected number of edges between them.

The URV Email data set contains 1,133 users including faculty, researchers, technicians, managers, administrators, and graduate students at Rovira i Virgili University of Tarragona, Spain, and 5451 user-to-user (address-to-address) links to interchange. In the network, each edge represents that at least one email was sent by a campus user. As in lots of network test data, there is lack of ground truth information. We apply most popular algorithm BGLL [3] and run out a result in Email network, the result will represent as ground-truth information with highest modularity 0.5660 within existing methods.

(3) We test the BSNMF-P method in Email network to evaluate communities number K with different initial dimension number of S.

Table 2. Community detection performance comparison on Email network (Ground-truth Q=0.566)

Method	AC	NMI	Q
NMF	0.501	0.522	0.490
SNMF	0.583	0.605	0.534
GNMF	0.503	0.524	0.501
GSNMF	0.595	0.616	0.534
BNMF	0.615	0.561	0.537
BSNMF-G	0.621	0.598	0.546
BSNMF-P	**0.670**	**0.650**	**0.558**

When K is set to the communities number 15, Table 2 show the detailed detecting AC, NMI and Q on the Email undirected network. From this result, we can see that both symmetric transformation and Bayesian inference promote

---

[1] http://deim.urv.cat/~alexandre.arenas/.

Table 3. Community number K_C inference with different initial K in BSNMF-P

	K	K_C	Modularity
1	12	12	0.5088
2	23	22	0.5363
3	34	30	0.5270
4	46	30	0.5302
5	57	29	0.5387
6	68	31	0.5368
7	80	**32**	**0.5402**
8	91	30	0.5316
9	102	36	0.5349
10	114	33	0.5275

the three indicators. The symmetric Bayesian method with Poisson likelihood achieves the best performance, especially in AC and NMI metric.

When run BSNMF-P method without initial communities numbers in Email network with 1133 nodes, Table 3 lists the different result K_C and relevant Modularty Q. We can see when the initial dimensionality K is large than 30, the result K_C will shrinkage into range between 30 and 35 on the Email undirected network. Best modularity Q is achieved when K is set to 80 from 1131 nodes, so without loss of generality, we may set the initial count from $1/10$ to $1/15$ of total nodes n.

4   Conclusions

In this paper, we study a network community detection problem using Bayesian symmetric NMF. We propose a model that considerate Bayesian inference process with different noise model into symmetric NMF, and derive the updating rules and conduct experiments to valid our model. We also apply Automatic Relevance Determination methods with sparse constrain to estimate the community count of a network. Our experimental evaluations for community detection tasks show that the proposed algorithm is effective and achieves the state-of-the-art performance.

Acknowledgments. This work was supported by NSFC (no. 61272247), the Science and Technology Commission of Shanghai Municipality (Grant No. 13511500200), the Arts and Science Cross Special Fund of Shanghai Jiao Tong University under Grant 13JCY14, the European Union Seventh Framework Programme (Grant NO. 247619).

References

1. Airoldi, E.M., Blei, D.M., Fienberg, S.E., Xing, E.P.: Mixed membership stochastic blockmodels. J. Mach. Learn. Res. **9**, 1981–2014 (2008)
2. Bishop, C.M.: Pattern Recognition and Machine Learning (Information Science and Statistics). Springer, New York (2008)

3. Blondel, V.D., Guillaume, J.L., Lambiotte, R., Lefebvre, E.: Fast unfolding of communities in large networks. J. Stat. Mech. Theory Exp. **2008**(10), P10008 (2008)

4. Cai, D., He, X., Wu, X., Han, J.: Non-negative matrix factorization on manifold. In: Proceedings of the 8th IEEE International Conference on Data Mining (ICDM 2008), 15–19 December 2008, Pisa, Italy, pp. 63–72 (2008). http://dx.doi.org/10.1109/ICDM.2008.57

5. Cemgil, A.T.: Bayesian inference for nonnegative matrix factorisation models. Comput. Intell. Neurosci. **2009**, 1–17 (2009)·

6. Ding, Y.: Community detection: topological vs. topical. J. Informetr. **5**(4), 498–514 (2011)

7. Fevotte, C., Idier, J.: Algorithms for nonnegative matrix factorization with the beta-divergence. Neural Comput. **23**(9), 2421–2456 (2011)

8. Guimerà, R., Danon, L., Díaz Guilera, A., Giralt, F., Arenas, À.: Self-similar community structure in a network of human interactions. Phys. Rev. E **68**(6), 065103-1–065103-4 (2003)

9. He, Y.C., Lu, H.T., Huang, L., Shi, X.H.: Non-negative matrix factorization with pairwise constraints and graph laplacian. Neural Process. Lett. **42**(1), 167–185 (2015)

10. He, Z., Xie, S., Zdunek, R., Zhou, G., Cichocki, A.: Symmetric nonnegative matrix factorization: algorithms and applications to probabilistic clustering. IEEE Trans. Neural Netw. **22**(12), 2117–2131 (2011)

11. Kuang, D., Park, H., Ding, C.H.: Symmetric nonnegative matrix factorization for graph clustering. In: SDM, vol. 12, pp. 106–117. SIAM (2012)

12. Lai, D., Wu, X., Lu, H., Nardini, C.: Learning overlapping communities in complex networks via non-negative matrix factorization. Int. J. Mod. Phys. C **22**(10), 1173–1190 (2011)

13. Lee, D., Seung, H.: Algorithms for non-negative matrix factorization. In: Advances in Neural Information Processing Systems, vol. 13 (2001)

14. Lee, D., Seung, H., et al.: Learning the parts of objects by non-negative matrix factorization. Nature **401**(6755), 788–791 (1999)

15. Leskovec, J., Lang, K.J., Mahoney, M.: Empirical comparison of algorithms for network community detection. In: Proceedings of the 19th International Conference on World Wide Web, pp. 631–640. ACM (2010)

16. Li, T., Ding, C.: The relationships among various nonnegative matrix factorization methods for clustering. In: Sixth International Conference on Data Mining, ICDM 2006, pp. 362–371. IEEE (2006)

17. Liu, Y., Tennant, D.A., Zhu, Z., Heath, J.K., Yao, X., He, S.: Dime: a scalable disease module identification algorithm with application to glioma progression. PloS one **9**(2), e86693:1–e86693:17 (2014)

18. Mørup, M., Hansen, L.K.: Automatic relevance determination for multi-way models. J. Chemometr. **23**(7–8), 352–363 (2009)

19. Newman, M.E.J.: Coauthorship networks and patterns of scientific collaboration. Proc. Natl. Acad. Sci. **101**(Suppl. 1), 5200–5205 (2004)

20. Newman, M.E.J., Girvan, M.: Finding and evaluating community structure in networks. Phys. Rev. E **69**, 026113:1–026113:15 (2004)

21. Newman, M.E.: Modularity and community structure in networks. Proc. Natl. Acad. Sci. **103**(23), 8577–8582 (2006)

22. Plantie, M., Crampes, M.: Survey on social community detection. In: Ramzan, N., van Zwol, R., Lee, J.-S., Clüver, K., Hua, X.-S. (eds.) Social Media Retrieval. CCN, pp. 65–85. Springer, London (2013)

23. Psorakis, I., Roberts, S., Ebden, M., Sheldon, B.: Overlapping community detection using bayesian non-negative matrix factorization. Phys. Rev. E **83**(6), 066114 (2011)
24. Radicchi, F., Castellano, C., Cecconi, F., Loreto, V., Parisi, D.: Defining and identifying communities in networks. Proc. Natl. Acad. Sci. USA **101**(9), 2658–2663 (2004)
25. Schmidt, M.N., Laurberg, H.: Nonnegative matrix factorization with gaussian process priors. Comput. Intell. Neurosci. **2008**, 3 (2008)
26. Shi, M., Yi, Q., Lv, J.: Symmetric nonnegative matrix factorization with beta-divergences. IEEE Signal Process. Lett. **19**(8), 539–542 (2012)
27. Shi, X., Lu, H., He, Y., He, S.: Community detection in social network with pairwisely constrained symmetric non-negative matrix factorization. In: Proceedings of the 2015 IEEE/ACM International Conference on Advances in Social Networks Analysis and Mining, ASONAM 2015, pp. 541–546. ACM, New York (2015)
28. Tan, V.Y.F., Fevotte, C.: Automatic relevance determination in nonnegative matrix factorization with the beta-divergence. IEEE Trans. Pattern Anal. Mach. Intell. **35**(7), 1592–1605 (2013)
29. Tang, L., Liu, H.: Community detection and mining in social media. Synth. Lect. Data Min. Knowl. Discov. **2**(1), 1–137 (2010)
30. Tang, X., Xu, T., Feng, X., Yang, G.: Uncovering community structures with initialized bayesian nonnegative matrix factorization. PLoS ONE **9**(9), e107884 (2014)
31. Wang, D., Li, T., Zhu, S., Ding, C.: Multi-document summarization via sentence-level semantic analysis and symmetric matrix factorization. In: Proceedings of the 31st Annual International ACM SIGIR Conference on Research and Development in Information Retrieval, pp. 307–314. ACM (2008)
32. Wang, F., Li, T., Wang, X., Zhu, S., Ding, C.: Community discovery using non-negative matrix factorization. Data Min. Knowl. Discov. **22**(3), 493–521 (2011)
33. Wu, M., Scholkopf, B.: A local learning approach for clustering. Adv. Neural Inf. Process. Syst. **19**, 1529 (2007)
34. Xie, J., Kelley, S., Szymanski, B.K.: Overlapping community detection in networks: the state-of-the-art and comparative study. ACM Comput. Surv. **45**(4), 43:1–43:35 (2013)
35. Xu, W., Liu, X., Gong, Y.: Document clustering based on non-negative matrix factorization. In: Proceedings of the 26th Annual International ACM SIGIR Conference on Research and Development in Informaion Retrieval, pp. 267–273. ACM (2003)
36. Yang, J., Leskovec, J.: Defining and evaluating network communities based on ground-truth. In: Proceedings of the ACM SIGKDD Workshop on Mining Data Semantics, pp. 3:1–3:8 (2012)
37. Yang, J., Leskovec, J.: Overlapping community detection at scale: a nonnegative matrix factorization approach. In: Proceedings of the Sixth ACM International Conference on Web Search and Data Mining, pp. 587–596. ACM (2013)
38. Zhao, Y., Levina, E., Zhu, J.: Community extraction for social networks. Proc. Natl. Acad. Sci. **108**(18), 7321–7326 (2011)

Maximizing the Influence Ranking
Under Limited Cost in Social Network

Xiaoguang Hong[1], Ziyan Liu[1], Zhaohui Peng[1(✉)], Zhiyong Chen[1,2], and Hui Li[1]

[1] School of Computer Science and Technology, Shandong University, Jinan, China
{hxg,pzh,chenzy,lih}@sdu.edu.cn, liuziyan@mail.sdu.edu.cn
[2] National Engineering Laboratory for ECommerce, Jinan, China

Abstract. Influence maximizing is an important problem which has been studied widely in recent years. There are many situations in which people are more concerned about influence ranking than influence coverage in competition network, because some times only the top-ranked users can win the rewards, while few researchers have studied this problem. In this paper, we consider the problem of selecting a seed set under limited cost to get as high influence ranking as possible. We show this problem is NP-hard and propose a Intelligent Greedy algorithm to approximately solve the problem and improve the efficiency based on the submodularity. Furthermore, a new Cost-Effective Multi-Step Influence Adjust algorithm is proposed to get high efficiency. Experimental results show that our Intelligent Greedy algorithm achieves better effectiveness than other algorithms and the Cost-Effective Multi-Step Influence Adjust algorithm achieves high efficiency and gets better effectiveness than Degree and Random algorithms.

Keywords: Social network · Influence propagation · Competition · Cost · Multi-step influence adjust

1 Introduction

With the popularity of online social network such as Facebook, Twitter, Sina Microblog, WeChat, etc., many researchers have studied the influence diffusion problem in social network, including the diffusion of ideas, news, new products, and so on. We collectively refer to these diffusions as influence propagation. In this field, much research has been related to influence maximization problem which is the problem of finding at most k nodes (seed nodes) in social network that could maximize the spread of influence. However, competition exists extensively in the real world, sometimes only the top-ranked users can win the rewards, such as in network voting, sales promoting competition, etc. In these situations, people are more concerned about influence ranking than influence coverage, so we address the influence ranking maximization problem in competitive network.

Some recent studies have considered competitive influence propagation. Most of them study the competitive influence maximizing problem [11], while some of

© Springer International Publishing Switzerland 2016
F. Li et al. (Eds.): APWeb 2016, Part I, LNCS 9931, pp. 220–231, 2016.
DOI: 10.1007/978-3-319-45814-4_18

them study the influence blocking problem [6,8]. In our problem, we not only need to expand our influence, but also need to block some competitors' influence simultaneously. Only when our influence is larger than that of competitors, we can achieve higher influence ranking in competitive social network. Further, different seed nodes often need different cost and the total budget is limited, so considering the total cost constraint of the seed set is more meaningful than only considering the size constraint of the seed set. The problem of how to select a seed set under limited cost to get as high influence ranking as possible is a very meaningful and challenging problem. However, to our best knowledge, few research has paid attention to this problem.

In this paper, we formulate the problem of maximizing the influence ranking under limited cost in competitive social network(MRLC). We prove that the problem under the competitive independence model(CIC) is NP-hard. Firstly, we propose the Intelligent Greedy algorithm to approximately solve this problem. As it is inefficient, we improve it's efficiency base on the submodularity. The improved Intelligent Greedy algorithm runs obviously faster than original algorithm. Furthermore, to meet the need of solving the problem efficiently in large social network, we propose an efficient algorithm called Cost-Effective Multi-Step Influence Adjust. Its effectiveness is better than Degree and Random algorithms while the efficiency is much better than Intelligent Greedy algorithm.

The rest of this paper is organized as follows: in Sect. 2, we review related work. We formulate this problem and prove it is NP-hard in Sect. 3. In Sect. 4, we present the Intelligent Greedy algorithm and Improved Intelligent Greedy algorithm. In Sect. 5, we propose the Multi-Step Influence Adjust algorithm. In Sect. 6, we experimentally compare the effectiveness and efficiency of those algorithms. Finally, we conclude the paper in Sect. 7.

2   Related Work

Influence maximization is the problem of finding at most k nodes in social network that could maximize the spread of influence. This problem is first modeled as discrete optimization problem in [1] and two basic influence propagation models are proposed, independent cascade model (IC) and linear threshold model (LT). Based on the two models, [1] proposes a greedy algorithm to solve the problem. A number of studies have tried to solve this problem more efficiently [2,4,5,7]. But those research do not consider the competition problem.

The competitive influence propagation has been captured by a number of research in recent years [7,8]. Bharathi, S extends the IC model to competitive influence propagation and gives approximation algorithm for computing the best response to an opponent's strategy [3]. Influence blocking maximizing problem has been studied in [6,8]. However, they do not consider that we need to expand our influence and block competitors' influence simultaneously. Goyal, Amit considers minimizing budget problem in social network but he does not address the competition problem [9]. Liu, Z considers to expand our influence and block competitors' influence simultaneously but they only consider one competitor [10].

When there are multiple competitors, we need to choose a more intelligent algorithms to solve this problem. Sometimes we need to maximize our influence and block competitors' influence simultaneously to get higher ranking. Further for the budget is limited, we have to get as high ranking as possible under limited cost. We're trying to solve this problem in this paper.

3   Model and Problem

In this section, we define the CIC model and MRLC Problem. Then, we prove MRLC Problem is NP-Hard.

3.1   CIC Model

Independence cascade model(IC) is a basic diffusion model proposed in [1], which has been used in many works [2,4,5]. We now define the competitive independence model (CIC) [3], every node in social network has multiple states. Influenced by S(it means it is activated by our selected seed Set S), influenced by C_i (it means the node is activated by the i-th competitor's selecting seed Set C_i), or not be influenced. At step 0, all nodes in seed set S were influenced by S, all nodes in seed set C_i were influenced by C_i. At any step $t > 0$, if vertex v is influenced by S or C_i in step t-1, at step t, v has one chance to influence each current inactive out-neighbor u with probability p_{vu}. If v succeeds, then u becomes the same state with v. If there are two or more nodes trying to influence u at the same step, their attempts are sequenced in arbitrary order.

3.2   Maximizing Influence Ranking Under Limited Cost Problem

In our problem, we use a directed graph $G = (V, E)$ to represent the social network and use the CIC model as the propagation model. We suppose every competitor i has selected a seed set C_i in V and then try to find seed set S under limited cost to get as high influence ranking as possible. As competitors will not change the seed sets, the seed set S is the only variable, and the influence of S and competitors will change with the change of S. Let $InfS(S)$ represents the number of users influenced by S in V and $InfC_i(S)$ represents the number of users influenced by C_i. When seed set S changes, $InfS(S)$ and $InfC_i(S)$ will change at the same time. As every set has an influence value, we can rank the sets' influence. The problem we want to solve is to find the set S under limited cost such that the influence of S will rank as high as possible. The formal problem is defined below.

Definition 1 (MRLC). *The input of the problem includes the directed graph $G(V, E)$, the seed set C_i which the competitor i has selected, the influence propagation probability P, the $cost_v$ of node $v \in V$ and the limited cost LC. We use $cost(S)$ to represent the total cost of seed set S.*

$$cost(S) = \sum_{v \in S} cost_v \tag{1}$$

The target is to find a seed set S^* that can achieve higher ranking in the competition under limited cost. That is

$$S^* = \arg\max_{cost(S)<=LC} Rank(S) \tag{2}$$

Theorem 1. *The MRLC problem is NP-hard under the CIC model.*

Proof. Considering an instance of Set Cover problem, $U = \{u_1, u_2, \ldots, u_n\}$ is the universal set. $S = \{S_1, S_2, \ldots, S_m\}$ represents the collection of sets whose union equals the universe U. The Set Cover Problem is to identify whether there exist k of the sets whose union equals the universe set U. We can encode the Set Cover instance as an special instance of MRLC problem.

Given an arbitrary Set Cover problem instance. We define a corresponding directed bipartite graph with $2n + m + k$ nodes representing a special MRLC problem. We partition the nodes into three parts, in which the first part has n nodes and node i corresponds to u_i in U, and the second part has m nodes with node j corresponding to the S_j in S, and the third part has $n + k$ nodes and not connecting to the first part and second part. If $u_i \in S_j$, there is a directed edge from node j to node i with a propagation probability $p_{ji} = 1$. We assume each node's cost is 1, and the limited cost is k and only one competitor who has selected the $n + k$ nodes in the third part as his seed set C. As he has influenced $n + k$ node in universe set U, if we want to get higher ranking, we must influence at least $n + k$ nodes in the universe set U with at most k seed nodes. The Set cover problem is equivalent to the problem of if there is a set S of at most k nodes with $InfS(S) > n + k$. Suppose we find a solution set S. The solution is also a solution to Set Cover problem. Since the set cover is NP-hard, MRLC problem is NP-hard too.

4   Intelligent Greedy Algorithm

In this section, we try to use greedy algorithm to approximately solve MRLC Problem. The most extensively studied algorithm in influence propagation problem is greedy algorithm and always has a better effectiveness than other algorithms. So we firstly try to use greedy algorithm to approximately solve the problem.

Algorithm 1 gives the details of Intelligent Greedy algorithm. $InfS(S)$ represents the number of users influenced by seed set S and $InfC_i(S)$ represents the number of users influenced by seed set C_i which is the seed set selected by the i-th competitor. C_B represents the seed set which is selected by competitors that we want to block. C represent all competitors' seed node. $InfC_B(S)$ represents the number of users influenced by seed set C_B. In this algorithm, we must intelligently select competitors to block. We firstly compute each competitor's influence when there is no blocking, then we choose the competitor whose influence is the most similar to S to block, and name it's seed set as C_B.

The number of users influenced by node v can be represented as:

$$Inf(v) = InfS(S \cup v) - InfS(S) \tag{3}$$

Algorithm 1. Intelligent Greedy algorithm

Input:
  $G(V, E)$, seed sets C selected by competitors, $cost_v$, Limited Cost LC
Output:
  Seed set S
1: initialize S=∅
2: **while** $cost(S) < LC$ **do**
3:   **for** each $v \in V \setminus (S \cup C)$ **do**
4:     $P(v) = \frac{Inf(v) + Block(v)}{cost_v}$
5:   **end for**
6:   select $u = \arg\max_v \{P(v) | v \in V \setminus (S \cup C)\}$
7:   $S = S \cup u$
8: **end while**
9: **return** S

The number of users blocked by node v can be represented as:

$$Block(v) = InfC_B(S) - InfC_B(S \cup v) \tag{4}$$

Then the cost performance of node v can be defined as:

$$P(v) = \frac{Inf(v) + Block(v)}{cost_v} \tag{5}$$

In Algorithm 1, we use Monte Carlo simulation to approximately calculate the influence of C_i and S under the CIC model. We use $P(v)$ to represent the cost performance of v and every round we select the node u with the highest cost performance until $cost(S \cup u) > LC$.

Because the Monte Carlo simulation needs to run too many times to obtain accurate estimation, the running time is very long especially in large social network. We improve the efficiency of Intelligent Greedy algorithm based on the submodularity. A fuction f(.) is submodular if for all $S \subseteq T$ and $v \notin S$, it has the following property: $f(S \cup v) - f(S) \geq f(T \cup v) - f(T)$. This means the margin gain of adding v to a set S is at least as high as that of adding v to the superset of S.

Theorem 2. $InfS(\cdot)$ *is submodular under the CIC model.*

Proof. Since influence spreading in $G(V, E)$ is a stochastic process under the CIC model. Considering that a newly influenced node v attempts to influence his neighborhood w with probability p_{vw}. We can view that this stochastic outcome has been determined by flipping a coin with bias p_{vw}. We assume that at the beginning of this process, we pre-flip all the coins to determine which edges are live and store the result. Let $G'(V', E')$ represents the pre-flip result. The E' represents the set of live edges in E, and V' represents the nodes in V. We now prove that for every pre-flip $G'(V', E')$, the $InfS_{G'}(\cdot)$ is submodular.

Let $S1$ and $S2$ be two set of nodes and $S1 \subseteq S2$, Let $s(v, w)$ represents the shortest graph distance from v to w in G' and $g(T, w)$ represents the shortest

graph distance from any node in set T to w in G'. C represents all competitors' seed nodes and never changes. If $s(v, w) <= g(C, w)$ and v is influenced by S, v will successfully influence w, and w becomes the same state with v. Let $L(v) = \cup\{w|(w \in V')\&s(v, w) <= g(C, w)\}$
$L(v)$ represents the node set which shortest graph distance to v is shorter than other node in set C. Then $InfS(S) = |\cup_{u \in S} L(u)|$.
$InfS_{G'}(S)$ represents the number of users influenced by seed set S in G'.
$Inf_{G'}(v)$ represents the number of users influenced by seed v in G'.
$Inf_{G'}(v) = InfS_{G'}(S \cup v) - InfS_{G'}(S)$. As $S1 \subseteq S2$, $\cup_{u \in S1}\{S(u)\} \subseteq \cup_{u \in S1}\{S(u)\}$. It is clear that the number in L(v) not in $\cup_{u \in S1}\{S(u)\}$ is at least as larger as the number in S(v) not in $\cup_{u \in S2}\{S(u)\}$, it means
$InfS_{G'}(S1 \cup v) - InfS_{G'}(S1) \geq InfS_{G'}(S2 \cup v) - InfS_{G'}(S1)$.
It means $InfS_{G'}(\cdot)$ is submodular.
$InfS(S) = \sum_{all\ G'} Prob[G'] \cdot InfS_{G'}(S)$.
Since all $InfS_{G'}(\cdot)$ are submodular, so $InfS(\cdot)$ is submodular too.

Theorem 3. $InfC_B(\cdot)$ *is submodular under the CIC model.*

As the provement of this theorem is similar to that of Theorem 2, the detail of this movement is omitted here due to space constraints.

We use p_v^i to represent the cost performance of node v and use S^i to represent the seed set which has been selected in previous i rounds. We use $P_j(v)$ to represent the cost performance of v and S^j to represent the seed set which has been selected in previous j rounds. If $j > i$ then $S^i \subseteq S^j$. As $InfS(\cdot)$ and $InfC_B(\cdot)$ is submodular, $InfS(S^i \cup v) - InfS(S^i) \geq InfS(S^j \cup v) - InfS(S^j)$
$InfC_B(S^i) - InfC_B(S^i \cup v) \geq InfC_B(S^j) - InfC_B(S^j \cup v)$
$P_i(v) = \frac{InfS(S^i \cup v) - InfS(S^i) + InfC_B(S^i) - InfC_B(S^i \cup v)}{cost_v}$
$P_j(v) = \frac{InfS(S^j \cup v) - InfS(S^j) + InfC_B(S^j) - InfC_B(S^j \cup v)}{cost_v}$
If $j \geq i$ then $P_i(v) \geq P_j(v)$.
If $P_j(v) \geq P_i(u)$ then $P_j(v) \geq P_i(u)$ and in the j round we need not to recompute $P_j(v)$, because node u's cost performance must less than v and can not be added to the seed S in this round.

Algorithm 2 gives the details of improved Intelligent Greedy algorithm, which is similar to the origin algorithm. In the first round, it also need to compute the cost performance of all nodes. In the next round, if the cost performance of a node in the last round is not better than the best cost performance having been computed in this round, there is no need to compute it again.

5   Cost-Effective Multi-step Influence Adjust Algorithm

Although we have improved the Intelligent Greedy algorithm, the running time is still fairly long especially in large network. We may need to find some high efficiency algorithm to solve this problem more efficiently. Degree is a simple heuristic which in every round selects the node with the largest degree. Degree is frequently used to select seed set in influence propagation problem, but it does

Algorithm 2. Improved Intelligent Greedy algorithm

Input:
$G(V, E)$, seed sets C selected by competitors, $cost_v$, Limited Cost LC
Output:
Seed set S
1: initialize S=$\emptyset$
2: **for** each $v \in V$ **do**
3:    $P(v) = 0$
4: **end for**
5: **while** $cost(S) < LC$ **do**
6:    $bestPerformance = 0$
7:    **for** each $v \in V$ **do**
8:       **if** $(P(v) > bestPerformance || P(v) == 0)$ **then**
9:          $P(v) = \frac{Inf(v) + Block(v)}{cost_v}$
10:          **if** $P(v) > bestPerformance$ **then**
11:             $bestPerformance = P(v)$
12:          **end if**
13:       **end if**
14:    **end for**
15:    **if** (S is empty) **then**
16:       rank $v \in V \setminus (S \cup C)$ by their cost performance
17:    **end if**
18:    select $u = \arg\max_v \{P(v)|v \in V \setminus (S \cup C)\}$
19:    $S = S \cup u$
20: **end while**
21: **return** S

---

not consider the neighbors state and influence. If a node v's out-neighbor u is influenced, node v can't influence u anymore and if u is an influenced in-neighbor of v, u may influence v in the next round. Node v's out-neighbors often have different influence, so we should treat them differently.

We propose a new heuristic called Multi-Step Influence Adjust Algorithm, which achieves better effectiveness than Degree and its running time is much shorter than Greedy. We use $O(v)$ to represents the out-neighbors of v and $I(v)$ represents the in-neighbors of v. C_B represents the seed set which is selected by competitors that we want to block. In this algorithm, we first compute every competitors' cost, then we find the competitor whose cost is the most similar to the limited cost to block, and select its seed set as C_B. In Multi-Step Influence Adjust Algorithm, $Inf(v)$ represents the influence of node v.

$$Inf(v) = 1 + \sum_{u \in O(v) \setminus (S \cup C_B)} Inf(u) * p_{vu} \tag{6}$$

$$Inf(v) = Inf(v) * \prod_{u \in I(v) \cap C_B} (1 + p_{uv}) * \prod_{u \in I(v) \cap S} (1 - p_{uv}) \tag{7}$$

Algorithm 3 gives the details of Cost-Effective Degree Adjust algorithm. First, We compute $Inf(v)$ through K times iteration with Eq. 6. The we adjust

---

Algorithm 3. Cost-Effective Multi-Step Influence Adjust Algorithm

---

Input:
  $G(V, E)$, seed sets C selected by competitors, $cost_v$, Limited Cost LC
.**Output:**
  Seed set S
1: initialize S=$\emptyset$
2: **for** each $v \in V$ **do**
3:    $Inf(v) = 1$
4: **end for**
5: **for** each $v \in C \cup S$ **do**
6:    $Inf(v) = 0$
7: **end for**
8: **for** $i = 0$ to K **do**
9:    **for** each $v \in V \setminus S \cup C$ **do**
10:       $Inf(v) = 1 + \sum_{u \in O(v) \setminus (S \cup C_B)} Inf(u) * p_{vu}$
11:    **end for**
12: **end for**
13: **for** each $v \in V \setminus S \cup C$ **do**
14:    $Inf(v) = Inf(v) * \prod_{u \in I(v) \cap C_B} (1 + p_{uv}) * \prod_{u \in I(v) \cap S} (1 - p_{uv})$
15:    $P(v) = \frac{Inf(v)}{cost_v}$
16: **end for**
17: **while** $cost(S) < LC$ **do**
18:    select $u = \arg\max_v \{P(v) | v \in V \setminus (S \cup C)\}$
19:    $S = S \cup u$
20:    **for** each $v \in V \setminus S \cup C$ **do**
21:       recompute $P(v)$ as before
22:    **end for**
23: **end while**
24: **return** S

$Inf(v)$ using Eq. 7. Then at each round we find the node v with the biggest $P(v)$ and recompute $P(v)$ of each node until $cost(S \cup v) > LimitCost$.

6   Experiments

To test the efficiency and effectiveness of our algorithms for solving MRLC problem, we use three real datasets to compare the probability of get a high ranking and running time of different algorithms. In the experiments, we use Monte Carlo simulation 10000 times to obtain competitors ranking. We suppose that there are five competitors and each competitor random selects five seeds. We count every ranking result of each simulation and calculate the probability of getting different ranking. As the rankings in every simulation are often different, we use the probability of getting higher ranking to compare different algorithm's effect. It is more accurate than computing the average influence ranking. It is obvious that there is a position correlation between the cost of node and it's out-neighbor number. In the experiment we design two kinds of cost function to represent different relations between the cost and the out-neighbor number of the node.

Linear cost Function:

$$cost(v) = 100 + v.outdegree \tag{8}$$

Exponential cost Function:

$$cost(v) = 100 + 1.01^{v.outdegree} \tag{9}$$

$v.outdegree$ represents the v's out-neighbor number. We set the influence propagation probability P = 0.01. We run the simulation under the CIC model on the network for 10000 times and count the influence and ranking result of each time.

6.1   Experimental Setup

The real datasets used in experiments are got from Stanford Large Network Dataset Collection[1]. The details of the three datasets are listed in Table 1.

Table 1. Details of datasets

Name	Nodes	Edges	Description
Slashdot	7.7K	176.5K	Slashdot social network
Epinions	7.6K	238.1K	Who-trusts-whom network of Epinions.com
Facebook	4.0K	88.5K	Facebook social network

The code is written in C++ and runs on a Linux server with 2.00 GHz Intel Xeon E5-2620 and 32 G memory. We compared the algorithms as follows:

- Random : randomly selecting node v and adding it to S until $cost(S \cup v) > LC$.
- Degree : a simple heuristic that every round selecting a seed v with the largest out-degree until $cost(S \cup v) > LC$.
- Improved Cost-Effective Greedy: (Algorithm 2) in every round adding the node v with the best cost performance to S until $cost(S \cup v) > LC$.
- Cost-Effective Multi-Step Influence Adjust: (Algorithm 3) computing the $P(v)$ for each node v in V at each round and adding the node v with the largest $P(v)$ to seed set S until $cost(S \cup v) > LC$.

6.2   Experimental Results

In the experiment, we mainly test the running time and probabilities of getting top ranking of different algorithms. Figures 1, 2, 3 and 4 show the results of

[1] https://snap.stanford.edu/data/.

using Linear Cost Function (Eq. 8). Figures 5, 6, 7 and 8 show the results of using Exponential Cost Function (Eq. 9).

Figure 1 shows the probabilities of gaining top ranking of different algorithms on the Slashdot dataset under the CIC model. The vertical axis represents the probability of gaining top ranking. The horizontal axis represents the limited cost of S. From Fig. 1 we can see that the effectiveness of Random algorithm is the worst and Cost-Effective Multi-Step Influence Adjust algorithm's effectiveness is much better than Degree and Random algorithms. The effectiveness of Improved Cost-Effective Greedy is better than other algorithm. Figures 2 and 3 show the experimental results in Epinions and Facebook datasets. From Figs. 2 and 3 we can see that the effectiveness of different algorithms are similar to that in Fig. 1.

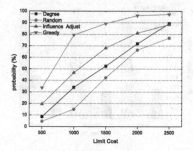

Fig. 1. Probabilities on the Slashdot graph (linear cost function).

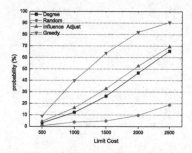

Fig. 2. Probabilities on the Epinions graph (linear cost function).

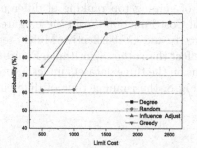

Fig. 3. Probabilities on the Facebook graph (linear cost function).

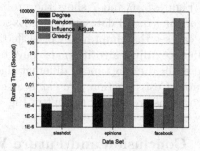

Fig. 4. Running time on all datasets (linear cost function).

Figure 4 shows the running time of different algorithms on the three datasets under the CIC model when the limited cost is 2000. The vertical axis represents the running time and the horizontal axis represents different datasets. From this figure we can see that the running time of Degree, Random and Cost-Effective Multi-Step Influence Adjust algorithm are approximately similar, while Improved Intelligent Greedy algorithm runs very slowly comparing to other algorithms even we have used submodularity to accelerate computation.

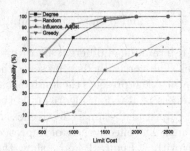

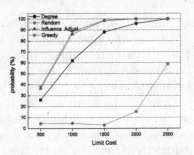

Fig. 5. Probabilities on the Slashdot graph (exponential cost function).

Fig. 6. Probabilities on the Epinions graph (exponential cost function).

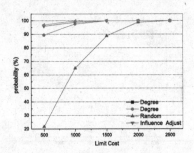

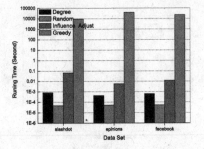

Fig. 7. Probabilities on the Facebook graph (exponential cost function).

Fig. 8. Running time on all datasets (exponential cost function).

Figures 5, 6 and 7 show the probability of gaining top ranking of different algorithms on all datasets under Exponential Cost Function. From these figures we can see that Improved Cost-Effective Greedy and Cost-Effective Multi-Step Influence Adjust algorithm also have better effect than other algorithms. Figure 8 shows the running time of different algorithms on the three datasets under Exponential Cost Function. We can see that in Fig. 8, the running time of different algorithms are similar to that in Fig. 4.

7   Conclusion and Future Work

Maximizing the influence ranking under limited cost is a very useful problem in the real word application. However, to our best knowledge, this problem is ignored by former works. In this paper, we define this problem and give some algorithms to solve this problem. Experiments in real datasets show that our Intelligent Greedy algorithm achieves better effectiveness and Cost-Effective Multi-Step Influence Adjust algorithm achieves high efficiency and get better effectiveness than Degree and Random algorithm. Future work will focus on the problem of optimizing the competitor blocking strategy to get better effectiveness.

Acknowledgments. This work is supported by NSF of China (No. 61303005), NSF of Shandong, China (No. ZR2013FQ009), the Science and Technology Development Plan of Shandong, China (No. 2014GGX101047, No. 2014GGX101019).

References

1. Kempe, D., Kleinberg, J., Tardos, E.: Maximizing the spread of influence through a social network. In: KDD, pp. 137–146 (2003)
2. Wei, C., Yajun, W., Siyu, Y.: Efficient influence maximization in social networks. In: SIGKDD, pp. 199–208 (2009)
3. Bharathi, S., Kempe, D., Salek, M.: Competitive influence maximization in social networks. In: Deng, X., Graham, F.C. (eds.) WINE 2007. LNCS, vol. 4858, pp. 306–311. Springer, Heidelberg (2007)
4. Chen, W., Wang, C., Wang, Y.: Scalable influence maximization for prevalent viral marketing in large-scale social networks. In: KDD, pp. 1029–1038 (2010)
5. Jung, K., Heo, W., Chen, W.: Irie: scalable and robust influence maximization in social networks. In: ICDM, pp. 918–923 (2012)
6. He, X., Song, G., Chen, W., et al.: Influence blocking maximization in social networks under the competitive linear threshold model. In: SDM (2012)
7. Borodin, A., Filmus, Y., Oren, J.: Threshold models for competitive influence in social networks. In: Saberi, A. (ed.) WINE 2010. LNCS, vol. 6484, pp. 539–550. Springer, Heidelberg (2010)
8. Budak, C., Agrawal, D., Abbadi, A.E.: Limiting the spread of misinformation in social networks. In: WWW, pp. 665–674 (2011)
9. Zhang, P., Chen, W., Sun, X., et al.: Minimizing seed set selection with probabilistic coverage guarantee in a social network. In: KDD, pp. 1306–1315 (2014)
10. Liu, Z., Hong, X., Peng, Z., et al.: Minimizing the cost to win competition in social network. In: Web Technologies and Applications, pp. 598–609 (2015)
11. Li, H., Bhowmick, S.S., Cui, J., GetReal, et al.: Towards realistic selection of influence maximization strategies in competitive networks. In: SIGMOD, pp. 1525–1537 (2015)

Towards Efficient Influence Maximization for Evolving Social Networks

Xiaodong Liu$^{(\boxtimes)}$, Xiangke Liao, Shanshan Li, and Bin Lin

National University of Defense Technology, Changsha 410073, China
{liuxiaodong,xkliao,shanshanli,binlin}@nudt.edu.cn

Abstract. Identifying the most influential individuals can provide invaluable help in developing and deploying effective viral marketing strategies. Previous studies mainly focus on designing efficient algorithms or heuristics to find top-K influential nodes on a given static social network. While, as a matter of fact, real-world social networks keep evolving over time and a recalculation upon the changed network inevitably leads to a long running time, significantly affecting the efficiency. In this paper, we observe from real-world traces that the evolution of social network follows the preferential attachment rule and the influential nodes are mainly selected from high-degree nodes. Such observations shed light on the design of IncInf, an incremental approach that can efficiently locate the top-K influential individuals in evolving social networks based on previous information instead of calculation from scratch. In particular, IncInf quantitatively analyzes the influence spread changes of nodes by localizing the impact of topology evolution to only local regions, and a pruning strategy is further proposed to effectively narrow the search space into nodes experiencing major increases or with high degrees. We carried out extensive experiments on real-world dynamic social networks including Facebook, NetHEPT, and Flickr. Experimental results demonstrate that, compared with the state-of-the-art static heuristic, IncInf achieves as much as 21× speedup in execution time while maintaining matching performance in terms of influence spread.

1 Introduction

Influence maximization (IM) is one fundamental and important problem which aims to identify a small set of influential individuals so as to develop effective viral marketing strategies in large-scale social networks [7]. As a matter of fact, real-world social networks keep evolving over time. For example, in Facebook, new people might join while old ones might withdraw, and people might make new friends with each other. Moreover, real-world social networks are evolving in a rather surprising speed; it is reported that as much as 1 million new accounts are created in Twitter every day. Such massive evolution of network topology, on the contrary, may lead to a significant transformation of the network structure, thus raising a natural need of efficient reidentification.

Existing researches on influence maximization focus mainly on developing effective and efficient algorithms on a given static social network. Although one

© Springer International Publishing Switzerland 2016
F. Li et al. (Eds.): APWeb 2016, Part I, LNCS 9931, pp. 232–244, 2016.
DOI: 10.1007/978-3-319-45814-4_19

could possibly run any of the static influence maximization methods, such as [5,6,12], to find the new top-K influential individuals when the network is updated, this approach has some inherent drawbacks that cannot be neglected: (1) the running time of a specific static method can be extremely long and unacceptable especially on large-scale social networks, and (2) whenever the network topology is changed, we need to recalculate the influence spreads for all the nodes which leads to very high costs. Can we quickly and efficiently identify the influential nodes in evolving social networks? Can we incrementally update the influential nodes based on previously known information instead of frequently recalculating from scratch?

Unfortunately, the rapidly and unpredictably changing topology of a dynamic social network poses several challenges in the reidentification of influential users. On one hand, the interconnections between edges in real-world social graphs are rather complicated; as a result, even one small change in topology may affect the influence spreads of a large number of nodes, not to mention the massive changes in large-scale social networks. It is very difficult to efficiently compute the changes of influence spreads for all the nodes after the evolution. On the other hand, since there are a great many nodes in large-scale social networks, how to effectively limit the range of potential influential nodes and reduce the amount of calculation as much as possible is a very challenging problem.

To well address these challenges, we investigate the dynamic characteristics exhibited during the evolution of real-world social networks. Through tests on three real-world dataset traces, Facebook [15], NetHEPT [2] and Flickr [14], we observe that, first, the growth of social network is mainly based on the preferential attachment principle [3], that is the new-coming edges prefer to attach to nodes with higher degree, which naturally leads to the "rich-get-richer" phenomena; and second, the top-K influential nodes are mainly selected from those high-degree nodes. Inspired by such observations, we know that the influence changes of some nodes will have no impact on the top-K selection, and thus can be pruned to reduce the amount of calculation. Motivated by this, we propose IncInf, an incremental method to identify the top-K influential nodes in evolving social networks instead of recalculating from scratch, thus significantly improves the efficiency and scalability to handle extraordinarily large-scale networks. To summarize, the main contributions of IncInf are as follows:

First, we design an efficient approach to quantitatively analyze the influence spread changes from topology evolution by adopting the idea of localization. A tunable parameter is provided to tradeoff between efficiency and effectiveness.

Second, we propose a pruning strategy which could effectively narrow the search space into nodes only experiencing major increases or with high degrees based on the changes of influence spread and the previous top-K information.

Third, we conduct extensive experiments on three real-world social networks. Compared with the state-of-the-art algorithm, IncInf achieves up to 21× speedup in execution time while providing matching influence spread. Moreover, IncInf provides better scalability to scale up to extraordinarily large-scale networks.

2   Preliminaries and Problem Statement

Social Network. A social network is formally defined as a directed graph $G = (V, E, P)$ where node set $V = \{v_1, v_2, \cdots, v_n\}$ denotes entities in the social network. Each node can be either active or inactive, and will switch from being inactive to being active if it is influenced by others nodes. Edge set $E \subset V \times V$ is a set of directed edges representing the relationship between different users. Take Twitter as an example. A directed edge (v_i, v_j) will be established from node v_i to v_j if v_i is followed by v_j, which indicates that v_j may be influenced by v_i. P denotes the influence probability of edges; each edge $(v_i, v_j) \in E$ is associated with an influence probability $p(v_i, v_j)$ defined by function $p : E \rightarrow [0, 1]$.

Independent Cascade (IC) Model. IC model is a diffusion model that has been well studied in [5,11,12]. Given an initial set S, the diffusion process of IC model works as follows. At step 0, only nodes in S are active, while other nodes stay in the inactive state. At step t, for each node v_i which has just switched from being inactive to being active, it has a single chance to activate each currently inactive neighbor v_j, and succeeds with a probability $p(v_i, v_j)$. If v_i succeeds, v_j will become active at step $t+1$. If v_j has multiple newly activated neighbors, their attempts in activating v_j are sequenced in an arbitrary order. Such a process runs until no more activations are possible [11]. We use $\sigma(S)$ to denote the influence spread of set S, which is defined as the expected number of active nodes at the end of influence propagation.

IM problem in Evolving Networks. An evolving network $\zeta = (G^0, G^1, \cdots, G^t)$ is defined as a sequence of network snapshots evolving over time, where $G^t = (V^t, E^t, P^t)$ is the network snapshot at time t. $\Delta G^t = (\Delta V^t, \Delta E^t, \Delta P^t)$ denotes the structural change of network G^t from time t to $t + 1$. Obviously, we have $G^{t+1} = G^t \bigcup \Delta G^t$. The influence maximization problem is defined as follows:

Given: Social network G^t at time t, the top-K influential nodes S^t in G^t, and the structural evolution ΔG^t of graph G^t.

Objective: Identify the influential nodes $S^{t+1} \subset V^{t+1}$ of size K in G^{t+1} at time $t + 1$, such that $\sigma(S^{t+1})$ is maximized at the end of influence diffusion.

3   Observations of Social Network Evolution

3.1   Preferential Attachment Rule

Understanding the pattern of the network topology evolution is of primary importance to design efficient influence maximization algorithms for evolving social networks. We first study the preferential attachment rule [3], or in other words, the "rich-get-richer" rule [8], which postulates that when a new node joins the network, it creates a number of edges, where the destination node of each edge is chosen proportional to the destination's degree. This means that new edges are more likely to connect to nodes with high degree than ones with

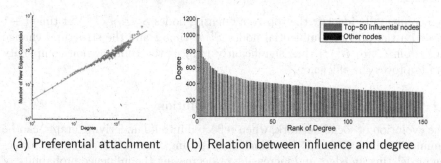

(a) Preferential attachment   (b) Relation between influence and degree

Fig. 1. Network topology evolution pattern on Facebook.

low degree. This is reasonable in reality; Lady Gaga gains 30,000 new followers on average every day which can never image for any common individual. The results on the Facebook dataset are demonstrated in Fig. 1(a) where the x axis is the degree of different nodes and the y axis is the average number of new edges attached to nodes of different degree. Note that both the x and y axis are in log scale. From Fig. 1(a) we can see that the degree of users in Facebook is linearly correlated with the number of new links created. This suggests that high-degree nodes get super-preferential treatment. Consequently, the influence spread change should be considerably great for the influential nodes, while there may be only small or even no change for ordinary people.

3.2 Relation Between Influence and Degree

Examining the relation between the influence and the degree of node can help us understand the effect of degree changing on the influence spread of nodes. For this reason, we run the static MixGreedy algorithm [5] on the final graph and identify the top-50 influential nodes. The results on the Facebook dataset are illustrated in Fig. 1(b) where the x axis is the rank of degrees of different nodes (we only show the top 150). Obviously, all the selected influential nodes have a large degree. In particular, among the 50 nodes, 48 nodes rank in top 100 of the whole 61,096 nodes in terms of degree, and the other two nodes rank 102 and 111 respectively. While on the NetHEPT and Flickr datasets, the top-50 influential nodes are selected from the top 1.79 % and 0.84 % nodes in degree, respectively. This demonstrates that the top-K influential nodes are mainly selected from those with large degrees.

4   IncInf Design

In this section, we present the detailed design of IncInf, an incremental approach to solve the influence maximization problem on dynamic social networks. The main idea of IncInf is to take full use of the valuable information that is inherent in the network structural evolution and previous influential nodes, so as to substantially narrow the search space of influential nodes. In this way IncInf is able to incrementally identify the top-K influential nodes S^{t+1} of G^{t+1} at time $t+1$ based on the previous influential nodes S^t at time t and the structural change ΔG^t from G^t to G^{t+1}, thus significantly reduces the computation complexity and improves the efficiency.

4.1   Basic Operations of Topology Evolution

The evolution of social network, when reflected into its underlying graph, can be summarized into six categories, which are inserting or removing a node, introducing or deleting an edge, and increasing or decreasing the influence probability of an edge. We denote the six types of topology change as $addNode$, $removeNode$, $addEdge$, $removeEdge$, $addWeight$, $decWeight$. The detailed descriptions and their effects on influence spread are shown in Table 1.

Table 1. Details of six types of basic operation

Operation	Description	Impact on influence spread
$addNode(u)$	add a new node u into the current network	the influence spread of u is set to 1
$removeNode(u)$	delete an existing node u from the network	the influence spread of u is set to 0
$addEdge(u,v,w)$	introduce a new edge (u,v) with $p(u,v) = w$	the influence spread of all the nodes that can reach u may be increased
$removeEdge(u,v)$	remove an existing edge (u,v) from the network	the influence spread of all the nodes that can reach u may be decreased
$addWeight(u,v,\Delta w)$	increase $p(u,v)$ by Δw	the influence spread of all the nodes that can reach u may be increased
$decWeight(u,v,\Delta w)$	reduce $p(u,v)$ by Δw	the influence spread of all the nodes that can reach u may be decreased

It should be noted that only after the $addNode$ operation can node u establish links ($addEdge$) or sever links ($removeEdge$) with other nodes, and node u can only be removed when all its associated edges are deleted. Moreover, the weight operation can be equivalently decomposed into two edge operations. For example, $addWeight(u,v,\Delta w)$ can be divided into $removeEdge(u,v)$ and $addEdge(u,v,w+\Delta w)$, supposing the previous weight of edge (u,v) is w.

4.2   Influence Spread Changes

As discussed above, whenever an edge (u,v) is introduced into or removed from the social network, the influence spread of all the nodes that can reach node u

may be changed. However, as a matter of fact, the real-world social networks exhibit small-world network characteristics and the connections between nodes are highly complicated. As a result, even one small change in topology, such as an edge addition or removal, may affect the influence spread of a large number of nodes, thus introducing massive recalculations. In order to reduce the amount of computation, we design an approach to efficiently calculate the changes on the influence spread of nodes which adopts the localization idea [6].

The main idea of localization is to use the local region of each node to approximate its overall influence spread. In particular, we use the maximum influence path to approximate the influence spread from node u to v. Here the maximum influence path $MIP(u, v, G)$ from node u to v in graph G is defined as the path with the maximum influence probability among all the paths from node u to v, and can be formally described as follows:

$$MIP(u, v, G) = \arg \max_{p \in P(u,v,G)} \{prob(p)\} \tag{1}$$

where $prob(p)$ denotes the propagation probability of path p and $P(u, v, G)$ denotes all the paths from node u to v in graph G. For a given path $p = \{u_1, u_2, \cdots, u_m\}$, the propagation probability of path p is defined as follows:

$$prob(p) = \prod_{i=1}^{m-1} p(u_i, u_{i+1}) \tag{2}$$

Moreover, an influence threshold θ is set to tradeoff between accuracy and efficiency. During the propagation process, we only consider paths whose influence probability are larger than θ while ignoring those with smaller probability. By doing this, the influence is effectively restricted to the local region of each node.

Similarly, in our proposal we localize the impact of topology changes on influence spread into local regions, and thus reduce the amount of computation. Among six types of topology change, $addNode$ (or $removeNode$) is the most straightforward since it simply sets the influence spread of the node to 1 (or 0); $addWeight$, $decWeight$ as well as $removeEdge$ are methodologically similar to $addEdge$. Consequently, in the following we take $addEdge$ as an example to show which nodes' influence spread need to be updated and how to determine those changes when a new edge is added into the graph.

Consider the case when a new edge $e = (u, v, w)$ is introduced between two existing node u and v. We denote the graph before and after such a topology change as G^t and $G^{t'}$, and the current seed set is S. The detailed algorithm is described in Algorithm 1. According to the principle of localization [6], if the propagation probability w is smaller than the specified threshold θ, or not bigger than the probability of $MIP(u, v, G^t)$, edge e can be simply neglected and there is no need to update any node's influence spread (lines 1–3). Otherwise, the newly-added edge e would become the $MIP(u, v, G^{t'})$. As a result, each node i whose maximum influence path to u has an influence probability larger than θ is likely to experience a rise in terms of influence spread (line 4) because node i may influence more nodes through the new edge e. So, we then check the probability

---

Algorithm 1. Edge addition

Input: a new edge $e = (u, v, w)$, graph G^t.
Output: The influence spread changes of nodes in $G^{t'}$.
 1: **if** $w < \theta$ or $w \leq prob(MIP(u, v, G^t))$ **then**
 2:     return;
 3: **end if**
 4: **for** each node i with $prob(MIP(i, u, G^t)) > \theta$ **do**
 5:     **for** each node j with $prob(MIP(v, j, G^t)) > \theta$ **do**
 6:         **if** $prob(MIP(i, j, G^t)) < \theta$ and
             $prob(MIP(i, j, G^{t'})) > \theta$ **then**
 7:             $deltaInf[i]+ = prob(MIP(i, j, G^{t'})) \times (1 - prob(j, S))$
 8:         **end if**
 9:         **if** $prob(MIP(i, j, G^t)) > \theta$ and
             $prob(MIP(i, j, G^{t'})) > \theta$ **then**
10:             $deltaInf[i]+ = (prob(MIP(i, j, G^{t'})) - prob(MIP(i, j, G^t))) \times (1 - prob(j, S))$
11:         **end if**
12:     **end for**
13: **end for**

---

of the maximum influence path from i to v and its successors in G^t and $G^{t'}$. Based on the two probabilities, we divide the problem into two small cases:

The first case is when the probability of maximum influence path from i to j in G^t is smaller than θ while that in $G^{t'}$ is larger than θ (lines 5–6). Here j denotes the node whose probability of $MIP(v, j, G^t)$ is larger than θ. In such a case, node i build a new path to j through the new edge e which increases the influence spread of i by $prob(MIP(i, j, G^{t'})) \times (1 - prob(j, S))$ (line 7). Here $prob(j, S)$ is the probability of that node j is influenced by the current seed set S, which is defined as follows:

$$prob(j, S) = \begin{cases} 1, & \text{if } j \in S \\ 1 - \prod_{w \in n(j)} 1 - prob(w, S) \cdot p(w, j), & \text{if } j \notin S \end{cases}$$

Here $n(j)$ denotes the in-neighbour set of j.

The second case is when the probability of maximum influence path from i to j is larger than θ in both G^t and $G^{t'}$ (lines 9–11). In this case, the influence increase of node i is $(prob(MIP(i, j, G^{t'})) - prob(MIP(i, j, G^t))) \times (1 - prob(j, S))$.

We treat the network dynamics from G^t to G^{t+1} as a finite change stream $c_1, c_2, \cdots, c_i, \cdots$ where each change c_i is one of the six topology changes we described above. When all the changes in the change stream are processed, we can obtain the influence spread change for all the nodes.

4.3   Potential Top-K Influential Users Identification

From the preferential attachment rule, we know that the influence spread changes of those high-degree nodes should be much greater than the ordinary nodes. Moreover, according to the power-law distribution, such high-degree nodes only account for a small part of the whole nodes. Consequently we can pick out nodes only experiencing major increases or with high degrees because these nodes are of great potential to become the top-K influential nodes in G^{t+1}. Then we only calculate the actual influence spread for these selected nodes while ignoring the others. In this way, a large percent of nodes are pruned and the search space is largely narrowed. It should be noted that a smart pruning strategy is of key importance since a poor selection might either affect the efficiency or reduce the accuracy in terms of influence spread. We describe the details of our pruning strategy as follows:

(1) In the ith iteration, if the influence spread of the previous influential node S_i^t increases in G^{t+1}, the chosen nodes are those with a larger influence spread change than $deltaInf[S_i^t]$;

In most cases, the influential nodes will attract a good many of new nodes and establish new links. Thus, their influence spreads will increase drastically. In such a case, the nodes whose influence spread changes are smaller than the influential nodes are completely impossible to become the most influential node in G^{t+1}. Therefore, when the influence spread of the previous influential nodes increase, we only select those whose influence spread changes are larger than the influential nodes in G^t. According to the preferential attachment rule, such a pruning method can greatly narrow the search space and reduce the amount of computation.

(2) In the ith iteration, if the influence spread of the previous influential node S_i^t decreases in G^{t+1}, in addition to item (1), the nodes are further selected to hold a sufficiently large degree or experience a sufficiently great increase. In order to formally define "large degree" and "great increase", here we set an threshold η to tradeoff between running time and influence spread. Here the nodes with sufficiently large degrees (or great increase) are defined as the set of node v_j whose degree (or degree increase ratio) is among the top η percent of all nodes in G^{t+1}. The degree increase ration of v_j is defined as $degree_j^{t+1}/degree_j^t$ where $degree_j^t$ denotes the degree of node v_j in graph G^t.

It should be noted that although the case the influence spread of a previous influential node decreases during the evolution rarely happens, we consider it here for completeness. In this case, the amount of nodes satisfying item (1) is relatively large which leads to mass computation and need to be further pruned. In order to select only the most potential nodes, we additionally select the nodes with large degree or large increase because a node with small degree has only very low probability to become an influential node in reality. Consequently, the search space is strictly circumscribed and the computational complexity is greatly reduced.

After the potential nodes are selected, we calculate the actual influence spread of these nodes in G^{t+1} and select the one with the maximum influence spread in

---

Algorithm 2. IncInf

Input: G^t, S^t, and G^{t+1}.
Output: the top-K influential nodes S^{t+1} in G^{t+1}.
 1: Initialize $S^{t+1} = \emptyset$;
 2: **for** $i = 1$ to K **do**
 3:     **for** each topology change c_j from G^t to G^{t+1} **do**
 4:         calculate the influence spread change $deltaInf[\cdot]$;
 5:     **end for**
 6:     select a set of potential nodes as pn according to pruning strategy;
 7:     **for** each node $v_l \in pn$ **do**
 8:         calculate the marginal influence spread $\sigma_{S^{t+1}}(v_j)$;
 9:     **end for**
10:     select $v_{max} = \arg\max_{v_j \in pn} (\sigma_{S^{t+1}}(v_j))$;
11:     $S^{t+1} = S^{t+1} \cup v_{max}$;
12: **end for**

---

each iteration. Algorithm 2 outlines the design of our proposed algorithm IncInf. IncInf iterates for K round (line 2) and in each round select one node providing the maximum marginal influence spread. Lines 3–5 calculate the influence spread change of each node caused by the topology evolution. Nodes with great potential to become top-K influential are selected (line 6) and their influence spread are computed in G^{t+1} (lines 7–9). Then the node providing the maximal marginal gain will be selected and added to the S^{t+1} (lines 10–11).

5   Experiments

5.1   Experimental Setup

We use three real-world social networks Facebook [15], NetHEPT [2], and Flickr [14], and each dataset includes multiple snapshots of different time stamps. Table 2 summarizes the statistical information of these datasets. We compare our algorithm with four static algorithms: **MixGreedy**, **ESMCE**, **MIA** and **Random**. MixGreedy is an improved greedy algorithm proposed by Chen et al. in [5]. ESMCE is a power-law exponent supervised estimation approach proposed in [12]. MIA is a heuristic that uses local arborescence structures of each

Table 2. Summary information of the real-world social networks

Datasets	Nodes			Edges		
	Initial Number	Final Number	Growth	Initial Number	Final Number	Growth
Facebook	12,364	61,096	394 %	73,912	905,665	1125 %
NetHEPT	5,802	29,555	409 %	57,765	352,807	511 %
Flickr	1,620,392	2,570,535	58.6 %	17,034,807	33,140,018	94.5 %

node to approximate the influence propagation [6]. Random is a basic heuristic that randomly selects K nodes from the whole datasets.

The propagation probability of the IC model is selected randomly from 0.1, 0.01, and 0.001 for each network snapshot, and we run simulations 10000 times and take the average influence spread. The pruning threshold η is set to 5 % and we aim to find the top 50 influential nodes from each dataset. The experiments are conducted on a PC with Intel Core i7 920 CPU @2.67 GHz and 6 GB RAM.

5.2 Efficiency Study

The time costs of different algorithms are illustrated in Fig. 2 where we record the total time cost for each snapshot of the three datasets. The experimental results show that the time costs of our algorithm on each snapshot are obviously less than those of static algorithms. Obviously, MixGreedy takes the longest time among four kinds of influence maximization algorithms. It takes MixGreedy more than as much as 6 h to identify the top 50 influential nodes on the final NetHEPT dataset, while the time is even longer on the larger dataset Facebook. Moreover, MixGreedy is not feasible to run on the largest dataset Flickr due to the unbearably long running time. ESMCE, benefiting from its sampling estimation method, runs much faster than MixGreedy, but it still takes as much as 3511 s on average to run on the five snapshots of Flickr. Compared with two greedy algorithms, the heuristic MIA performs much better. It only takes MIA 23.8 s to run on the final Facebook graph. When running on the Flickr dataset with as much as 2.5 M nodes and 33 M edges, however, its speedup is far from satisfactory, since it still needs more than 45 min to finish. While our proposed algorithm, IncInf, outperforms all the static algorithms in terms of efficiency. In particular, IncInf is almost four orders of magnitude faster than the MixGreedy algorithm on the Facebook dataset. While compared with the MIA heuristic, the speedup of IncInf is 8.41× and 6.94× on the Facebook and NetHEPT datasets, respectively; What's more, when applied on the largest dataset Flickr, IncInf can achieve as much as 20.65× speedup on average. This is because IncInf only

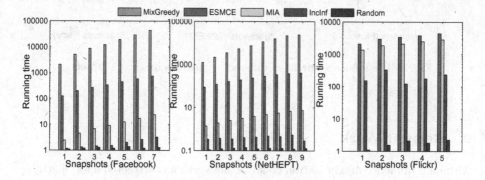

Fig. 2. The time costs of different algorithms on three real-world datasets.

computes the incremental influence spread changes and adaptively identifies the influential nodes based on the previous influential nodes and the current influence spread changes. The experimental results clearly validate the efficiency advantage of our incremental algorithm IncInf. Without doubt, Random runs the fast among all the algorithms. However, as we will show in Sect. 5.3, its accuracy is much worse and unacceptable when developing real-world viral marketing strategies.

5.3   Effectiveness Study

Figure 3 shows the experimental results. MixGreedy outperforms all the other algorithms in terms of influence spread. However, the efficiency issue limits its application to large-scale dataset such as Flickr. The performance of ESMCE, MIA and IncInf almost match MixGreedy on the Facebook dataset, while on NetHEPT, the gaps become larger but remain acceptable (only 3.4 %, 4.7 % and 5.1 % lower than MixGreedy on average). When applied to the Flickr dataset, ESMCE performs the best since ESMCE strictly control the error threshold by iterative sampling. Compared with MIA, IncInf shows very close performance and is only 2.87 % lower on average of all five snapshots, which demonstrates the effectiveness of our proposal. Random, as the baseline heuristic, clearly performs the worst on all the graphs. The influence spread of Random is only 15.6 %, 12.1 % and 10.9 % of that of IncInf on Facebook, NetHEPT and Flickr, respectively.

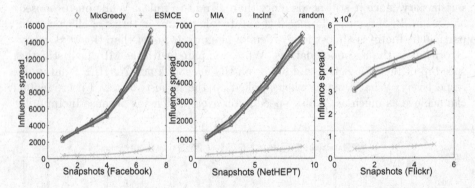

Fig. 3. The influence spread of different algorithms on three datasets.

6   Related Work

Although Influence maximization on static networks has attracted a lot of attentions [5,6,13], studies on dynamic social networks still remains largely unexplored to date. Habiba et al. [10] propose a dynamic social network model which is different from ours. In their proposal, the network keeps evolving during the process of influence propagation, and their goal is to find the top-K influential nodes over such a dynamic network. Chen et al. [4] extend the IC model to incorporate the time delay aspect of influence diffusion among individuals in social networks, and consider time-critical influence maximization, in which one wants to maximize influence spread within a given deadline. While in [9], the authors consider a continuous time formulation of the influence maximization problem in which information or influence can spread at different rates across different edges. Charu Aggarwal et al. [1] try to discover influential nodes in dynamic social networks and they design a stochastic approach to determine the information flow authorities with the use of a globally forward approach and a locally backward approach.

7 Conclusion

In this paper, we consider the influence maximization problem in evolving social networks, and propose an incremental algorithm, IncInf, to efficiently identify top-K influential nodes. Taking advantage of the structural evolution of networks and previous information on individual nodes, IncInf substantially reduces the search space and adaptively selects influential nodes in an incremental way. Extensive experiments demonstrate that IncInf significantly reduces the execution time of state-of-the-art static influence maximization algorithm while maintaining satisfying accuracy in terms of influence spread.

Acknowledgments. This research was supported by NSFC under grant NO. 61402511. The authors would like to thank the anonymous reviewers for their helpful comments.

References

1. Aggarwal, C., Lin, S., Yu, P.: On influential node discovery in dynamic social networks. In: SDM, pp. 636–647. SIAM, California, USA (2012)
2. ArXiv NetHEPT dataset (2003). http://www.cs.cornell.edu/projects/kddcup/datasets.html. Accessed 18 June 2013
3. Barabasi, A.-L., Albert, R.: Emergence of scaling in random networks. Science **286**(5439), 509–512 (1999)
4. Chen, W., Lu, W., Zhang, N.: Time-critical influence maximization in social networks with time-delayed diffusion process. In: AAAI, Toronto, Canada (2012)
5. Chen, W., Wang, Y., Yang, S.: Efficient influence maximization in social networks. In: SIGKDD, pp. 199–208. ACM, Paris, France (2009)
6. Chen, W., Wang, C., Wang, Y.: Scalable influence maximization for prevalent viral marketing in large-scale social networks. In: SIGKDD, pp. 1029–1038. USA (2010)
7. Domingos, P., Richardson, M.: Mining the network value of customers. In: SIGKDD, pp. 57–66. ACM, San Francisco, CA, USA (2001)
8. Easley, D., Kleinberg, J.: Power laws and rich-get-richer phenomena. In: Networks, Crowds, and Markets: Reasoning about a Highly Connected World (2010)

9. Gomez-Rodriguez, M., Scholkopf, B.: Influence maximization in continuous time diffusion networks. In: ICML. IEEE, Edinburgh (2012)
10. Habiba, T.B.W., Berger-Wolf, T.Y.: Maximizing the extent of spread in a dynamic network. DIMACS TR: 2007–20 (2007)
11. Kempe, D., Kleinberg, J., Tardos, E.: Maximizing the spread of influence through a social network. In: SIGKDD, pp. 137–146. ACM, Washington, D.C., USA (2003)
12. Liu, X., Li, S., Liao, X., Wang, L., Wu, Q.: In-time estimation for influence maximization in large-scale social networks. In: SNS, pp. 1–6, Switzerland (2012)
13. Liu, X., Li, M., Li, S., Peng, S., Liao, X., Lu, X.: IMGPU: GPU-accelerated influence maximization in large-scale social networks. In: TPDS, pp. 136–145 (2014)
14. Mislove, A., Koppula, H.S., Gummadi, K.P., Druschel, P., Bhattacharjee, B.: Growth of the Flickr social network. In: SNS, pp. 25–30, USA (2008)
15. Viswanath, B., Mislove, A., Cha, M., Gummadi, K.P.: On the evolution of user interaction in Facebook. In: SNS, pp. 37–42, Spain (2009)

Detecting Community Pacemakers
of Burst Topic in Twitter

Guozhong Dong[1], Wu Yang[1(✉)], Feida Zhu[2], and Wei Wang[1]

[1] Information Security Research Center,
Harbin Engineering University, Harbin, China
`yangwu@hrbeu.edu.cn`
[2] Singapore Management University, Singapore, Singapore

Abstract. Twitter has become one of largest social networks for users to broadcast burst topics. Influential users usually have a large number of followers and play an important role in the diffusion of burst topic. There have been many studies on how to detect influential users. However, traditional influential users detection approaches have largely ignored influential users in user community. In this paper, we investigate the problem of detecting community pacemakers. Community pacemakers are defined as the influential users that promote early diffusion in the user community of burst topic. To solve this problem, we present DCPBT, a framework that can detect community pacemakers in burst topics. In DCPBT, a burst topic user graph model is proposed, which can represent the topology structure of burst topic propagation across a large number of Twitter users. Based on the model, a user community detection algorithm based on random walk is applied to discover user community. For large-scale user community, we propose a ranking method to detect community pacemakers in each large-scale user community. To test our framework, we conduct the framework over Twitter burst topic detection system. Experimental results show that our method is more effective to detect the users that influence other users and promote early diffusion in the early stages of burst topic.

Keywords: Twitter · Burst topic · User graph model · Community pacemakers

1 Introduction

With the development of social media, Twitter has been an important medium for providing the rapid spread of burst topic. When breaking news or events occur, influential users can post tweets about breaking news and share with their friends. Due to large number of people that have different user interests participating in conversation and discussion, some tweets spread among Twitter users and become the source of burst topics. As such, the main cause of burst topic is the information diffusion in user community. Figure 1 illustrates the

© Springer International Publishing Switzerland 2016
F. Li et al. (Eds.): APWeb 2016, Part I, LNCS 9931, pp. 245–255, 2016.
DOI: 10.1007/978-3-319-45814-4_20

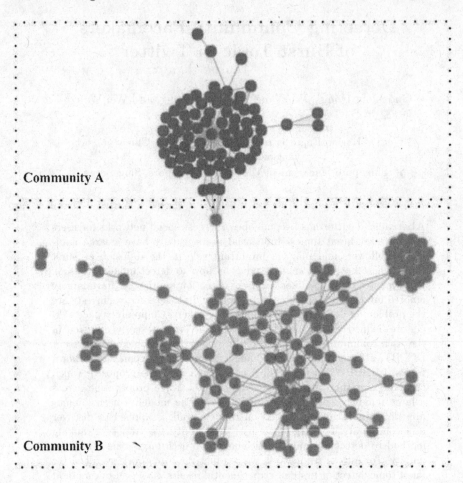

Fig. 1. Example of user community of burst topic

user community of burst topics detected by CLEar system[1], in which two main user communities are marked in different colors. So far, plenty of works focus on the influential users who are popular or famous in burst topics. However, these famous influential users are not the early spreader of burst topic that also influence their followers to spread the topic. In quite a lot of scenario, it is more important to detect the cause of burst topic diffusion in different user communities, which are called community pacemakers in this paper.

Unfortunately, detecting community pacemakers in burst topic has not been solved by the existing works. In this paper, we propose DCPBT (Detecting Community Pacemakers in Burst Topics) framework and implement the framework on CLEar system. When new burst topics are detected by CLEar system, DCPBT applies burst topic user graph construct algorithm to conduct user graph for each

---

[1] http://research.pinnacle.smu.edu.sg/clear/.

burst topic. Based on burst topic user graph, a user community detection algorithm based on random walker is proposed to detect user community in burst
topic, which can adjust the number of user community adaptively and select
large-scale user community. For large-scale user community, we propose a ranking method to detect community pacemakers in each large-scale user community.
To summarize, the contributions of our work are listed as follows:

(1) We propose a burst topic user graph model which can represent the topology
    structure of burst topic propagation across a large number of Twitter users.
    In the burst topic user graph, nodes represent the burst topic users and
    edges represent the follower/followee relationship between users.
(2) A community pacemakers detection algorithm is proposed to detect com-
    munity pacemakers in each large-scale user community of burst topic, which
    is more effective to detect the users that influence other users and promote
    early diffusion in the early stages of burst topic.
(3) We implement DCPBT framework on CLEar system, which can demonstrate
    the effectiveness of DCPBT framework.

The rest of the paper is organized as follows. Section 2 reviews the related work.
Section 3 presents the framework of DCPBT. Section 4 describes the experimental results. Finally, we conclude our work in Sect. 5.

2   Related Work

The study of burst topic [1–8] and user influence [9–17] have been studied in
the last decade. As there are numerous research works focusing on it, here we
introduce the ones most related to our work.

Burst Topic Detection: Prasadet et al. [1] propose a framework to detect
emerging topics through the use of dictionary learning. They determine novel
documents in the stream and subsequently identify topics among the novel documents. Agarwal et al. [2] model emerging events detection problem as discovering
dense clusters in highly dynamic graphs and exploit short-cycle graph property
to find dense clusters efficiently in microblog streams. Alvanaki et al. [3] present
the "en Blogue" system for emergent topic detection. En Blogue keeps track of
sudden changes in tag correlations and presents tag pairs as emergent topics.
Takahashi et al. [4] apply a recently proposed change-point detection technique
based on Sequentially Discounting Normalized Maximum Likelihood (SDNML)
coding to detect abnormal messages and detect the emergence of a new topic
from the anomaly measured through the model. Wang, Liu et al. [5] propose a
system called SEA to detect events and conduct panoramic analysis on Weibo
events from various aspects. Xie et al. [6,7] present a real-time system to provide
burst event detection, popularity prediction, and event summarization. Shen
et al. [8] analyze different burst patterns and propose real-time burst topics
detection oriented Chinese microblog stream. The method detect burst entities

and cluster them to burst topics without requiring Chinese segmentation, which can obtain related messages and users at the same time.

User Influence: Cha et al. [9] analyze the influence of Twitter users by employing three measures that capture different perspectives: indegree, retweets, and mentions. They find that influence is not determined by single factor, but through many factors. Lee et al. [10] propose a method to find influentials by considering both the link structure and the temporal order of information adoption in Twitter. Weng et al. [11] propose an extension of PageRank algorithm to measure the influence of users in Twitter, which measures the influence taking both the topical similarity between users and the link structure. Brown et al. [12] investigate a modified k-shell decomposition algorithm based on user relationship to compute user influence on Twitter. Fang et al. [13] develop a novel Topic-Sensitive Influencer Mining (TSIM) framework in interest-based social media networks to find topical influential users and images. Saez-Trumper et al. [14] propose a ranking algorithm to detect trendsetters in information networks. The algorithm can identify persons that spark the process of disseminating ideas that become popular in the network.

Note that previous studies mainly aim at detecting burst topics and influential users. Different from other works, we consider the role of user community in burst topic diffusion and propose the problem of detecting community pacemakers in burst topics. We focus on detecting influential users that promote early diffusion in the user community of burst topic.

3   Framework of DCPBT

The framework of DCPBT that construct on CLEar system is shown in Fig. 2, which contains three functional layers, namely Data Layer, Model Layer and Presentation Layer.

The Data Layer provides two databases for efficient data storage and data query. The first one is to store burst topics detected by CLEar system, and provide query operation for new burst topic monitor module (NBTM) in Model Layer. The second one is to store Twitter stream data, which stores necessary data involved in burst topics. The Model Layer utilizes several important modules to detect community pacemakers in burst topics. NBTM monitors new burst topics via polling burst topic database. Once new burst topics are detected, NBTM sends burst topic data collect command to burst topic data collect module (BTDC). BTDC retrieves burst topic data from Hadoop cluster, constructs burst topic user graph for further processing. In order to detect pacemakers in burst topic, community pacemakers detection algorithm based on burst topic user graph is proposed. The Presentation Layer presents the user engagement series and pacemakers detected by DCPBT with a user-friendly interface.

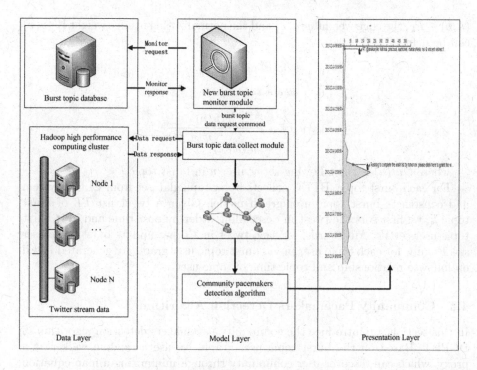

Fig. 2. The framework of DCPBT

4 Methods

In this section, we introduce some important models employed by DCPBT.

4.1 Burst Topic User Graph Model

Once a user posts a tweet related to the burst topic in Twitter, the tweet can spread to the user's followers, then followers who are interested in the burst topic may post or retweet the message. In order to represent the topology structure of burst topic propagation across a large number of Twitter users, in the burst topic user graph model, nodes represent the burst topic users and edges represent the follower/followee relationship between users.

The burst topic user graph of burst topic k can be formally defined as $G_k = <V_k, E_k, T_k>$. In detail, $V_k = \{u, \cdots, v, \cdots\}$ is the set of Twitter users over burst topic k, E_k represents the set of edges among Twitter users, in which a directed edge (u, v) means that u is the follower of v. $T_k = \{t(u), \cdots, t(v), \cdots\}$ is the earliest post time set of users over burst topic k.

By considering time information, the directed edges in topic user graph model can represent the direction of information flow and play a key role in detecting pacemakers, so we include time information in the edge weight. For each

$(u, v) \in E_k$, the edge weight $w(u, v)$ and the normalization of edge weight $W(u, v)$ can be defined as follows

$$w(u, v) = \begin{cases} e^{-\frac{t(u)-t(v)}{\alpha}}, & if\ t(v)>0\ and\ t(v)<t(u),\ \alpha>0 \\ 0, & otherwise \end{cases} \tag{1}$$

$$W(u, v) = \frac{w(u, v)}{\sum\limits_{m \in Out(u)} w(u, m)} \tag{2}$$

where $Out(u)$ is the following set of user u in burst topic k.

For each burst topic, BTDC collects burst topic dataset from CLEar system and constructs burst topic user graph model. Given a tweet list TL of burst topic k, we first sort the tweet in descending order by post time and in it burst topic user set V_k. Afterwards, for each tweet in TL, we update burst topic user set. Finally, for each topic user in V_k, burst topic user graph are generated based on followee relationship and topic time of topic user.

4.2 Community Pacemakers Detection Algorithm

In this section, we introduce the community pacemakers detection algorithm in DCPBT. Based on the burst topic user graph, we use a random walker as a proxy, which can discover user community through minimizing a map equation over burst topic user graph. The map equation is first introduced in ref. [18]. Given a burst topic user graph with n users, the conditional probability that the random walker steps from user u to user v is given by the edge weight:

$$p_{u \to v} = W(u, v) \Big/ \sum\nolimits_v W(u, v) \tag{3}$$

To ensure the independent of the random walker starts in burst topic user graph, we use smart teleportation scheme and only record steps along links [19]. The stationary distribution is given by p_u^*, which can be expressed:

$$p_u^* = (1 - \tau) \sum_v p_v^* p_{v \to u} + \tau \frac{\sum_v W(u, v)}{\sum_{u,v} W(v, u)} \tag{4}$$

The unrecorded visit rates on edge $q_{v \to u}$ and nodes p_u can now be formalized as follows:

$$q_{v \to u} = p_v^* p_{v \to u} \tag{5}$$

$$p_u = \sum_v q_{v \to u} \tag{6}$$

We use C to denote the community partition of burst topic user graph into m modules, with each node u assigned to a community i. m community codebooks and one index codebook are used to describe the random walker's movements within and between communities. The community transition rates $q_{i \leftarrow}$ and $q_{i \to}$

represent that the random walker enter and exit community i, which can be expressed by unrecorded visit rates on edge:

$$q_{i\leftarrow} = \sum_{u\in j\neq i, v\in i} q_{u\to v} \tag{7}$$

$$q_{i\to} = \sum_{u\in i, v\in j\neq i} q_{u\to v} \tag{8}$$

The map equation that can measure the per-step theoretical lower limit of a modular description of a random walker on user graph is given by:

$$L(M) = q_{\leftarrow}H(Q) + \sum_{i=1}^{m} R_i H(P^i) \tag{9}$$

Below we explain the terms of the map equation in detail. $L(M)$ represents the per-step description length for community partition, $q_{\leftarrow}$ represents the total probability that the random walker enters any of the m communities, which can be expressed:

$$q_{\leftarrow} = \sum_{i=1}^{m} q_{i\leftarrow} \tag{10}$$

$H(Q)$ represents the frequency-weight average length of codewords in the index codebook, which is given by:

$$H(Q) = -\sum_{i=1}^{m} (q_{i\leftarrow}/q_{\leftarrow}) \log(q_{i\leftarrow}/q_{\leftarrow}) \tag{11}$$

R_i represents the rate at which the community codebook i is used, which is given by:

$$R_i = \sum_{u\in i} p_u + q_{i\to} \tag{12}$$

$H(P^i)$ represents the frequency-weight average length of codewords in community codebook i, which is given by:

$$H(P^i) = -(q_{i\to}/R_i) \log(q_{i\to}/R_i) - \sum_{u\in i} (p_u/R_i) \log(p_u/R_i) \tag{13}$$

With the map equation, the burst topic user graph can be divided into different user communities. First, each user node is assigned to its own community. Then, in random order, each user node is moved to the neighboring community that results in the largest decrease of the map equation. If no move results in a decrease of the map equation, the user node stays in its original community. This procedure is repeated, each time in a new random order, until no move generates a decrease of the map equation. Then the network is rebuilt, with the communities of the last level forming the nodes at this level, and, exactly as at the previous level, the nodes are joined into communities. This hierarchical rebuilding of the network is repeated until the map equation cannot be reduced further.

At last, each community C_i represents user community in burst topic. The large scale user community set, denoted by C_l, is selected by:

$$C_l = \{C_i \,||C_i| \geq n/5\} \tag{14}$$

The community detection algorithm can adjust the number of user community adaptively, in which community number parameter is not needed. For large-scale user community, we propose a ranking method to detect community pacemakers in each large-scale user community. The pacemaker weight of user v in each large scale user community $C_l \in C$, denoted by $PM_{C_l}(v)$, is given by

$$PM_{C_l}(v) = dD(v) + (1 - d) \sum_{u \in IN_{C_l}(v)} PM_{C_l}(u)W(u,v), \, 0 \leq d \leq 1 \tag{15}$$

where d is the damping factor, $IN_{C_l}(v)$ is the follower set of user v in user community C_l and $D(v)$ is a probability distribution over C_l. The distribution is topic dependent and is set to $1/|C_l|$ for all $v \in C_l$. The community pacemakers of user community C_l are the top N_l pacemaker weight of users in user community C_l.

5 Experiments

In order to test the advantage of community pacemakers detection algorithm, we have implemented and conducted a set of experiments on CLEar system. In this section, we first describe the dataset used in the experiments, and then present the experimental evaluation. The goal of experiment is to prove that our framework is more efficient than other approaches. In each experiment, we compare our PM ranking with TS ranking [14], and traditional PageRank(PR). The parameters are set through a large number of experiments and applied with $\alpha = 1800\,s$ in Eq. 1 and $d = 0.2$ in Eq. 15.

5.1 Dataset

We collected burst topic dataset from CLEar system. The system can detect and summarize burst topics in Singapore Twitter stream as soon as they emerge in real-time, which is convenient for us to collect burst topic features, tweet data and users data involved in burst topics. The collected burst topic dataset covered the period from November 1 to November 30 in 2015. Furthermore, in order to conduct burst topic user graph, the follower/followee relationships of burst topic users were also collected.

5.2 Influenced Followers Ratio

In this section, we compare the influence of the top users in each ranking approach. To evaluate this, we create a simple indicator called Influenced Followers Ratio for a burst topic k, IFR_k, defined as the fraction of followers of top N users in burst topic that post the tweets related to burst topic k. In the three ranking approaches, the value of N is determined by PM ranking, which is given by:

$$N = \sum N_l(0 < l \le |C_l|) \tag{16}$$

Table 1 shows the average Influenced Followers ratio (IFR) of PM ranking, TS ranking and traditional PageRank (PR) in our dataset. As shown in Table 1, Influenced Followers Ratio in PM is bigger than TS and PR, which shows that top users in PM ranking influence more their followers to spread burst topics than other ranking approaches.

Table 1. Influenced followers ratio

Approaches	IFR
PM	0.141
TS	0.098
PR	0.071

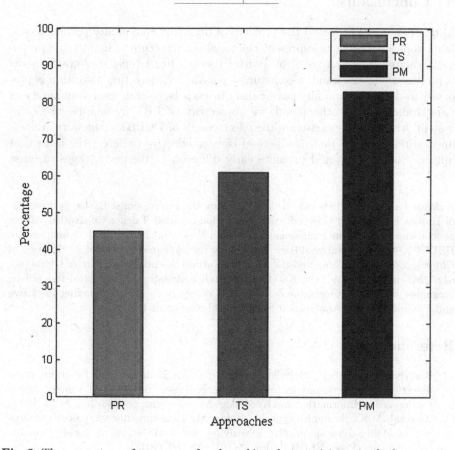

Fig. 3. The percentage of top users of each ranking that participate in the burst topic before the burst

5.3 Promoting Early Diffusion in the Early Stages

To compare the ability of top users in promoting early diffusion in the early stages of burst topic, we first obtained the detecting time of each burst in burst topic in our dataset. Due to different burst patterns, we formalize the median of detecting times of each burst in burst topic k as B_k. Next, we have to compare it with the time of top N users of each ranking that participates in the burst topic, where $T_k(r)$ represents the participation time of the user that rank r in burst topic k. If $B_k - T_k(r) < 0$, this means that the user participates in the burst topic before the burst. Finally, in our burst topic dataset, we compute the percentage of top N users of each ranking that participates in the burst topic before the burst, which is shown in Fig. 3. As shown in Fig. 3, the percentage of top users of PM ranking is larger than other approaches. More than 80 % of the top users participate in the burst topic before the burst, which indicates that our approach is more effective to detect the users that promote early diffusion in the early stages of burst topic.

6 Conclusions

In this paper, we proposed the problem of detecting community pacemakers in burst topics. In order to represent the topology structure of burst topic propagation across a large number of Twitter users, a burst topic user graph model is proposed. On one hand, a community pacemakers detection algorithm is proposed to detect community pacemakers in each large-scale user community of burst topic. On the other hand, we implement DCPBT framework on CLEar system, which can demonstrate the effectiveness of DCPBT framework. Experimental results show that our method is more effective to detect the users that influence other users and promote early diffusion in the early stages of burst topic.

Acknowledgment. This work is supported by the International Exchange Program of Harbin Engineering University for Innovation-oriented Talents Cultivation, China Scholarship Council, the Fundamental Research Funds for the Central Universities (no. HEUCF100605), the National High Technology Research and Development Program of China (no. 2012AA012802) and the National Natural Science Foundation of China (no. 61170242, no. 61572459); the National Research Foundation, Prime Ministers Office, Singapore under its International Research Centres in Singapore Funding Initiative and Pinnacle Lab for Analytics at Singapore Management University.

References

1. Kasiviswanathan, S.P., Melville, P., Banerjee, A., Sindhwani, V.: Emerging topic detection using dictionary learning. In: Proceedings of the 20th ACM International Conference on Information and Knowledge Management, pp. 745–754. ACM (2011)
2. Agarwal, M.K., Ramamritham, K., Bhide, M.: Real time discovery of dense clusters in highly dynamic graphs: identifying real world events in highly dynamic environments. Proc. VLDB Endow. 5(10), 980–991 (2012)

3. Alvanaki, F., Sebastian, M., Ramamritham, K., Weikum, G.: EnBlogue: emergent topic detection in Web 2.0 streams. In: Proceedings of the 2011 ACM SIGMOD International Conference on Management of data, pp. 1271–1274. ACM (2011)
4. Takahashi, T., Tomioka, R., Yamanishi, K.: Discovering emerging topics in social streams via link anomaly detection. In: IEEE 11th International Conference on Data Mining (ICDM), pp. 1230–1235. IEEE (2011)
5. Wang, Y., Liu, H., Lin, H., Wu, J., Wu, Z., Cao, J.: SEA: a system for event analysis on Chinese tweets. In: Proceedings of the 19th ACM SIGKDD International Conference on Knowledge Discovery and Data Mining, pp. 1498–1501. ACM (2013)
6. Xie, W., Zhu, F., Jiang, J., Lim, E.P., Wang, K.: Topicsketch: real-time bursty topic detection from Twitter. In: IEEE 13th International Conference on Data Mining (ICDM), pp. 837–846. IEEE (2013)
7. Xie, R., Zhu, F., Ma, H., Xie, W., Lin, C.: CLEar: a real-time online observatory for bursty and viral events. Proc. VLDB Endow. **7**(13), 1637–1640 (2014)
8. Shen, G., Yang, W., Wang, W.: Burst topic detection oriented large-scale microblogs streams. J. Comput. Res. Dev. **52**(2), 512–521 (2015). (in Chinese)
9. Cha, M., Haddadi, H., Benevenuto, F., Gummadi, K.P.: Measuring user influence in Twitter: the million follower fallacy. In: Fourth International AAAI Conference on Weblogs and Social Media (ICWSM 2010), pp. 10–17. AAAI Press (2010)
10. Lee, C., Kwak, H., Park, H., Moon, S.: Finding influentials based on the temporal order of information adoption in Twitter. In: Proceedings of the 19th International Conference on World Wide Web, pp. 1137–1138. ACM (2010)
11. Weng, J., Lim, E.P., Jiang, J., He, Q.: Twitterrank: finding topic-sensitive influential twitterers. In: Proceedings of the Third ACM International Conference on Web Search and Data Mining, pp. 261–270. ACM (2010)
12. Brown, P.E., Feng, J.: Measuring user influence on Twitter using modified K-shell decomposition. In: Fifth International AAAI Conference on Weblogs and Social Media, pp. 18–23. AAAI Press (2011)
13. Fang, Q., Sang, J., Xu, C., Rui, Y.: Topic-sensitive influencer mining in interest-based social media networks via hypergraph learning. IEEE Trans. Multimedia **16**(3), 796–812 (2014)
14. Saez-Trumper, D., Comarela, G., Almeida, V., Baeza-Yates, R., Benevenuto, F.: Finding trendsetters in information networks. In: Proceedings of the 18th ACM SIGKDD International Conference on Knowledge Discovery and Data Mining, pp. 1014–1022. ACM (2012)
15. Wu, Y., Hu, Y., He, X., Deng, K.: Impact of user influence on information multi-step communication in a microblog. Chin. Phys. B **23**(6), 5–12 (2014)
16. Bakshy, E., Hofman, J.M., Mason, W.A., Watts, D.J.: Everyone's an influencer: quantifying influence on Twitter. In: Proceedings of the Fourth ACM International Conference on Web Search and Data Mining, pp. 65–74. ACM (2011)
17. Liu, D., Wu, Q., Han, W.: Measuring micro-blogging user influence based on user-tweet interaction model. In: Tan, Y., Shi, Y., Mo, H. (eds.) ICSI 2013, Part II. LNCS, vol. 7929, pp. 146–153. Springer, Heidelberg (2013)
18. Rosvall, M., Bergstrom, C.T.: Maps of random walks on complex networks reveal community structure. Proc. Natl. Acad. Sci. **105**(4), 1118–1123 (2008)
19. Lambiotte, R., Rosvall, M.: Ranking and clustering of nodes in networks with smart teleportation. Phys. Rev. E **85**(5), 056107(1–9) (2012)

Dynamic User Attribute Discovery on Social Media

Xiu Huang, Yang Yang(✉), Yue Hu, Fumin Shen, and Jie Shao

School of Computer Science and Engineering,
University of Electronic Science and Technology of China, Chengdu, China
{huangxiu,huyue}@std.uestc.edu.cn, dlyyang@gmail.com,
fumin.shen@gmail.com, shaojie@uestc.edu.cn

Abstract. Social media service defines a new paradigm of people communicating, self-expressing and sharing on the Web. Users in today's social media platforms often post contents, inferring their interests/attributes, which are significant for many Web services such as social recommendation, personalized searching and online advertising. User attributes are temporally dynamic along with internal interest changing and external influence. Based on topic modeling, we present a probabilistic method for dynamic user attribute discovery. Our method automatically detects user attributes and models the dynamics using time windows and decay function, thereby facilitating more accurate recommendation. Evaluation on a Sina Weibo dataset shows the superiority in terms of precision, recall and F-measure as compared to baselines, such as static user attribute modeling.

Keywords: Dynamic user attribute · Topic model · Time window

1 Introduction

In recent years, we have witnessed dramatic growth of social media services such as Twitter[1] and Pinterest[2], where people can publish, share and consume instant information. In China, as one of the leading microblogging service providers, Sina Weibo[3] receives significant attention from research area. Launched by Sina Corporation in August 2009, Sina Weibo has approximately 500 million registered users by December 2012, on which more than 4.6 million users are active on a daily basis, generating 100 million microblogs per day. On Sina Weibo, it allows people to create concise microblogs with a limitation of 140 characters, made up with a mix of Chinese and English characters as well as self-defined hashtags (e.g. # and @) and external URLs. Thus, users' activities in real time from Sina Weibo stream enable us to automatically discover user attributes by dynamically monitoring users' status, which would help us to timely detect and analyze users' opinions, sentiments and preferences.

---

[1] http://www.twitter.com/.
[2] http://www.pinterest.com/.
[3] http://www.weibo.com/.

© Springer International Publishing Switzerland 2016
F. Li et al. (Eds.): APWeb 2016, Part I, LNCS 9931, pp. 256–267, 2016.
DOI: 10.1007/978-3-319-45814-4_21

Nevertheless, on account of the sparsity and noise of content in short text, diverse and fast changing topics, and large data volume, it is challenging to dynamically discover user attributes. As a result, addressing the specialty and uncertainty of microblogs is crucial for us to analyze changing tendency in user attributes and behaviors. In previous works [5,9,13,20], users attributes or interests were constructed by using language models in a static manner. Actually, as newly-emerging elements keep occurring on the Web, user attributes may be temporally dynamic, i.e., some interests will be out-of-date while others may become popular attributes that are likely to better reflect current user requirements. Therefore, it is necessary to explore user attributes from the very recent Web contents.

In this work, a novel dynamic user attribute model (DUAM) is proposed to overcome the shortcomings of static attribute model. In particular, we leverage a topic model by the name of Biterm Topic Model (BTM) [19], which is capable of addressing the sparsity of content in short text. BTM is an extending model of Latent Dirichlet Allocation (LDA) [6], which generates topics over microblogs through modeling the biterms directly. As defined in [19], a biterm that is an unordered word-pair co-occurring in a short context can model the word co-occurrence patterns, and it also can avoid the problem of sparse patterns through aggregating word co-occurrence patterns so as to discover topics, thus distinguishing from adding external knowledge to content. Besides, since microblogs are normally input as a rapidly growing stream of prohibitively large volume, they require the user attribute model to dynamically update with the continuously arrived new data. Inspired by this analysis, we introduce a decay function over time windows to model the dynamics in user interests. We assume that microblog documents arrive in a batch mode and in our experiments we divide the whole dataset by a fixed time window (e.g., three months). In this way, our proposed model only needs to store a small part of microblog data online, which can be much more efficient than static attribute discovery. From the experiments on real dataset crawled from Sina Weibo, the dynamic property of user attributes can be detected according to DUAM, which outperforms static user attribute models. The major contributions of this work are summarized as follows:

1. We propose a Dynamic User Attribute Model (DUAM) based on a topic model named Biterm Topic Model (BTM), which can effectively address the sparsity of short text content and significantly overcome the shortcoming of static attribute.
2. Our model dynamically establishes topic-attribute mapping by introducing a decay function over time windows, and detects the shift over user attributes based on the microblog stream.
3. We construct a Sina Weibo microblog dataset by manually labeling user attributes. The promising results demonstrate that our proposed approach significantly outperforms static user profiles.

The rest of the paper is organized as follows. Section 2 illustrates the related work about user attribute modeling. In Sect. 3, we discuss how to generate user attribute dynamically. Section 4 demonstrates the experimental results and evaluation. Finally, conclusions and future work are given in Sect. 5.

2 Related Work

In this section, we summarize some related work about user profiling and indicate the differences in relation to our own work.

Researchers have long been interested in studying mining user interests, which are established by extracting users' characteristics and preferences from posted content on social media [10,12,17,18]. Most previous studies attempted to exploit external knowledge (e.g. Wikipedia, DBpedia) for semantic linking to enrich the presentation of microblogging. For instance, Abel et al. [1] analyzed methods for contextualizing Twitter activities via connecting Twitter posts with the related news articles. The proposed method semantically represents individual Twitter activities through extracting from tweets and the related news articles. Lim et al. [11] proposed a method that can automatically classify the relative interests of Twitter users with a weighting in relation to their other interests using information from Wikipedia. Besides, Ding et al. [8] studied user biographies from Twitter to indicate user interests and analyzed the extracted interest tags from biographies to enrich the information of tweets. However, their work relies heavily on the availability of users' biographies. In those works, exploiting external knowledge, which is widely leveraged for enriching semantics of microblogging, is only effective when auxiliary data are in a close correlation to the original data. On the contrary, our method exploits statistics of word co-occurrence in the microblog corpus with no need to infer external knowledge.

Besides, some previous works also attempted to exploit cross-OSN content to extract user interests. Abel et al. [2] studied form-based user profiles in social web services, e.g. Twitter, Facebook, and also investigated tag-based user profiles based on user tagging activities in some other social systems, e.g. Flickr, StumbleUpon and Delicious, in order to explore the benefits of building user profiles between different systems. Ottoni et al. [14] studied behavior and interests of users, whose accounts are associated Twitter with Pinterest. However, the majority of the existing works focus on tackling the sparse and noisy user-generated data, which represent static user attribute, leading to inconsistence with users' actual status.

Our work employs BTM algorithm to address the sparsity of short texts in microblogs, so that generate topics from microblog content. Then, we propose a novel DUAM model to dynamically discover user attributes over time. Yet, there are some prior works exploiting user attributes by extracting topics, which base on LDA-like model utilizing various inference algorithms. For example, Rosen-Zvi et al. [15] presented the author-topic model, which extends Latent Dirichlet Allocation (LDA) and is a generative model for authorship information and documents. Then, Xu et al. [18] proposed a twitter-user model, which is a modified author-topic model by using a latent variable to indicate author's interests, instead of constructing a "bag-of-words" user profile. Bhattacharya et al. [4] proposed a novel mechanism named Labeled LDA, which aims to generate topics of interest for individual users on Twitter. Overall, Our work distinguishes from the above researches in that we focus on exploiting user dynamic attribute by automatically modeling topic-attribute mapping in time windows, thus overcoming the shortcoming of static user attribute.

3   The Proposed Approach

In this section, we first present the overall framework of dynamic user attribute discovery. Then, we elaborate how to utilize the Biterm Topic Model (BTM) to extract topics of user attributes. Finally, we formally present the algorithm of inferring Dynamic User Attribute Model (DUAM) in detail.

As depicted in Fig. 1, in the data collection process, we crawl microblogs of randomly selected users according to several different topics from Sina Weibo. After filtering out noisy microblogs, we conduct the following preprocessing steps: (1) We remove links from the microblogs; (2) We eliminate non-Chinese characters and self-defined characters (e.g. "@"); (3) We segment the crawled microblog documents into words; and (4) We remove stop words and non-sense words of high frequency in the microblogs. Subsequently, we leverage the BTM model to extract topics of attributes from the users' microblogs and employ DUAM to dynamically learn attributes for general users, through which we can capture the changing tendency of attributes from an individual user.

3.1   Biterm Topic Model

Biterm Topic Model (BTM) is a probabilistic topic model extending LDA. The underlying idea in BTM is that if two word co-occurrences appear more frequently in the same microblog document, there is a better chance for them to belong to the same topic. As defined in [19], a biterm denotes an unordered

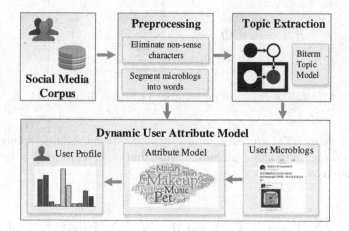

Fig. 1. The overall framework of dynamic user attribute discovery model.

word pair, which is composed of any two different words in a microblog document. For instance, if there are three distinct words in a microblog document, we can generate three biterms:

$$(w_1, w_2, w_3) \Rightarrow \{(w_1, w_2), (w_2, w_3), (w_1, w_3)\}.$$

where $(\cdot, \cdot)$ is unordered combination. BTM considers the whole corpus of microblogs as a mixture of attributes (topics), where any pair of words are drawn from a specific topic independently and a topic submits to the topic mixture distribution over the whole microblog corpus. Particularly, the topics extracted from microblogs of a specific user indicate the user attributes, characteristics and preferences.

Suppose single-valued hyperparameters α and β are Dirichlet priors for θ and ϕ_k, respectively. The specific generating process of BTM can be described as below:

1. Draw topic proportions $\theta \sim$ Dirichlet(α);
2. For each topic k, where $k = 1, 2, ..., K$
   draw word probability $\phi_k \sim$ Dirichlet(β);
3. For each biterm $b_i \in B$
   draw topic $z_i \sim$ Multinomial(θ), and
   draw biterm $(w_{i,1}, w_{i,2}) \sim$ Multinomial(ϕ_{z_i});

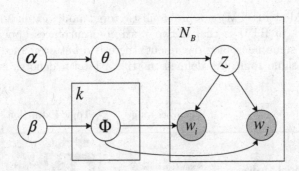

Fig. 2. The graphical presentation of Biterm Topic Model.

Figure 2 shows BTM graphical representation. Given the single-valued hyperparameters α and β, the joint probability distribution of a biterm $b_i = (w_{i,1}, w_{i,2})$ can be written as:

$$P(b_i|\alpha, \beta) = \sum_{k=1}^{K} P(w_{i,1}, w_{i,2}, z_i = k|\alpha, \beta) = \int \int \sum_{k=1}^{K} \theta_k \phi_{k,w_{i,1}} \phi_{k,w_{i,2}} d\Theta d\Phi \quad (1)$$

Thus we can get the likelihood in the whole microblog corpus, where $B = \{b_i\}_{i=1}^{N_B}$, N_B referring the number of biterms in documents.

$$P(B|\alpha,\beta) = \prod_{i=1}^{N_B} \int \int \sum_{k=1}^{K} \theta_k \phi_{k,w_{i,1}} \phi_{k,w_{i,2}} d\Theta d\Phi \tag{2}$$

We can see that BTM directly uses the word co-occurrence patterns as an unit revealing the latent semantics of attributes (topics), rather than a single word. In addition, BTM assigns a topic for every biterm in order to learn a global topic distribution. We can obtain the topic proportions of a document by follows:

$$P(z_i = k|d) = \sum_{b=b_i}^{N_B} P(z_i = k|b = b_i) P(b = b_i|d) \tag{3}$$

3.2 Inference Parameters

To perform approximate inference for Θ and Φ, we adopt Gibbs sampling, which is introduced detailedly in [3]. For BTM algorithm, in order to infer the topics, we are required to sample the topic assignment z for each biterm according to its conditional distribution, thus obtaining the following conditional probability:

$$P(z_i = k|z_{-b}, B, \alpha, \beta) \propto$$
$$(n_{-b,k} + \alpha) \frac{(n_{-b,w_i|k} + \beta)(n_{-b,w_j|k} + \beta)}{[\sum_{w=1}^{U}(n_{-b,w|k} + \beta) + 1][\sum_{w=1}^{U}(n_{-b,w|k} + \beta)]} \tag{4}$$

where z_{-b} means the topic assignments for biterms except b, and n_{-b} denotes the number of biterms assignment over topic k except b, as well as $n_{-b,w|k}$ is the number of times for a word w assignment over topic k excluding b.

Finally, the counts of the topic assignments of biterms and word occurrences are used to infer the distributions Φ of topic-word and global topic distribution Θ as follows:

$$\Phi_{k,w} = \frac{n_{w|k} + \beta}{\sum_{w=1}^{U} n_{w|k} + U\beta} \tag{5}$$

$$\Theta_k = \frac{n_k + \alpha}{N_B + K\alpha} \tag{6}$$

3.3 Dynamic User Attribute Model

DUAM model keeps user attribute dynamic where new microblog data arrive continuously. We assume microblog documents are divided by time windows, therefore, we can dynamically generate user attribute using topic-attribute mapping in time windows. By leveraging BTM algorithm, we can obtain m topics/attributes presenting m clusters of microblogs. At the same time, we use top-k method to choose n words with highest scores as the representative keywords of a topic. The top n words' scores are presented by a n-dimension vector

$s = [s_1, \ldots, s_n]$. Relevantly, a user's attribute is represented by an m-dimension vector $a(u, t)$. Here, in time window t, our model generates attribute vector $a(u, t)$ for user u by using obtained attributes in the previous time windows. Therefore, we utilize both previous data and new data to produce user attributes, which can better reflect up-to-date preferences of users.

For general users, after data preprocessing, we calculate the frequency of keywords f_m corresponding to the mth topic. Then, the product between f_{mi}, the frequency of ith keyword in topic m, and s_{mi}, the weight of ith keyword in topic k, denotes the weight of the topic m for a single user.

$$a_m = \sum_{i=1}^{k} f_{mi} s_{mi} \tag{7}$$

After normalizing a_m, we can obtain the relative value $a'_m = a_m / \sum_{i=1}^{m} a_i$, which is used to decide whether a user have the topic/attribute.

During the topic-attribute mapping, we set a threshold θ for mapping rules, i.e., if the matching number is larger than the given threshold θ, we consider that the topic successfully maps to user attribute.

After topic-attribute mapping, users' static attributes during various periods of time is obtained. However, the obtained attributes are restricted to the specific time, and the long-term attributes that do not appear in the specific time are ignored. User attributes have a continuous changing tendency, which attributes in every period of time have mutual connections. Inspired by [7,16], in order to invoke user attribute in last time window, we define a decay coefficient $0 < \lambda < 1$ to infer the influence of prior attribute as follows:

$$\lambda(t) = 1 - \mu t^v, \tag{8}$$

where $0 < \mu < 1$ and $v > 0$ are two decay parameters. $a(u, t_i)$, a m-dimension vector, indicates m attributes in time t_i. Then, we estimate the user attribute vector $a(u, t_i)$ as below:

$$a(u, t_i) = \sum_{j=1}^{i} \lambda(t_j) a(u, t_j). \tag{9}$$

Here, $a(u, t_0) = [a_{1_0}, a_{2_0}, \ldots, a_{10_0},]$

Thus, we can obtain the latest attributes revealing users' current status, and by analyzing changing tendency of user attribute over time windows, we are able to make a prediction on users attributes in the near future.

4   Experiments

In this section, the collecting process of our experimental dataset from Sina Weibo is introduced firstly. Then, we demonstrate how to implement our experiment in details and give the evaluation metric. Finally, compared with static method with no time windows, we illustrate the effectiveness of our proposed DUAM model.

4.1 Experimental Dataset

We build up our dataset via crawling information streams over a one-year period from January to December, 2015, published by randomly selected 2100 users about 640 000 microblogs from Sina Weibo. In order to get sufficient text data, we filter out those who posted less than 200 microblogs. As our work aims to model the dynamic attribute on Sina Weibo for a single user, we employ a simple and effective means of user-selection through randomly selecting 100 active users who generate more than 5000 microblogs as the training set to obtain topics of attributes. Then, we also randomly select 100 active users to evaluate the experimental performance compared with static method.

After data collection and noise removal, we take the following preprocessing steps: (1) We remove links from the micrologs; (2) We eliminate non-Chinese characters and self-defined characters (e.g. "@"); (3) We segment the crawled microblog documents into words; and 4) We remove stop words and non-sense words of high frequency in the micrologs.

4.2 Implementation Details

In BTM algorithm, we set $\alpha = 50/K$, and $\beta = 0.01$ empirically and use training dataset to generate $m = 10$ topics of attributes, including $n = 20$ top words as keywords in each topic. In such case, we can obtain 10 topics over the whole microblog document and the keywords are with scores s_{mi} over the corresponding topic. Showed in Table 1, we present top 5 keywords with scores for every topic. We can see that the keywords are closely related to the corresponding topics. In DUAM, we divide the microblog documents by 3 months as a time window and empirically set $\mu = 0.56$, $v = 0.06$ separately.

To show the effectiveness of DUAM, we compare it with the static method without time windows. For the static method, we conduct the experiment with fixed time in the first three months to generate user static attributes. While, in our proposed model, we slide the time window and leverage decay function to generate fresh attributes of users according to new data arriving continuously.

Table 1. 10 attributes and corresponding top 5 keywords extracted by BTM.

健身	动作(0.0231)	健身(0.0083)	马甲(0.0075)	教程(0.0072)	瑜伽(0.0068)
美食	美食(0.0171)	做法(0.0097)	吃货(0.0079)	美味(0.0065)	好吃(0.0064)
数码	手机(0.0174)	苹果(0.0083)	小米(0.0076)	摄像头(0.0068)	处理器(0.0063)
体育	篮球(0.0101)	曼联(0.0100)	比赛(0.0074)	范加尔(0.0068)	科比(0.0067)
美妆	化妆(0.0173)	教程(0.0173)	妆容(0.0104)	美妆(0.0074)	技巧(0.0061)
旅游	旅行(0.0224)	攻略(0.01881)	旅途(0.0102)	见闻(0.0060)	旅游(0.0057)
军事	中国(0.0096)	航母(0.0060)	美国(0.0057)	海军(0.0048)	南海(0.0045)
音乐	一首歌(0.0140)	音乐厅(0.0127)	翻唱(0.0104)	私人(0.0077)	演唱会(0.0077)
萌宠	主人(0.0110)	狗狗(0.0105)	汪星(0.0070)	主子(0.0062)	猫咪(0.0043)
游戏	游戏(0.0295)	玩家(0.0106)	发售(0.0067)	本作(0.0065)	地址(0.0052)

4.3   Evaluation Metric

In the dataset collection process, we also crawl the tags labeled by the authors themselves. Due to the crawled tags, which are lack of complete information, we manually annotate 100 users' tags. Similar to clustering, we define a topic of attribute as a cluster C. To evaluate the experimental result, we compare the obtained attributes from each microblog document with that provided both by authors and the manual work. The clustering performance is measured by frequently used evaluation methods, precision, recall and F-measure analysis.

Here, F-measure is the average value of recall and precision and used in our experimental evaluation as a measure of accuracy. Higher F-measure value reflects the algorithm is better. Higher the precision implies better quality of the algorithm in prediction as recall indicates quantitative analysis.

4.4   Experimental Results

We employ the proposed DUAM to generate dynamic user attribute. Based on BTM algorithm, we can obtain 10 attributes. Figure 3 shows quite different set of attributes for each of the randomly selected 3 users. Apparently, user 1 is most interested in makeup and food, and user 2 has an affection on fitness, while user 3 shows special preference to military. Due to the diversity of different users, we analyze a randomly selected user to see the changing tendency of user's attributes in a time span. As displayed in Fig. 4, we can see that in different time period the user has different preferences, which represent long-term attributes and short-term attributes. As we can see, the user scarcely has interest in sport, electronics, military and music. However, there is a rising trend for him or her on food and travel.

Compared with the static method, the precision, recall and F-measure values of the proposed DUAM based on Sina Weibo are showed in Table 2. As we can see there is a significant increase in precision, recall about 8.9 % and 18.1 %

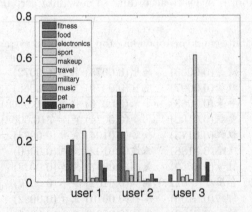

Fig. 3. The distribution of attributes from 3 randomly selected users.

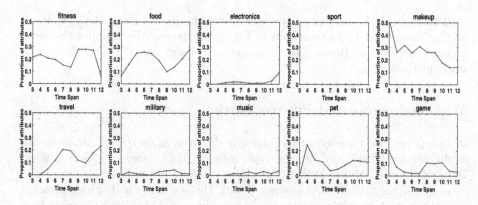

Fig. 4. The changes of a randomly selected user over 10 attributes.

Table 2. Precision, recall and F-measure.

	Precision	Recall	F-measure
the static method	0.4412	0.6317	0.5195
DUAM	0.5300	0.8125	0.6415

separately on the DUAM over time windows. The average value of precision and recall is also higher on DUAM. Accordingly, the result of DUAM is better for us to predict user attribute in the near future and consequently deliver personalized recommendation in line with users' current preferences.

Obviously, different thresholds exert great influence on our model to generate attributes in accordance with users. To visually evaluate performance in the proposed DUAM and the static method, we utilize ROC curve which is typically used to evaluate binary classifier output quality and can also be applied to assess our model.

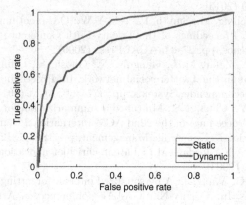

Fig. 5. ROC curve of the static method and DUAM.

The accuracy is denoted as the area under the ROC curve (AUC), which is ranging from 0 to 1. As presented in Fig. 5, the larger AUC implies the more accurate prediction. Hence, our proposed model significantly outperforms the static method.

5   Conclusions and Future Work

In this paper, we targeted at dynamically discovering user attribute on social media service. Based on Biterm Topic Model (BTM), we proposed a novel Dynamic User Attribute Model (DUAM) to analyze changing tendency of user attribute on Sina Weibo. As compared with the static method, which presents user attribute in a static process, our proposed model leveraged time windows and a decay function to describe fresh attributes that better meet user current demands. Extensive experiments on our crawled dataset from Sina Weibo showed the effectiveness of our model.

In future work, we will further research the problem with multi-data sources adding images or short videos. For our initial exploration, we only focus on the content-based data. However, in order to set up a real application running on a variety of social media platforms, we should further investigate how to automatically discover user attribute through user-generated contents, images and short videos. On all accounts, our proposed method has a great potential to stimulate future research in social network.

References

1. Abel, F., Gao, Q., Houben, G.-J., Tao, K.: Semantic enrichment of Twitter posts for user profile construction on the social web. In: Antoniou, G., Grobelnik, M., Simperl, E., Parsia, B., Plexousakis, D., Leenheer, P., Pan, J. (eds.) ESWC 2011, Part II. LNCS, vol. 6644, pp. 375–389. Springer, Heidelberg (2011)
2. Abel, F., Herder, E., Houben, G.J., Henze, N., Krause, D.: Cross-system user modeling and personalization on the social web. User Model. User-Adap. Inter. **23**(2–3), 169–209 (2013)
3. Asuncion, A., Welling, M., Smyth, P., Teh, Y.W.: On smoothing and inference for topic models. In: Proceedings of the Twenty-Fifth Conference on Uncertainty in Artificial Intelligence. pp. 27–34. AUAI Press (2009)
4. Bhattacharya, P., Zafar, M.B., Ganguly, N., Ghosh, S., Gummadi, K.P.: Inferring user interests in the Twitter social network. In: Proceedings of the 8th ACM Conference on Recommender Systems, pp. 357–360. ACM (2014)
5. Bian, J., Yang, Y., Chua, T.S.: Multimedia summarization for trending topics in microblogs. In: Proceedings of the 22nd ACM International Conference on Conference on Information & Knowledge Management, pp. 1807–1812. ACM (2013)
6. Blei, D.M., Ng, A.Y., Jordan, M.I.: Latent Dirichlet allocation. J. Mach. Learn. Res. **3**, 993–1022 (2003)
7. Chen, J., Wang, C., Wang, J.: A personalized interest-forgetting Markov model for recommendations. In: Twenty-Ninth AAAI Conference on Artificial Intelligence (2015)

8. Ding, Y., Jiang, J.: Extracting interest tags from Twitter user biographies. In: Jaafar, A., Mohamad Ali, N., Mohd Noah, S.A., Smeaton, A.F., Bruza, P., Bakar, Z.A., Jamil, N., Sembok, T.M.T. (eds.) AIRS 2014. LNCS, vol. 8870, pp. 268–279. Springer, Heidelberg (2014)

9. Gao, Q., Abel, F., Houben, G.J., Tao, K.: Interweaving trend and user modeling for personalized news recommendation. In: Proceedings of the 2011 IEEE/WIC/ACM International Conferences on Web Intelligence and Intelligent Agent Technology, vol. 1, pp. 100–103. IEEE Computer Society (2011)

10. Geng, X., Zhang, H., Song, Z., Yang, Y., Luan, H., Chua, T.S.: One of a kind: user profiling by social curation. In: Proceedings of the ACM International Conference on Multimedia, pp. 567–576. ACM (2014)

11. Lim, K.H., Datta, A.: Interest classification of Twitter users using Wikipedia. In: Proceedings of the 9th International Symposium on Open Collaboration, p. 22. ACM (2013)

12. Michelson, M., Macskassy, S.A.: Discovering users' topics of interest on Twitter: a first look. In: Proceedings of the Fourth Workshop on Analytics for Noisy Unstructured Text Data, pp. 73–80. ACM (2010)

13. He, W., Liu, H., He, J., Tang, S., Du, X.: Extracting interest tags for non-famous users in social network. In: Proceedings of the 24th ACM International on Conference on Information and Knowledge Management, pp. 861–870. ACM (2015)

14. Ottoni, R., Las Casas, D.B., Pesce, J.P., Meira Jr., W., Wilson, C., Mislove, A., Almeida, V.: Of pins and tweets: investigating how users behave across image-and text-based social networks (2014)

15. Rosen-Zvi, M., Griffiths, T., Steyvers, M., Smyth, P.: The author-topic model for authors and documents. In: Proceedings of the 20th Conference on Uncertainty in Artificial Intelligence, pp. 487–494. AUAI Press (2004)

16. Sen, W., Xiaonan, Z., Yannan, D.: A collaborative filtering recommender system integrated with interest drift based on forgetting function. Int. J. u- and e- Serv. Sci. Technol. 8(4), 247–264 (2015)

17. Wang, T., Liu, H., He, J., Du, X.: Mining user interests from information sharing behaviors in social media. In: Pei, J., Tseng, V.S., Cao, L., Motoda, H., Xu, G. (eds.) PAKDD 2013, Part II. LNCS, vol. 7819, pp. 85–98. Springer, Heidelberg (2013)

18. Xu, Z., Lu, R., Xiang, L., Yang, Q.: Discovering user interest on Twitter with a modified author-topic model. In: 2011 IEEE/WIC/ACM International Conference on Web Intelligence and Intelligent Agent Technology (WI-IAT), vol. 1, pp. 422–429. IEEE (2011)

19. Yan, X., Guo, J., Lan, Y., Cheng, X.: A biterm topic model for short texts. In: Proceedings of the 22nd International Conference on World Wide Web, pp. 1445–1456. International World Wide Web Conferences Steering Committee (2013)

20. Yin, H., Cui, B., Chen, L., Hu, Z., Huang, Z.: A temporal context-aware model for user behavior modeling in social media systems. In: Proceedings of the 2014 ACM SIGMOD International Conference on Management of Data, pp. 1543–1554. ACM (2014)

The Competition of User Attentions Among Social Network Services: A Social Evolutionary Game Approach

Jingyuan Li[1(✉)], Yuanzhuo Wang[1], Yuan Lu[2], Xueqi Cheng[1], and Yan Ren[3]

[1] CAS Key Lab of Network Data Science and Technology,
Institute of Computing Technology, Chinese Academy of Sciences,
Beijing, China
lijingyuan@ict.ac.cn
[2] School of Automation and Electrical Engineering,
University of Science and Technology Beijing, Beijing, China
[3] National Computer Network Emergency Response Technical
Team Coordination Center of China, Beijing, China
ry@cert.org.cn

Abstract. As the total amount of users in the social network services approach the total amount of netizens, social network service providers have to compete with each other for the attention of the existing users, rather than attracting totally new users without any social network service using experiences. Most of the current game theoretical studies on social network services focus on the evolution of the cooperation/defection between users, and the evolution of the structure of the networks for a single social networks, and fail to consider the competition of multiple social networks over a same and stable users group and their user attentions. In this paper, we propose a competitive social evolutionary game model to describe the competition of user attentions among social network services. We introduce the concept of user attention, and the popularity of social networks to describe the local and general user attention distributions, respectively. Our simulation of the competition between two social networks with different initial network structures and cooperation/defection utilities shows that a greater reputation awareness can suppress the influence of the defect temptation value.

Keywords: Social network · User attention · Competitive social evolutionary game

1 Introduction

Online social network services, or SNs for short, have definitely been the super star of the era, winning almost everyones fondness all around the world. During the past decade, SNs, including Microblogs, Social Network Sites, Wechat etc., expand in an astonishing speed of over 100 percent per year [1], with millions of new users joining the SNs. Moreover, a great percentage of the people who have already been using some of the SNs tend to spend more time in the SNs they

© Springer International Publishing Switzerland 2016
F. Li et al. (Eds.): APWeb 2016, Part I, LNCS 9931, pp. 268–279, 2016.
DOI: 10.1007/978-3-319-45814-4_22

have joined in, or to join new and more popular SNs. We call this the developing phase of social network services.

However, as the total amount of users involved in SNs approach the total amount of netizens in the world, getting new users, or obtaining more using time from the existing users is becoming more and more difficult. We call this the developed phase of social network services. Now that online SNs have entered the developed phase, and the new resources are dwindling, SN providers are forced to compete with each other, to try to turn over more users from other competing SNs, so as to maintain their position in the industry.

Many of the previous research findings on the evolution of social networks have been focused on the describing of the behavior of social network service users (or SU for short), and the corresponding network structures. Social evolutionary game (SEG) model proposed by Yu [2,3] considers two main factors: the **short term gain** and the **long term reputation** of the SUs. The short term gain controls the behavior of users' information exchange actions, and are commonly described by cooperation games. If a user chooses to accept and rebroadcast a neighbor's status, it is considered an act of cooperation C. If a user fails to do so, it is considered an act of defection D. Two users both obtain rewards R upon mutual cooperation and P upon mutual defection. A defector exploits a cooperator with temptation T and exploited cooperator receives S. The usual setting is $T > P > R > S$. The long term reputation controls the structure of the network. If a user always fails to cooperate with other users, to a certain time, its neighbors would have the opportunity to end the relationship. More factors could be introduced to make the model more realistic, such as considering mixed-strategies in the information exchange game [4].

However, not much of the work has been done to describe the behaviors and strategies of SNs themselves, which is of great importance to analyze multiple SN providers in the developed phase. Wu [5] studied the relation between SNs and SUs on the subject of privacy issues in an evolutionary game model. Though for the information dissemination issue, SN providers cannot directly interfere with the play's actions, they can change the rules of the services, such as encouraging more cooperation strategies by adding more payoffs to some strategies. The initial state of the network structures may also influence the evolution of the social networks [6]. Nevertheless, there is of little study on how can we describe the evolution of two or more SNs under a same set of SUs', especially when there are a stable number of users and a relatively stable amount of time of the users to spend on SNs.

In this paper, we propose a competitive social evolutionary game model ($CSEG$) to describe multiple social network services under the same group of users as depicted in Fig. 1, so as to see the influences among different kinds of SNs' initial statues and information dissemination rules, and how they affect the amount of times the SUs' spend in each SNs. We suppose that one netizen can only spend a fix amount of time in all of the SNs it has joined in. Therefore if it spends more time in SN_1, than it will spend less time in other SNs, where the netizens spend their time to repost information and to add or remove neighbors. We introduce the concept of **user attention** to denote the amount

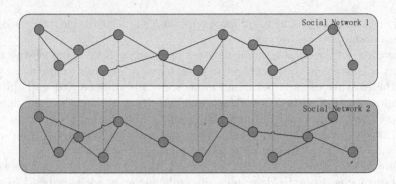

Fig. 1. Two social networks under a same group of users.

of time of SUs to spend in SNs, and we suppose that the competition of SN providers is intrinsically to compete the user attentions of the SUs. Hence our goal will be to analyze the user attentions under different initial conditions and strategies, including the different initial network structures, cooperation game's payoff matrices, and neighbor adjusting strategies, etc..

The contributions of our work include:

- We propose $CSEG$, a competitive social evolutionary game model to describe the competition of SNs over the user attentions, which considers both the information dissemination behaviors of the SUs and the structure changes of the SNs. It can be a comprehensive and efficient tool to study SNs in the developed phase of social network services.
- By our simulation, we have found out that a greater reputation awareness p_r can suppress the influence of the defect temptation value b, and the importance of the initial p_r of the small world network is more important to the stableness of the evolutionary process than the initial p_r in the regular network.

The rest of the paper is organized as follows. Section 2 gives a brief introduction of the evolutionary game studies on the social network services. In Sect. 3, we propose our $CSEG$ model, describing the entities in the game as well as the main factors to be considered in the evolution dynamics. Section 4 shows the comparisons of a user set with different initial network structures and cooperation settings, and the competitive results of the user attentions between two SNs. Finally, we conclude the paper in Sect. 5.

2   Related Work

Evolutionary game theory was firstly presented by John Maynard Smith and George R. Price, which is originally to describe the evolving populations of lifeforms in biology. Later on, researchers realized that, when adding certain mechanisms to describe the competition of minds or humans thoughts, evolutionary game can also been applied to human related research topics, such as in the cooperation phenomenon of human society [7], in the detection of macroscopic-behavioral dynamics of civil violence [8], and in the association of wireless communication networks [9].

One of the most important application aspect of evolutionary game theory is about the study of online social networks, which has a small-world feature [10], and can evolve very quickly due to the easy-access online features and the huge number of users. It is of great importance to study the evolution of online social network services in order to find the fundamental rules or patterns of it, and so we can quickly adapt to the changes to get value from it. The research is mainly two folds: (1) about information dissemination and (2) about the change of network structure. For information dissemination, Traag [11] believed it is basically a study of cooperation and defection and can therefore be described by prisoner's dilemma game. For the change of networks, Jiang [12] proposed an evolutionary game theoretic framework to model the dynamic information diffusion process in social networks, and compared his model under three kinds of networks: complete, uniform degree, and nonuniform degree networks.

Some of the work considers both the collaborative behaviors of the players, and the evolution of the social network structure [13,14]. Yu [2,3] presented a social evolutionary games methodology to describe the information dissemination behavior and the corresponding influences on the network structure through a reputation mechanism, which is the closest of our work. However, previous work did not fully consider the situation where multiple social network services are built upon a same users set, where the policies or regulations of one SN may seriously influence the SU's cohesion of other SNs. The model we have presented in this paper make a detailed study on this subject, which can be very useful to analyze the current social network services' ecology where few new users are joining in, and SN's have to compete with other SNs to survive.

3   The Competitive Social Evolutionary Game Model

In this section, we present the framework of our competitive social evolutionary game $(CSEG)$ for multiple social networks upon a same group of players, where each social networks, with the group of players, forms a SEG. Each player has two concerns, namely, its short-term utility and long-term reputation. Based on players' concerns, there are two updating mechanisms, strategy updating for better utility and partnership adjusting for better reputation environment. The factor that connects the two social networks is the user attention factor $\mathbf{A}$ of all players about the preference of one social network service over the others.

3.1   The Game

A competitive social evolutionary game is represented as a 6-tuple set of $G = (V, \mathcal{E}, I, U, R, \mathbf{A})$, where:

- $V = \{v_1, v_2, \ldots, v_n\}$ is the set of players.

- $\mathcal{E} = \{E_1, E_2, \ldots, E_m\}$ is the set of edge sets of the m social networks, where $E_k = \{e_{ij}|v_i, v_j \in V\}$ is the set of partnerships among the players in SN_k, $k = 1, 2, \ldots, m$.
- I is the set of actions which are performed by the players based on their corresponding strategies, including both the actions about short-term utility and long-term reputation.
- U is the set of the short term utility functions of the players.
- R is the set of the long term reputation functions of the players.
- $\mathbf{A_{mn}} = \{\mathbf{a_1}, \mathbf{a_2}, \ldots, \mathbf{a_n}\}$ is the user attentions of the n players in m social networks.

The Denomination About Directed and Undirected Graph. If we consider a SN as directed networks as in Microblogs, then a player i will have two types of neighbors, namely, IN-neighbors and OUT-neighbors. A player j is i's IN-neighbor, if there is an edge e_{ji} which is directed as $i \leftarrow j$. Reversely, i is j's OUT-neighbor. Let $\mathcal{N}_i^I$ denote the set of i's IN-neighbors, and $\mathcal{N}_i^O$ be the set of i's OUT-neighbors. Player i's neighbors can then be denoted as $\mathcal{N}_i = \mathcal{N}_i^I \cup \mathcal{N}_i^O$. d_i^I, d_i^O are player i's indegree and outdegree, respectively, and $d_i = d_i^I + d_i^O$ is its degree.

If we consider a SN as undirected graph as in Social Network Sites, then there will be only the definitions of player i's neighbors $\mathcal{N}_i$ and its degree d_i.

The User Attention of the Players. As defined in Sect. 1, the attention of a player is all of the time spent by the player to play in all of the SNs. Therefore at one time for one SN, the player will only spend part of the time. An because of the presumption we made that social network services have been in a developed phase, both of the number of players and the attention of players are stable.

More formally, we define **user attention matrix $\mathbf{A_{mn}}$** as:

$$\mathbf{A_{mn}} = \begin{bmatrix} a_{11} & a_{12} & a_{13} & \cdots & a_{1n} \\ a_{21} & a_{22} & a_{23} & \cdots & a_{2n} \\ a_{31} & a_{32} & a_{33} & \cdots & a_{3n} \\ \cdots\cdots\cdots\cdots\cdots\cdots \\ a_{m1} & a_{m2} & a_{m3} & \cdots & a_{mn} \end{bmatrix} \tag{1}$$

which denotes the attentions of n players in m social networks. $a_{ij} \in A_{mn}$ is the percentage of the time for player v_j to play in social network SN_i. The sum of each column $\sum_{i=1}^{m} a_{ij} = 1$ for all $j \in n$, which is the total amount of play time of each player.

We also define the popularity a_i^* to be the successfulness of the SN_i. It is simply the average of the players' attention in SN_i: $a_i^* = \frac{1}{n}\sum_{k=1}^{n} a_{ik}$. Hence the popularity vector of all SNs can be denoted as $\mathbf{p} = [a_1^*, a_2^*, \ldots, a_m^*]$.

From a SN's point of view, the attention of a player for this SN is called the local attention, and the attentions spent on other SNs are call the remote attentions. From a player's point of view, when playing in a SN, the attention of the player itself is considered as the neighbor attention, and the average attention of the entire SN is called the popularity of the SN.

The Utility of Players. In a $CSEG$ as well as a SEG, the utility of player i is the cumulative payoffs obtained from its opponents in a single round. It is players' short-term concern, which affects players' choices when they imitate others' strategies. The game can be set as a matrix game of prisoner's dilemma, where the payoffs of the matrix could either be a value according to the strategy settings of the SN provider, or they could be a series of utility functions that consider the neighbor attentions of this SN and other SNs as well.

The Reputation of Players. In a SEG, the reputation of player i is the opinion of its partners on it, which is the results of its partners's evaluation upon i's behavior history. It is players's long-term concern, which affects players' choices when they seek new partners. The reputation of player i at time t is formulated as

$$R_i(t) = \sigma R_i(t-1) + \Delta R_i(t) \tag{2}$$

where $\Delta R_i(t)$ is the increment of reputation at time t, σ is the memory decaying rate of reputation.

$\Delta R_i(t) = \sin(\frac{\pi}{2}\frac{n_i^C}{d_i})$, where n_i^C represents the number of C-strategy neighbors of i. sin() function is used to suppress the rise rate of reputation.

The reputation of a $CSEG$ is a little bit different than SEG in that the user attentions can influence the reputation of players, or the adding/removing strategies of the neighbors.

3.2   The Cross Social Network Coevolutionary Dynamics

At each time step t, players take actions in every social networks in an asynchronous way, an example of the coevolutionary process of three social networks is shown in Fig. 2. Three major actions are processed: strategy updating, partnership updating, or prisoner's dilemma game (PDG) playing. The two updating strategies are performed according to the asynchronous update [14]: at time t_{ij} (which denotes the SN_j's actions at time step i), a randomly selected player performs updating operation. The chosen player updates its strategy with probability $1/(1+W)$, otherwise it adjusts its partnership. W is the ratio between the time scale of strategy updating τ_e and that of partnership adjustment τ_p, $W = \tau_e/\tau_p$. The frequency of partnership adjusting increases with W.

Updating Strategies. When updating strategies, the chosen player changes its strategy. We believe that the variation is an intrinsic attribute of the players, and therefore should not be influenced by the user attentions.

When a player i is determined to update its strategy, it chooses an player j from its partners who satisfies

$$j = \arg \max_{l \in \mathcal{N}_i^O} \{u_l > u_i\} \tag{3}$$

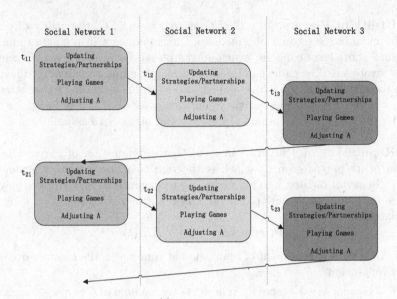

Fig. 2. The cross social network coevolutionary dynamics.

and then imitates j's strategy. The imitation process can be expressed by Fermi update rule [15]:

$$\omega(s_i \leftarrow s_j) = \frac{1}{1 + exp[\beta(u_j - u_i)]} \qquad (4)$$

where β represents the extent of imitation noise which can be considered as the willingness of players to imitate others' strategies.

Updating Partnerships. When updating partnerships, the chosen player who adjusts its partnership list can only be aware of the local reputation information about its neighbors and its neighbors' neighbors. At each round, the player removes some of its neighbors who have the lowest reputation, and add some of its neighbors' neighbors with the highest reputation. Because the goal of $CSEG$ is to study the developed phase of social network services, the ratio between adding and removing partnership should be around 1.

Player i dismisses the partnership with its one OUT-neighbor based on the minimum reputation rule and establishes a new one with one of its IN-neighbors and next-nearest OUT-neighbors based on the maximum reputation update rule. Specifically, with probability p_s, player i removes the partnership with j, satisfying

$$j = \arg \min_{l \in \mathcal{N}_i^O} \{R_l(t) < R_i(t)\} \qquad (5)$$

And with probability p_r, player i chooses player k as its new partner, satisfying

$$k = \arg \max_{\mathcal{N}_i^I \cup_{l \in \mathcal{N}_i^O}(\{i\} \cup \mathcal{N}_i^O)} \{R_l(t) > R_i(t)\} \qquad (6)$$

Otherwise with probability $1 - p_r$, player i randomly selects player k' from its IN-neighbors and next-nearest OUT-neighbors.

Playing Game. After the process of the updating strategies, for each SN, a player a_j plays prisoner's dilemma game with all its neighbors one by one over the influences of the user attention matrix $\mathbf{A}$. a_j's utility is the cumulative payoffs it obtains from its opponents, formulated by

$$u_i = \sum_{j \in \mathcal{N}_i} \mathbf{s_i^T M s_j} \tag{7}$$

where $\mathcal{N}_i$ is the set of i's neighbors, $\mathbf{s_i}$ and $\mathbf{s_j}$ are the strategies adopted by i and j, respectively, where $\mathbf{C} = [0,1]^{\mathbf{T}}$ stands for the cooperation strategy, and $\mathbf{D} = [0,1]^{\mathbf{T}}$ stands for the defection strategy. $\mathbf{M}$ is a rescaled 2×2 payoff matrix

$$\mathbf{M} = \begin{bmatrix} 1 & 0 \\ b & 0 \end{bmatrix} \tag{8}$$

where players are rewarded by 1 for mutual cooperation and tempted by $1 \leq b \leq 2$ to defect when facing cooperators.

Adjusting Attentions. All the players adjust their user attentions at the end of each time step. Player i adjusts its attention vector $\mathbf{a_i} = [a_{1i}, a_{2i}, \ldots, a_{mi}]^{\mathbf{T}}$ according to the neighbor cooperation ratios $f_{ki}(C) = n_{ki}^C/d^{ki}$ of all the social networks, $k = 1, 2, \ldots, m$. If

$$f_{ki}(C) \leq \frac{1}{m} \sum_{j=1}^{m} f_{ji}(C) \tag{9}$$

then a_{ki} should be set to a smaller value than the previous round. Otherwise, a_{ki} should be set to a greater value. Player i's current utilities in different SNs can also influence its attention.

After calculating the user attention matrix $\mathbf{A_{mn}}$, we can calculate the popularity vector $\mathbf{p} = [a_1^*, a_2^*, \ldots, a_m^*]$, which can be used to compare the successfulness among the SNs.

4   Simulations

We simulated our proposed $CSEG$ under the situations where two social networks competed with each other for the user attentions of a group of users. The initialization of the networks were regular network and small world network.

4.1   The Calculation of User Attention for Two Competing Social Networks

The Influence of User Attention to Strategy Updating. The utilities of players are the intrinsic gains that are decided by the SN providers, and therefore should not be influenced by user attentions. However, the importance of reputation can be influenced by user attentions. For example, if a player spends more time in SN_1 than SN_2, its reputation in SN_1 will be more important than in SN_2, which means that p_r in SN_1 should be in increased.

We assume a competitive situation of two SNs in our simulations, therefore the updating of player k's p'_r can be defined as

$$
p'_r = \begin{cases} p_r + \delta_{max}(a_{1k} - a_{2k}), & 0 \le p'_r \le 1 \\ 0, & p'_r < 0 \\ 1, & p'_r > 1 \end{cases} \tag{10}
$$

where δ_{max} is the maximum change of user attention in one time step.

The Updating of User Attentions. The updating of user attentions in our simulations is related to two factors: the utility of players in this time step in the two social networks, and the cooperation/defection ratio of players. Player k's new attention in SN_1 can be calculated as

$$
a'_{1k} = \alpha a_{1k} + \beta \frac{u_{1k}}{u_{1k} + u_{2k}} + \gamma \frac{f_{1k}(C)}{f_{1k}(C) + f_{2k}(C)} \tag{11}
$$

where α, β, γ are the relative importance coefficients, and $\alpha + \beta + \gamma = 1$. Equation 11 guarantees that $a_{1k} + a_{2k} = 1$ for the two SNs' situation.

4.2 The Simulation Settings and Assumptions

The number of players in our simulation was 100, the initial user attention matrix $\mathbf{A}_{2 \times 100} = \{a_{ij} = 0.5 | \forall i \in \{1, 2\}, j \in \{1, 2, \dots, 100\}\}$. We generated the edges of the graphs using igraph [16], where the small world network was generated by the Watts-Strogatz model, and the degree of the generated networks was 4 on average. The maximum number of evolutionary time steps were 20,000, during which the parameters of α, β and γ were set to be 0.5, 0.4 and 0.1, respectively. If one of the network's popularity a_i^* was less than 0.1 or greater than 0.9, then the simulation stopped, and the ending time step was recorded.

4.3 Simulation Results

We generated a regular network SN_1 and a small world network SN_2, and compared the evolution of user attentions over time with thermograms of the ending time steps as shown in Fig. 3. The ending steps in the red area are larger than that in the blue area, which means that the user attention coevoluntionary process is slower in the red area than in the blue area.

We can see from Fig. 3 that the diagonals from bottom-left to top-right of the four subfigures are hotter than other areas, which means that if the defect temptation values b_1 and b_2 in $\mathbf{M}$ are similar, then the coevoluntionary process will be long. On the contrary, the top-left and bottom-right areas are cooler, meaning the coevoluntionary processes here will be quicker. That is to say, SN

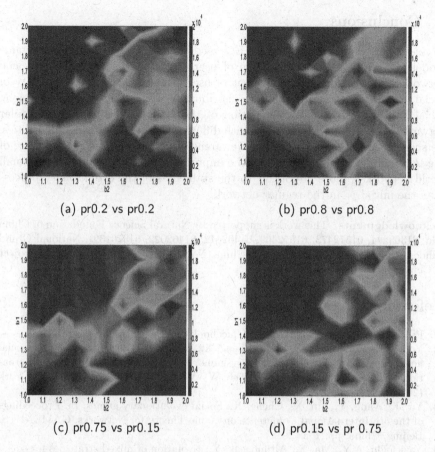

(a) pr0.2 vs pr0.2

(b) pr0.8 vs pr0.8

(c) pr0.75 vs pr0.15

(d) pr0.15 vs pr 0.75

Fig. 3. Regular network v.s. small world network. (Color figure online)

providers can increase user attentions by giving a smaller b than its opponent to encourage more cooperation strategies.

Figure 3a and b show the situation where the regular network and small world network have a same initial p_r of 0.2 and 0.8 respectively. We can see that Fig. 3a has a sharper temperature change than Fig. 3b, which means that the importance of reputation in the SN can also influence the coevolutionary process, where a greater p_r will suppress the influence of b.

In Fig. 3c and d, we simulate the situation where the regular network and small world network have a very different initial p_r. We observe that a small world network which values the reputation factor more than a regular network will have a quite uneven evolutionary process time, or we can say that the importance of the initial p_r for the small world network is more important to the stableness of the evolutionary process.

5   Conclusions

In this paper, we have proposed $CSEG$: a competitive social evolutionary game model for studying the competition of user attentions among multiple social networks. We have given the concept of user attention, and the popularity of social networks to describe the time distribution of social network users among social network services. By simulation of the competition between a regular network and a small world network with different cooperation/defection utilities, we showed that a greater reputation awareness p_r can suppress the influence of the tempted defection value b, and the importance of the initial p_r of the small world network is more important to the stableness of the evolutionary process than the initial p_r in the regular network.

Acknowledgments. This work is supported by Natural Science Foundation of China (No. 61303244, 61572473, 61572469, 61402442, 61402022, 61370132), National Grand Fundamental Research 973 Program of China (No. 2014CB340401), and the 242 Project of China (No. 2016F107).

References

1. Facebook Statistics (2016). https://facebook.com/press/info.php?statistics
2. Yu, J., Wang, Y., Jin, X., Li, J., Cheng, X.: Evolutionary analysis on online social networks using a social evolutionary game. In: Proceedings of the 23rd International Conference on World Wide Web, WWW 2014, pp. 415–416. ACM, New York (2014)
3. Yu, J., Wang, Y., Jin, X., Cheng, X.: Social evolutionary games. In: Proceedings of the 5th International Conference on Game Theory for Networks, GAMENETS, Beijing, China, pp. 1–5 (2014)
4. Yazicioglu, A.Y., Ma, X., Altunbasak, Y.: Evolution of mixed strategies for social dilemmas on structured networks. In: Proceedings of the 2011 International Conference on Networking, Sensing and Control, Delft, The Netherlands, pp. 175–180 (2011)
5. Wu, L., Chen, X.: Modeling of evolutionary game between SNS and user: from the perspective of privacy concerns. In: Proceedings of the 21st International Conference on Management Science & Engineering, Helsinki, Finland, pp. 115–119 (2014)
6. Wang, Y., Nakao, A.: On cooperative and efficient overlay network evolution based on a group selection pattern. IEEE Trans. Syst. Man Cybern. Part B Cybern. **40**(3), 656–667 (2010)
7. Axelrod, R., Hamilton, W.D.: The evolution of cooperation. Science **211**(4489), 1390–1396 (1981)
8. Quek, H.Y., Tan, K.C., Abbass, H.A.: Evolutionary game theoretic approach for modeling civil violence. IEEE Trans. Evol. Comput. **13**(4), 780–800 (2009)
9. Han, K., Liu, D., Chen, Y., Chai, K.K.: Energy-efficient user association in Het-Nets: an evolutionary game approach. In: Proceedings of the 4th International Conference on Big Data and Cloud Computing, BDCloud 2014, Sydney, Australia, pp. 648–653 (2014)
10. Zhang, S., Song, Z., Wang, X., Zhou, W.: Emergence of small-world networks via local interaction using prisoner's dilemma game. In: Proceedings of the 2007 IEEE Congress on Evolutionary Computation, CEC 2007, pp. 3706–3710 (2007)

11. Traag, V.A., Dooren, P.V., Nesterov, Y.: Indirect reciprocity through gossiping can lead to cooperative clusters. In: Proceedings of the IEEE Symposium on Artificial Life, pp. 154–161 (2011)
12. Jiang, C., Chen, Y., Liu, K.J.R.: Evolutionary dynamics of information diffusion over social networks. IEEE Trans. Signal Process. **62**(17), 4573–4586 (2014)
13. Santos, F.C., Pacheco, J.M., Lenaerts, T.: Cooperation prevails when individuals adjust their social ties. PLoS Comput. Biol. **2**(10), 1284–1291 (2006)
14. Fu, F., Hauert, C., Nowak, M.A., Wang, L.: Reputation-based partner choice promotes cooperation in social networks. Phys. Rev. E **78**(2), 026117 (2008)
15. Szabo, G., Toke, C.: Evolutionary prisoner's dilemma game on a square lattice. Phys. Rev. E **58**(1), 69 (1998)
16. igraph: the Network Analysis Package (2016). http://www.igraph.org/

Mechanism Analysis of Competitive Information Synchronous Dissemination in Social Networks

Yuan Lu[1,2(✉)], Yuanzhuo Wang[1], Jianye Yu[3], Jingyuan Li[1], and Li Liu[2]

[1] Insitute of Computing Technology, Chinese Academy of Science, Chengdu, China
819222971@qq.com
[2] School of Automation and Electical Engineering, University of Science and Technology Beijing, Beijing, China
[3] School of Information, Beijing Wuzi University, Beijing, China

Abstract. Different group of information, such as advertising and product promotion, compete with each other as they diffuse over social networks. Most of the existing methods analyze the dissemination mechanism mainly upon the information itself, without considering human characteristics. This paper uses a framework of social evolutionary game to simulate the dissemination and adjusts utility function and updating mechanism based on coordination game. We find that individuals consider more about their own reputation and more communication between them, individuals are more cautious in the face of strategy choice. When the benefit of competitive information is nearly 1.2 times of the original one, it can make up the loss of reputation caused by changing strategy. For the specific network environment based on simulation, the actual data on Sina Weibo strongly verify this rule and shows that factor of reputation promotes the cooperation and users won't easily change their information.

Keywords: Social network · Social evolutionary game · Coordination game · Information dissemination

1 Introduction

Social network is a kind of social structure, which is formed by the connection between the individual or groups in the information networks. Different from the traditional media, the new social network is based on the communication mode of "user-to-user". Through the interactive behavior, such as publishing, comments, recommendations, forwarding and praise, etc., the users can interact frequently and information can be disseminated wide and fast. Information dissemination in social networks brings mutual influences as well as competitive conflicts within one marketing group, such as advertising. In the case of information in the same field spreading in social networks, it will make mutual influence and reflect the characteristic of competition among them. It is necessary to study rules of information propagation, including key factors and evolutionary rule, in order to manipulate spreading process to maximize or minimize

© Springer International Publishing Switzerland 2016
F. Li et al. (Eds.): APWeb 2016, Part I, LNCS 9931, pp. 280–291, 2016.
DOI: 10.1007/978-3-319-45814-4_23

the outcome of information dissemination. At present, the study of basic laws of information dissemination is still in a primary stage, mainly involves network structure and information characteristics, while individual factors have not yet been considered. Moreover, synchronous information competition, a common dissemination model in networks, which describes a pair of rival information propagating at the same time. Aiming at the characteristics of large scale and fast dynamic changing of data in social network, especially considering the personality factors of networks' participants, evolutionary game is considered as a suitable method for solving the problem of information dissemination. And information diffusion and strategy selection can be regarded as the network dynamic behavior [1]. Evolutionary game theory is able to well describe the dynamic characteristics of information communication on social networks and reflect the influence of user's behavior. Social evolutionary game (SEG) [2] is introduced to model the dynamic characteristics between network structure and users' behaviors. We propose co-evolutionary mechanisms upon SEG to study the evolutionary dynamics of information dissemination, where SEG is to effectively model the phenomenon that the more users receive the information, the more profit the producer of the information will get. Through model simulation, we analyze the variation of the cooperation rate in the process of competition and find that the individual reputation plays an important role in the dissemination. The actual data of "Let Red Packet Fly" activity on Sina Weibo strongly verify the feasibility of the model.

2 Related Work

The model of information dissemination based on network structure mainly includes independent cascade model [3], linear threshold model [4] and extended model. The model based on population state includes Susceptible-Infective(SI), Susceptible-Infective-Susceptible(SIS), Susceptible-Infective-Removed(SIR) [5]. The model of multi source information transmission [6] and competition SI1-SI2(1,2 represent two infective individuals) [7] are aimed at information characteristics.

Also many scholars have carried on the research to the influence factor of the information dissemination. Kate et al. found that strong connectivity increases reputation between individuals which plays an important role in reaching a consensus [8]. Scott et al. also believe that the higher density and more connection can reduce the uncertainty and contribute to information dissemination [9].

In recent years, information dissemination on social networks has been deeply explored. Some mature communication models is presented focusing on the topics of the information cascades, viral marketing and product penetration, however, all of these works are limited in single virus models and independent dissemination. In addition, there are more works in limiting the spread of competition information. Pathak et al. looked at a two-virus SIS model on graphs [10]. Beutel et al. investigated the competition of mutual exclusion of infection and proposed that infecting one virus can reduced the chance of infecting other viruses, or even got immunity. But this model can't describe the dynamic characteristics well [11]. On the basis of Twitter, Wang et al. has studied the problem of information dissemination, but only from the point of view of data analysis, they didn't consider the possibility of coexistence of information cooperation [12]. There is still a lack of a complete model and method for the multi information dissemination. Usually users will choose the same behavior or technology from their neighbors to gain the profit. This kind of multi information diffusion competition attracts researchers from the perspective of game theory to study the manner of the competition information. Kempe et al. began to pay attention to the sudden and widespread adoption of various strategies in games and the effects of "word of mouth" in the promotion of new products [13]. But his model still didn't consider the influence of the individuals of the network on the spread. In work [14], motivated by applications to marketing, Domingos and Richardson et al. posed a fundamental algorithmic problem for such systems while the probability models failed to describe the character of the process of the information dissemination. In recent years, game-theoretic models, which are well studied in economic, sociology, ecology etc. are adopted by computer scientists for analyzing network behaviors. Kostka et al. [15] used the concept of game theory and location theory examined the dissemination of competing rumors in social network. Zinoviev D et al. [16, 17] adopted game theoretic models to understand human aspects of information dissemination taking into account personalities of individuals. The most interesting is in work [18], Yu et al. proposed a game-theoretic model for competitive information dissemination in social network and tried to understand human impacts on competitive information dissemination, however, the model and the method still needed to be improved.

3 Modeling and Mechanism Analysis

3.1 Game Model

A social evolutionary game is represented as a unweighted graph $G = (V, E; I, U, R)$ [19], where $V = \{v_1, v_2, \cdots, v_n\}$ is the set of individuals; $E = \{e_1, e_2, \cdots, e_n\}$ is the set of partnerships between individuals; I is the set of interactions performed by individuals based on their corresponding strategies; U is the set of utility functions of individuals, the cumulative payoffs obtained from his opponents in a single round indicates the short-term concern, which affects his choices when imitating others' strategies; R is the set of reputation functions indicating the opinion of his partners on him, which is the results of his partners' evaluation upon his behavior history and indicates the long-term concern, affects his choices when he seeks new partners. Participants have two alternative strategies: cooperation $C(s = [1, 0]^T)$ and defection $D(s = [0, 1]^T)$. In the directed network, an individual i have two types of neighbors IN-neighbors and OUT-neighbors. If there is an edge e_{ji} that directions from individual j to individual i, individual j will be i's IN-neighbor and i is j's OUT-neighbors. N_i^I indicates the set of i's IN-neighbors and N_i^O is the set of i's OUT-neighbors. The set of i's neighbors is indicated as N_i. The individual i's indegree and outdegree are d_i^I and d_i^O, respectively, d_i denotes the individual's degree which is the sum of his indegree and outdegree.

Considering the network effect of social networks, individual usually not only pursue the benefits of information influence, but also consider the impact on their reputation for the information dissemination. Coordination game theory can well describe the information competition under the influence of the multiple factors. In a coordination game, the more widely spread of the certain information over social networks, the higher users would accept it and also users are less likely to accept the alternative information if they have already accepted a similar one. Individuals benefit from coordinating their actions on a common goal, benefit will increase if the number of cooperators increases, however, the cost of defecting original strategy would be very high once participants stick to their strategies. Individuals strengthen mutual trust by choosing Pareto efficiency strategy in coordination game. Both persisting in the same strategy or defecting to adopt a new strategy are both pure strategy Nash equilibrium: the payoff-dominant equilibrium and the risk-dominant equilibrium [20], which depended on the utility and participants' behavior expectation and perception of risk and benefit, which make information dissemination uncertain. We assume that the game is defined by a 2×2 matrix as Fig. 1.

	A	B
A	a	c
B	d	b

Fig. 1. Payoff matrix of the coordination game.

More precisely, we have $a > d$ and $b > c$. Individuals can choose strategy A or strategy B. In this matrix, (A,A) and (B,B) are both pure strategy Nash equilibrium: (A, A) is the payoff-dominant equilibrium and (B,B) is the risk-dominant equilibrium. Utility is one of the factors that affect the coordination game equilibrium and the equilibrium is also dependent on the participants' behavior expectation and the perception of risk and benefit, so that make the information transmission uncertain. (A,A) and (B,B) two kinds of pure Nash equilibrium points are pursued by individuals in competitive information dissemination. Therefore, the coordination game can describe the characteristics of competitive dissemination in social network.

Each participant coordinates with all his direct partners whose utilities are cumulative payoffs obtained from their opponents:

$$u_i = \sum_{j \in N_i} s_i^T M s_j \tag{1}$$

Where N_i is the set of participant i's neighbors choosing cooperation or defection. In a pure coordination games with a rescaled 2×2 payoff matrix $M = \begin{pmatrix} 1 & 0 \\ 0 & b \end{pmatrix}$. The Nash equilibrium in pure strategies corresponds to the diagonal elements in matrix M when the two players adopt the same strategies(cooperation or defection) in the game. When players adopt different strategies, both of their benefits are 0 and the non diagonal elements in M are set to 0, because of this is not the Nash equilibrium state. Participants are rewarded by 1 when coordinating on the same strategy and tempted by $0 \leq b \leq 2$ to defect. When $0 \leq b < 1$, participants obtain the higher payoff from adopting the unanimous strategies, named "payoff-dominant equilibrium". In "risk-dominant equilibrium" when $1 < b \leq 2$, the temptation of defection is even higher.

In an asynchronous update mechanism [6], a randomly selected participant updates his strategy with probability $\frac{1}{1+W}$ at each time step, otherwise adjusting his partnership. $W = \frac{\tau_e}{\tau_p}$ is the ratio between time scale of strategy updating τ_e and the partnership adjustment τ_p [21, 22]. The frequency of partnership adjusting increases with W. At each evolutionary time step, each participant's reputation $R_i(t)$ is renewed as

$$R_i(t) = \sigma R_i(t-1) + \Delta R_i(t) \tag{2}$$

Where $\Delta R_i(t) = \sin\left(\frac{\pi}{2}\frac{n_i^c}{d_i}\right)$ and n_i^C is the number of C-strategy neighbors of participant i. Choosing more admitted information is benefit to participants' reputation.

Strategy Updating: The focal participant i chooses $j = \arg\max_{l \in N_i^o}\{u_l > u_i\}$ from his partners and then imitates j's strategy. The imitation process can be expressed by Fermi update rule [21, 23]:

$$W(s_i \leftarrow s_j) = \frac{1}{1 + \exp[\beta(u_j - u_i)]} \tag{3}$$

where s_i, s_j and u_i, u_j are the adapted strategies and utilities of participants i and j. β represents the extent of imitation noise considered as the willingness to imitate others' strategies.

Partnership Adjusting: Participant i searches j, satisfying:

$$j = \text{agr } \min_{l \in N_i^o}\{R_l(t) > R_i(t)\} \tag{4}$$

with the lowest reputation in his direct partners and sever their partnership with probability p_s. And also participant i searches k, satisfying:

$$k = \text{agr } \max_{N_i^l \cup_{l \in N_i^o} N_i^o(\{i\} \cup N_i^o)}\{R_l(t) > R_i(t)\} \tag{5}$$

with probability p_r as his new partner with the maximum reputation from his direct partners and partners' partners, otherwise randomly select k' with probability $1 - p_r$, p_r denotes the degree of participants' attention on their reputation.

3.2   Simulation Results

In our simulation, we built a random regular network with each participant's degree $k = 4$ and the total number of participants was $N = 10^3$, all participants initially choose to cooperate or defect randomly with equivalent probability. A set of Monte Carlo simulations were conducted with different p_r and W for analyzing the presented mechanism. p_r is the value degree of reputation and W indicates the frequency of partnership strategy adjusting. The sampling interval of pr and W is 0.1 and the range is from 0 to 1. The fraction of cooperators f_c describes the evolution of cooperation, the percentage of original information in networks. The small fluctuation of f_c means the coevolution is evolving into a relative stable state, or it runs a relatively long time, 2×10^5 Monte Carlo(MC) time steps. f_c is averaged over the following 10^3 MC time steps. We set $\beta = 0.01$, $p_s = 0.01$, the memory decaying rate of reputation is 1. The income of original information is 1. b is denoted as the income of the information added into competition later which range is from 0 to 2 and the interval is 0.04. Two competitive information begin to spread in the random network, the higher f_c means information A is dominant in the dissemination, the lower f_c means information B is more popular.

In Fig. 2(a), $W = 0.8$ means the participants in the network communicate with each other relatively frequently. When reputation-based partners adjust p_r at a low level and $b < 1.2$, the cooperation emerge to a high level. While $b > 1.2$ means the cooperators

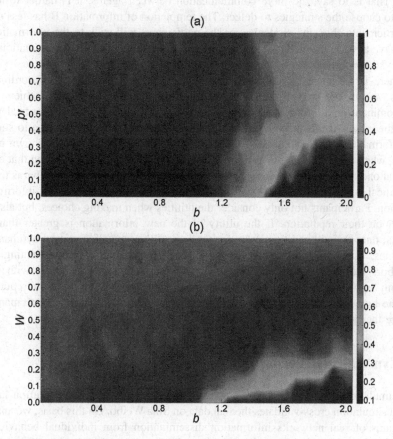

Fig. 2. (a) Relative stationary states of fc with variation of b and pr when $W = 0.8$. (b) Relative stationary states of fc with variation of b and W when $pr = 0.2$. In temperature chart, the changing of temperature color indicates the changing of fc. fc describes the evolution of cooperation, the percentage of original information in networks. The higher temperature(red) means more cooperation behaviors between users, the lower temperature(blue) means more defection behaviors. (Color figure online)

are hard to survive. The cooperation level decreases with b increasing, and increases with p_r increasing. The more temptation for new information, the easier it is to attract participants to adopt it. There exists a relative high area of cooperation when p_r at a high level, the fluctuation of f_c is not obvious because of the reputation makes individual's strategies choice more cautious, the effect of temptation of new information is not significant.

When p_r remain constant and the value of b and W change at the same time, the fraction of cooperators is shown in Fig. 2(b). We set $p_r = 0.2$, which shows that the participants in the networks consider less reputation when making a strategic decisions. When W is at a low level, the change of b value has a great effect on f_c and the f_c deceases with the increase of b. Communications become less among participants, cooperation or defection will rest with temptation. When W is relatively high, the temptation of defection is hard to effect the f_c, the cooperation is maintained at a high level. That is to say, the more communication between agents, it is harder to tempt them to choose the strategies to defect. The temptation of information B has less effect on participants choosing strategies, so the participants will stick to the information A they have already taken. The more communication, the greater role of reputation will be played by participants in the change of information.

There is an interesting phenomenon in the above charts. Because coordination games reach the payoff-dominant equilibrium when $b < 1$ and achieve the risk-dominant equilibrium when $b > 1$. So $b = 1$ can be regarded as a critical point from theoretic. At this point, each participant has 50 % probability choose to stick to the information they have already taken or switch to a new one. But as shown in the charts, when the benefit of new information is 1.2 times more attractive than that of the original one, participants begin to switch to choose the new information. That is to say, the critical point has a certain extension. Reputation makes an impact on information selection. Participants not only consider the utilities when making choices, but also the impact on their reputation. If the utility of the new information is greater than the original one, but still can't make up for the loss of the reputation caused by changing strategies, users will still adhere to the original information. Communicating with neighbors means they take their reputation seriously and make choices with more caution. So we get an important conclusion here: The more attention to the reputation and the higher frequency of communication causes much trouble for participants to change their choices.

4   Experimental Verification

The simulation give a further guide to analyze the actual data. The evolution law of model simulation cross-validates the real data on Sina Weibo. On this basis, we analyze the status of real networks information dissemination from individual behavior. In order for the accuracy and validity of the model for our analysis and verification, we selected the most influential online social application Sina Weibo as an experiment platform for this test. All the data we collected is filtered, to ensure the fairness of the data. We considered the focal effect among users: the crowd's attention tends to focus on the information resources which they have interest in common and users prefer to recommend the related information of the same movie. This feature is well consistent with coordination game strategy and describes this phenomenon of social networks.

During 2015 Spring Festival, two representative electrical companies *Micoe* and *Midea* participated into the "Let Red Packet Fly" activity on Sina Weibo. Discussions of them can be regarded as a pair of competitive information diffusing on social networks at the same time. The number of red packet distributed by two companies as well as each search volumes and forwarding amount of two companies' message can be the utility b and f_c of the evolution. We collected the relevant data about the amount of forwarding of *Micoe* and *Midea* during from Feb. 02 to Feb. 28 in the year of 2015. The data set included more than 700,000 users' statuses in February 2015 from Sina Weibo, an undirected network with two-way interactions and low reputation between users.

In Fig. 3, the forwarding amount reflects how popular *Micoe* and *Midea* are among users. We can find that the forwarding amount of them both present a certain fluctuation trend in Fig. 3. It can be seen clearly in the figure that the forwarding amount of the two companies alternately holds the upper hand and get to a peak in a specific time period. That is to say the two key words of two companies sharp compete on the microblog platform and individuals are strong concerned by the network user groups in a certain period of time. Both companies have taken effective marketing tools and distributed lots of red packets in "Let Red Packet Fly" activity to increase users' interest.

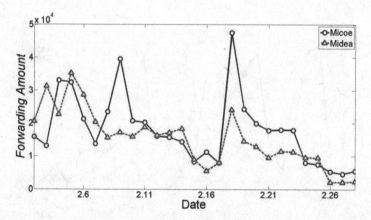

Fig. 3. Fluctuation of forwarding amount. Red line and blue line separately represent related terms of *Midea* and *Micoe* spreading degree. The higher value means the greater amount of forwarding of the company's message. (Color figure online)

In this competition, f_c reflects the percentage of the relevant information about *Micoe* spreaded on microblog in the competition with the topic of *Midea*. The fluctuation of f_c shows the evolution process of the two competitive information transmission. When $f_c > 0.5$, users are biased in favor of the competitive information A and more inclined to stick with the strategy they have already taken, which means the *Midea* is more popular in this information competition. On the contrary, $f_c < 0.5$, users tend to abandon the information A and turn to accept the new information B, which means the *Micoe* make more influence on users. It can be regarded as the two kinds of information that participate in competition at the same time, the variation of their own utilities caused the change of the users' attention. In the process of information accepted, users will also have a preference, which leads to the differences of the spreading on the microblogs.

Attraction of two companies on users occupied the advantage alternatively in Fig. 4, where the utilities of *Midea* and *Micoe* reflect the attractiveness of users. The more search volume and distributed red packets, the greater appeal of the company to users and the high utility of the information in the game evolution will be. In the early of February, the red line is basically above the blue line, which means the utility of *Midea* is more dominant and users are likely to search its related information. Subsequently, the red line swung below the blue line, the utility of *Micoe* has exceeded the utility of *Midea*, users on networks began to extensive search the information of *Micoe* and the effect of distributing red packets of Micoe on users began to appear. During the Spring Festival, the microblog search volume of *Micoe* got certain advantages, but *Midea* successfully increased users' interest and the search volume. While in the end of Februrary, *Midea* carried out a series of activities and successfully increased users' interest and search volume in this period of time. The number of red packets distributed by two companies affected the users' initiative of forwarding the related topics to a large extent.

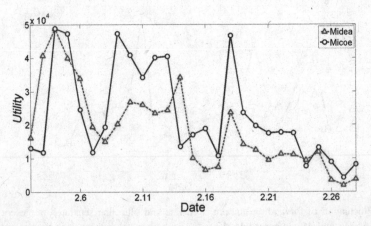

Fig. 4. Fluctuation of utility. Red line and blue line represent propaganda power of *Midea* and *Micoe*. The higher value means the more red packets distributed in the competition of the company and getting the greater attractiveness of the users. (Color figure online)

In Fig. 5, the simulation curve with high level of p_r fits with actual evolutionary curve to a certain extent. We set the utility of the *Midea* information as standard 1 and the utility of the *Micoe* information is set as a variable b fluctuating according to standard 1. And f_c reflects the percentage of the relevant information about *Micoe* spreaded on microblog in the competition with the topic of *Midea*. There are some

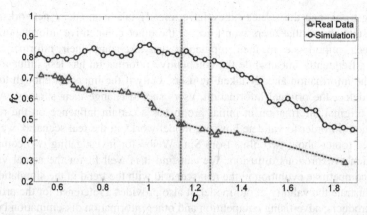

Fig. 5. Comparative analysis of evolution process. Red line is real data curve and blue line is simulation curve. The simulation curve selects the simulation data in Fig. 2(a) with $pr = 0.7$, $W = 0.2$. (Color figure online)

differences between two curves due to the fact that there are other companies distributed the red packets in the sales activity to attract the users and also some uncertainty factors in actual situation. Our work mainly describe the trend of competitive information dissemination by simulation and the similar fluctuation trends of two curves verify the feasibility of our model. The more accurate model validation can be achieved through parameter setting.

When $b < 1$ and $f_c > 0.5$, it shows that the Weibo users are not willing to recommend the *Midea* on their own initiative if the red packets of the company is not enough to attract them. They will choose to insist on discussing the related topic of *Micoe* and share its red packets. f_c decreases with the increasing of b because of well marketing and large amount distributed red packets of *Midea* attracting public attention and users begin to actively search for its topics instead of *Micoe* they once interested in. The real data curve falls down to 0.5 approximately at $b = 1.2$. This phenomenon well verifies the impact of reputation factor on the diffusion of competitive information. In a word, the attraction of information must be high enough to draw the public attention, which can make up for the loss of reputation caused by changing information or strategy. Only in this way can information takes advantage in information competition and attracts more users.

5   Conclusion

In this paper, we introduce a framework of social evolutionary game (SEG) and adopt the coordination game strategy to model and simulate the synchronous competitive information dissemination on social networks. Participants in a SEG care about the short-term utility and the long-term reputation, which are not only the basis of strategy updating and partnership adjusting, but also an important factor to influence the diffusion of competitive information. Coordination game we used in this paper can describe the communication and competition well, and imitate users' behavior that unwilling to change the strategy they have already taken on social networks. Simulation results shows that users won't accept the other competitive information, when they attach importance to their reputation and exchange their information with neighbors frequently, meanwhile the competitive information has less influence compare to the information already taken by users. Only if the impact is enough to attract them to defect the original information, users should change their strategy. Also the scale of original information in initial group has a certain influence on the result of competition. In order to validate how our model works in the real scenario, we collect the related topics about two films from Sina Weibo for investigating two competitive information synchronous diffusion. We can find it is well fit for the actual variation trend of competitive evolution in the real scenario with the trend in the simulation case, which illustrate the validity of our model. It also provides a reference for the promotion of new products, advertising competition and other information dissemination behavior on social networks. In the further research work, we will continue to improve the model in multi-participant game and analyze the factor more accurate.

Acknowledgments. This work is supported by National Grand Fundamental Research 973 Program of China (No. 2014CB340401), National Natural Science Foundation of China (No. 61572469, 61173008, 61303244, 61402442, 61402022, 61370132, 61303049).

References

1. Wang, Y., Yu, J., Qu, W., Shen, H., Cheng, X., Lin, C.: Evolutionary game model and analysis methods for network group behavior. Chin. J. Comput. **38**(2), 282–300 (2015)
2. Yu, J., Wang, Y., Jin, X., et al.: Evolutionary analysis on online social networks using a social evolutionary game. In: Proceedings of the 23rd International Conference on World Wide Web Companion, pp. 415–416. ACM (2014)
3. Goldenberg, J., Libai, B., Muller, E.: Talk of the network: a complex systems look at the underlying process of word-of-mouth. Mark. Lett. **12**(3), 211–223 (2001)
4. Granovetter, M.: Threshold models of collective behavior. Am. J. Sociol. **83**, 1420–1443 (1978)
5. Kermack, W.O., McKendrick, A.G.: A contribution to the mathematical theory of epidemics. In: Proceedings of the Royal Society of London A: Mathematical, Physical and Engineering Sciences, Vol.115, No.772, pp. 700–721. The Royal Society (1927)
6. Myers, S.A., Zhu, C., Leskovec, J.: Information diffusion and external influence in networks. In: Proceedings of the 18th ACM SIGKDD International Conference on Knowledge Discovery and Data Mining, pp. 33–41. ACM (2012)
7. Beutel, A., Prakash, B.A., Rosenfeld, R., et al.: Interacting viruses in networks: can both survive?. In: Proceedings of the 18th ACM SIGKDD International Conference on Knowledge Discovery and Data Mining, pp. 426–434. ACM (2012)
8. Ten Kate, S., Haverkamp, S., Mahmood, F., et al.: Social network influences on technology acceptance: a matter of tie strength, centrality and density. In: BLED 2010 Proceedings Paper, p. 40 (2010)
9. Scott, J.: Social network analysis: developments, advances, and prospects. Soc. Netw. Anal. Min. **1**(1), 21–26 (2011)

10. Pathak, N., Banerjee, A., Srivastava, J.: A generalized linear threshold model for multiple cascades. In: 2010 IEEE 10th International Conference on Data Mining (ICDM), pp. 965–970. IEEE (2010)

11. Prakash, B.A., Beutel, A., Rosenfeld, R., et al.: Winner takes all: competing viruses or ideas on fair-play networks. In: Proceedings of the 21st International Conference on World Wide Web, pp. 1037–1046. ACM (2012)

12. Weng, L., Flammini, A., Vespignani, A., et al.: Competition among memes in a world with limited attention. Nat. Sci. Rep. 2(7391), 1–8 (2012)

13. Kempe, D., Kleinberg, J., Tardos, É.: Maximizing the spread of influence through a social network. In: Proceedings of the Ninth ACM SIGKDD International Conference on Knowledge Discovery and Data Mining, pp. 137–146. ACM (2003)

14. Richardson, M., Domingos, P.: Mining knowledge-sharing sites for viral marketing. In: Proceedings of the Eighth ACM SIGKDD International Conference on Knowledge Discovery and Data Mining, pp. 61–70. ACM (2002)

15. Kostka, J., Oswald, Y.A., Wattenhofer, R.: Word of mouth: rumor dissemination in social networks. In: Shvartsman, A.A., Felber, P. (eds.) SIROCCO 2008. LNCS, vol. 5058, pp. 185–196. Springer, Heidelberg (2008)

16. Zinoviev, D., Duong, V.: A game theoretical approach to broadcast information diffusion in social networks. In: Proceedings of the 44th Annual Simulation Symposium, pp. 47–52. Society for Computer Simulation International (2011)

17. Zinoviev, D., Duong, V., Zhang, H.: A game theoretical approach to modeling information dissemination in social networks. Comput. Sci. Game Theor. I, 407–412 (2010). arXiv preprint

18. Yu, J., Wang, Y., Li, J., et al.: Analysis of competitive information dissemination in social network based on evolutionary game model. In: 2012 Second International Conference on Cloud and Green Computing (CGC), pp. 748–753. IEEE (2012)

19. Yu, J., Wang, Y., Jin, X., Cheng, X.: Social evolutionary games. In: 2014 5th International Conference on GameNets, pp. 1–5. IEEE (2014)

20. Andrea, M., Amin, S.: The spread of innovations in social networks. J. Proc. Nat. Acad. Sci. 107(47), 20196–20201 (2010)

21. Fu, F., Hauert, C., Nowak, M.A., et al.: Reputation-based partner choice promotes cooperation in social networks. Phys. Rev. E 78(2), 026117 (2008)

22. Santos, F.C., Pacheco, J.M., Lenaerts, T.: Cooperation prevails when individuals adjust their social ties. PLoS Comput. Biol. 2(10), 1284–1291 (2006)

23. Szabo, G., Toke, C.: Evolutionary prisoner's dilemma game on a square lattice. Phys. Rev. E 58(1), 69 (1998)

A Topic-Specific Contextual Expert Finding Method in Social Network

Xiaoqin Xie[✉], Yijia Li, Zhiqiang Zhang, Haiwei Pan, and Shuai Han

College of Computer Science and Technology, Harbin Engineering University,
Harbin 150001, China
xiexiaoqin@hrbeu.edu.cn

Abstract. Expert retrieval is a widely studied problem. However, most existing expert finding methods focus on social network which contains topic-irrelevant users and interactions. This results in that the expert results are not topic-specific and practical because many users need to find experts for certain topic. Furthermore, contextual factors of social network also affect the accuracy of expert finding and are seldom concerned comprehensively in existing approaches. To solve above problems, in this paper, we propose a topic-specific contextual expert finding method. At first, we define a topic-specific contextual feature model (TSCFM) which consists of a topic-aware model (TAM) for topical feature and a context-aware model (CAM) for contextual feature. TAM uses LDA and HITS to extract topical feature, and CAM evaluates social relation, time and location factors to extract contextual features. Then based on TSCFM, we learn an expert scoring function which synthetically concerns topical and contextual features using SVM algorithm and rank the experts. The experiments on two datasets demonstrate that our proposed expert finding method is feasible and can improve the accuracy.

Keywords: Social network · Expert finding · Topic-aware · Context-aware

1 Introduction

Users in social network service, such as Twitter, can create content on many topics like shopping, music, etc., and the content can influence other users. With the popularity of social networks, more and more people tend to ask authority friends in social network for advice. And it is common that users want to find experts for a specific topic. Social relationship among users in social network are always related to various topics and mixed together, and these facts issue many challenges to expert finding problem. Furthermore, in addition to topic, contextual factors of social network like social relation, location and so on also affect user's choices for expert users.

Authority-based method is popular approach for expert finding. Some other expert finding researches are based on analysis for contextual factors in social network. However, although authority-based methods like PageRank and HITS [1] are effective for expert finding, these methods focus on topic-independent or topic-dependent social network which includes many irrelevant users and relationships, and there are not

© Springer International Publishing Switzerland 2016
F. Li et al. (Eds.): APWeb 2016, Part I, LNCS 9931, pp. 292–303, 2016.
DOI: 10.1007/978-3-319-45814-4_24

enough topic-specific information for expert finding. For instances, HITS is used for expert finding [5, 15], but the network like Java Forum and large online help-seeking community they analyzed is not topic-specific, which includes topic-irrelevant users.

Besides, topic model is also feasible for expert finding. A topic modeling approach for expert finding is proposed in [10], and this approach works well for expert results by using LDA. While, data sparsity influences the accuracy of expert finding method that only concern topic aspects. Some other expert finding researches concern topic aspects and analyze contextual factors of social network to reduce the influence of the data sparsity. But they only concern one of the contextual factors in social network. For instances, the proposal methods for expert finding in [6, 11] just analyze location factor. Less work for expert finding based on topic analysis comprehensively concern social relation, temporal and location factors of social network.

In order to solve aforementioned problems, this paper proposes a topic-specific contextual expert finding method TSCEFM, which concerns both topical and contextual factors. Firstly, we propose a topic-specific contextual feature model TSCFM. The TSCFM uses a topic-aware model TAM to extract topical feature and a context-aware model CAM to extract contextual feature. Secondly, we design and realize the algorithms for our expert finding method. LDA and HITS [17] is applied to TAM to get topical feature of users. Evaluating functions [16] for social relation, time and location factors are used in CAM to get contextual features of users. We use SVM algorithm to learn an expert scoring function which synthetically concerns topical and contextual features and rank experts by their authority value calculated by the function. Finally, we conduct experiments on two datasets to evaluate our method. The results show that our method is feasible and achieves better accuracy than existing methods.

In the remainder of this paper, Sect. 2 is about related works. Section 3 presents problem definition. Section 4 describes the topic-specific contextual expert finding method. Section 5 discusses the experimental design and results. Section 6 concludes the paper.

2   Related Works

Expert finding problem attracts many researchers' attention. Popular expert finding methods include authority-based methods like HITS, topic modeling methods like LDA and contextual analysis methods and so on.

Authority-based methods like HITS are commonly used for expert finding. HITS were used for e-mail expert ranking analysis in [7]. Yeniterzi, R. et al. [17] proposed a variation of HITS to provide a more effective topic-specific method for modeling authority network of expert users. Yang, R.R. et al. [12] proposed a weighted HITS algorithm to find experts in online knowledge community.

Some other approaches used topic model like LDA for expert finding. Liu, J. et al. [4] used a topic-centric candidate priors (TCCP) model to improve the expert finding system and LDA is used to extract topics and Gibbs sampling is used to estimate parameters. Naumann, F. et al. [2] extracted probability of topic by LDA to represent authority of expert candidates. Yang, L. et al. [3] also used LDA for expert finding.

These above expert finding methods only concerned topic aspects. Besides, some expert finding methods based on topic model also concerned some contextual factors. Macedo, A.Q. et al. [16] considered several contextual signals for event recommendation to choose the events that best fit users' interests. Adomavicius, G. et al. [8] used contextual information to generate relevant recommendations. Daud, A. et al. [9] identified a user as expert for different time periods by a novel Temporal-Expert-Topic (TET) approach based on Semantics and Temporal Information. Lin, L. et al. [14] proposed a topical and weighted factor graph model to find experts.

To conclude, many existing researches have made efforts on proposing expert finding methods. However, most existing methods concern social network which includes topic-irrelevant users and interactions. Furthermore, less work concern the social relation, time and location contextual factors and topical factor in a unified way. In this paper, we propose a topic-specific contextual expert finding method to concern topical factor and contextual factor synthetically.

3   Problem Definition

In this section, we present related terms definitions and problem formation.

Definition 3.1.   Social network is denoted as $G = (V, E)$, where $V = \{v_1, v_2,..., v_n\}$ means user nodes, $E = \{e_1, e_2,..., e_n\}$, e_j represents the friendship links between v_i.

Definition 3.2.   Expert Nodes. Each user node can be given an authority value calculated by function $f(x)$ that can be referred to Sect. 4. Those nodes whose authority value is greater than a boundary value are defined as expert nodes.

Definition 3.3.   User Topic Feature Matrix (UTFM). $UTFM = (b_{ik})_{n \times d}$, where b_{ik} means the probability of user v_i belonging to the kth topic and d is the number of topics.

Definition 3.4.   Topic-Relevant Nodes (TRN). Those nodes whose probability on the given topic is greater than a boundary value, are defined as topic-relevant nodes.

Definition 3.5.   Topic-Relevant Root Set (TRRS). $TRRS = G_r(V_r, E_r)$, where $V_r = \{v_1, v_2,..., v_n\}$ means the set of all topic-relevant nodes, $E_r = \{e_1, e_2,..., e_n\}$ represents the set of all friendship links between v_i.

PROBLEM 1.   Topic-Relevant Root Set (TRRS) Construction Problem. Given a social network $G = (V, E)$, the task is to find the topic-relevant nodes, then add these nodes and the links between them into the topic-relevant root set.

Definition 3.6.   Topic-Relevant Base Set (TRBS). $BS = G_b(V_b, E_b)$, where $V_b = \{v_1, v_2, ..., v_n\}$ means the set of topic-relevant users and those users that have direct links to topic-relevant users, E_b represents friendship links between v_i, $E_b = \{e_1, e_2,..., e_n\}$.

PROBLEM 2. Expert Finding Problem. Given a social network $G = (V, E)$, assume that there is a user u, who want to find some experts in a given topic t, our task is to find the expert nodes.

4   Topic-Specific Contextual Expert Finding Method

In this section, we propose a topic-specific contextual expert finding method (TSCEFM) which considers both topical feature and contextual feature of social network.

4.1   Method Framework

The architecture of TSCEFM method is shown in Fig. 1. The left part is the input, which includes a social network and a topic t required by user. The output shown in right part is expert users. The middle two parts are critical components of our method: feature extraction and expert finding. At first we extract both topical and contextual feature of social network and construct a topic-specific contextual feature model (TSCFM). Then expert finding problem is transformed into an expert scoring function learning problem, and both types of features are used as input of the learning algorithm.

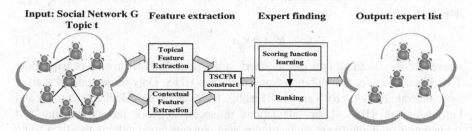

Fig. 1. Topic-specific contextual expert finding method architecture

In the next sections, we describe detailed model and algorithms in TSCEFM.

4.2   Topic-Specific Contextual Feature Model

The topic-specific contextual feature model (TSCFM) consists of topic-aware model (TAM) and context-aware model (CAM) parts, and can be presented as: TSCFM = <TAM, CAM >.

Yeniterzi, R. et al. [17] prposed an HITS model to construct an authority network. Based on their idea, we propose TAM that using LDA and the HITS to extract topical feature. Macedo, A.Q., et al. [16] proposed evaluating functions for several contextual factors. Based on their idea, we design a CAM for contextual features using evaluating functions. For a given user u, TAM will set a topical feature and CAM will set three features on three types contextual factors involving social relation, temporal interval and location.

Assume that we use U to denote the set of users, and use L, T and D to respectively denote the set of features on social relation, temporal interval and location three types of contextual factors. And use P to denote the set of topical feature of users, the critical step in expert finding task is to find a scoring function which can consider both contextual factors and topical factors, and it can be stated as the following form:

$$\hat{f}: U \times L \times T \times D \times P \to E \tag{1}$$

where E means the set of scores for different experts. Thus, the top-k expert finding problem can be transferred to optimized problem, which can be formalized as follows:

$$top - k(u, L_u, T_u, D_u, P_u) := \underset{u \in U}{\arg \max}^k \hat{f}(u, L_u, T_u, D_u, P_u) \tag{2}$$

where $L_u \in L$, $T_u \in T$, $D_u \in D$ and $P_u \in P$.

TAM can be represented as TAM = \<value\> and we calculate authority values of users based on HITS to get the topical feature. Unlike traditional HITS, we expand TRRS into TRBS instead of expanding RS into BS and extract topic-relevant nodes TRN for a certain topic according to the probability in a user topic feature matrix UTFM. We get the probabilities of user on topics by a LDA model which based on Gibbs sampling method. These probability values can form UTFM. The top N nodes ranked by probabilities can be extracted as TRN and added to TRRS. The number of TRN can be calculated as follows:

$$N = \partial \times n \tag{3}$$

where $0 < \partial < 1$, and n is nodes numbers in the network. The process of expanding TRRS into TRBS is also different from traditional HITS. Only those nodes that have topic-relevant links to TRRS are used. Finally, TAM iteratively calculates authority and hub values of HITS. User interactions in social network include reading and commenting, and intuitively, being commented presents more authoritative than being read. The final topical feature score can be calculated as follows:

$$behaWeight = readNum + 2 \times commentNum \tag{4}$$

$$P_u = finalAuthority^\lambda \times behaWeight^\beta + b \tag{5}$$

where *readNum* and *commentNum* represent the reading number of times and commenting number of times separately, b is the probability value of user in UTFM, $\lambda + \beta = 1$ and $\lambda < \beta$. The Score value is the topical feature of user and can evaluate the authority of a user.

CAM includes social relation, location and temporal factors and can be represented as: CAM = $\langle L_u, D_u, T_u \rangle$. L_u, D_u and T_u denote the scores for three factors respectively. Besides, CAM can generate contextual feature scores (L_u, D_u, T_u) of a user:

Social Relation. L_u means the social relation feature, which is calculated by a group frequency model. Users who interact with others more frequently are more likely to be expert. The L_u can be represented as a value and can be calculated as follows:

$$L_u = \frac{|E_{u,g}|}{|E_U|} \tag{6}$$

where $E_{u,g}$ denotes the set of microblogs that user u interact with other users, E_u denotes the set of microblogs of all users.

Location Factor. D_u means the location feature for a given user. A kernel-based density estimation approach is applied to get D_u:

$$d(l) = \frac{1}{|L_u|} \sum_{l' \in L_u} K_H(l - l') \tag{7}$$

$$K_H(x) = \frac{1}{\sqrt{2\pi|H|}} e^{-\frac{xx^T}{2\sqrt{H}}} \tag{8}$$

$$D_u = f(l) \tag{9}$$

where L_u is the set that contained location coordinates of all users, l is the location coordinate of the target calculating user.

Temporal Factor. Intuitively, user is more likely to choose expert who surf net at the same time. Each microblogs of user u is denoted as a time vector e. User u can be represented as follows:

$$\vec{u} = \frac{1}{|E_u|} \sum_{e \in E_u} \vec{e} \tag{10}$$

T_u is calculated as follows:

$$T_u = \sum_{\vec{e'} \in e_u} \cos(\vec{u}, \vec{e'}) \tag{11}$$

where e_u is the set of all microblogs denoted as time vector of users.

4.3 Topic-Specific Contextual Expert Finding Algorithm

Based on TSCFM, the topic-specific contextual expert finding algorithm is described as follows:

Algorithm 4.1. topic-specific contextual expert finding
Input: Social network $G = (V, E)$, a certain topic t;
Output: Expert ranking list l_f;
1: $A \leftarrow constructFM(G, t)$; //construct TSCFM to extract feature vector A including topical and contextual features;
2: $l_e \leftarrow findExpert(A, G)$; //use A as input and find experts in social network;
3: $l_f \leftarrow Rank(l_e)$;//rank experts in l_e to generate expert ranking list;
4: save and return the expert ranking list l_f;

As depicted in Algorithm 4.1, we construct TSCFM to extract topical and contextual features and get a feature vector A in step 1. Then, in step 2 we use *findExpert()* algorithm to find experts in social network. The *Rank()* algorithm ranks these experts to generate final ranking list.

The *constructFM()* algorithm we call in Algorithm 4.1 is described as follows:

Algorithm 4.2. *constructFM()* algorithm

Input: Social network $G= (V, E)$, a certain topic t;
Output: a feature vector $A=(l_o, l_c)$, where l_o is topical feature and l_c is contextual feature;
1: $l_o \leftarrow TAM(G, t)$; // extract users' topical feature of social network G;
2: $l_c \leftarrow CAM(G)$; // extract users' contextual feature of social network G;
3: $A \leftarrow (l_o, l_c)$; //expanding TRRS G_r into TRBS G_b;
4: return A;

In Algorithm 4.2, we extract users' topical feature list in step 1 and extract contextual feature list in step 2. In step 3, topical and contextual features are combined into a unified vector A.

The task of *findExpert()* algorithm in step 2 of Algorithm 4.1 is to learn a scoring function $f(x)$ and predict whether a user is expert or not to generate expert list. We represent each user x as a feature vector, including topical feature and contextual feature. The full list of features can be found in Table 1. We train the dataset using support vector machine (SVM) [13] to learn decision function, and we also treat the decision function as scoring function $f(x)$ to evaluate users' authority.

Table 1. List of features used in *findExpert()* algorithm.

Topical feature	The topical score P_u of user x
Contextual feature	The social relation score L_u of user x
	The location score D_u of user x
	The temporal score T_u of user x

The *Rank()* algorithm in Algorithm 4.1 ranks users by the authority scores calculated by the scoring function learned in *findExpert()* algorithm.

5   Experiments

5.1   Dataset

There are two datasets tested in our experiments, Citeseer site and Twitter. We crawl about 1000 papers from Citeseer, and we extract the title, abstract, citation relationship to form the Citeseer dataset. Twitter dataset is the subset of 2009 Twtitter user dataset, including user ID, posts, reading and commenting information. We consider the authors or users as nodes, and consider their citation relationships or user reading and commenting relationships as edges.

5.2 Experimental Design

We conduct three experiments to demonstrate the accuracy and feasibility of our method. MAP (mean average precision) and NDCG (Normalized Discounted Cumulative Gain) are used as the evaluation metrics.

We label the users whose number of followers ranks top k as real experts.

Experiment 1: Comparison of Influence of Parameter α to Results. The parameter α of formula (3) decides the number of TRN we select. We conduct this experiment to find the best α value for the accuracy of our expert finding method.

Experiment 2: Comparison of Expert Finding Quality of Different Methods. To demonstrate that the expert finding results of our method conform to the topic that the end users need and the contextual situation including social relation, time and location three aspects that users need, we compare the accuracy of our method with accuracy of existing expert finding methods.

We compare TSCEFM with CFLA [6], SAR [6] and TWG [4] which concerned both topical and contextual factors, and HITS [15], TSM [17] and ERPR [12] which just concerned topical factor, and E-CA [16] which just concerned contextual factors.

Experiment 3: Comparison of Different Influential Factors. Our proposed TSCFM concern both topical and contextual influential factors for expert finding. To demonstrate that both topical and contextual influences are necessary in our method, we compare the accuracy of TSCFM with accuracy of TAM of TSCFM and CAM of TSCFM.

Besides, the TAM of TSCFM expands TRRS into TRBS. To demonstrate this expanding pattern is necessary for expert finding, we compare the accuracy of TSCEFM based on TAM with accuracy of TSCEFM based on HITS which expands RS into BS.

5.3 Results and Discussion

Analysis for Experiment 1. Figure 2 shows the accuracy of TSCEFM as α varies. As depicted in Fig. 2, we find that TSCFM performs more accurately when α in formula (3) is between 0.3 and 0.4.

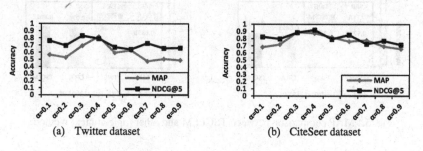

(a) Twitter dataset   (b) CiteSeer dataset

Fig. 2. MAP and NDCG of TSCEFM for different α

Analysis for Experiment 2. The accuracy of TSCEFM is shown in Figs. 3 and 4.

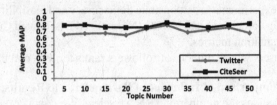

Fig. 3. MAP of TSCEFM in Twitter and CiteSeer dataset

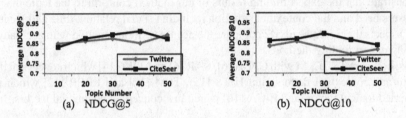

Fig. 4. NDCG of TSCEFM in Twitter and CiteSeer dataset

As shown in Figs. 3 and 4, the average MAP values of TSCEFM are between 0.643 and 0.829, and the average NDCG values are between 0.768 and 0.912. Figures 5 and 6 indicate that our TSCEFM achieves better accuracy than other state-of-the-art expert finding methods concern both topic and context. The average MAP values of SAR and TWG are 13.31 % lower than the values of TSCEFM. The NDCG values of CFLA are 12.37 % less than the values of TSCEFM. Figures 7 and 8 demonstrate that our method has clear advantage in accuracy over other methods concerning single aspect. The MAP values of TSCEFM are 15.32 % higher than the MAP values of these methods.

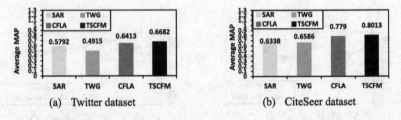

Fig. 5. MAP comparison between TSCEFM and other topic-context method

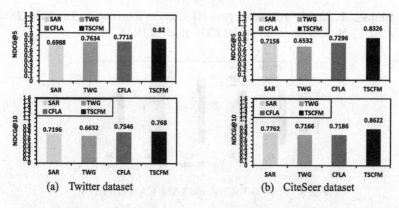

Fig. 6. NDCG comparison between TSCEFM and other topic-context method

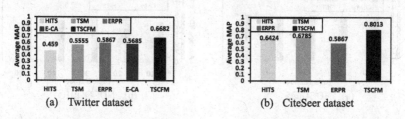

Fig. 7. Average MAP comparison between TSCEFM and single aspect method

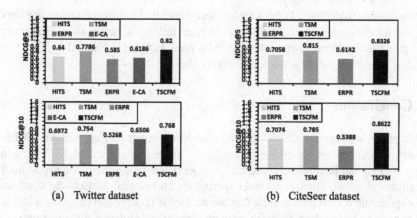

Fig. 8. NDCG comparison between TSCEFM and single aspect method

We can conclude that more accurate experts which conform to topic that users input and contextual situation that users need can be found by our proposed TSCEFM.

Analysis for Experiment 3. As shown in Figs. 9 and 10, TSCFM achieves higher MAP and NDCG value compared to TAM. And TSCFM is obviously more accurate than CAM. TSCFM achieves 16.82 % bigger MAP than TAM, and 29.28 % higher MAP than

CAM. Besides, TAM performs better than HITS under most topics. The NDCG values of TAM are 17.59 % higher than the NDCG values of HITS.

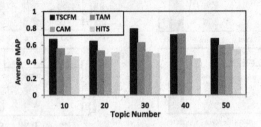

Fig. 9. MAP of TSCFM and TAM, CAM and HITS

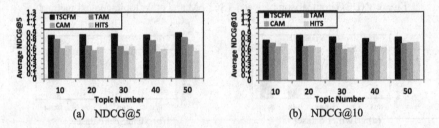

Fig. 10. NDCG of TSCFM and TAM, CAM and HITS

We can conclude that: (1) Both topical and contextual influential factors are important and necessary for expert finding. (2) Topical influential factor affect the accuracy of expert finding more than contextual influential factor. (3) Expanding TRRS into TRBS is necessary for the accuracy of expert finding.

6   Conclusions

In this paper, we propose a topic-specific contextual expert finding method TSCEFM. At first, we designed a topic-specific contextual feature model called TSCFM, which can extract topical feature and contextual feature. Then we described the expert finding algorithms in detail. Finally, several experiments on two real datasets are conducted. The experiment results demonstrate that our method is feasible and can more accurately find experts conforming to a certain topic and contextual situation that users need.

Our future work will consider the dynamic condition of social network.

Acknowledgements. This work is supported by the National Natural Science Foundation of China (No. 61202090, 61370084, 61272184), the Science and Technology Innovation Talents Special Fund of Harbin under grant (No. 2015RQQXJ067), the Fundamental Research Funds for the Central Universities under grant (No. HEUCF10060), the Nature Science Foundation of Heilongjiang Province under Grant (No. F2016005).

References

1. Kleinberg, J.: Authoritative sources in a hyperlinked environment. J. ACM **46**(5), 604–632 (1999)
2. Momtazi, S., Naumann, F.: Topic modeling for expert finding using latent Dirichlet allocation. J. Wiley Interdisc. Rev. Data Min. Knowl. Discov. **3**(5), 346–353 (2013)
3. Yang, L., Zhang, W.: Combining distance and sequential dependencies in expert finding. In: IEEE International Conference on Intelligent Computing and Intelligent Systems 2009, pp. 491–495. IEEE (2009)
4. Liu, J., Li, B., Liu, B., Li, Q.: Topic-centric candidate priors for expert finding models. In: Yang, Y., Ma, M., Liu, B. (eds.) ICICA 2013, Part I. CCIS, vol. 391, pp. 253–262. Springer, Heidelberg (2013)
5. Campbell, C.S., Maglio, P.P., Cozzi, A.: Expertise identification using Email communications. In: The Twelfth International Conference, pp. 528–531 (2003)
6. Neshati, M., Beigy, H., Hiemstra, D.: Expert group formation using facility location analysis. J. Inf. Process. Manag. **50**(2), 361–383 (2014)
7. Dom, B., Eiron, I., Cozzi, A.: Graph-based ranking algorithms for e-mail expertise analysis. In: Proceedings of the 8th ACM SIGMOD Workshop on Research Issues in Data Mining and Knowledge Discovery, pp. 42–48. ACM (2003)
8. Adomavicius, G., Tuzhilin, A.: Context-aware recommender systems. In: ACM Conference on Recommender Systems, pp. 2175–2178. ACM (2008)
9. Daud, A., Li, J., Zhou, L.: Temporal expert finding through generalized time topic modeling. J. Knowl.-Based Syst. **23**(6), 615–625 (2010)
10. Saeedeh, M., Felix, N.: Topic modeling for expert finding using latent Dirichlet allocation. J. Data Min. Knowl. Discov. **3**(5), 346–353 (2013)
11. Zhao, K., Cong, G., Yuan, Q.: SAR: a sentiment-aspect-region model for user preference analysis in geo-tagged reviews. In: International Conference on Data Engineering, pp. 675–686. IEEE (2015)
12. Yang, R.R., Wu, J.H.: Study on finding experts in community question-answering system. J. Appl. Mech. Mater. **513**, 1760–1764 (2014)
13. Metzler, D., Croft, W.B.: Linear feature-based models for information retrieval. J. Inf. Retrieval **10**(3), 257–274 (2010)
14. Lin, L., Xu, Z., Ding, Y.: Finding topic-level experts in scholarly networks. J. Scientometrics **97**(3), 797–819 (2013)
15. Zhang, J., Ackerman, M.S., Adamic L.: Expertise networks in online communities: structure and algorithms. In: International Conference on World Wide Web, pp. 221–230. ACM (2007)
16. Macedo, A.Q., Marinho, L.B., Santos, R.L.T.: Context-aware event recommendation in event-based social networks. In: ACM Conference on Recommender Systems, pp. 123–130. ACM (2015)
17. Yeniterzi, R., Callan, J.: Constructing effective and efficient topic-specific authority networks for expert finding in social media. In: International Workshop on Social Media Retrieval & Analysis, pp. 45–50. ACM (2014)

Budget Minimization with Time and Influence Constraints in Social Network

Peng Dou, Sizhen Du, and Guojie Song[(⊠)]

Key Lab of Machine Perception(MOE), Peking University, Beijing, China
billydou20@gmail.com, {sizhen,gjsong}@pku.edu.cn

Abstract. The problem of influence maximization in social network gains increasing attention in recent years. The target is to seek a seed set that maximizes the influence coverage with given budget and infinite time space. However, in many real-world applications, people are eager to achieve the desired influence coverage with limited time and the smallest budget, where we want to find a minimum seed set with constrained time and influence. We refer to the problem as Budget Minimization (BM). The BM problem is much more challenging compared with traditional influence maximization w.r.t. the following reasons: (1) it requires to find both the minimum size of the seed set and the most influential nodes simultaneously, leading to complex and expensive optimization procedure; (2) the estimation of the influence coverage of a given seed set should be accurate enough, since we have to decide whether it can reach the influence threshold exactly. In this paper, we propose an Extended Simulated Annealing on Budget Minimization (ESABM) method to efficiently find the smallest seed set in the considered BM problem. The ESABM method extends the traditional Simulated Annealing (SA) algorithm by importing the 'delete' and 'insert' operators in addition to the 'replace' operator, which is used in the traditional SA algorithm. Based on the operators, some operator selection techniques are proposed with detailed theoretical guarantees. Moreover, since we have to estimate the influence coverage in the ESABM algorithm, we further propose an efficient layered-graph based influence coverage estimation method. Experimental results conducted on five real world data show that our proposed method outperforms the existing state-of-the-art methods in terms of both accuracy and efficiency.

1 Introduction

The problem of influence maximization in social networks is to find an influential seed set that maximizes their influence coverage, which is originally motivated from the application in viral marketing. This problem was first formulated as a discrete optimization problem by [1], referred as influence maximization, and has been well studied in the literature. The representative method for solving the influence maximization problem is the greedy approximation algorithm, which has been shown to achieve $(1 - \frac{1}{e} - \varepsilon)$ accuracy [1]. This seminal work motivated extensive researches on influence maximization in the past decade, including the

© Springer International Publishing Switzerland 2016
F. Li et al. (Eds.): APWeb 2016, Part I, LNCS 9931, pp. 304–315, 2016.
DOI: 10.1007/978-3-319-45814-4_25

cost-effective lazy forward (CELF) optimization [2], arborescence based influence coverage approximation (MIA) [3], and community based greedy (CGA) [4] etc.

The existing methods for solving influence maximization problem assume that there are not any constraints on time and budget. However, in many real world applications, some conditions that specify user preferences, such as time limit and expected influence coverage. These constraints have to be taken into account for better applied for real life. For example, in commercial analysis, the companies expect to reach a desired influence coverage within limited time when they promote their products, therefore the problem of selecting an influential seed set with minimum size, indicating less budget needed, should be well solved. And we call this problem to be Budget Minimization (BM). In this paper, unlike the influence maximization and other related problems, we consider the budget minimization problem with both time and influence constraints.

The considered BM problem is much more challenging compared with the traditional influence maximization w.r.t. the following reasons: (1) in the BM problem, the budget minimization and the influential nodes selection have to be considered simultaneously, while in the traditional influence maximization problem, the budget is fixed. In other words, in addition to the influential nodes selection, the budget is unknown and we have to minimize it as well. Unfortunately, an efficient algorithm for solving the BM problem is absent hitherto; (2) enough accuracy is necessary for the estimation of the influence coverage from a seed set, since we have to determine whether a seed set is qualified for a feasible solution and as a consequence it reaches the influence coverage threshold within t steps of propagation. However, in the traditional influence maximization problem, the steps of propagation are unlimited.

In this paper, we propose an Extended Simulated Annealing on Budget Minimization (ESABM) method to cope with the BM problem. In the traditional Simulated Annealing (SA) algorithm, only the replacement operation on the nodes is allowed, therefore the size of a seed set can not be changed. In the ESABM method, we introduce the 'delete' and 'insert' operators and the associated mechanism of solution selection to create the neighborhood of the seed set. The probability of selecting which operator is given according to the quality of the seed set, as well as the gap between the influence coverage of the seed set and the threshold, are regarded as the main factor when selecting the operator. In the ESABM algorithm, an accurate influence estimation is required, because it determines whether a seed set is qualified for a feasible solution compared with the given influence threshold η. Therefore, we propose a novel scalable method to estimate the influence coverage, where a layered-graph is used to accelerate the diffusion process. This method considers all the possible propagations between any node pairs, while maintains high estimation accuracy as the Monte-Carlo method does, however, without expensive Monte Carlo simulations. In order to evaluate the proposed ESABM method, we experiment on five real-world datasets, and the results demonstrate that the proposed method outperforms the state-of-the-art methods in terms of both accuracy and efficiency.

1.1 Related Work

In [1], the influence maximization problem is first formulated as a discrete optimization problem. This seminal work has motivated a large body of research on influence maximization in the past decades [2–7]. However, all the aforementioned works assume the budget is given and manage to maximize the influence. By taking time constraints on the influence maximization into account, Chen et. al. proposed a time-critical influence maximization method [8] in social networks based on IC-M model and a Continuous-Time IC model is also studied by [9]. On the other side, the influence of the seed set can be another constraint in influence maximization problem. Goyal et al. consider the MINTSS problem [10], which aims to find the minimized budget with a given influence coverage threshold. Wang et. al. propose a seed minimization with probabilistic coverage guarantee (SM-PCG) [11]. However, all these methods never give a consideration about time and influence constraints simultaneously, moreover, they only provide a naive time-consuming greedy based algorithms to solve it.

Our work is also closely related to the influence coverage estimation problem. The Monte-Carlo algorithm [1] can tackle the problem by simulating many times, which, however, results in low efficiency. Another related algorithm, named as SP1M, [12] is based on short-path, but it is not related to the propagation probabilities. Moreover, the PMIA algorithm [6], which is based on the maximum influence path, may ignore some potential diffusions and will lead to inaccurate approximation. Du et al. proposed a randomized algorithm on continuous-time IC model [9] and it is not suitable for the traditional IC model.

2 Preliminary

2.1 Independent Cascade (IC) Model

We consider the IC model [1] as the information diffusion process as follows: Let $G(V, E)$ be a directed graph with edge labels $pp : E \rightarrow [0, 1]$. At step 0, there is a seed set that consists of several nodes that are activated at the beginning. When a node u is activated at step t, the node v is possibly to be activated depended on the condition whether there exists an edge between nodes u and v with the probability $pp(u, v)$ at step $t + 1$. For convenience, we assume that the probabilities defined for the edges are equivalent, denoted by pp.

2.2 Simulated Annealing (SA)

The SA algorithm [13] is a heuristic algorithm that simulates the process of metal annealing and approximates the optimal solutions of a number of NP-hard problems. It works as follows.

1. First, it creates an initial solution S_i and system temperature T_0, then the algorithm calculates the fitness of the solution $f(S_i)$, which stands for the initial energy of the system;

2. Second, it consequently searches the neighborhood of the current solution and creates a new solution S_j. If $\Delta f = f(S_j) - f(S_i)$ is negative, then S_j achieves a better solution and S_i is replaced with S_j; otherwise, S_j will replace S_i with possibility $p_{i,j} = \exp(\frac{-\Delta f}{T})$ (Metropolis criterion [13]), where T is the current system temperature. After a number of iterations, the system temperature is decreased by $T = T * \gamma$, where γ is the cooling rate, as the solution is modified. When the temperature reaches the termination T_f, the algorithm stops.

3   The BM Problem

Given a directed social network $G(V, E)$ with pp defined on each edge, the time limit t and the influence coverage threshold η, we aim to find the smallest set $S \subseteq V$, such that the influence coverage of seed S within t time, i.e. $\sigma(S, t)$, is no less than η. Generally, the BM problem can be formulated as follows:

$$S_{min} = argmin_{\sigma(S,t) \geq \eta} |S| \tag{1}$$

Theorem 1. *The BM problem is NP-hard for the IC model.*

Proof. We show the Set Cover problem, which is a well-known NP-hard problem, is a special case of the proposed BM problem. When we set $t = 1$ and $\eta = |V|$, the resulted problem it is equivalent to the problem of finding the smallest seed set with every other node lying in the neighborhood of the seed set. Therefore, BM is NP-hard. □

From Theorem 1, we know that no polynomial algorithm exists for solving the BM problem. Therefore, we propose an efficient algorithm ESABM for BM problem rather than greedy algorithm with low efficiency and a method to estimate the influence coverage because the influence estimation for the ESABM algorithm requires an accurate estimation.

4   Proposed Method

We propose the ESABM algorithm and the layered-graph based influence coverage estimation method for solving the BM problem.

4.1   The ESABM Algorithm for BM

Suppose we are given an initial set $A = \{v_1, v_2, \ldots, v_k\}$ with pre-specified budget k. During each iteration, we get the neighbor solution A' of A by replacing one node in set A with a node in $V - A$. If $\Delta(f) = \sigma(A') - \sigma(A)$ is positive, where $\sigma(A)$ is the fitness function of a solution set A, A' is accepted. Otherwise, A' is accepted if $min\{1, exp(\Delta(f)/T)\} > random[0, 1]$. The above process is repeated until convergence.

As introduced previously, the SA algorithm only finds the solution with the size k predefined. However, in the BM problem, k is unknown and has to be learned. Therefore the traditional SA algorithm can not be used directly. To solve the hard BM problem, the following problems have to be settled.

1. **Operator Extension**: Operations in the SA algorithm need to enable changes of the seed set size when creating new neighbors in addition to replacement;
2. **Operator Selection**: After the employment of new operators, it is necessary to develop strategies to determine which operator should be chosen when creating a neighbor solution set;
3. **Solution Substitution**: Once a neighbor solution set is created, the acceptance of the newly produced seed set should be determined.

Next, we show how to solve these problems.

4.1.1 Operator Extension

We introduce two new operators, i.e. the 'delete' and 'insert' operators.

Definition 1 (ESABM Operators). *Given the seed set $S = \{x_1, x_2, ..., x_m\}$, the new seed set S' can change with the following operators conducted:*

- *Replace: $S' = \{x_1, ...x_{i-1}, x'_i, x_{i+1}..., x_m\}$, where x_i is replaced by x'_i;*
- *Delete: $S' = \{x_1, ..., x_{i-1}, x_{i+1}, ..., x_m\}$, where x_i is deleted from S;*
- *Insert: $S' = \{x_1, ..., x_m, x_{m+1}\}$, where x_{m+1} is inserted into S.*

From the definition, the 'replace' operator can improve the quality of the current solution with larger influence coverage and keep the size of the seed set fixed. Once we need to optimize the size of the seed set to approach the final solution, the 'delete' and 'insert' operators should be used.

4.1.2 Operator Selection

We show how to select these operators when generating a new solution set. Intuitively, we can choose the operators randomly, implying that the probability of choosing each operator is $1/3$. However, this method ignores the use of the current solution S and will lead to low efficiency even though it will converge to the same solution. Actually, we can obtain sufficient conditions for selecting these operators based on the current solution S.

Lemma 1 (Sufficient condition for using 'delete' operator). *It is sufficient to delete a node from S to get S', whose influence coverage is still over the threshold, if the seed set S satisfies:*

$$\sigma(S,t) > \eta/(1 - 1/|S|). \tag{2}$$

Proof. It is obvious that there is at least one node x in S, of which the marginal influence is no more than $\sigma(S,t)/|S|$. When we delete x from S, the influence of the newly-generated seed set S' should exceed the threshold, which can be expressed as follows:

$$\sigma(S\setminus\{x\}, t) > \sigma(S,t) - \sigma(S,t)/|S| > \eta \tag{3}$$

Thus, we get

$$\sigma(S,t) > \eta/(1 - 1/|S|) \tag{4}$$

$\square$

Lemma 2 (Sufficient condition for using 'insert' operator). *A node is guaranteed to be inserted into S, i.e. the optimum size of a feasible seed set is more than $|S|$, if the seed set S satisfies:*

$$\sum_{0<i<|S|} \sigma(x_i, t) < \eta. \tag{5}$$

Proof. Assuming the budget is given as $|S|$, without considering the overlap between nodes, the influence coverage of $|S|$ nodes, is the sum of the top $|S|$ nodes of which the influence coverage is maximum. Provided the sum is less than η, the influence coverage of the seed set cannot be over η.   □

Theorem 2. *Given the seed set S in the current solution, the probability for choosing the operators defined in Definition 1 during the learning process is defined as follows:*

- if $\sigma(S,t) > \eta/(1 - 1/|S|)$, $P_{del} = 1$,
- if $\eta < \sigma(S,t) < \eta/(1 - 1/|S|)$,
  - $P_{del} = (\sigma(S,t) - \eta)/(\sigma(S,t)/|S|)$,
  - $P_{rep} = 1 - (\sigma(S,t) - \eta)/(\sigma(S,t)/|S|)$,
- if $sum_{0<i<|S|}\sigma(x_i,t) > \eta$ and $\sigma(S,t) < \eta$,
  - $P_{ins} = (\eta - \sigma(S,t))/(\sigma(S,t)/|S|)$,
  - $P_{rep} = 1 - (\eta - \sigma(S,t))/(\sigma(S,t)/|S|)$,
- if $sum_{0<i<|S|}\sigma(x_i,t) < \eta$, $P_{ins} = 1$,

where P_{del}, P_{ins} and P_{rep} represents probability of deleting a node, the probability of inserting a node, and the probability of replacing a node respectively.

Proof. Firstly, according to Lemmas 1 and 2, we are able to decide whether we should delete or insert given the current seed set. If the influence of the seed set is close to the threshold so that we cannot determine whether to delete or insert a node. Assuming the influence of the seed set is over the threshold, the redundancy is known as $\sigma(S,t) - \eta$, and the loss of delete a node can be estimated by $\sigma(S,t)/|S|$. The ratio between these two is from 0 to 1. When the ratio is large, which means there is much redundancy. Otherwise, we choose to replace one node rather than deleting. Therefore, the ratio $(\sigma(S,t) - \eta)/(\sigma(S,t)/|S|)$ represents the probability of deleting. Similarly, the ratio $(\eta - \sigma(S,t))/(\sigma(S,t)/|S|)$ represents the probability of inserting.   □

4.1.3   Solution Substitution

After choosing the operators to generate a new solution, we have to decide whether this solution is acceptable depended on the quality of this solution according to the Metropolis Criterion.

Lemma 3. *For a new seed set S' generated from the seed set S by using the 'replace' operator, the probability of accepting S' can be computed as follows:*

- *if $\sigma(S',t) \geq \sigma(S,t)$, we accept S' with $p = 1$;*
- *otherwise, $p = exp((\sigma(S',t) - \sigma(S,t))/T)$.*

Lemma 4. *For a new seed set S' generated from the seed set S by using the 'delete' operator, the probability of accepting S' can be computed as follows:*

- *if $\sigma(S',t) \geq \eta$, we accept S' with $p = 1$;*
- *otherwise, $p = exp((\sigma(S',t) - \sigma(S,t) - \eta)/T)$.*

Lemma 5. *For a new seed set S' generated from the seed set S by using the 'insert' operator, the probability of accepting S' can be computed as follows:*

- *if $(\sigma(S',t) \geq \eta$ and $\sigma(S,t) < \eta)$ or $\sigma(S',t) < \eta$, we accept S' with $p = 1$;*
- *if $\sigma(S',t) < \eta$ and $\sigma(S,t) < \eta$, $p = exp((\sigma(S',t) - \sigma(S,t) - \eta)/T)$.*

Next, we prove that the convergence of ESABM algorithm can be guaranteed by Theorem 3.

Theorem 3. *The Extended Simulated Annealing on Budget Minimization algorithm converges.*

Proof. Applying the Metropolis Criterion, when the temperature is cooled logarithmical, the annealing process will lead the convergence of the whole process, which is proven by Geman [14]. It will converge to the solution which minimizes the energy function. Therefore, both Metropolis Criterion and an energy function to meet the requirement of BM should be both taken into account. The goal of BM is to find the minimal seed set size while the influence coverage is over η. Considering the conditions above, we propose the energy function in our algorithm:

$$E(S) = |S| * \eta - \sigma(S,t) + (\eta - |S|) * \eta * sgn(\sigma(S,t) - \eta). \tag{6}$$

$E(S)$ is determined by $|S|$ and $\sigma(S,t)$. Provided $|S|$ is a feasible solution, when $|S|$ is the smallest set and $\sigma(S,t)$ is large enough, $E(S)$ can get to the minimum. Note that $|\sigma(S,t) - \sigma(S',t)| < \eta$, so the change of the size has the superior effect compared with the influence coverage change of the seed set. According to Metropolis Criterion, the probability that accepts the new solution S' is $min\{1, exp((E(S') - E(S))/T)\}$ and the probability of solution substitution is derived from the criterion. Now, we complete the proof.   □

4.2   Layered-Graph Based Influence Estimation

In this section, we propose a method to calculate the influence coverage $\sigma(S,t)$ based on a layered-graph with both high efficiency and accuracy.

4.2.1 Construction of the Layered-Graph

Given a directed graph $G(V, E)$ and a seed set S, we build a layered-graph $G^T(V^T, E^T)$ as follows. Let V_t be the set of nodes that is possible to be activated at step t, and the node v is denoted as v_t if v is possible to be activated at step t. We define $V^T = V_0 \cup V_1 \cup ... \cup V_T$, where the seed set is V_0. For each edge (u, v) in E, if $u \in V_{t-1}$ and $v \in V_t$, then the edge (u_{t-1}, v_t) is included in set E_t with the probability pp. Without loss of generality, we define $E^T = E_0 \cup E_1 \cup ... \cup E_T$. Note that the time cost for the activation of a node in a single diffusion process is certain, however a node in V_t can be activated in time t. We define $P(v, t)$ as the probability that node v is activated exactly in time t.

4.2.2 Influence Estimation

Since the goal to estimate $\sigma(S, T)$, we initially find the relationship between $\sigma(S, T)$ and $P(v, t)$ where $\sigma(S, T) = \sum_{v \in V} \sum_{t \leq T} P(v, t)$. Therefore, the key point is to calculate $P(v, t)$ efficiently and accurately. Indeed, we can estimate $P(v, t)$ for each node v at $t \leq T$ recursively as presented by the following theorem.

Theorem 4. *Given any node v in G and any time $t \leq T$, the activation probability $P(v, t)$ can be estimated as follows:*

1. if $t = 0$,

$$P(v, t) = \begin{cases} 1 & v \in S, \\ 0 & v \notin S, \end{cases}$$

2. else if $t > 0$,

$$P(v, t) = \sum_{t' < t} P(v, t') + (1 - \sum_{t' < t} P(v, t'))$$

$$* [1 - \prod_{j=1}^{n} (1 - P(u_j, t-1))] * pp(u, v). \tag{7}$$

Proof. We can easily prove the accuracy of the case when $t = 0$. For $t > 0$, once we have already estimated $P(v, t-1)$ for each node $v \in V_{t-1}$, we then compute $P(v, t)$ for each node $v \in V_{t-1}$. Assuming node $v \in V_{t-1}$ has n predecessors, denoted by $u_1, u_2, ..., u_n$. We then estimate $P(v, t)$ by using $P(u_1, t-1), P(u_2, t-1), ..., P(u_n, t-1)$. The probability that node v is not activated until time t is $1 - \sum_{t' < t} P(v, t')$. Recall that the probability that one node can be activated is denoted as pp, and each node can be activated only once, we obtain Formula 7. $\square$

From the above theorem, for every $t \in \{1, 2, ..., T\}$, we compute $P(v, t)$ for each $v \in V_t$. Finally we can obtain $\sigma(S, T)$ after computing $P(v, t)$. Since each edge in the layered-graph represents a possible propagation, we have to traverse all the edges when the process starts. Therefore, the time complexity for estimating $\sigma(S, T)$ is $O(|V^T| + |E^T|)$, where V^T and E^T are the vertex set and edge set in the layered-graph G^T, respectively.

5 Experiments

In this section, we conduct experiments on five real-world datasets. We compare the ESABM algorithm with the greedy algorithm MINTSS [10], the degree algorithm, the TIM algorithm, Two-phase Influence Maximization (TIM), an algorithm that aims to bridge the theory and practice in influence maximization [7], and a counterpart of the proposed ESABM algorithm, i.e. the ESABM_RDM algorithm. The proposed algorithm for influence estimation is also compared with Monte-Carlo and SP1M [6].

5.1 DataSets

We evaluate all the algorithms on five real-world datasets [15], which are commonly studied in the literature of the time-critical influence maximization problem [8]. All the networks are directed graphs. The details of these datasets are listed in Table 1.

Table 1. Statistics of the real-world datasets.

Dataset	HepTh	AstroPh	Epinions	Amazon	Orkut
Node	9,877	18,772	75,879	262,111	3,072,441
Edge	25,988	396,160	508,837	1,234,877	117,185,083

5.2 Setting of the Parameters

In the experiments, we set the parameters according to Table 2. The default setting for the parameters are highlighted. To compare the influence estimation of all the algorithms, we run the Monte-Carlo simulations 10000 times.

Table 2. Setting of the parameters.

Para.	HepTh	Amazon	Epinions	AstroPh	Orkut
t	1, 2, **3**, 4, 5, 6	1, **2**, 3		1, **2**, 3, 4	1, **2**, 3
η	200, **500**, 1000		500, **1000**, 2000		**20000**
p	0.01, **0.05**, 0.1				

5.3 Results and Analysis

In this part, we report the results and compare the learning performance of the algorithms by varying the time limit, and the influence coverage threshold.

5.3.1 Varying the Time Limit

Figure 1 shows the needed budget (seed size) when the time limit increases. From Fig. 1, we see that the ESABM algorithm outperforms all the other algorithms. When the influence coverage threshold is large, the ESABM algorithm achieves much better performance compared with other algorithms, especially on the Amazon and Hepth data. For example, when t equals to 6 on the Hepth data, the seed set size learned by the ESABM algorithm is 30 % smaller than the seed set size learned by the degree algorithm. On the other hand, when the minimized budget becomes smaller, the ESABM algorithm shows less improvements.

5.3.2 Varying the Influence Coverage Threshold

Table 3 shows the performance of different algorithms when the influence coverage threshold changes. From the result, we observe that our algorithm outperforms other algorithms in all of the cases. Compared with greedy algorithm, the seed set size learned by ESABM is at least 20 % smaller. For example, in the hepth data, the seed set size learned by the ESABM algorithm is 14 % smaller than TIM, 28 % smaller than degree, 31 % smaller than greedy. In the Astroph data, the seed set size learned by the ESABM algorithm is 28 % smaller than TIM, 42 % smaller than degree and 57 % than greedy.

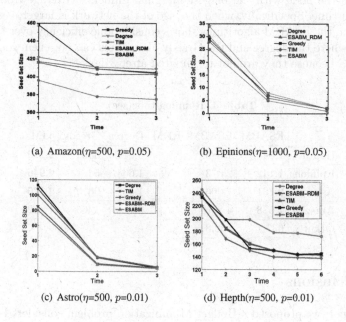

(a) Amazon(η=500, p=0.05)   (b) Epinions(η=1000, p=0.05)

(c) Astro(η=500, p=0.01)   (d) Hepth(η=500, p=0.01)

Fig. 1. Seed set size with time step (t)

Table 3. Seed set size with threshold.

	ESABM	ESABM_RDM	Degree	Greedy	TIM
Hepth	**150**	161	193	197	171
Epinion	**53**	58	62	67	60
Astroph	**14**	17	20	22	18
Amazon	**367**	393	429	441	415
Orkut	**259**	332	–	–	–

5.3.3 Running Time

Table 4 shows the running time of different algorithms with the default setting of the parameters. We can see that the ESABM algorithm is faster than all the compared algorithms. The running time of the ESABM algorithm and the ESABM_RDM algorithm are almost equivalent, because they are equipped with the same parameters. When the size of the considered network is small, e.g. on the hepth data, the ESABM algorithm, the degree algorithm and the TIM algorithms are very efficient and can be completed within only several seconds. In contrast, the greedy algorithm is very computational expensive because it has to find the node with the biggest marginal influence coverage from all the nodes each time. Specifically, when the size of the network is larger, e.g. on the Orkut data, the ESABM algorithms shows considerable efficiency over the other methods, where the degree and the Greedy algorithms can not even work on the Orkut data because they are not terminated after 3 days.

Table 4. Running time(sec).

	ESABM	ESABM_RDM	Degree	Greedy	TIM
Hepth	1.548	1.086	5.611	336	2.219
Epinion	1.142	2.073	1.664	54	2.946
Astroph	2.181	1.606	1.582	90	3.175
Amazon	7.638	4.835	66	1365	149.26
Orkut	637.1	552.8	–	–	–

6 Conclusions

In this paper, we proposed a Budget Minimization problem considered in social network analysis. In order to solve the hard BM problem, we developed an Extended Simulated Annealing on Budget Minimization, i.e. ESABM, algorithm. Experimental results show that the proposed algorithm is efficient and effective for solving the BM problem compared with the existing methods.

Acknowledgement. This work was supported by the National Science and Technology Support Plan(2014B AG01B02), the National Natural Science Foundation of China (61572041, 61402020), the National High Technology Research, Development Program of China (2014AA015103), and Beijing Natural Science Foundation (4152023) and the Ph.D. Programs Foundation of Ministry of Education of China (No. 20130001120021).

References

1. Kempe, D., Kleinberg, J., Tardos, É.: Maximizing the spread of influence through a social network. In: ACM SIGKDD International Conference on Knowledge Discovery Data Mining, pp. 137–146 (2003)
2. Kieffer, A., Paboriboune, P., Crpey, P., Flaissier, B., Souvong, V., Steenkeste, N., Salez, N., Babin, F.X., Longuet, C., Carrat, F.: Cost-effective outbreak detection in networks. In: ACM SIGKDD International Conference on Knowledge Discovery and Data Mining, pp. 420–429 (2007)
3. Chen, Y.C., Peng, W.C., Lee, S.Y.: Efficient algorithms for influence maximization in social networks. Knowl. Inf. Syst. **33**(3), 577–601 (2012)
4. Wang, Y., Cong, G., Song, G., Xie, K.: Community-based greedy algorithm for mining top-K influential nodes in mobile social networks. In: ACM SIGKDD International Conference on Knowledge Discovery and Data Mining, Washington, DC, USA, pp. 1039–1048, July 2010
5. Jung, K., Heo, W., Chen, W.: Irie: scalable and robust influencemaximization in social networks. In: 2012 IEEE 12th International Conference on Data Mining (ICDM), pp. 918–923 (2012)
6. Chen, W., Wang, C., Wang, Y.: Scalable influence maximization for prevalent viral marketing in large-scale social networks. In: Proceedings of KDD 2010, pp. 1029–1038 (2010)
7. Tang, Y., Xiao, X., Shi, Y.: Influence maximization: near-optimal timecomplexity meets practical efficiency, pp. 75–86 (2014)
8. Chen, W., Lu, W., Zhang, N.: Time-critical influence maximization in social networks with time-delayed diffusion process. Chin. J. Eng. Design. **19**(5), 340–344 (2012)
9. Du, N., Song, L., Gomez-Rodriguez, M., Zha, H.: Scalable influence estimation in continuous-time diffusion networks. Adv. Neural Inf. Process. Syst. **12**(3), 418–420 (2013)
10. Goyal, A., Bonchi, F., Lakshmanan, L.V.S., Venkatasubramanian, S.: On minimizing budget and time in influence propagation over social networks. Soc. Netw. Anal. Min. **3**(2), 179–192 (2012)
11. Zhang, P., Chen, W., Sun, X., Wang, Y., Zhang, J.: Minimizing seed set selection with probabilistic coverage guarantee in a social network, pp. 1306–1315 (2014)
12. Kimura, M., Saito, K.: Tractable models for information diffusion in social networks. In: Fürnkranz, J., Scheffer, T., Spiliopoulou, M. (eds.) PKDD 2006. LNCS (LNAI), vol. 4213, pp. 259–271. Springer, Heidelberg (2006)
13. Metropolis, N., Rosenbluth, A.W., Rosenbluth, M.N., Teller, A.H., Teller, E.: Equation of state calculations by fast computing machines. J. Chem. Phys. **21**(6), 1087–1092 (2004)
14. Geman, S., Geman, D.: Stochastic relaxation, gibbs distributions, and the bayesian restoration of images. IEEE Trans. Pattern Anal. Mach. Intell. **6**(6), 721–741 (1984)
15. Leskovec, J.: Snap. http://snap.stanford.edu/

B-mine: Frequent Pattern Mining and Its Application to Knowledge Discovery from Social Networks

Fan Jiang, Carson K. Leung$^{(\boxtimes)}$, and Hao Zhang

University of Manitoba, Winnipeg, MB, Canada
kleung@cs.umanitoba.ca

Abstract. As an important data mining task, frequent pattern mining has drawn attention from many researchers. This has led to the development of many frequent pattern mining algorithms, which include Apriori-based, tree-based, and hyperlinked array structure-based algorithms, as well as vertical mining algorithms. Although these algorithms are efficient and popular, they also suffer from some drawbacks. To tackle these drawbacks, we present in this paper an alternative algorithm called *B-mine* that uses a bitwise approach to mine frequent patterns. Evaluation results show the space- and time-efficiency of B-mine for frequent pattern mining, as well as the practicality of B-mine for social network analysis and knowledge discovery from social networks.

Keywords: Data mining · Knowledge discovery · Frequent patterns · Bitmap · Social network analysis

1 Introduction and Related Works

Data mining aims to discover implicit, previously unknown, and potentially useful knowledge from data. As an important data mining task, *frequent pattern mining* finds frequently co-occurring items, events, or objects (e.g., frequently purchased merchandise items in shopper market basket, frequently collocated events). Since the introduction of the research problem of frequent pattern mining [1], numerous frequent pattern mining algorithms [2,9,16] have been proposed. For instance, the Apriori algorithm [1] applies a generate-and-test paradigm in mining frequent patterns in a level-wise bottom-up fashion. Specifically, the algorithm first generates candidate patterns of cardinality k (i.e., candidate k-itemset) and tests if each of them is frequent (i.e., tests if its support or frequency meets or exceeds the user-specified *minsup* threshold). Based on these frequent patterns of cardinality k (i.e., frequent k-itemsets), the algorithm then generates candidate patterns of cardinality $k+1$ (i.e., candidate $(k+1)$-itemsets). This process is applied repeatedly to discover frequent patterns of all cardinalities. A disadvantage of the Apriori algorithm is that it requires K database scans to discover all frequent patterns (where K is the maximum cardinality of discovered patterns).

© Springer International Publishing Switzerland 2016
F. Li et al. (Eds.): APWeb 2016, Part I, LNCS 9931, pp. 316–328, 2016.
DOI: 10.1007/978-3-319-45814-4_26

To address this disadvantage of the Apriori algorithm and to improve efficiency, the FP-growth algorithm [3] uses an extended prefix-tree structure called Frequent Pattern tree (FP-tree) to capture the content of the transaction database. Unlike the Apriori algorithm, FP-growth scans the database *twice*. The key idea of FP-growth is recursively extract relevant paths from the FP-tree to form projected databases (i.e., collection of transactions containing some items), from which subtrees (i.e., smaller FP-trees) capturing the content of relevant transactions are built. While FP-growth avoids the generate-and-test paradigm of Apriori (because FP-growth uses the divide-and-conquer paradigm), a disadvantage of FP-growth is that many smaller FP-trees (e.g., for $\{a\}$-projected database, $\{a, b\}$-projected database, $\{a, b, c\}$-projected database,...) need to be built during the mining process. In other words, FP-growth requires lots of memory space.

To avoid building and keeping multiple FP-trees at the same time during the mining process, some other algorithms (e.g., TD-FP-Growth [15], H-mine [10]) have been proposed. Unlike the FP-growth (which mines frequent patterns by traversing the global FP-tree and subtrees in a *bottom-up* fashion), the TD-FP-Growth algorithm traverses only the global FP-tree and in a *top-down* fashion. During the mining process, instead of recursively building sub-trees, TD-FP-Growth keeps updating the global FP-tree by adjusting tree pointers. Along this direction, the H-mine algorithm uses a new data structure—namely, a hyperlinked-array structure called H-struct. Like FP-growth and TD-FP-Growth, the H-mine algorithm also scans the database twice, but it captures the content of the transaction database in the *hyperlinked-array structure* (instead of the tree structure). During the mining process, the H-mine algorithm also recursively updates links in the H-struct. During the mining process, some array entries in the H-struct may contain K hyperlinks, one hyperlink for each cardinality. A disadvantage of TD-FP-Growth and H-mine is that many of the pointers/hyperlinks need to be updated during the mining process.

While the aforementioned algorithms mining frequent patterns "horizontally" (i.e., using a transaction-centric approach to find what k-itemset is supported by or contained in a transaction), frequent patterns can also be mined "vertically" (i.e., using an item-centric approach to count the number of transactions supporting or containing the patterns). Two notable vertical frequent pattern mining algorithms are VIPER [13] and Eclat [20]. Like the Apriori algorithm, Eclat also uses a levelwise bottom-up paradigm. With Eclat, the database is treated as a collection of item lists. Each list for an item x keeps IDs of transactions containing x. The length of the list for x gives the support of 1-itemset $\{x\}$. By taking the intersection of lists for two frequent itemsets α and β, we get the IDs of transactions containing $(\alpha \cup \beta)$. Again, the length of the resulting (intersected) list gives the support of the pattern $(\alpha \cup \beta)$. Eclat works well when the database is sparse. However, when the database is dense, these item lists can be long.

Alternatively, VIPER represents the item lists in the form of bit vectors. Each bit in a vector for a domain item x indicates the presence (bit "1") or absence (bit "0") of transaction containing x. The number of "1" bits for x gives the support of 1-itemset $\{x\}$. By computing the dot product of vectors for two frequent itemsets α and β, we get the vector indicating the presence of transactions containing $(\alpha \cup \beta)$. Again, the number of "1" bits of this vector gives the support of the resulting pattern $(\alpha \cup \beta)$. VIPER works well when the database is dense. However, when the database is sparse, lots of space may be wasted because the vector contains lots of 0s.

In this paper, we present an alternative frequent pattern mining algorithm. One of our *key contributions* of this paper is such a new space- and time-efficient frequent pattern mining algorithm, which requires only one database scan because (i) it only needs to keep a global structure for capturing the content of the original database, and (ii) it also captures the contents of subsequent projected databases without complicated update operations on pointers/hyperlinks. Our second *key contribution* of this paper is an application of our mining algorithm to social network analysis. This contribution not only shows the effectiveness of our algorithm, but also shows its ability and practicality of handling real-life problems.

The remainder of this paper is organized as follows. Section 2 presents our frequent pattern mining algorithm called B-mine, and Sect. 3 shows its practical application in social network analysis. Evaluation and conclusions are given in Sects. 4 and 5, respectively.

2   Our B-mine Algorithm and Its Associated B-table Structure

To mine frequent patterns from transactional database, we capture the contents of the transactions in a **bitwise table (B-table)**. In addition, we also use the following auxiliary structures during the mining process:

1. A **vertical index list (VI-List)** for a frequent k-itemset X, which is a two-dimensional data structure capturing column indices (indicating the first "1" columns in the suffix of the transactions containing X) and their corresponding row indices (indicating the IDs of transactions with that column index).
2. A **horizontal index counter (HI-Counter)** for a frequent k-itemset X, which is a vector capturing column sums of the B-table. The sum for column c represents the support of $(X \cup \{c\})$.

2.1   Construction of the B-table, HI-Counters and VI-Lists

Given a transactional database, we construct a B-table by applying the following steps to capture important contents of the database so that frequent patterns can then be discovered from the constructed B-table:

1. For each row in the database, we create a bitmap row with size M, where M is the number of domain items. Each column (represented by a bit) in the row represents the items in each transaction. We put (i) a "1" in the i-th bit (where $1 \leq i \leq N$) if the i-th item is present in the transaction or (ii) a "0" otherwise (i.e., when the i-th item is absent from the transaction). When we repeat the aforementioned actions in this step, we get multiple bitmap rows to form a B-table.

2. For each row r of the B-table, we create a Level-0 VI-List by recording the column index (i.e., the column index for the first occurrence of the "1" bit in row r).

3. For each column c of the B-table, we create a Level-0 HI-Counter by counting the number of 1s in column c and put the count in the c-th entry of the HI-Counter.

Note that all the above steps can be completed in only *one scan* of the database. See the following examples for more details.

Example 1. Consider an illustrative database with $N = 6$ transactions about $M = 6$ domain items: $t_1 = \{b, e\}$, $t_2 = \{a, c\}$, $t_3 = \{a, e\}$, $t_4 = \{b, c, e\}$, $t_5 = \{a, b, c\}$ and $t_6 = \{a, b, c, e\}$. The B-table, as shown in Fig. 1(a), can be constructed by scanning the database only once to capture the contents of the transactional database. Each row in the table represents a transaction. Note that each 1 or 0 entry takes only one bit of memory in the actual implementation. □

Observation 1. *Recall that, two scans are required to build an H-struct [10]. As each entry in the H-struct is a hyperlink and H-struct contains a high number of hyperlinks, H-struct requires lots of memory space. In contrast, only one database scan is required to build a B-table. Moreover, the B-table does not contain any hyperlinks. Each entry in the B-table takes only 1 bit in memory.*

Based on this B-table, we then construct its associated Level-0 VI-List and HI-Counter. Specifically, the Level-0 VI-List contains several rows of the first appearance of an item. For example, in Fig. 1(b), the VI-List captures the information: (i) rows t_2, t_3, t_5 and t_6 contain four transactions that start with item a; (ii) rows t_1 and t_4 contain two transactions that start with item b (but not item a). The size of

		a b c d e f
(a) B-table	t_1	0 1 0 0 1 0
	t_2	1 0 1 0 0 0
	t_3	1 0 0 0 1 0
	t_4	0 1 1 0 1 0
	t_5	1 1 1 0 0 0
	t_6	1 1 1 0 1 0
(c) Level-0 HI-Counter		4 4 4 0 4 0

(b) Level-0 VI-List

column index	row indices
column a (i.e., $\{a\}$)	rows t_2, t_3, t_5, t_6
column b (i.e., $\{b\}$)	rows t_1, t_4

Fig. 1. B-table and its Level-0 VI-List & HI-Counter

each row index in the VI-List is bounded above by $N = 6$ (when all $N = 6$ transactions start with the same item), where N is the total number of transactions. Given $M = 6$ domain items and Level $L = 0$, the number of rows in the Level-0 VI-List is bounded by $M - L = 6$ rows (when each transaction with a different item), where M is the number of domain items.

Observation 2. *Each row index appears at most once in the VI-List because each represents a transaction.*

As for the Level-0 HI-Counter, it stores the count of each column of the B-table. For example, as shown in Fig. 1(c), the count of "1" in column a of HI-Counter is 4, which means that item a appears in four transactions. Each count is a non-negative integer ranging from 0 to N, where N is the total number of transactions (e.g., each count in Fig. 1(c) is at most 6 because there are only $N = 6$ transactions). Given $M = 6$ domain items and Level $L = 0$, the size/length of this Level-0 HI-Counter vector is $M - L = 6$ entries, where M is the number of domain items.

2.2 Mining of Frequent Patterns from B-table

Once the B-table is constructed, our B-mine algorithm can mine frequent patterns from the B-table. The mining process involves two major procedures: VIListComb() and HICounterIncr(). See Figs. 2 and 3 for skeletons of these procedures.

VI-List-Combo(currLevelVI-List, preLevelVI-List){
    currRow ← currLevelVI-List.lastRow
    if (not all rows in prevLevelVI-List has been processed) **then**
        for each row r in prevLevelVI-List **do**
            if r.ColumnIndex = currRow.ColumnIndex
            then append currRow.RowIndex to r.RowIndex
}

Fig. 2. The VI-List-Combo() procedure for combining two VI lists.

HI-Counter-Incr(B-table, HI-Counter, VI-List){
    for $i \leftarrow 1$ to M **do** HI-Counter$[i] \leftarrow 0$
    for each row index r in VI-List **do**
        for each tID in r **do**
            for each corresponding column index c in VI-List **do**
                if B-table$[tID][c] = 1$ **then** HI-Counter$[c]$++
        return HI-Counter
}

Fig. 3. The HI-Counter-Incr() procedure for incrementing the HI-Counter.

Our B-mine algorithm first finds frequent 1-itemsets by checking the Level-0 HI-Counter. Items with values $\geq$ *minsup* are frequent. Based on the contents of the Level-0 VI-List, frequent non-singletons can then be mined. From each row r (representing an itemset X) of the Level-0 VI-List, our B-mine algorithm extracts the corresponding rows (based on the row index for r) in B-table to obtain the projected database for X. Afterwards, Level-1 HI-Counter for X can then be computed. Again, if the value for an item y in this HI-Counter $\geq$ *minsup*, then $X \cup \{y\}$ is considered a frequent 2-itemset, from which Level-2 VI-List is then computed. This mining process is repeated until Level-k VI-List is empty (i.e., no more row indices), meaning all relevant transactions have been examined and all frequent patterns have been discovered from the database. Example 2 illustrates how B-mine algorithm finds all frequent patterns.

Example 2. To explain the mining process, let us continue with Example 1. Let *minsup* be set to 2. The mining process starts from the first row of Level-0 VI-List as shown in Fig. 1(c): Column a (i.e., $\{a\}$) in rows t_2, t_3, t_5 & t_6. By using the column and row indices, B-mine extracts the relevant rows from the B-table to obtain appropriate projected databases, from which frequent k-itemsets can be discovered. Specifically, from the first row of VI-List, B-mine extracts the relevant transactions t_2, t_3, t_5 and t_6 based on the row indices (as shown in Fig. 4). For readability, irrelevant portions of the B-table were "faded out" in the table. However, *it is important to note that, the contents of the B-table reminds unchanged throughout the entire mining process.*

Based on the relevant portions of the B-table showing the $\{a\}$-projected database, B-mine forms Level-1 VI-List and HI-Counter for the $\{a\}$-projected database. See Fig. 4(b) and (c). □

Observation 3. *The total number of row indices in VI-List for the $\{a\}$-projected database (as shown in Fig. 4(b) is 4, which is the same as the number of row indices in the first row of Level-0 VI-List (as shown in Fig. 1(b)) representing transactions t_2, t_3, t_5, t_6 containing itemset $\{a\}$. The row indices t_2, t_3, t_5 & t_6 are now subdivided into three subgroups (i) t_5 & t_6, (ii) t_2, and (iii) t_3.*

As shown in Fig. 4, the newly constructed Level-1 HI-Counter records all the column sums of the B-table that are relevant to $\{a\}$. These sums are frequency

(a) B-table

	a b c d e f
t_1	
t_2	0 1 0 0 0
t_3	0 0 0 1 0
t_4	
t_5	1 1 0 0 0
t_6	1 1 0 1 0

(c) Level-1 HI-Counter | 2 3 0 2 0

(b) Level-1 VI-List

column index	row indices
column b (i.e., $\{a, b\}$)	rows t_5, t_6
column c (i.e., $\{a, c\}$)	rows t_2
column e (i.e., $\{a, e\}$)	rows t_3

Fig. 4. B-table for the $\{a\}$-projected database and its Level-1 VI-List & HI-Counter

support values of all frequent patterns with prefix 1-itemset $\{a\}$. In this example, we have frequency support values for frequent patterns $\{a,b\}, \{a,c\}$ and $\{a,e\}$, which appear in 2, 3 and 2 transactions, respectively. Note that, although the support counts are 0 for items d and f (i.e., $\{a,d\}$ and $\{a,f\}$) are shown in Fig. 4(c) for the $\{a\}$-projected database, B-mine does not actually count the support for itemsets $\{a,d\}$ and $\{a,f\}$ because these itemsets are known to be infrequent. As B-mines can find out which patterns are frequent and which are not from Level-0 HI-Counter, B-mine ignores these infrequent itemsets in subsequent mining process (including the formation of Level-1 HI-Counter for the $\{a\}$-projected database).

Observation 4. *Recall from Sect. 2.1 that the size of the Level-L HI-Counter is $(M-L)$, where M is the number of domain items. So, at this stage of the mining process (when $L = 1$), size of the Level-1 HI-Counter (as shown in Fig. 4(c)) is $6-1=5$.*

Based on this Level-1 VI-List, B-mine extracts relevant transactions for the $\{a,b\}$-projected database (see Fig. 5(a)) to form Level-2 VI-List and HI-Counter for the $\{a,b\}$-projected database (as shown in Fig. 5(b) and (c)).

Based on the Level-2 VI-List for the $\{a,b\}$-projected database, B-mine finds frequent 3-itemset $\{a,b,c\}$ with support 2. Afterwards, as no more frequent patterns can be discovered from this $\{a,b\}$-projected database, the B-mine algorithm backtracks to the previous level. The algorithm checks if the Level-1 VI-List has more rows. If so, it (i) checks if all other rows in the Level-1 VI-List has the same column index as the current VI-List rows and (ii) updates the new row in the Level-1 VI-List. By doing so, the resulting Level-1 VI-List for the $\{a\}$-projected database becomes the one shown in Fig. 6(b) while both B-table and Level-1 HI-Counter are similar to those shown in Fig. 4(a) and (c) with column b "faded out".

From this updated Level-1 VI-List for the $\{a\}$-projected database, B-mine applies the same procedure as described above to find frequent pattern $\{a,c\}$ with support 3. The Level-2 VI-List and Level-2 HI-Counter for the $\{a,c\}$-projected database shows that no more frequent patterns containing $\{a,c\}$ can be found. B-mine, once again, updates the Level-1 VI-List for the $\{a\}$-projected

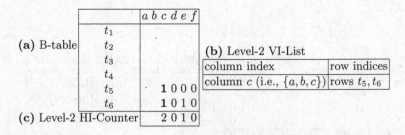

(b) Level-2 VI-List

column index	row indices
column c (i.e., $\{a,b,c\}$)	rows t_5, t_6

Fig. 5. B-table for the $\{a,b\}$-projected database and its Level-1 VI-List & HI-Counter

(a) B-table

	a b c d e f
t_1	
t_2	1 0 0 0
t_3	0 0 1 0
t_4	
t_5	1 0 0 0
t_6	1 0 1 0
(c) Level-1 HI-Counter	3 0 2 0

(b) Updated Level-1 VI-List

column index	row indices
column c (i.e., $\{a,c\}$)	rows t_2, t_5, t_6
column e (i.e., $\{a,e\}$)	rows t_3

Fig. 6. B-table for the $\{a\}$-projected database and its Level-1 updated VI-List & HI-Counter

(a) B-table

	a b c d e f
t_1	
t_2	0 0 0
t_3	0 1 0
t_4	
t_5	0 0 0
t_6	0 1 0
(c) Level-1 HI-Counter	0 2 0

(b) Updated Level-1 VI-List

column index	row indices
column e (i.e., $\{a,e\}$)	rows t_3, t_6

Fig. 7. B-table for the $\{a\}$-projected database and its Level-1 updated VI-List & HI-Counter

database (as shown in Fig. 7(b)) while both B-table and Level-1 HI-Counter are similar to those shown in Fig. 4(a) and (c) with columns b & c "faded out".

Afterwards, B-mine applies similar steps to find the remaining frequent patterns $\{b,c\}, \{b,c,e\}, \{b,e\}$ and $\{c,e\}$.   □

Observation 5. *Observed from Example 2 that our B-mine algorithm applies a divide-and-conquer (or depth-first search) approach for mining (cf. Apriori [1] applies a levelwise or breadth-first search approach). Unlike FP-growth [3] or TD-FP-Growth [15], our B-mine algorithm only constructs VI-Lists and HI-Counters (which require much less memory than the constructions of sub-trees in FP-growth or TD-FP-Growth).*

3   Application of Our B-mine to Social Network Analysis

In the previous section, we described our B-mine algorithm and applied it for frequent pattern mining. In this section, we apply our B-mine in a real-life application—namely, *social network analysis* [11,21].

Social networks are generally made of social entities (e.g., individuals, corporations, collective social units, or organizations) that are linked by some specific types of interdependency (e.g., kinship, friendship, common interest, beliefs, or financial ex-change). A social entity is connected to another entity as his next-of-kin, friend, collaborator, co-author, classmate, co-worker, team member, and/or business partner. Today, we have various social networking websites or services-such as Facebook, Flickr, Google+, LinkedIn, Twitter, and Weibo [12,18,19]. For instance, Facebook is an online social networking website that allows users to create a personal profile, add other Facebook users.as friends, and exchange messages. In addition, a Facebook user can also join common-interest user groups. Similarly, LinkedIn is a business-oriented social networking service that allows users to create a professional profile, establish connections to other LinkedIn users, and exchange messages. In addition, a LinkedIn user can also join common-interest user groups and tag his or her connections according to some categories, which may or may not be overlapped (e.g., colleagues, classmates, and friends). As a popular online social networking and blogging service, Twitter allows its users to read the tweets of other users by following them. As such Twitter users are linked by "following" relationships (e.g., a user follows another user).

Over the past few years, social computing and its applications have become an emerging research area in the fields of computer science. Numerous works have been presented. Examples include clustering and classification of social media data [17], mining and analysis of co-authorship networks [5], as well as visualization of social networks [4]. Moreover, there have been works on discovery of "following" patterns [6], diverse social entities [8], detection of leaders [11], influential and strong friends [7,14]. In contrast, we apply B-mine in social network analysis for finding *active participants in multiple common-interest user groups*. See Example 3.

Example 3. Let us consider an illustrative portion of a social network. Such a portion consists of $N = 6$ common-interest user groups about $M = 6$ social entities: Group 1 = {Bob, Eva}, Group 2 = {Alice, Carol}, Group 3 = {Alice, Eva}, Group 4 = {Bob, Carol, Eva}, Group 5 = {Alice, Bob, Carol}, and Group 6 = {Alice, Bob, Carol, Eva}. Our B-mine algorithm first builds a B-table and its VI-Lists & HI-Counters as shown in Fig. 8. Then, B-mine finds the following sets of social entities actively participating in at least two common-interest user groups: {Alice}, {Bob}, {Carol}, {Eva}, {Alice, Bob}, {Alice, Carol}, {Alice, Eva}, {Alice, Bob, Carol}, {Bob, Carol}, {Bob, Carol, Eva}, {Bob, Eva} and {Carol, Eva}.   □

		Alice	Bob	Carol	Don	Eva	Frank
	Group 1	0	1	0	0	1	0
(a) B-table	Group 2	1	0	1	0	0	0
	Group 3	1	0	0	0	1	0
	Group 4	0	1	1	0	1	0
	Group 5	1	1	1	0	0	0
	Group 6	1	1	1	0	1	0
(c) Level-0 HI-Counter		4	4	4	0	4	0

(b) Level-0 VI-List

col idx	row indices
{Alice}	Groups 2, 3, 5, 6
{Bob}	Groups 1, 4

Fig. 8. B-table and its Level-0 VI-List & HI-Counter for social network analysis

4   Evaluation

Memory Usage Analysis. We first analyzed the memory space consumption of our B-mine algorithm (and its associated structures) when compared with related works (e.g., FP-growth). Recall that the FP-growth algorithm is the tree-based divide-and-conquer approach to mine frequent itemsets from shopper market basket datasets. The algorithm first builds a global FP-tree to capture important contents of the dataset in the tree. The number of tree nodes is theoretically bounded above by the number of occurrences of all items (say, $O(N_{occurrence})$) in the dataset. Practically, due to tree path sharing, the number of tree nodes (say, $O(N_{tree}) < O(N_{occurrence})$) is usually smaller than the upper bound. However, during the mining process, multiple smaller sub-trees need to be constructed. Specifically, for a global FP-tree with depth L, it is not usual for $O(L)$ sub-trees to coexist with the global tree, for a total of $O(L \times N_{tree})$ nodes. In contrast, on the surface, our B-table may appear to take up more space (due to the lack of tree path sharing). The B-table contains $O(N_{occurrence})$ entries. However, it is important to note that each entry in our B-table is just *a single bit*, instead of an integer for an item ID. In other words, the B-table requires $\frac{O(N_{occurrence})}{8}$ bytes of space. Moreover, unlike FP-growth, we do not need to build any sub-B-table. In other words, the same B-table is used throughout the entire mining process. Thus, our B-table requires $\frac{O(N_{occurrence})}{8}$ bytes $\ll O(L \times N_{tree})$ bytes in FP-growth.

For any interesting frequent patterns of length L, we need to create $O(L)$ VI-Lists and HI-Counters. Note that FP-growth also creates $O(L)$ header tables for the $O(L)$ sub-trees. In other words, the amount of space required by VI-Lists and HI-Counters is similar to that by sub-trees. However, our B-table requires much less space than all FP-trees.

Frequent Pattern Mining Evaluation. We then compared the performance of our B-mine algorithm with related works by using (i) sparse IBM synthetic datasets (with 10 K transactions and 100 K transactions) and (ii) a dense online retail dataset from UCI Machine Learning Repository[1] (with 541,909 transactions). Five existing frequent pattern mining algorithms were used compare with our proposed B-mine algorithm: FP-growth [3], TD-FP-Growth [15], H-mine [10], VIPER [13], and Eclat [20]. All experiments were run in a time-sharing environment in a 1 GHz machine. The reported figures are based on the average of multiple runs. Figure 9 shows the evaluation result of our proposed B-mine against five existing algorithms over different minimum support threshold (*minsup*) values on different datasets. Furthermore, the results on the dense dataset show that our proposed B-mine gave the best performance over other five existing algorithms because our B-mine algorithm takes the advantage of using B-table capture the information in bitwise data structure (especially, for dense dataset). The results on the sparse datasets show that when *minsup* is low, algorithms such as H-mine and VIPER did not perform very well. Eclat showed its advantages

---

[1] https://archive.ics.uci.edu/ml/.

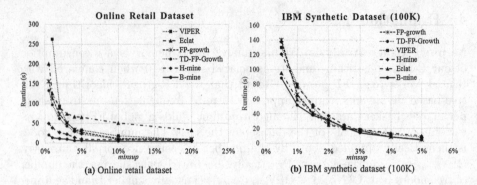

Fig. 9. Experimental results on frequent pattern mining

when handling sparse datasets. Moreover, our proposed B-mine performed better than the other five.

Social Network Mining Evaluation. Recall from Sect. 4 that our B-mine can be applicable to social network analysis. To evaluate this real-life application, we experimented with two datasets—namely, Stanford SNAP Facebook dataset and Stanford SNAP Twitter dataset[2]. The SNAP Facebook dataset contains 4,039 social entities and 88,234 connections between these social entities. The SNAP Twitter dataset contains 81,306 social entities and 1,768,149 connections between these social entities. The results are shown in Fig. 10.

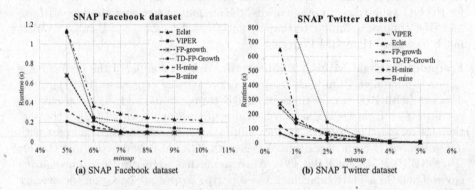

Fig. 10. Experimental results on social network analysis

5 Conclusions

As frequent pattern mining is one of the most important data mining tasks, many algorithms have been proposed over the past decades. While these algorithms are showing their advantages and benefit our research, they also suffer

---

[2] http://snap.stanford.edu/data/.

from a few disadvantages. In this paper, we present an alternative frequent pattern mining algorithm, namely B-mine, which uses (i) a bitwise table (B-table) to capture transactions in databases, (ii) vertical index lists (VI-Lists) to capture information for mining extensions of frequent patterns, and (iii) horizontal index counter (HI-Counters) to compute support values of patterns. Evaluation results analytically show the space-efficiency of our B-table and its associated VI-Lists & HI-Counters during the mining process; they empirically show the time-efficiency of our B-mine algorithm when mining frequent patterns. Moreover, they also practically show the applicability of B-mine in social network analysis for knowledge discovery from social networks.

Acknowledgements. This project is partially supported by NSERC (Canada) and University of Manitoba.

References

1. Aggarwal, R., Srikant, R.: Fast algorithms for mining association rules. In: Proceedings of VLDB, pp. 487–399 (1994)
2. Cuzzocrea, A., Jiang, F., Lee, W., Leung, C.K.: Efficient frequent itemset mining from dense data streams. In: Chen, L., Jia, Y., Sellis, T., Liu, G. (eds.) APWeb 2014. LNCS, vol. 8709, pp. 593–601. Springer, Heidelberg (2014)
3. Han, J., Pei, J., Yin, Y.: Mining frequent patterns without candidate generation. In: Proceedings of ACM SIGMOD, pp. 1–12 (2000)
4. Leung, C.K., Carmichael, C.L.: Exploring social networks: a frequent pattern visualization approach. In: Proceedings of IEEE SocialCom, pp. 419–424 (2010)
5. Leung, C.K.-S., Carmichael, C.L., Teh, E.W.: Visual analytics of social networks: mining and visualizing co-authorship networks. In: Schmorrow, D.D., Fidopiastis, C.M. (eds.) FAC 2011. LNCS (LNAI), vol. 6780, pp. 335–345. Springer, Heidelberg (2011)
6. Leung, C.K., Jiang, F., Pazdor, A.G.M., Peddle, A.M.: Parallel social network mining for interesting 'following' patterns. Concurrency Comput. Pract. Experience (2016). doi:10.1002/cpe.3773
7. Leung, C.K., Tanbeer, S.K., Cameron, J.J.: Interactive discovery of influential friends from social networks. Soc. Netw. Anal. Min. **4**(1), 13 (2014). doi:10.1007/s13278-014-0154-z. Article no. 154
8. Leung, C.K., Tanbeer, S.K., Cuzzocrea, A., Braun, P., MacKinnon, R.K.: Interactive mining of diverse social entities. KES J. **20**(2), 97–111 (2016)
9. Lin, J.C., Gan, W., Fournier-Viger, P., Hong, T.: Mining weighted frequent itemsets with the recency constraint. In: Cheng, R., Cui, B., Zhang, Z., Cai, R., Xu, J. (eds.) APWeb 2015. LNCS, vol. 9313, pp. 635–646. Springer, Heidelberg (2015)
10. Pei, J., Han, J., Lu, H., Nishio, S., Tang, S., Yang, D.: H-mine: hyper-structure mining of frequent patterns in large databases. In: Proceedings of IEEE ICDM, pp. 441–448 (2001)
11. Rahman, Q.M., Fariha, A., Mandal, A., Ahmed, C.F., Leung, C.K.: A sliding window-based algorithm for detecting leaders from social network action streams. In: Proceedings of IEEE/WIC/ACM WI-IAT, vol. 1, pp. 133–136 (2015)

12. Schaal, M., O'Donovan, J., Smyth, B.: An analysis of topical proximity in the Twitter social graph. In: Aberer, K., Flache, A., Jager, W., Liu, L., Tang, J., Guéret, C. (eds.) SocInfo 2012. LNCS, vol. 7710, pp. 232–245. Springer, Heidelberg (2012)

13. Shenoy, P., Bhalotia, J.R., Bawa, M., Shah, D.: Turbo-charging vertical mining of large databases. In: Proceedings of ACM SIGMOD, pp. 22–33 (2000)

14. Tanbeer, S.K., Leung, C.K., Cameron, J.J.: Interactive mining of strong friends from social networks and its applications in e-commerce. J. Organ. Comput. Electron. Commer. (JOCEC) **24**(2–3), 157–173 (2014)

15. Wang, K., Tang, L., Han, J., Liu, J.: Top down FP-growth for association rule mining. In: Chen, M.-S., Yu, P.S., Liu, B. (eds.) PAKDD 2002. LNCS (LNAI), vol. 2336, pp. 334–340. Springer, Heidelberg (2002)

16. Tong, W., Leung, C.K., Liu, D., Yu, J.: Probabilistic frequent pattern mining by PUH-Mine. In: Cheng, R., Cui, B., Zhang, Z., Cai, R., Xu, J. (eds.) APWeb 2015. LNCS, vol. 9313, pp. 768–780. Springer, Heidelberg (2015)

17. Xu, H., Yang, Y., Wang, L., Liu, W.: Node classification in social network via a factor graph model. In: Pei, J., Tseng, V.S., Cao, L., Motoda, H., Xu, G. (eds.) PAKDD 2013, Part I. LNCS (LNAI), vol. 7818, pp. 213–224. Springer, Heidelberg (2013)

18. Yang, X., Ghoting, A., Ruan, Y., Parthasarathy, S.: A framework for summarizing and analyzing Twitter feeds. In: Proceedings of ACM KDD, pp. 370–378 (2012)

19. Yuan, Q., Cong, G., Ma, Z., Sun, A., Magnenat-Thalmann, N.: Who, where, when and what: discover spatio-temporal topics for Twitter users. In: Proceedings of ACM KDD, pp. 605–613 (2013)

20. Zaki, M.J.: Scalable algorithms for association mining. IEEE TKDE **12**(3), 372–390 (2000)

21. Zhang, Y., Pang, J.: Distance and friendship: a distance-based model for link prediction in social networks. In: Cheng, R., Cui, B., Zhang, Z., Cai, R., Xu, J. (eds.) APWeb 2015. LNCS, vol. 9313, pp. 55–66. Springer, Heidelberg (2015)

An Efficient Online Event Detection Method for Microblogs via User Modeling

Weijing Huang[1], Wei Chen[1(✉)], Lamei Zhang[2], and Tengjiao Wang[1]

[1] Key Laboratory of High Confidence Software Technologies (Ministry of Education),
EECS, Peking University, Beijing, China
huangwaleking@gmail.com, {pekingchenwei,tjwang}@pku.edu.cn
[2] Baidu Inc., Beijing, China
citlmzhang@163.com

Abstract. Detecting events in microblog is important but still challenging. As tweet stream is a mixture of user interests and external events, its expensive to distinguish them. Existing methods are ineffective since they ignore user interests or only model interests and events on a fixed dataset without scalability. In this paper, we introduce an online learning model User Modeling Based Interest and Event Topic Model (UMIETM). UMIETM (1) exploits user modeling's information to discover events, which usually capture attentions from users with different interests, and (2) treats the arriving data as stream and run the detection in online learning style. Furthermore, UMIETM can handle dynamic increased vocabulary in tweet stream. The UMIETM is verified on the real dataset which spans one year and contains 16 million tweets, and it outperforms state-of-the-art models in quantitative.

Keywords: Event detection · Online learning model · Microblog stream · User modeling

1 Introduction

Detecting events in microblog is very important as the microblog has become one of the most popular sites for users to publish and get recent news. However it is still a challenging task, for the reason that tweets are (1) in large scale and (2) in a mixture of user interests and external events. Users post tweets not only for reporting breaking news, but also talking about interest related affairs. Existing methods are insufficient in this scenario, such as [1] which is ineffective when ignoring user interests, and [2] which lacks scalability when considering user interests but on a fixed dataset.

In another way, user profile or user description is vital in social media. And the existing event detection lacks the exploiting of them. Users describe themselves in profiles to show their interests, locations, occupations or identifications.

This research is supported by the Natural Science Foundation of China (Grant No. 61300003, 61572043), and the Specialized Research Fund for the Doctoral Program of Higher Education (Grant No. 20130001120001).

© Springer International Publishing Switzerland 2016
F. Li et al. (Eds.): APWeb 2016, Part I, LNCS 9931, pp. 329–341, 2016.
DOI: 10.1007/978-3-319-45814-4_27

These factors are stable and the high generalization of user's characteristics. One example is Biz Stone[1] whose profile is *Co-founder of Twitter, Medium, and now Co-founder and CEO of Askjelly.com.* We read his recent 100 tweets, and found 7 tweets talking about Twitter and 36 about Jelly. This example suggests that user profile is stable to reflect user's interests. As another example, the profile of a Chinese microblog user @hadoopchina[2] is *#Cloud, #YARN, #Spark, #Big_Data, #Hadoop.* He mainly tweets IT news until external events happen, such as celebrating spring festival. This suggests profile can help to distinguish events from user interest-related tweets. Although some accounts do not have the accurate self-descriptions, they can be augmented by gathering their followings' self-descriptions. And the user profile information can be utilized to model users' long term interests.

In this paper, we introduce an online learning model, the *User Modeling Based Interest and Event Topic Model* (UMIETM) to discover the bursty events, taking users' profiles into consideration which indicate long term interests. In UMIETM, the bursty events can be better confirmed when the events receive the attentions from users with different interests.

It's also very interesting when we treat the event detection in microblogs as a kind of crowd sourcing service from different twitterers. As the old saying "Birds of a feather fly together", there must be something happened when all birds fly. We take user profiles as the strong signal to indicate user's behavior or interests. When different twitterers with different interests post or retweet the same thing, they are telling us that we should pay attention to what they have paid attention to. We take this intuition into our event detection in microblogs.

The main contributions in our paper can be summarized as follows.

1. The proposed model UMIETM exploits user modeling's information to discover bursty events, which usually captures attention from users with different interests.
2. The UMIETM is further scaled up to treat the arriving data as stream and run the event detection in online learning style, and is able to handle the dynamic increased vocabulary in tweet stream.
3. We conduct the experiment on the real microblog dataset which spans one year and contains 16 million tweets, and verify that the better user interest modeling can lead to better performance of event detection.

2   Related Work

Event Detection. Study of event detection on text stream can be divided into three ways: word frequency based, text similarity based and topic model based.

Several word frequency methods have been developed for event detection such as Discrete Fourier Transform [3], wavelet analysis [4]. They treat the word's document frequency along timeline as time series and do the analysis in frequency

---

[1] https://twitter.com/biz.

[2] http://weibo.com/hadoopchina.

domain. DFT method suffers the problem that it can not locate the time point for bursty. The complexity of wavelet analysis based method used by EDCoW [4] is very high, so its scalability is limited.

Text similarity based online event detection methods [5,6] suffer from the lexical variation which means different words describe the same events. Similarity based method can successfully detect the tweet which is retweeted by many times, but fails to find out the event which is described from many different perspectives. As a result, many events are duplicately detected due to their popularities, which may bury other events and overwhelm users with duplicated unwanted content.

In contrast, topic model can handle the lexical variation problem with word co-occurrence [7]. As many events are highly related to topics, a number of methods based on topic model have been proposed for event detection, including online detection and offline detection. Lau [1] introduces an online topic model to track emerging events in microblogs. It can deal with a massive of tweets, but it doesn't filter out the tweets related to user interests. Diao [2,8] show that event detection can benefit from filtering out user interest related tweets. And Yan [9] models the bursty topic by incorporating burstiness of biterms as prior knowledge. But they are different from ours. As these models need the whole dataset as the input, they are offline detections, which is not scalability for large dynamic dataset such as micorblogs. Instead, we gather user's self-description and the followings' self-description as user profile which is stable to characterize user interests. Based on this fine-grained user modeling information, user's long term interest related tweets and short term bursty event related tweets can be distinguished in online style, and can be efficiently applied on microblogs.

User Modeling. Since user modeling has the significant impact on users' activities on microblogs, the usage of user modeling has drawn attention in many research fields. [10] exploits user modeling information and network topology to infer user's role in social network. [11] also introduces user modeling to community detection where some edges are not observable, but user profile can provide additional information. User modeling can be carried out to obtain web users' demographics information [12] (e.g., gender, age, ethnicity, education, income, etc.). Compared with demographics information, user interests' modeling can be verified more easily [13]. Different from the above work, we do not only treat the user modeling result as an additional feature, but also an important representative factor for users, who are the source of information in microblogs.

3   Method

3.1   Problem Formulation

User Profile. User usually describes herself or himself by a piece of short text on microblog platform. This short text can be a continuous string or an array of hash tags, e.g., "*Co-founder of Twitter, Medium, and now Co-founder and CEO of Askjelly.com*" of Biz Stone or "*#Cloud, #YARN*" of @hadoopchina. Though it's easy to estimate Biz Stone's interests by his self-description text, it's not always capable of doing so because some users' are very short. To overcome this limitation, we define **user profile** p_u as combining user u's self-description text with the texts provided by u's followings. Taking Biz Stone as an example, he follows 696 accounts, in which there are 60 founders, 27 CEOs, 21 Google related, and 9 medium related accounts, etc. This example demonstrates that user profile p_u can be augmented by gathering the followings' information.

User Modeling. User modeling is used to capturing user's long term interests. For example, Biz Stone's long term interests can be inferred as *Social Media*, *Business*, and *Technology* from his user profile *Medium*, *CEO*, and *Google* respectively.

The notations used in this paper are summarized in Table 1(a). We consider u's user timeline as the triple $\{uid, p_u, w_u\}$, where p_u represents user u's profile and $w_u = \{(tweetid, t_{ud}, w_{ud})\}$ means the set of tweets posted by user u. The element of w_u is a triple of tweet id, time stamp t_{ud} and tweet content w_{ud}.

Event. We define the event in the given time window t as the set of tweets denoted by $\{w_{te}\}$. The event related tweets in set $\{w_{te}\}$ hold two properties: (1) they are posted by users with different interests (2) they are similar within the set. The task of event detection is to find out all the events in corpus. Different methods treat the above properties in different ways: LSH based methods [5] treat the difference and similarity in word vector space; while the topic model based methods [2] take the semantic meaning into consideration. Under the framework of topic model, we divide the topics into user interest related topics and bursty event related topics, and further promote an online learning model UMIETM.

3.2   Model Descriptions

UMIETM is motivated by two observations: (I)user profiles are more stable to reflect users' interests than tweets; and (II)external events draw global attention in short time.

To capture Observation I, we enhance the association between profile topic and tweet topic which is inspired by [14]. More particularly, the model generates observed data for user u in two phases as shown in Fig. 1. The first phase generates the hidden topic s_{un} from user interest topic distribution θ_u, then generates profile token p_{un}. When $y_{ud} = 0$, the second phase generates tweet topic z_{ud} from associated profile topics $\{s_{u1}, ..., s_{un}\}$ uniformly. To overcome the deficiency of user profile words, we add the smoothing factor κ for each topic to s_{un}. The second phase makes a closer correlation between profile topics and tweet topics.

For Observation II, we introduce time dependent event distribution η_t and switcher y_{ud} to distinguish users long term interests from short term responses to external events. Only if switcher $y_{ud} = 1$, tweet topic z_{ud} will be sampled from multinomial event distribution η_t.

We introduce UMIETM's generative process and leave the detail of online learning to Sect. 3.4. We assume that there are K latent topics corresponding

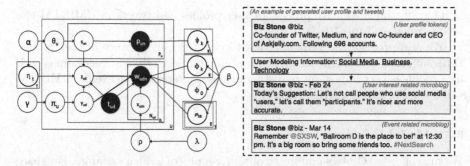

Fig. 1. Illustration of UMIETM (left), and an example user Biz Stone's tweets about his long term interests and bursty event (SXSW is a music, film and interactive conference and festival hosted on March 11–20) (right).

to all users' interests in corpus. ψ_k and ϕ_k are profile word distribution and tweet word distribution on k-th interest topic respectively. We also assume a background word distribution ϕ_0 to filter out common words. We set E events as word distributions $\{\varphi_{te}\}$ in each time window t.

Table 1. UMIETM's Notations and Generative Process

(a) Notations

T	number of time windows
E	number of event-related topics in each time window
K	number of interest-related topics
U	number of users
P_u	number of tokens appeared in user u's profile
D_u	number of tweets published by user u
α, β	priors of Dirichlet distributions
θ_u	K dimension vector indicating user u's interest distribution
p_{un}	user u's n-th profile token
s_{un}	the hidden topic of user u's n-th profile token
ψ_k	the user profile token's distribution on k-th profile topic
η_t	the events' distribution in time window t
π_u	the preference of user u to participate the discussion of global events
y_{ud}	the type of user u's d-th tweet ($y_{ud} = 0$ indicates interest-related, $y_{ud} = 1$ event-related)
t_{ud}	the discrete value between 1 and T indicating the timestamp of user u's d-th tweet
z_{ud}	the topic of user u's d-th tweet , $z_{ud} \in \{1,...K\}$ if $y_{ud} = 0$; $z_{ud} \in \{1,...,E\}$ if $y_{ud} = 1$
ϕ_k	tweet token's distribution on k-th interest-related topic
ϕ_{te}	tweet token's distribution on e-th event-related topic in time window t
x_{udn}	the boolean indicator of n-th token in user u's d-th tweet: background word if $x_{udn} = 0$; non-background word if $x_{udn} = 1$
w_{udn}	the n-th token in user u's d-th tweet: may chosen from ϕ_0, ϕ_k or ϕ_{te}
ρ	the Bernoulli parameter for boolean indicator x_{udn}
ϕ_0	tweet token's distribution on background topic
γ, λ	the priors for Bernoulli distributions

(b) Generative Process

$\rho \sim Beta(\lambda)$, $\phi_0 \sim Dir(\beta)$
for k=1 to K:
  $\psi_k \sim Dir(\beta)$, $\phi_k \sim Dir(\beta)$
for t=1 to number of time windows:
  $\eta_t \sim Dir(\alpha)$
  for e=1 to E:
    $\varphi_{te} \sim Dir(\beta)$
for each user u:
  $\pi_u \sim Beta(\gamma)$
  user interest $\theta_u \sim Dir(\alpha)$
  for n=1 to number of u's profile tokens:
    $s_{un} \sim Multinomial(\theta_u)$
    $p_{un} \sim Multinomial(\psi_{s_{un}})$
  for tweet d=1 to D_u:
    $y_{ud} \sim Multinomial(\pi_u)$
    if y_{ud}=0:
      $z_{ud} \sim Uniform(\{s_{u1},...,s_{un}\})$
    else:
      $z_{ud} \sim Multinomial(\eta_{t_{ud}})$
    for n=1 to N_{ud}:
      $x_{udn} \sim Bernoulli(\rho)$
      if x_{udn}=0:
        $w_{udn} \sim Multinomial(\phi_0)$
      else:
        if y_{ud}=0:
          $w_{udn} \sim Multinomial(\phi_{z_{ud}})$
        else:
          $w_{udn} \sim Multinomial(\phi_{t_{ud}.z_{ud}})$

Overall, the generative process of user profiles and tweets in UMIETM can be described as Table 1(b).

We also propose the variant IETM to check the significance of user profile. Different from UMIETM, IETM models user interest related tweets directly. If user profile is very important to distinguish user interests from events, UMIETM will outperform IETM, vice versa.

3.3   Model Inference

We run collapsed Gibbs sampling to obtain samples of hidden variables. For space limit, we omit the detail of inference and only list the conditional distribution of each hidden variable. As there is a coupling between profile topics and tweet topics, the Gibbs sampling should be divided into two phases, first for profiles, second for tweets.

In first inference phase (shown in Algorithm 1 line 2 to line 6) we sample user profile's hidden topic s_{un} as standard LDA's collapsed Gibbs sampling [15], where $c_{uk}^{(p)}$ is the number of user u's profile words assigned to topic k, $c_{u,.}^{(p)}$ is the total number of profile words of u and $c_{kv}^{(p)}$ is the times of profile word v assigned to topic k.

$$p(s_{un} = k | s_{\neg un}, \boldsymbol{p}, \alpha, \beta) \propto \frac{c_{uk}^{(p)} + \alpha}{c_{u,.}^{(p)} + K\alpha} \frac{c_{kv}^{(p)} + \beta}{c_{k,.}^{(p)} + V\beta} \tag{1}$$

After the convergence of the first Gibbs sampling phase, we start the second (shown in Algorithm 1 line 7 to line 21). We joint sample for y_{ud} and z_{ud} using Eqs. (2) and (3) where $c_{kv}^{(0)}$ is the number of tweet word v assigned to k-th interest topic, and $n_{kv}^{(0)}$ is the number of word v in d-th tweet assigned to k-th interest topic. c_u^0 and c_u^1 denote the number of user u's tweets labeled as interest related and event related respectively. In order to avoid the deficiency of user profile tokens, we add smoothing parameter κ to $c_{uk}^{(p)}$ in Eq. (2).

$$p(y_{ud} = 0, z_{ud} = k | \boldsymbol{y}_{\neg ud}, \boldsymbol{z}_{\neg ud}, \boldsymbol{t}, \boldsymbol{w}, \boldsymbol{s}, \alpha, \beta, \gamma)$$

$$\propto \frac{c_u^0 + \gamma}{c_u^1 + c_u^0 + 2\gamma} \frac{c_{uk}^{(p)} + \kappa}{c_{u,.}^{(p)} + K\kappa} \frac{\prod_{v=1}^{V} \prod_{b=0}^{n_{kv}^{(0)} - 1} (c_{kv}^{(0)} + \beta + b)}{\prod_{b=0}^{n_{k,.}^{(0)} - 1} (c_{k,.}^{(0)} + V\beta + b)} \tag{2}$$

$$p(y_{ud} = 1, z_{ud} = e | \boldsymbol{y}_{\neg ud}, \boldsymbol{z}_{\neg ud}, \boldsymbol{t}, \boldsymbol{w}, \boldsymbol{s}, \alpha, \beta, \gamma)$$

$$\propto \frac{c_u^1 + \gamma}{c_u^1 + c_u^0 + 2\gamma} \frac{c_{t,e}^{(1)} + \alpha}{c_{t,.}^{(1)} + E\alpha} \frac{\prod_{v=1}^{V} \prod_{b=0}^{n_{tev}^{(1)} - 1} (c_{tev}^{(1)} + \beta + b)}{\prod_{b=0}^{n_{te,.}^{(1)} - 1} (c_{t,e,.}^{(1)} + V\beta + b)} \tag{3}$$

Finally we filter out common words from semantic meaningful words by Eqs. (4) and (5). Here the hidden variable $x_{udn} = 0$ indicates that w_{udn} is a common word. $c_v^{(B)}$ is the times of tweet word v assigned to the background topic. M_0^ρ is the total number of common words in corpus and $M_0^\rho + M_1^\rho$ equals the total number of tokens.

$$p(x_{udn} = 0 | x_{\neg und}, w_{udn} = v, w_{\neg udn}, \boldsymbol{y}, \boldsymbol{z}, \boldsymbol{t}, \alpha, \beta, \gamma, \lambda)$$

$$\propto \frac{M_0^p + \lambda}{M_0^p + M_1^p + 2\lambda} \frac{c_v^{(B)} + \beta}{\sum_{v=1}^V c_v^{(B)} + V\beta} \tag{4}$$

$$p(x_{udn} = 1 | x_{\neg und}, w_{udn} = v, w_{\neg udn}, \boldsymbol{y}, \boldsymbol{z}, \boldsymbol{t}, \alpha, \beta, \gamma, \lambda)$$

$$\propto \frac{M_1^p + \lambda}{M_0^p + M_1^p + 2\lambda} \left(\frac{c_{k,v}^{(0)} + \beta}{c_{k,.}^{(0)} + V\beta} \right)^{I(y_{ud}=0)} \left(\frac{c_{t,e,v}^{(1)} + \beta}{c_{t,e,.}^{(1)} + V\beta} \right)^{I(y_{ud}=1)} \tag{5}$$

3.4   Online Learning on Tweet Stream

Gibbs Sampling on fixed large dataset is very expensive both in memory and time. Each Gibbs sweep need to maintain 12 statistics such as $c_{k,v}^{(p)}$, $c_{u,k}^{(p)}$ appeared in Eqs. (1) to (5). Generally, the complexity of the Gibbs sampling is $O(I_1 K |P| + I_2(K+E)|W|)$ where I_1, I_2 are iteration numbers for profiles and tweet tokens, K is topic number, E is event number, $|P|$ is number of profile tokens and $|W|$ is number of tweet tokens. More important, we have to maintain all tweets in memory for batch learning but it is unacceptable.

We propose the online learning method shown in Algorithm 2, and denote the previous learned model as $\mathcal{M}$, previous trained users $\mathcal{U}$. We update $\mathcal{M}$ increasingly by tweets in each time window.

There are two tricks in our online method. The first one is line 6 and line 8 in Algorithm 2 which runs the batch sampling for all users' profile. User profile can reflect user's interests better than tweets and the user profile token number $|P|$ is much smaller than $|W|$. The second one is line 10 which run the sampling only on the tweets in current time window. It can distinguish interest-related tweets by user profile, and detect event-related tweets. Overall, the complexity of online learning is reduced to $O(I_1 K |P| + I_2(K+E)|W_t|)$, where $|W_t|$ is number of tweet tokens appeared in time window t, which is significantly smaller than $|W|$.

Finally, we handle the dynamic increased vocabulary in line 4. Take $\phi_{k,v}$ as an example, when we meet the word v unseen, $\phi_{k,v}$ in current time window can be initially estimated as $\beta/(c_{k,.}^{(0)} + V'\beta)$, and other words appeared in previous time windows can be estimated as $(c_{k,v}^{(0)} + \beta)/(c_{k,.}^{(0)} + V'\beta)$. And V' is the size of increased vocabulary.

4   Experiment

Here we present the effectiveness of our proposed algorithm UMIETM and the efficiency of its online performance. We evaluate the effectiveness by perplexity, precision for event detection. We check the time cost and complete likelihood for efficiency.

Algorithm 1. UMIETM batch learning algorithm

1   initiate the topic label and the statistics
2   **for** $i = 1 : I_1$ **do**
3   **for** u *in user set* U **do**
4   **for** $n = 1 : P_u$ **do**
5   sample profile's hidden topic s_{un} by (1)
6   update s_{un}, $c_{u,k}^{(p)}$ and $c_{k,v}^{(p)}$
7   **for** *iteration* $i = 1 : I_2$ **do**
8   **for** $t = 1 : T$ **do**
9   **for** u *in user set* U_t **do**
10   **for** $d = 1 : D_u$ **do**
11   sample y_{ud} and z_{ud} by (2), (3)
12   **if** $y_{ud} = 0$ **then**
13   update z_{ud}, y_{ud}, $c_u^{(0)}$, $c_{u,k}^{(0)}$, $c_{k,v}^{(0)}$
14   **else**
15   update z_{ud}, y_{ud}, $c_u^{(1)}$, $c_{t,k}^{(1)}$, $c_{t,k,v}^{(1)}$
16   **for** n *in* $1, \cdots, N_{ud}$ **do**
17   sample x_{udn} by (4), (5)
18   **if** $x_{udn} = 0$ **then**
19   update x_{udn}, M_0^ρ, $c_v^{(B)}$
20   **else**
21   update x_{udn}, M_1^ρ, $c_{k,v}^{(0)}$, $c_{t,k,v}^{(1)}$

Algorithm 2. UMIETM online learning algorithm

1   **for** all $u \in \mathcal{U}$, load $\boldsymbol{p_u}$ and $\boldsymbol{w_u}$
2   **for** all $u \in \mathcal{U}$, k, v, load M_0^ρ, M_1^ρ, $c_{u,k}^{(p)}$, $c_{k,v}^{(p)}$, $c_u^{(0)}$, $c_u^{(1)}$, $c_{u,k}^{(0)}$, $c_{k,v}^{(0)}$, $c_v^{(B)}$ from trained Model $\mathcal{M}$.
3   **for** $t = 1 : T$ **do**
4   update the vocabulary for profile and tweet
5   **for** *iteration* $i = 1 : I_1$ **do**
6   **for** u *in user set* $\mathcal{U} \cup U_t$ **do**
7   do operation as line 5 to line 6 in Algorithm 1
8     $\mathcal{U} = \mathcal{U} \cup U_t$
9   **for** *iteration* $i = 1 : I_2$ **do**
10   **for** u *in user set* U_t **do**
11   do operation as line 10 to line 21 in Algorithm 1.

Weibo is a popular Chinese microblogging service[3]. We crawl weibo data by its public API[4] from Jan 2012 to Dec 2012. To improve the quality of analyzing on tweets, we do necessary pre-processing: (1) splitting dataset by week, (2) segmenting Chinese words, (3) removing stop words and low frequency words whose document frequency in its time window is less than 3, (4) removing tweets whose token number is less than 3. To model user interests better, we remove users from dataset who has less than 2 hashtags in profile. After pre-processing we get 252 thousand users, 16 million tweets and 251 million tweet tokens listed in Table 2(a).

We compare our model UMIETM with twitterLDA [16], timeUserLDA [2], Author-LDA (aggregating tweets into long pseudo-document as [17]), and our model's variant IETM (mentioned in Subsect. 3.2). Author-LDA combines the tweets posted by same author into a single document, then run standard LDA on the assembled tweets. TwitterLDA is designed for topic modeling on twitter. We compare with Author-LDA and TwitterLDA for confirming the significance of distinguishing user interests from events in microblog. TimeUserLDA [2] is designed for retrospective event detection in microblog, and considers to distinguish events from user interests. We compare with timeUserLDA to show the impact of user profile. The variant IETM (*Interest and Event Topic Model*) models user's interests and events without the help of user profiles.

---

[3] http://en.wikipedia.org/wiki/Sina_Weibo/.
[4] http://open.weibo.com/.

Table 2. Statistics of processed dataset

(a) Weibo dataset

	#user	#tweet
whole year	252,369	16,421,167
week1	9,785	31,503
week2	29,721	242,554
week3	30,891	254,698
...	...	...

(b) Metrics of event detection

	precision	recall
UMIETM	0.894	0.913
UMIETM(-)	0.847	0.697
IETM	0.824	0.536
LSH[5]	0.394	0.913
EDCoW[4]	0.731	0.435

(c) Held out perplexity

Author-LDA	twitterLDA	timeUserLDA	IETM	UMIETM
20422.25	6027.47	4810.92	3926.76	3107.83

We set the asymmetric α and symmetric $\beta = 0.01$ for UMIETM, where α will be optimized by Gibbs EM algorithm [18]. $\alpha = 0.1$, $\beta = 0.01$ are set for all remaining models. After cross validation we find that UMIETM and IETM perform best on $K = 90$ and $E = 30$, $\kappa = 0.01$. To compare equally, we set the same topic number for Author-LDA, twitterLDA, timeUserLDA.

To verify the role of user profiles played, we set UMIETM(-) as the degradation of UMIETM, which take symmetric prior α for user profile topic inference. We further choose the event topics from the set of $\{\phi_{te}\}$ learned by UMIETM as detected event. The selection criteria is that whether y_{ud} is stable for the tweets that $z_{ud} = e$ after burn-in period (a.k.a., after 5 gibbs sweeps in the experiment).

4.1 Effectiveness

In this subsection, we illustrate the performance of UMIETM in which user profile is considered.

Quantitative Measure. We initialize UMIETM and IETM by batch learning on data from first week to third week, then run them in online learning way from fourth week to ninth week. On each week we calculate their perplexities [19], where $perplexity(D_{test}) = \exp\{-\frac{\sum_{u=1}^{U}\sum_{d=1}^{D_u}\log p(w_{ud})}{\sum_{u=1}^{U}\sum_{d=1}^{D_u}N_{ud}}\}$ and $p(w_{ud}) = (1 - \pi_u)\sum_{k=1}^{K}\theta_{uk}\prod_{n=1}^{N_{ud}}(\phi_{s,w_{udn}}(1-\rho) + \phi_{k,w_{udn}}\rho) + \pi_u\sum_{k=1}^{K}\eta_{t,k}\prod_{n=1}^{N_{ud}}(\phi_{s,w_{udn}}(1-\rho) + \phi_{k,w_{udn}}\rho)$. The other models are also trained from first week to third week, and the held out perplexities are calculated on data from fourth week to ninth week. TimeUserLDA and IETM's perplexities are smaller than Author-LDA, twitterLDA as they both consider distinguishing user interests from events.

In Table 2(c) the perplexity of UMIETM is 3107.8, and much smaller than others. It demonstrates that user profile is significant for tweet stream's modeling.

In Table 2(b), we evaluate the events detected by models. As mentioned in the related work section, there are mainly three types of methods to detect events. We compare UMIETM with the text similarity based method LSH [5], word frequency based method EDCoW [4]. As existing topic model based methods [2,8,9] do not scale well on tweet stream, they are not listed in Table 2(b). LSH [5] detects events as the cluster of tweets, and use the cosine similarity without considering semantic meaning of words. So LSH based method may split the same event into clusters of similar tweets and generate duplicate events. We asked the annotators to label the event with score 1, and non-event related topic as 0. The precision and recall is illustrated in Table 2(b), where UMIETM(-) performs slightly better than IETM. A reasonable analysis is that UMIETM uses the profile information sufficiently by asymmetric priors [20]. In this way, we also prove that the well exploiting of user profile information is important to model user interests and events on tweets.

Case Study. Some events detected by our UMIETM model are shown in Table 3. Comparing with UMIETM, timeUserLDA fails to discover the *shoddy construction* event in the second week, while IETM reports this event as *bi, women, elegant, adoption, engineering, reed*. Obviously IETM fails to distinguish this event from user interests. UMIETM filters users' interests like *bi, women, elegant, adoption* using their profile #baby, #women, and detect the event.

Table 3. Example Events detected by UMIETM

Time window	Top words of example events	Example events
The first week of 2012	Japan, earthquake, occur, the first day, January, 7.0, 2012	In January 1 of 2012, a magnitude-7 earthquake occurred in Japan.
	New, year, happy, 2012, New Years Day, healthy, blessing, happiness	Everyone blesses happy new year in the first day of 2012
The second week of 2012	Reed, steel, appearance, engineering, shoddy construction, criminal	In an accident, a car crashed through the guardrail into the river. People found that, the guardrail was built with reed which should be built with steel bar.
The third week of 2012	Apple, 4S, iPhone, line up, scalper, Sanlitun	Many scalpers lined up to buy the Apple iPhone 4S when it started to sell at Sanlitun Apple store in January 13th

4.2 Efficiency

We implement our methods on mallet[5], and run them on Linux server with 8 cores(2.00 GHz) and 64 GB memory. In this subsection, we verify the convergence and online performance of our online learning algorithm.

Convergence. The convergence of algorithm is vital in real time streaming environment. UMIETM needs several Gibbs sweeps to find out the optimal parameters of model. But the size of data is usually large even spitted into time windows. The time complexity of UMIETM is $O(I_1 K|P| + I_2 K|W| + I_2 E|W|)$ given in Sect. 3.4. It suggests that the more iterations Gibbs sampling have to do, the less efficiently our model performs. Fortunately UMIETM is very economic to converge to its stable state and does not need so much iterations. As suggested by [20], we choose LDA with asymmetric priors as baseline, which is much more comparable than standard LDA on microblog dataset. We run UMIETM on the first three time windows. Correspondingly we only use the tweets in these time windows for LDA with asymmetric priors. Burn-in period is set to 20 and optimize interval also 20, which means the priors are optimized every 20 iterations as the red curve in Fig. 2(a). One of the criteria for stopping Gibbs sampling is that the complete log likelihood is stabilized. We stop UMIETM after 50 iterations since the complete log likelihood only increases 0.02 % in last iteration. It will take LDA(asymmetric priors) 300 min to run 1000 iterations, and 20 min per 10 iterations for UMIETM. There is a trade-off between the effectiveness by more iterations and efficiency by less, we set the iteration number as 10 in online setting. After 10 iterations the UMIETM's complete log likelihood will increase no more than 0.2 %.

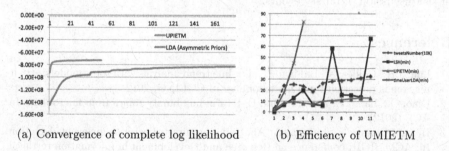

(a) Convergence of complete log likelihood   (b) Efficiency of UMIETM

Fig. 2. (a) The convergence of complete log likelihood of UMIETM and LDA(with asymmetric priors). The x-axis represents the round of iteration and y-axis shows the complete log likelihood. (b) Efficiency of UMIETM. The x-axis represents the time window, and y-axis shows the duration of processing or the number of tweets in each corresponding time window.

---

[5] http://mallet.cs.umass.edu/dist/mallet-2.0.7.tar.gz.

Online Performance. As verified in the above subsection *Convergence*, we run 10 Gibbs sweeps for each time window in online learning phase. When UMIETM is implemented in online way, it doesn't have to revisit the tweets in previous windows. So its time cost in each window is proportional to the data size of current window as shown in Fig. 2(b). In average, online UMIETM can process five thousand tokens per second.

One problem of LSH based event detection method is that the bad design may lead to unstable performance. The tweets are mapped into LSH defined space in a skew way which is illustrated in Fig. 2(b). The other comparison is timeUserLDA [2] which is designed for retrospective event detection. The purple line in Fig. 2(b) goes up straightly as timeUserLDA has to revisit previous tweets to make a decision whether the tweet is event related. The perplexity of online UMIETM on the 11 time windows in Fig. 2(b) is 3036.44 ± 397.14. This indicator demonstrates that UMIETM can perform well in online learning way.

5   Conclusions

Microblog is mixed with user interests and external events. And the external bursty events can be better detected by well user modeling, as the bursty events usually capture the attention from users with different interests. Since user profile can help to identify user's long term interest, our proposed model UMIETM exploit this vital information into user modeling and event detection. We further treat the arriving data as stream and run the detection in online learning style. The experiments demonstrate that our method is effective and efficient for online event detection in microblogs. As future work, it would be interesting to discover which groups with specific long term interests are more positive to participate in the discussion of bursty events.

References

1. Lau, J.H., Collier, N., Baldwin, T.: On-line trend analysis with topic models: #twitter trends detection topic model online. In COLING, 2012
2. Diao, Q., Jiang, J., Zhu, F., Lim, E.-P.: Finding bursty topics from microblogs. In: ACL (2012)
3. He, Q., Chang, K., Lim, E.-P.: Analyzing feature trajectories for event detection. In: ACM SIGIR conference on Research and Development in information retrieval, pp. 207–214. ACM (2007)
4. Weng, J., Lee, B.-S.: Event detection in twitter. In: ICWSM (2011)
5. Petrović, S., Osborne, M., Lavrenko, V.: Streaming first story detection with application to twitter. In: HLT-NAACL (2010)
6. McCreadie, R., Macdonald, C., Ounis, I., Osborne, M., Petrovic, S.: Scalable distributed event detection for twitter (2013)
7. Blei, D.M., Ng, A.Y., Jordan, M.I.: Latent dirichlet allocation. JMLR (2003)
8. Diao, Q., Jiang, J.: A unified model for topics, events and users on twitter. In: EMNLP (2013)

9. Yan, X., Guo, J., Lan, Y., Jun, X., Cheng, X.: A Probabilistic Model for Bursty Topic Discovery in Microblogs. AAAI, pp. 353–359 (2015)
10. Zhao, Y., Wang, G., Yu, P.S., Liu, S., Zhang, S.: Simon: inferring social roles and statuses in social networks. In: SIGKDD. ACM (2013)
11. Yoshida, T.: Toward finding hidden communities based on user profile. J. Intell. Inf. Syst. **40**(2), 189–209 (2013)
12. Culotta, A., Kumar, N.R., Cutler, J.: Predicting the demographics of twitter users from website traffic data. In: AAAI, pp. 72–78 (2015)
13. Faralli, S., Stilo, G., Velardi, P.: Large scale homophily analysis in twitter using a twixonomy. In: IJCAI, pp. 2334–2340. AAAI Press (2015)
14. Blei, D.M., Jordan, M.I.: Modeling annotated data. In: SIGIR (2003)
15. Griffiths, T.L., Steyvers, M.: Finding scientific topics. PNAS (2004)
16. Zhao, W.X., Jiang, J., Weng, J., He, J., Lim, E.-P., Yan, H., Li, X.: Comparing twitter and traditional media using topic models. In: Clough, P., Foley, C., Gurrin, C., Jones, G.J.F., Kraaij, W., Lee, H., Mudoch, V. (eds.) ECIR 2011. LNCS, vol. 6611, pp. 338–349. Springer, Heidelberg (2011)
17. Quan, X., Kit, C., Ge, Y., Pan, S.J.: Short and sparse text topic modeling via self-aggregation. IJCAI, pp. 2270–2276 (2015)
18. Wallach, H.M.: Structured topic models for language. Doctoral dissertation, Univ. of Cambridge (2008)
19. Wallach, H.M., Murray, I., Salakhutdinov, R., Mimno, D.: Evaluation methods for topic models. In: ICML (2009)
20. Wallach, H.M., Mimno, D.M., McCallum, A.: Rethinking lda: Why priors matter. In: NIPS, vol. 22, pp. 1973–1981 (2009)

Man-O-Meter: Modeling and Assessing the Evolution of Language Usage of Individuals on Microblogs

Kuntal Dey[1,2(✉)], Saroj Kaushik[1], Hemank Lamba[3], and Seema Nagar[2]

[1] Indian Institute of Technology (IIT), Delhi, India
saroj@cse.iitd.ac.in
[2] IBM Research, Bangalore, India
{kuntadey,senagar3}@in.ibm.com
[3] Carnegie Mellon University, Pittsburgh, USA
hlamba@cs.cmu.edu

Abstract. Language usage behavior of users evolves over time, as they interact on social media such as Twitter. We study the evolution of language usage behavior of individuals, across topics, on microblogs. We propose *Man-O-Meter*, a framework to model such evolution. We model the evolution using a combination of three dimensions: (a) time, (b) content (topics) and (c) influence flow over social relationships. We assert the goodness of our approach, by predicting ranks of experts, with respect to their influence in their respective expertise category, using the change in language used in time. We apply our framework on 2,273 influential microbloggers on Twitter, across 62 categories, spanning over 10 domains. Our work is applicable in predicting activity and influence, interest evolution, job change and community change expected to happen to a user, in future.

1 Introduction

Proliferation of online social sites like Facebook, Twitter, Youtube *etc.* has lead to tremendous growth in user generated content. Before Web 2.0 era, online content used to be structured; however, users on social websites today are defining varieties of linguistic expressions suiting their needs. Never before the study of linguistic aspects of user-generated content, has been possible at such a large scale.

Expression analysis studies show that microblog platforms, like Twitter, can be used as corpus for tasks, such as sentiment analysis and opinion mining [25]. Research shows that on Twitter, hashtags dynamically evolve, effectively serving as models to characterize linguistic form propagations [9]. Studies also predict emerging conversations, and show that norms of interactions change over time on social networks [12,17].

Several norms arise in group interactions, such as minimization of *joint effort* [7], and mechanisms such as *accommodation* [8] and *audience design* [4]. Same

© Springer International Publishing Switzerland 2016
F. Li et al. (Eds.): APWeb 2016, Part I, LNCS 9931, pp. 342–355, 2016.
DOI: 10.1007/978-3-319-45814-4_28

mechanisms are also in play in online social networks [16,28,30]. Mizil *et al.* [10] propose a framework to track the evolving linguistic norms of users, as they join and depart online communities. They explore how individual users react to individual community norms at different stages of their career. However, specific topical interests, and analysis of user's language evolution addressing topics, has not gained much research attention.

As participants actively interact on social media, the language of expression used by them evolves with time, specifically with respect to topics, evident by the fact that the hashtag symbol on Twitter has become a phenomenon [30]. This we believe can be attributed to the fact that as an individual's stays over time on a social media site, she tries to establish her identity as an opinion leader or an influencer or news spreader [6,19]. This process leads to the evolution of her interests over time as she obtains deeper insights into topics. Experience (or, expertise) level of users may change at different rates for different topics, which may reflect upon the language used. For example, a user's experience for tennis may change differently for her experience over calculus, reflecting on her language of expression.

We characterize this evolution on microblogs, by collecting social network platform messages, for given topical interests. We propose a novel framework, namely *Man-O-Meter*, to capture the evolution of language usage of individuals, over topics, using microblog content as corpus. We model the evolution using a combination of three broad dimensions: (a) time, (b) content (top0ics) and (c) influence flow over social relationships. We assert the goodness of our approach, by predicting ranks of experts, with respect to their influence in their respective expertise category, using the change in language used in time. We believe, ours is a first study of its kind, that involves all these three dimensions, namely time, topics and social influence flow.

We apply the *Man-O-Meter* framework on tweets of 3, 200 influential microbloggers on Twitter. We select 62 different categories across 10 different domains, obtaining data for top 50 influential microbloggers per category, leading to a data set having a total of 2, 273 microblogger records. For this, we used the online service provided by [13], to collect appropriate Twitter data. We observe the change of users' topical interests over time, and the varying rate of change for different individuals, across topics.

The main contributions of this paper are:

- We study the evolution of language usage behavior of users, for given topics.
- We propose the *Man-O-Meter* framework, to identify evolution of language usage of users, on microblogs, around given topics. We are the first ones to model this by combining three factors together: (a) time, (b) content (topics) and (c) influence flow over social relationships.
- We study how evolution of language change models across topics can be used to predict influence, and quantify the goodness of our approach using loss functions learned over the well-known support vector machine (SVM) based learning.
- We empirically evaluate our proposed framework with real-life Twitter data.

2   Related Work

Of late, proliferation of online social networks has inspired significant volumes and variety of research. Our work is inspired by three prior research topics.

Language Change Models: Researchers have for long investigated how the language used by people over the forums, blogs and social media evolves. Cristian *et al.* [10] studied the change of language of users in a community in 2 popular beer rating forums. The authors also used it to model the lifecycle of a user in the community. Kooti *et al.* [17] studied how the conventions like writing 'Retweet' as RT evolved over Twitter. Studies have tried to model the linguistic norms for emails [26] and Internet forums [23]. On Twitter, the spread of hashtags and content has also been taken as an indicator of linguistic norm acceptance [28,30]. Our work explores how the user evolves across various topics, with respect to the set of domain co-experts.

Studying Influence: Influence has been one of the major areas of research and has been defined in various ways. Cha *et al.* discussed 3 types of influence in the near-complete dataset of Twitter [6]. Bakshy *et al.* [3] studied influencing behavior in terms of cascade spread. They found that the past influence of users and the interestingness of content can be used to predict the influencers. Tang *et al.* [20] attempt to study social influence in large scale networks using a topical sum-product algorithm in the machine learning paradigm, and thereby investigates the impacts of topics in social influence propagation. Romero *et al.* [27] studied the role of passivity and proposed a Page Rank like measure to find influence over Twitter. Weng *et al.* too proposed a PageRank like measure to quantify influence on Twitter [31], the proposed measure was based on high link reciprocity and homophily. A lot of work has been done in studying influence on popular online social media websites. However, little has been done in studying how influence varies across time with change in language style of the author.

Studying Time: Our work also follows from the extensive work done by researchers in studying the change in user behavior and the role of language used across time. Much has been done in studying change in network structure [18,21], communities across time [2,11,15]. Similarly, evolution in user behavior has also been extensively studied. Morrison *et al.* proposed evolutionary clustering methods and analyzed changes in the role membership of users across time [24]. Angeletou *et al.* used statistical analysis and semantic models and rules for computing behavior in online communities [1]. Though a lot of work has been done in studying temporal evolution of various attributes of an online social media website, there is still dearth of work in studying the temporal evolution of 2 or more of these attributes collectively. We try to analyze the change in language and how it correlates with change in influence of the user.

3   Problem Overview and Our Approach

Social network/microblog users, change their way of expression, and thereby the underlying language of expression, as they age on the network. This could be potentially attributed to evolution of their interests, acquiring deeper insights about subject matters or topics and influence of other users. The underlying language change can also lead to a significant change in their influence behavior, and the number of their followers.

Objective: The objective is to study the language usage behavior of microblog users, focusing on the evolution of such usage over time, with respect to topics.

3.1 Creating the *Man-O-Meter* Model Framework

We design the (*Man-O-Meter*) model, capturing the temporal evolution of language used by individuals over topics. This model spans over the steps below.

1. We first generate a corpus document, using microblog posts made by users, with respect to a particular user community (for example, an academic community known to specialize on web science). We learn the topics present in the corpus, using a LDA topic modeler.
2. We temporally split the posts made by each user, into k uniformly distributed segments, over the time period under investigation. We assign each post to a particular topic with a certain level of confidence, based on our topic modeler classification.
3. For each time split, we measure the divergence of the topics in the posts made by each individual user within the split, for the topics present overall in the corpus within the same split.
4. We finally analyze these divergences over all splits for each user, to make our overall observations.

Having created our framework, we further propose the following:

- A characterization framework, involving (a) rate of change of interest of individual users for each given topic, (b) overall rate of change of these interests across all the topics and (c) average cross entropies across time segments for each given user.
- A prediction framework, that uses a machine learning method (a form of support vector machines) to predict influence, using the findings of our method as input, and qualitatively demonstrate the goodness of prediction with our method.

Generating ad-hoc Corpus and Learning Topics. We first generate the ad-hoc corpus. This corpus is subsequently used to learn the topics present in the corpus, using LDA topic modeling. For this, we collect the content generated by the user on the microblog, over time span $[T_{start}, T_{end}]$, and the generation timestamp t_i, for each piece of content (post) p_i. Since microblogs come with a message forwarding service (e.g., re-tweet in Twitter), we add a weightage for each p_i, to form a participant content P_j of the ad-hoc corpus. Let $|p_i|$ be the number of times that a message got forwarded. Each P_j, therefore, can be expressed as:

$$P_j \leftarrow \{p_i, |p_i|\} \tag{1}$$

The weighted participant contents form the corpus document as a set union of P_j, and sum of the weights.

$$C \leftarrow \cup_j \{P_j\} \tag{2}$$

We refer to the document thus generated as an ad-hoc corpus, because: (a) this comprises of the known concepts used by the participants, found on a best-effort basis, and (b) this is formed on an as-is basis, subject to data availability. This method of forming ad-hoc corpus, with certain topical focus, is intuitively scalable, and hence suitable for big data applications. Further, the ad-hoc corpus thus formed, is independent of user attributes, and spans over the complete activity of the users under consideration.

We attempt to learn the broader topic model, from the ad-hoc corpus, using LDA [5]. The learning process forms n topics omnipresent over C, represented as $Z \leftarrow \{z_1, z_2, \ldots, z_n\}$. Each topic z_i also has a corresponding weight w_i, denoting the significance of the topic within the overall corpus.

The Temporal Segmentation. We segment our dataset on a temporal basis, to study temporal evolution. For simplicity, we divide the entire time range, $[T_{start}, T_{end}]$, into k equal time segments:

$$T_{start}(= T_0), T_1$$

$$T_1, T_2$$

$$\ldots$$

$$T_{k-1}, T_{end}(= T_k)$$

Using the topics found earlier, we determine the presence of each topic within given time segment T_i, examining the content generated within the corresponding time segment. Let $C_1, C_2, \ldots, C_k$ denote the ad-hoc corpus formed at each respective time segment, considering all p_i, generated within time segment T_i. Clearly,

$$C = \cup_k \{C_k\} \tag{3}$$

Using the topic modeler, we provide the set of topics $Z \leftarrow \{z_1, z_2, \ldots, z_n\}$ found earlier, to determine the weight w_l of each topic z_l, within each C_i, in a weighted manner. This gives: $W_{C_i} \leftarrow \{w_{1,C_i}, w_{2,C_i}, \ldots, w_{n,C_i}\}$. Here,

$$\sum_{l=1}^{n} w_{l,C_i} = 1. \tag{4}$$

We measure the alignment of each user, with the C_i corresponding to time segment T_i, by investigating the alignment of the topics present within C_i, and the topics of the content generated by the user within T_i, using methods described subsequently.

Measuring and Analyzing Distances. Finally, for each time segment T_i, we measure the distances of each user's topical alignment, with each topic z_l, having weight w_{l,C_i}, in the corresponding ad-hoc corpus C_i.

We introduce the notion of accumulated topic set posted by a user u_i within T_i, as $Z_{u_i} \leftarrow \{z_1, z_2, \ldots, z_n\}$, having weights $W_{u_i} \leftarrow \{w_{1,u_i}, w_{2,u_i}, \ldots, w_{n,u_i}\}$,

We present an over-simplified example, to provide an intuitive understanding of the underlying process. In time segment T_i, let there be only 4 posts, namely $\{p_1, p_2, p_3, p_4\}$, posted by user u.

Let's say, a total of 20 posts were generated within T_i, namely $\{p_1, p_2, \ldots, p_{20}\}$ made by all users u. Thus, the overall corpus within T_i, C_i, will incorporate all of $\{p_1, p_2, \ldots, p_{20}\}$. Further, we assume there were 3 topics learned over the global corpus, namely $Z \leftarrow \{z_1, z_2, z_3\}$.

Using the ad-hoc corpus generation method above, we generate corpus C_{u_i}, from $\{p_1, p_2, p_3, p_4\}$, for this user u_i, in T_i. Using topics learned earlier from the overall data, we learn the association weights of the user corpus towards these topics: $W_{u_i} \leftarrow \{w_{1,u_i}, w_{2,u_i}, w_{3,u_i}\}$. Similarly, we generate a topic vector for the corpus, using the posts generated within T_i, as: $W_{C_i} \leftarrow \{w_{1,C_i}, w_{2,C_i}, w_{3,C_i}\}$.

The example is detailed in Table 1. Note that the sum of values of any given row is unity, as that is the total distribution of the posts, across all the topics.

Table 1. An example of topic association vector of user and corpus

	z_1	z_2	z_3
W_{u_i}	0.10	0.61	0.29
W_{C_i}	0.49	0.33	0.18

The divergence of a user's language model, for each topic z_l, in the corresponding ad-hoc corpus C_i, is measured using a multiplicative combination of three factors.

1. The relative weightage of a user towards this topic within the given time segment, from all the existing topics across all time segments. In the example shown in Table 1, this is given in the first row.
2. The measurement of the weightage of a corpus, formed within T_i, towards the given topic, for all existing topics. This appears on the second row of Table 1.
3. Cross-entropy (CE_{u_i}) measures cross-entropy [29] of the topics of the user, with respect to the topics in the corpus, within T_i. This is measured as the product of the user's topic weightage vector, and the set of topics present in the T_i corpus.

To complete the formation of our *Man-O-Meter* framework, we finally obtain the divergence of a user with respect to a topic within a given time segment, as:

$$D_{u,z_l,T_i} = w_{l,u_i} \times w_{l,C_i} \times CE_{u_i} \tag{5}$$

3.2 Characterization Metrics

To characterize the observations made from the *Man-O-Meter* model, we propose following 3 metrics per user: topical interest change rate for a topic $TIR(u,z_l)$, topical interest change rate for over all topics $TIRO(u)$ and average topical interest change rate for over all topics $AIRO(u)$.

1. TIR (u,z_l) measures the interest change rate for a user u, for a topic z_l. This quantifies the fluctuation in a user's language expression at topical level. TIR (u,z_l) is computed by summing up the first order norm of divergences corresponding two consecutive time segments for the user u, for the topic z_l, as:

$$TIR(u, z_l) = \sum_{i=2}^{k} |(D_{u,z_l,T_i} - D_{u,z_l,T_{i-1}})| \qquad (6)$$

2. $TIRO(u)$ measures the interest change rate for overall topics for a user u. This quantifies the fluctuation in a user's language expression across overall topics. $TIRO(u)$ is computed by summing up the first order norm of divergences corresponding two consecutive time segments for user u, as:

$$TIRO(u) = \sum_{i=2}^{k} |(CE_{u_i} - CE_{u_{i-1}})| \qquad (7)$$

3. $AIRO$ (u) measures the average cross entropies across time segments for a user u. This measure estimates, how well a user's language expression aligns with the overall corpus on an average. $AIRO(u)$ is computed by taking average of a user's cross entropies across time segments, as:

$$AIRO(u) = \frac{1}{k}(\sum_{i=1}^{k} CE_{u_i}) \qquad (8)$$

3.3 Rank Prediction

So far, we have characterized the deviation of users from their respective domain across topics over time. Based upon this, we propose the following hypothesis: the change in language used across time, can be used to predict rank of users, with respect to their influence in their respective expertise category. We validate our hypothesis with a supervised algorithm framework. We define *influence* of a user as the number of times, on average, the tweets authored by her were retweeted. Normalizing the retweet count by the number of statuses authored by a particular user removes activity bias, which exists when user is authoring a lot of tweets. We consider following types of features:

– **Language Change Features:** Change in language from all the other experts of the same domain can be critical and can catch followers' attention or lose it. Increase in attention can be due to the user talking about a new concept/innovation or a topic which is not being talked about her peers. However, it can also lead to decrease in attention since followers might not be interested in listening to the particular topic which the user is currently talking about and might be more interested in the usual topics. We use the features pertaining to the topic wise cross entropy feature of the user in the current time segment and the previous 3 time segments and changes in the topic wise cross entropy across the last 4 time segments.

- **Community Mention Feature:** A significant feature of Twitter is *mentions*, where another user has mentioned a given user in a tweet. Receiving a high number of mentions indicates that a given user is highly cited or sought after. It adds value when the users who want to engage are the co-experts in the domain. We take the number of mentions in the current and the previous 3 time segments, and the change in the number of community mentions across the past 4 time segments, as features.
- **Activity Feature:** The number of tweets denote how active a particular user is. We take the number of tweets posted by the user in the previous 3 time segments as the feature. We do not use the number of tweets posted in the current time segment as it also plays a part in our output variable and it can lead to a bias.
- **Retweet Feature:** The number of retweets a user gets on a tweet authored by her indicates the influence the user has in her network. We consider the total retweet count in the previous 3 time segments as a feature.

We use SVM^{Rank} [14], a rank list learning algorithm to build a model for influence prediction. SVM^{Rank} trains a Ranking SVM on the training set, taking as input the features of each attribute along with their ranking. Based on the learned model, the algorithm outputs a score for each attribute in each rank list in the testing set. This set of scores, when sorted in decreasing order, produces the predicted rank list. We measure the performance of the fitted model using the following 2 performance indicators:

- **Zero-One Loss:** The Zero-One Loss function yields 0 if the predicted list exactly matches the actual list, else yields 1. It is computed for every rank-list in the testing set, and summed over all such rank lists.
- **Average Loss:** This is the fraction of swapped pairs in the predicted list required to make it equivalent to the actual list in the testing set.

4 Experiments

4.1 Data Collection

We collect tweet history data from Twitter, along with timestamps, for top experts in different domains. To find experts over Twitter, we use a crowd-sourced method - Cognos [13]. For each query, the tool returns top 50 experts. We query the tool for 10 domains, and form queries under each category, shown in Table 2. Overall, we extract experts for 62 popular categories. We the past 3, 200 tweets for each expert from Twitter, between Wed Mar 07 06:33:27 2007

Table 2. Categories for which data was collected

News	News, Political News, Sports News, Entertainment News, Science News, Technology News, Business News
Journalists	Journalists, Politics Journalists, Sports Journalist, Entertainment Journalist, Science Journalist, Technology Journalist, Business Journalist
Politics	Politics, Conservative News, Liberal politicians, USA politicians, German politicians, Indian politicians, Brasilian politicians
Sports	Sports, F1, baseball, soccer, poker, tennis, NFL, NBA, Bundesliga, LA Lakers
Entertainment	Entertainment, celebrities, movie reviews, theater, music
Hobbies	hobbies, hiking, cooking, chefs, traveling, photography
Lifestyle	lifestyle, wine, dining, book clubs, health, fashion
Science	science, biology, computer science, astronomy, complex networks
Technology	technology, iPhone, mac, linux, cloud computing
Business	business, markets, finance, energy

and Fri Jan 03 23:10:01 2014. This gives a total of $6,953,303$ tweets across $2,273$ unique users. On an average, we find $112,150$ tweets per query and 48.3617 users per category. We retain only those categories for which we get tweets of more than 30 users, leaving us with 47 categories.

4.2 Experimental Setup

We set each time segment k is a period of 30 days, and put each tweet into a time segment based on its timestamp. We train the LDA model with the number of topics as 5, which is overserved to be a good balance between too general and too specific values, via trial-and-error.

4.3 Topic Detection

For each category, we generate an adhoc-corpus C from the tweets authored by the experts of the category and use MALLET [22] to find LDA-based topics. Without loss of generality, we set the number of topics as 5. An example of topics detected for one time-segment for query *computer science* is shown in Table 3.

We temporally split the posts made by each expert over 104 parts, each consisting of the posts authored in a given month. Using the corresponding category-learned topic-model, we compute topic distribution for each expert individually across each time segment. We generate an ad-hoc corpus for each time segment for each category using the posts authored by the experts in that time segment. We compute the topic distribution of each time segment's adhoc-corpus using the corresponding category-learned model. We subsequently compute cross entropy and divergence, for each expert, with respect to each time segment topic distribution.

Table 3. Three randomly selected topics, shown as representative examples

Topic ID	Concepts
1	apple iphone mobile apps store ios ipad free video features
2	digg ubuntu tinyurl www geek twitlonger cloud nasa
3	video women check computer aws make post 11 ow lt youtube
4	windows microsoft java ubuntu live day time surface sony pro pc phone
5	software week check online phone code tv event world great back web don

4.4   User Level Interest

We observe that experts' cross entropies and divergences change over time, and the rate of change varies across topics for different individuals. For each expert in each category, we first compute $TIR(u, z_l)$. We subsequently compute $TIRO(u)$, the measurement for the interest change rate for overall topics for a user u. As it is impractical to plot the cross entropy and divergence change rate for every expert, we present indicative results by picking the top 5 $TIRO(u)$ across categories, top 5 $TIR(u, z_l)$ across categories, across topics and plot them over time. Figures 1 and 2 show the indicative results. The top 5 $TIRO(u)$ are for categories NBA, cloud, music, sports journalist and politics journalist, which are all highly dynamic in terms of events happening and controversies.

From Fig. 1, it is evident that, for all the categories represented in the figure, there are wide fluctuations in the beginning but as time progresses, the five experts align their topical interests with the community. This leads to a significant empirical observation: experts *acclimate* with the *social language* as time elapses and participation increases, leading to *language usage stability*.

From Fig. 2, it is evident that what is happening at overall topic level is also happening at a particular topic level. As time progress, the divergences for the topics corresponding to top 5 $TIR(u, z_l)$ becomes stable. The categories corresponding to top 5 $TIR(u, z_l)$ are business, news, politics, tennis and theatre. All of these communities are highly dynamic in terms of topics being discussed over social media.

4.5   Category Level Interest

Each category is inherently different from other categories. Users in different categories behave differently. To investigate the rate of change of interest amongst experts at a category level, we plot $AIRO(u)$ of experts on a box-whiskers plot in Fig. 3. We observe that for each category there is wide distribution of $AIRO(u)$ values. This means that every category has users who align to category's language of expression to the users who do not align themselves so much at different point of times. We also observe that categories are of very different natures in terms of alignment of its user base with its language of expression. Categories ranges from highly volatile, characterized by high median and large range for

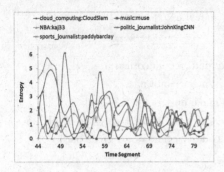

Fig. 1. Range of cross entropies for users corresponding top 5 $TIRO(u)$

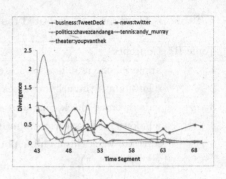

Fig. 2. Range of divergence for users and topics corresponding top 5 $TIR(u,z_l)$

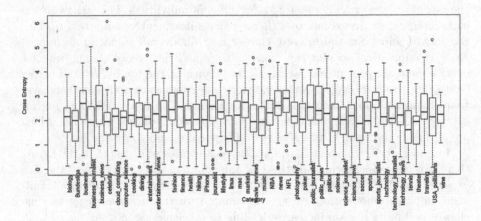

Fig. 3. Range of average cross entropies of users for different categories

$AIRO(u)$, to categories which are really stable, characterized by low median and small range for $AIRO(u)$.

4.6 Rank Prediction

We run the supervised SVM^{Rank} algorithm for each category separately, using a 50-50 training-testing random split, a linear kernel, termination a epsilon of 0.001, and a learning cate $C = 2.0$. The average number of rank lists (instances) per category us 60.693 and the loss function is given by the total number of swapped pairs.

We compare our approach across 2 naive baseline feature sets: community mention features and number of tweet features. The performance of the supervised algorithms, in terms of the two performance indicators, is shown in Figs. 4 and 5. Smaller values of loss function across all categories indicate better model fits. Clearly, our approach that includes the cross-entropy variables outperform

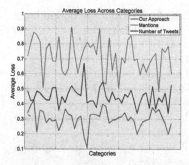

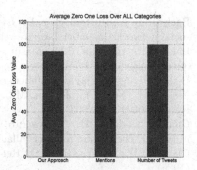

Fig. 4. Average loss function of model over different categories

Fig. 5. Average zero-one loss function of model over all categories

the other 2 baseline measures. This indicates the effectiveness of cross-entropy features across time segments and topics, for predicting the overall future influence of a user among her peers.

5 Conclusion

In this paper, we attempted to model the evolution of language usage behavior of social media users. We proposed the *Man-O-Meter* model, to track over: (a) time, (b) content (topics) and (c) influence flow over social relationships. We studied 2,273 influential microbloggers on Twitter, across 62 categories over 10 domains, for their entire Twitter life time, using our model. This provided insights into the change of topical interests of users over time. We observed that this rate of change varied across topics for different users. We asserted the goodness of our approach, by predicting ranks of experts, with respect to their influence in their respective expertise category, using the change in language used in time, using loss function to quantitatively measure the goodness. Our work is useful in predicting activity and influence, interest evolution, job change and community change expected for a user. Our analysis indicates that experts acclimate with the social language as time elapses and participation increases, leading to language usage stability. This can be used for a wide range of social applications, all the way from expert recommendation to different forms of targeted and personalized social advertising. In future, we propose improving our temporal segmentation method, to factor for participation skew of users, across time segments.

References

1. Angeletou, S., Rowe, M., Alani, H.: Modelling and analysis of user behaviour in online communities. In: Aroyo, L., Welty, C., Alani, H., Taylor, J., Bernstein, A., Kagal, L., Noy, N., Blomqvist, E. (eds.) ISWC 2011, Part I. LNCS, vol. 7031, pp. 35–50. Springer, Heidelberg (2011)

2. Backstrom, L., Huttenlocher, D., Kleinberg, J., Lan, X.: Group formation in large social networks: membership, growth and evolution. In: KDD (2006)
3. Bakshy, E., Hoffman, J.M., Mason, W.A., Watts, D.J.: Everyone's an influencer: quantifying influence on twitter. In: WSDM (2011)
4. Bell, H.: Language style as audience design. In: Language in society (1984)
5. Blei, D.M., Ng, A.Y., Jordan, M.I.: Latent Dirichlet allocation. J. Mach. Learn. Res. **3**, 993–1022 (2003)
6. Cha, M., Haddadi, H., Benevenuto, F., Gummadi, P.K.: Measuring user influence in twitter: the million follower fallacy. In: ICWSM (2010)
7. Clarke, H.H., Wilkies-Gibbs, D.: Referring as a collaborative process. In: Cognition (1986)
8. Coupland, J., Coupland, N., Giles, H.: Accommodation theory: communication, context and consequences. In: Contexts of Accommodation, Cambridge, pp. 1–68 (1991)
9. Cunha, E., Magno, G., Comarela, G., Almeida, V., Gonalves, M.A., Benevenuto, F.: Analyzing the dynamic evolution of hashtags on twitter: a language-based approach. In: Workshop on Language in Social Media - ACL, pp. 58–65 (2011)
10. Danescu-Niculescu-Mizil, C., West, R., Jurafsky, D., Leskovec, J., Potts, C.: No country for old members: user lifecycle and linguistic change in online communities. In: WWW (2013)
11. Ducheneaut, N., Yee, N., Nickell, E., Moore, R.: The life and death of online gaming communities: a look at guilds in world of warcraft. In: CHI (2007)
12. Garley, M., Hockenmaier, J.: Dissemination, diversity and dynamics of English borrowings in german hip hop forum. In: ACL (2012)
13. Ghosh, S., Sharma, N.K., Benevenuto, F., Ganguly, N., Gummadi, K.P.: Whom to follow? discover topic authorities on twitter! (2012). http://twitter-app.mpi-sws.org/whom-to-follow/
14. Joachims, T.: Training linear SVMs in linear time. In: KDD (2006)
15. Kairam, S., Wang, D., Leskovec, J.: The life and death of online groups: predicting group growth and longevity. In: WSDM (2012)
16. Kooti, F., Yang, M.C.H., Gummadi, K.P., Mason, W.A.: The emergence of conventions in online social networks. In: ICWSM (2012)
17. Kooti, F., Mason, W.A., Gummadi, K.P., Cha, M.: Predicting emerging social conventions in online social networks. In: CIKM (2012)
18. Kumar, R., Novak, J., Tomkins, A.: Structure and evolution of online social networks. In: KDD (2006)
19. Kwak, H., Lee, C., Park, H., Moon, S.B.: What is twitter, a social network or a news media? In: WWW (2010)
20. L. L, J. Tang, J. Han, M. Jiang, S. Yang.: Mining topic-level influence in heterogeneous networks. In: CIKM (2010)
21. Leskovec, J., Kleinberg, J., Faloustos, C.: Graph evolution: densification and shrinking diameters. In: TKDD (2007)
22. McCallum, A.K.: Mallet: a machine learning for language toolkit (2002). http://mallet.cs.umass.edu
23. Mizil, C.D.-N., Lee, L., Pang, B., Kleinberg, J.: Echoes of power: language effects and power differences in social interaction. In: WWW (2012)
24. Morrison, D., McLoughlin, I., Hogan, A., Hayes, C.: Evolutionary clustering and analysis of user behaviour in online forums. In: ICWSM (2012)
25. Pak, A., Paroubek, P.: Twitter as a corpus for sentiment analysis and opinion mining. In: LREC (2010)

26. Postmes, T., Spears, R., Lea, M.: The formation of group norms in computer-mediated communication. In: Human Communications Research (2000)

27. Romero, D.M., Galuba, W., Asur, S., Huberman, B.A.: Influence and passivity in social media. In: Gunopulos, D., Hofmann, T., Malerba, D., Vazirgiannis, M. (eds.) ECML PKDD 2011, Part III. LNCS, vol. 6913, pp. 18–33. Springer, Heidelberg (2011)

28. Romero, D.M., Meeder, B., Kleinberg, J.M.: Differences in the mechanics of information diffusion across topics: idioms, political hashtags, and complex contagion on twitter. In: WWW, pp. 695–704 (2011)

29. Rubinstein, R.Y.: The cross-entropy method for combinatorial and continuous optimization. Methodol. Comput. Appl. Probab. $1(2)$, 127–190 (1993)

30. Tsur, O., Rappoport, A.: What's in a hashtag?: content based prediction of the spread of ideas in microblogging communities. In: WSDM, pp. 643–652 (2012)

31. Weng, J., Lim, E., He, Q.: Finding topic-sensitive influential twitterers. In: WSDM (2011)

Research Full Paper: Modelling and Learning with Big Data

Forecasting Career Choice for College Students Based on Campus Big Data

Min Nie[1], Lei Yang[1], Bin Ding[1], Hu Xia[1], Huachun Xu[2], and Defu Lian[1(✉)]

[1] Big Data Research Center,
University of Electronic Science and Technology of China, Chengdu, China
nieminde@gmail.com, dove.ustc@gmail.com
[2] College of Teacher Education and Psychology,
Sichuan Normal University, Chengdu, China

Abstract. Career indecision is a difficult obstacle in front of adolescents. Traditional vocational assessment research measure it by means of questionnaires and diagnose the potential sources of career indecision. Based on the diagnostic outcomes, career consolers develop the treatment plans tailor to students. However, because of personal motives and the architecture of the mind, it may be difficult for students to know themselves, so that the outcome of questionnaires can not fully reflect their inner states and statuses. Self-perception theory suggest students' behavior could be used as clue for inference. Thus, we proposed a data-driven framework for forecast student career choice of graduation based on their behavior in and around the campus, playing an important role in supporting career counseling and career guiding. By evaluating on 10M behavior data of over four thousand students, we show the potential of this framework in these functionality.

Keywords: Campus big data · Career identity · Career prediction · Self-knowledge

1 Introduction

According to Erikson, the formation of a vocational identity is one of the main tasks of adolescence, and viewed as a part of the larger task of identity development [9]. Indicating the possession of a clear and stable picture of ones goal, interests and talents, vocational identity is possibly formed by sufficient career exploration and subsequent commitment at college [19]. During this period of vocational identity formation, many adolescents experience periods of indecision regarding their career [18,23]. Thus, career counseling services are essential at college, to help students make career decisions, so that special career counseling centers have even been established.

From the psychological perspective, career counseling for career indecision of college students is usually a cognitive based approach in which logical processes are employed in collecting, sifting and evaluating relevant career and personal

F. Li et al. (Eds.): APWeb 2016, Part I, LNCS 9931, pp. 359–370, 2016.
DOI: 10.1007/978-3-319-45814-4_29

information. Concretely, instruments such as Career Decision-Making Difficulties Questionnaire (CDDQ) [11] are first used to precisely diagnose the sources of students' career indecision, ranking from a lack of readiness to a lack of information about self, occupations and ways of obtaining information. Based on the diagnostic outcomes, career counselors are in a position to develop a "treatment" plan for intervening in the students career indecision.

In order to be able to engage in occupational decision making, students should first of all develop competence and skill in self-concept [24]. Because of personal motives and the architecture of the mind, it may be difficult for people to know themselves [31]. According to self-perception theory, inferring persons' internal states from their behavior is a major source of self-concept [4]. For example, if students notice that they are constantly late for class, they might rightly infer that they are not as conscientious as they thought. Since many of student behaviors are driven by internal states which are "weak, ambiguous, or uninterpretable", people can use students' behaviors as a clue to their hidden dispositions.

With the development of information technology, more and more advanced information management and monitor systems have been established in many colleges/universities, with the aims of making students life and study more convenient and efficient via smart cards. When students continuously interact within a cyber-physical space, their behaviors in and around the campus, such as having meals, shopping, borrowing books and taking courses, are accumulated in real time. These behavior data can capture different patterns that reflecting their unique habits, capability, preference and mental status, so its explosive growth in the amount has just created an opportunity for proposing a data-driven framework to help students to better know themselves.

To this end, in this paper, we propose a supervised career choice forecasting framework based on students' behavior data and career choices of graduation. Within this framework, we put forward behavior-based representative factors for affecting student career selection. These factors, supported by psychological study, include professional skills/abilities [1] learned from course-taking records, behavior order in the conscientiousness of big five personality [8], interest and preference for borrowing books, and family economic status [27] estimated by daily consumption from smart card usage. It is intuitive to cast career forecasting into a multi-class classification problem, so that algorithms such as KNN, Decision Tree or logistic regression could be used to predict his potential career choice in a determinant or probabilistic way. These multi-class classification algorithms essentially capture each college student' similarity/distance/divergence with graduates over those representative factors, rightly agreeing with the social comparison theory [10] in psychology. The central proposition of the social comparison theory is the "similarity hypothesis", which indicates that human evaluate their ability and limitations by comparing with similar others, particularly when objective and non-social means of evaluation are not available. Self-evaluation in this case probably becomes more stable and accurate. More importantly, people are especially likely to make upward comparisons, that is to evaluate themselves against successful individuals, higher probably leading to self-improvement ultimately.

In addition to supporting students to determine their career choice, in this framework, we will conduct correlation analysis between behavior-characterized factors and career choices, in order to discover the influent representative factors for affecting student career selection. Therefore, it is possible to leverage these knowledge to help students achieve their early-stated goals. For example, if we have observed the significant effect of English courses at students' career choice about going abroad for further study. Thus, students, stating this as their goal, should strive to acquire the language skills (reading, writing, listening) of English.

Finally, we evaluate the proposed framework on behavior data and career choice records of over 4,000 students. Grouping career choices of graduation into "abroad further study", "seeking jobs", "domestic further study", and "others", Micro F1-measure of the best multi-class classification algorithm could achieve 0.6 at the first semester and improve with the increase of semesters. According to the correlation analysis between behavior-based factors and career choices, we find that factors like professional skills, behavior regularity and economic status could significantly correlate with career choices.

2   Related Work

This paper forecasts career for college students based on campus big data, which could be related with both professional career mining and vocational counseling and guiding. The former one includes the prediction of expertise, skill and career movement, and recommendation of jobs while the latter one will summary the influence factors on vocational indecision and introduce some intervention techniques as well as matching theory of career selection.

2.1   Professional Career Mining

There are several directions of professional career mining, including the prediction of expertise, skills and career movement. For example, expertise can be predicted based on documents such as project descriptions, human resource databases, professional articles, program code [22], based on emails [2] due to its usage for communication about work topics, based on the use of social media [13], such as blogs, wikis, forums, microblogs, people tags and so on, and based on enterprise systems of record and data from the internal corporate social networking site as features [28]. Professional skills, being related to expertise, but having multi-label characteristics, were usually predicted by matrix completion based collaborative filtering with side information [29]. Given the volatile and unpredictable world of the work environment at present, There has been less opportunity and willingness for individuals to engage in a single organization for a lifetime [3]. Thus more and more employees begin to choose external, lateral or even downward job changes. The research topics related to career movement/mobility include uncovering a set of determinants, regularities and reproducible patterns behind career movement [7,14,30,32], ranging from move-ment propensity, brain circulation, vocational preference to spatio-temporal reg-ularity and the characteristic of stratification. In addition, how to recommend tailored jobs for users when there are potential career movements is an research topics about career movement. For example, a supervised machine learning algo-rithm are used to recommend jobs to people based on their past job histories [20]. Due to reciprocity of job seekers and recruiters, collaborative filtering and its boost with profiles of job seekers and descriptions and requirements of jobs have been applied [16,20]. The key part in collaborative filtering is to measure (career) similarity between job seekers for either themselves or recruiters, which can be the profile/self-description similarity, career path similarity or their combina-tion [33].

2.2 Vocational Guiding

Vocational selection is a key research topic in vocational psychology, starting with the talent matching approach developed by Parsons [21]. Matching the-ory was subsequently developed into the trait and factor theory of occupational choice, which necessarily measured individual talents and the attributes of par-ticular jobs. Although it is similar to job recommendation frameworks, it placed more emphasis on individual personalities, interests, aptitudes, or other explain-able and measurable characteristics by instrument tools [15]. Instead, job rec-ommendation strives to learn the traits and factors from behavior observations, which may suffers from low interpretability. Due to its criticism in a lack of adap-tiveness with the change of individual and occupational environments, develop-mental theories [26], career exploration theories [25], social learning/cognitive theory [17] and others was developed for determining and explaining vocational choices at different periods of life span. The goal of these theories is to help people to acquire evolving self-concept (interest, ability, motivation and need) and dynamic occupational environments, and to further make career decisions.

3 Career Choice Forecasting Based on Campus Big Data

Students' behavior in campus are continuously recorded, in cases of making payments, borrowing library books and taking courses. Forecasting career choice for students requires first to disaggregate these records into different evidence sets and then predict the career choices based on these evidence. In particular, course taking histories are used for extracting professional skills and learning mastery levels of these skills since lots of the skills equipped for future vacation is delivered by taking courses; consumption records are time-stamped so that they are leveraged for modeling the regularity of behaviors such as having breakfast and taking shower; students often borrow books for learning specific skills or expanding their knowledge, mining book borrowing preferences from their book-loan histories could benefit. Finally, since each consumption record could reflect the economic status of family [12], they are used for estimating economic status by extracting patterns, such as expenditure of each breakfast/lunch/dinner and monthly expenditure. Based on these four types of evidences, grouping career choice into four groups, i.e., "abroad further study", "seeking jobs", "domestic further study", and "others", we can leverage multi-class classification algorithm for career choice prediction.

Below, we assume that the career choices of M students $\mathcal{U} = \{u_1, \cdots, u_M\}$ and their four college years of behavior data are given. These students have borrowed N books $\mathcal{B} = \{b_1, \cdots, b_N\}$, and taken S courses $\mathcal{C} = \{c_1, \cdots, c_S\}$ in total. Each book has a category attribute, and there are T categories $\mathcal{A} = \{a_1, \cdots, a_T\}$ in total.

3.1   Learning Mastery Level of Professional Skills

As aforementioned, professional skills are extracted from course taking histories. Its each record gives students' score in the corresponding course. However, there are total thousands of courses in one university, if scores of each course are taken features, this feature representation will face with sparsity challenge. In addition, many professional skills may be determined by students' performance on several courses. For example, "Machine Learning" skills may depend on the performance on "probability and statistics", "linear algebra" and "mathematical analysis". Therefore, matrix factorization based dimension reduction algorithm will be applied for feature extractions. In particular, this algorithm takes a student-course scoring matrix as input, and maps students and courses onto the same joint latent space. Assume the scoring matrix is denoted as $R \in \mathbb{R}^{M \times S}$, whose each entry $r_{i,j}$ represents the grade of the student u_i on the course c_j, and that each student $u_i \in \mathcal{U}$ and each course $c_j \in \mathcal{C}$ are represented by points in the latent space of dimension K, denoted as a user latent factor $\mathbf{p}_i \in \mathbb{R}^K$ and a course latent factor $\mathbf{q}_j \in \mathbb{R}^K$ respectively. Each dimension of the latent space could be explained by skills according to [6], so that user latent factors represent mastery level of corresponding skills and course latent factor indicate the correlation of course with corresponding skills. The dot product between user latent factor and course latent factor approximate students' performance on courses. Student latent factors $\mathbf{P} = (\mathbf{p}_1, \cdots, \mathbf{p}_M)'$ and course latent factors $\mathbf{Q} = (\mathbf{q}_1, \cdots, \mathbf{q}_S)'$ are can be learned by optimizing the following objective functions,

$$\min_{\mathbf{P}, \mathbf{Q}} \sum_{i,j} I_{i,j}(r_{i,j} - \mathbf{p}_i'\mathbf{q}_j)^2 + \lambda(\sum_i \|\mathbf{p}_i\| + \sum_j \|\mathbf{q}_j\|), \tag{1}$$

where $I_{i,j}$ indicates whether a student i has taken a course j or not. In other words, students' performance over those courses without being taken by students is missing.

The learning of the parameters $\mathbf{p}_i$ and $\mathbf{q}_j$ could be achieved by alternative least square or stochastic gradient descent. Alternative least square makes use of the following formula for updating parameters:

$$\mathbf{p}_i = (\lambda \mathbf{I}_K + \sum_j \mathbf{I}_{i,j} \mathbf{q}_j \mathbf{q}_j')^{-1} (\sum_j \mathbf{I}_{i,j} r_{i,j} \mathbf{q}_j)$$

$$\mathbf{q}_j = (\lambda \mathbf{I}_K + \sum_i \mathbf{I}_{i,j} \mathbf{p}_i \mathbf{p}_i')^{-1} (\sum_i \mathbf{I}_{i,j} r_{i,j} \mathbf{p}_i) \qquad (2)$$

After the learning of parameters, user latent factors correspond to the features on the mastery level of professional skills.

However, when feeding the student-course score matrix into this algorithm, one important preprocessing step should be first finished. Since one course may be taught by several teachers, the grade of the same course taught by different teachers cannot probably be compared with each other due to the different teaching level. Therefore, we compute the averaging grade of each course w.r.t each teacher and then subtract it from his/her students' grade of the corresponding course. For example, a student A and a student B take a course taught by a teacher X and obtain grades r_a and r_b respectively, while a student C and student D take the same course taught by another teacher Y and obtain grades r_c and r_d respectively. The grades of the former two students are normalized as $r_a - (r_a + r_b)/2$ and $r_b - (r_a + r_b)/2$.

3.2 Modeling Behavior Regularity

Conscientious is an important personality trait and has been shown to be positively related to job/academic performance [8]. Conscientious people exhibit a tendency to show self-discipline, which just could be reflected by regularity of daily activities. Therefore, behavior regularity should be useful for help student to determine their future career choice. We particularly focus on the daily regularity of having breakfast, and going to library for the first time in each day and taking showers. Regularity of a behavior could be considered as repeatability, and will be measured by the entropy of probability that the behavior occur within specific time intervals. Assume there are n time intervals $\mathcal{T} = \{t_1, \cdots, t_n\}$, for any given student, the probability that a behavior $v \in \mathcal{V} = \{$"breakfast", "library", "shower"$\}$ will take place within time interval t_i is computed as

$$P_v(T = t_i) = \frac{n_v(t_i)}{\sum_i n_v(t_i)} \qquad (3)$$

where $n_v(t_i)$ is the occur frequency of the behavior v within the time interval t_i. Then the entropy of the behavior v is computed as

$$E_v = - \sum_{i=1}^{n} P_v(T = t_i) \log P_v(T = t_i) \qquad (4)$$

If the entropy of a behavior is higher, the probability over time intervals is more uniformly distributed and the regularity of this behavior is lower. When computing entropy, we assume that each time interval span half an hour with respect to all three behaviors. Since breakfast behavior is specified by time periods from

6 am to 10 am, thus the number of time intervals is 8, less than the other two cases (48 time intervals). In summary, there are three entropy features in total, to reflect students' regularity.

3.3 Mining Book Reading Interest

As introduced above, students borrow library books for learning skills outside of class to expand their knowledge. Therefore, the loan history of each student could reflect his/her interest, part of which may be correlated with future occupational choices. However, there are a great number of books which have been borrowed by students, but each student only borrow a few books among them. Thus if directly using loan frequency as evidence for borrowing preference will suffer from the sparsity challenge. One solution is to leverage dimension reduction techniques based on the loan history, similar to the extraction of professional skills. Fortunately, each book has a rich set of attributes, whose category, characterized by Chinese Library Classification[1] in this case, could be useful but more easy and more interpretable to identify the interest of students. To this end, we make statistics on the loan frequency of each category of books with respect to each student and normalize them to get a probabilistic distribution. Categories are organized as a several levels of hierarchy, but to balance between representative ability and sparse, we focus on the second level of categories in the hierarchy. Thus there are around two hundreds of categories in total.

3.4 Estimating Family Economic Status

It is possible to know family economic status by questionnaire, but students may overstate their situations to get better finical support. Therefore, it is appealing to estimate family economic status from students' consumption histories, as suggested in [12]. The consumption at different locations may play different important roles for this goal, and we put emphasis on the payment history at mess halls and supermarkets since the cost from them could occupy a very large portion of total cost. In particular, we calculate the cost of each meal and each shopping by summing multiple payments within short time interval (10 mins in this case) and compute expenditure of each day, and then convert consumption history into meal cost sequence, shopping cost sequence and daily expenditure sequence, by concatenating them by chronologically. Then we first exploit first order and second order *descriptive statistics*, including minimum, maximum, median, mean, interquartile range, standard deviation and kurtosis, as evidences for family economic status. Second, we compute the *ratio* of transaction amount per day on weekends to weekdays, and conduct Fast Four Transformation (FFT) and calculate *energy* as the sum of the squared magnitudes of each FFT component [5], which could capture consumption periodicity and thus provide another evidence for family economic status. However, in order to cancel out the effect of consumption level, the mean value of the sequence $[x_1, x_2, \cdots, x_n]$, of the length

---

[1] https://en.wikipedia.org/wiki/Chinese_Library_Classification.

n, should be subtracted from its each value. Based on the converted sequence $[\tilde{x}_1, \tilde{x}_2, \cdots, \tilde{x}_n]$ where $\tilde{x}_i = x_i - \sum x_i/n$, the energy is defined as,

$$\text{Energy} = \frac{\sum_{i=0}^{n-1} |F_i|^2}{n}$$
$$F_i = \sum_{i=0}^{n-1} e^{-j2\pi \frac{ki}{n}} \tilde{x}_i \qquad (5)$$

To summary, we have seven descriptive statistics, one ratio feature and one energy feature for each of three expense sequences, thus having 27 features in total.

3.5　Career Forecasting Algorithms

Due to the developmental characteristic of career identity, we should be able to predict career choice for each students in each semester. Thus, we organize these evidence by semesters. In other words, in each of the first six semesters, based on their behavior data within the semester, we extract all introduced features for each student. Then in a given semester, features of the proceeding semesters will be concatenated with the current one in an chronological order. Assume the number of features is F, the first semester includes F features, the second semester includes $2F$ features and the sixth semester includes $6F$ features. For each semester, based on the features and career choices of graduation, we train multi-class classification algorithms for career choice prediction. However, this paradigm could not enable the classifier of each semester to leverage weight of classifiers of proceeding semesters. Fortunately, these classification algorithms have relation with the additive models, such as AdaBoost. This motivates us to leverage a two-level ensemble framework for this problem. Its top level algorithm is AdaBoost with six learners, where each learner is trained on F features of each semester. Therefore, the first learner of AdaBoost is trained on features of the first semester and the last learner of AdaBoost is trained the features of the last semesters. Feature sets of different learner are disjoint with each other, we can also continue using AdaBoost algorithm for each learner. Base learners in the second AdaBoost are multi-class classification algorithms.

4　Experiments

In this section, we first introduce the dataset and the settings for evaluation. Following that, we report experimental results and deliver some discussions.

4.1　Dataset and Settings

The evaluation is conducted on a dataset from 4,246 student of the same grade. The total number of consumption records is 13,122,696, among which the records in the mess is 6,875,698. Within four years in the university, these students have borrowed 172,894 books, generating 336,238 book loan records, and taken 1,072 courses in total, generating 276,588 course-grade records. For assessing the performance of multi-class classification, we exploit widely-used Macro-F1 measure and Micro-F1 measure [34]. The former one gives equal weight to every class, regardless of its frequency, and is a per-category average of F1, and the latter one gives equal weight to every document, and is a per-document average of F1. For each of six semesters, 5-folds cross validation is performed. For the multi-class classification, we will conduct comparison between logistic regression (LR), SVM, Random Forest (RF) and Decision Tree (DT).

4.2 Results and Discussions

Comparison of Multiclass Classification Algorithms. We first show the performance comparison between four classification algorithms on the six semesters of data in Fig. 1. The performance comparison on other semesters show consistent result, and thus would not have shown here. We can see that Random Forest performs comparatively better than the others, although the difference is not large. Due to the efficiency of Random Forest, we choose it for subsequent usage.

Prediction Performance. Based on the aforementioned evaluation scheme, we study the change of prediction performance with the increase of semesters and show the results in Fig. 2. First, the results indicate that the performance of career choice prediction will improve with the increase of semesters. This may be because students will learn new skills, extent their reading interest and increase the mastering level of professional skills as time goes by. And the performance of prediction is significantly better than the Random guess (25 % by pure random and 44 % by the MostFreq class). This indicate the career choice could be

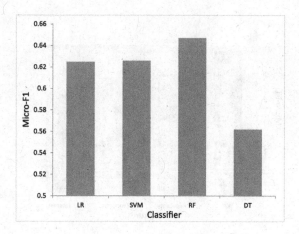

Fig. 1. The comparison of different classification algorithms

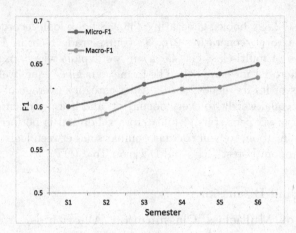

Fig. 2. Micro-F1 measure and Macro-Measure with the increasing semester

predictable from students' life and study behavior in and around the campus. In order to see the effect of different types of features, we study the feature importance for career choice prediction below.

Feature Analysis. By feed each type of features into the classification algorithm, we could obtain the performance of each type of features. The results of evaluating feature importance for career choice is shown in Fig. 3. From them, all features besides book reading interest have shown significant value for predicting career choice. And the effect of skill-based features is most salient. This should be intuitive, since one major goal that students come to university is to learn skills for preparing future occupation. Therefore, the mastery level of professional skills is directly correlated with, or determines students' career choices.

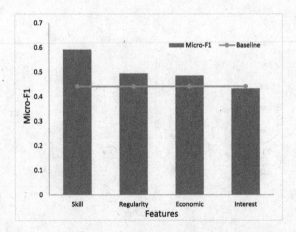

Fig. 3. Importance of four types of features by comparing them with

Behavior regularity and economic status also show significant effect for career choice prediction, confirming our aforementioned assumption. However, the reading interest reflecting in book borrowing preference is not even as well as random guess. This may lies in the sparsity of book loan history, so that it is not robust for such features to reflect students' reading interest. In the future work, it may be more appealing to leverage dimension reduction based algorithms.

5 Conclusion

In this paper, we studied career choice prediction based on students' campus behavior and proposed a data-drive framework for career choice prediction. Based on four types of behavior features, we evaluated the effectiveness of such a framework and found that the extracted professional skills, behavior regularity and economic status was significantly correlated with career choices.

In the future work, we should focus on efficient feature extraction (interest and skills) based supervised dimension reduction and further improve the performance of prediction. And we will also conduct further analysis about the relationship between student's behavior and salary.

References

1. Albion, M.J., Fogarty, G.J.: Factors influencing career decision making in adolescents and adults. J. Career Assess. **10**(1), 91–126 (2002)
2. Balog, K., de Rijke, M.: Finding experts and their details in e-mail corpora. In: Proceedings of WWW 2006, pp. 1035–1036. ACM (2006)
3. Baruch, Y.: Transforming careers: from linear to multidirectional career paths: organizational and individual perspectives. Career Dev. Int. **9**(1), 58–73 (2004)
4. Bem, D.J.: Self-perception theory (1973)
5. Chen, Y.-P., Yang, J.-Y., Liou, S.-N., Lee, G.-Y., Wang, J.-S.: Online classifier construction algorithm for human activity detection using a tri-axial accelerometer. Appl. Math. Comput. **205**(2), 849–860 (2008)
6. Desmarais, M.C.: Mapping question items to skills with non-negative matrix factorization. ACM SIGKDD Explor. Newslett. **13**(2), 30–36 (2012)
7. Deville, P., Wang, D., Sinatra, R., Song, C., Blondel, V.D., Barabási, A.-L.: Career on the move: geography, stratification, and scientific impact. Sci. Rep. **4**, 1–7 (2014)
8. Dudley, N.M., Orvis, K.A., Lebiecki, J.E., Cortina, J.M.: A meta-analytic investigation of conscientiousness in the prediction of job performance: examining the intercorrelations and the incremental validity of narrow traits. J. Appl. Psychol. **91**(1), 40 (2006)
9. Erikson, E.H.: Identity: Youth and Crisis, vol. 7. WW Norton & Company, New York (1994)
10. Festinger, L.: A theory of social comparison processes. Hum. Relat. **7**(2), 117–140 (1954)
11. Gati, I., Krausz, M., Osipow, S.H.: A taxonomy of difficulties in career decision making. J. Couns. Psychol. **43**(4), 510 (1996)
12. Guan, C., Lu, X., Li, X., Chen, E., Zhou, W., Xiong, H.: Discovery of college students in financial hardship. In: Proceedings of ICDM 2015, pp. 141–150. IEEE (2015)

13. Guy, I., Avraham, U., Carmel, D., Ur, S., Jacovi, M., Ronen, I.: Mining expertise and interests from social media. In: Proceedings of WWW 2013, pp. 515–526. International World Wide Web Conferences Steering Committee (2013)
14. Hadiji, F., Mladenov, M., Bauckhage, C., Kersting, K.: Computer science on the move: inferring migration regularities from the web via compressed label propagation. In: Proceedings of IJCAI 2015, pp. 171–177. AAAI Press (2015)
15. Holland, J.L.: Making Vocational Choices: A Theory of Vocational Personalities and Work Environments. Psychological Assessment Resources, Odessa (1997)
16. Hong, W., Li, L., Li, T., Pan, W.: iHR: an online recruiting system for Xiamen talent service center. In: Proceedings of KDD 2013, pp. 1177–1185. ACM (2013)
17. Krumboltz, J.D., Mitchell, A.M., Jones, G.B.: A social learning theory of career selection. Couns. Psychol. **6**(1), 71–81 (1976)
18. Lopez, F.G.: A paradoxical approach to vocational indecision. Pers. Guidance J. **61**(7), 410–412 (1983)
19. Marcia, J.E., Waterman, A.S., Matteson, D.R., Archer, S.L., Orlofsky, J.L.: Ego Identity: A Handbook for Psychosocial Research. Springer Science & Business Media, New York (2012)
20. Paparrizos, I., Cambazoglu, B.B., Gionis, A.: Machine learned job recommendation. In: Proceedings of RecSys 2011, pp. 325–328. ACM (2011)
21. Parsons, F.: Choosing a Vocation. Houghton Mifflin, Boston (1909)
22. Reichling, T., Wulf, V.: Expert recommender systems in practice: evaluating semi-automatic profile generation. In: Proceedings of CHI 2009, pp. 59–68. ACM (2009)
23. Savickas, M.L.: Identity in vocational development. J. Vocat. Behav. **27**(3), 329–337 (1985)
24. Savickas, M.L.: The transition from school to work: a developmental perspective. Career Dev. Q. **47**(4), 326–336 (1999)
25. Schein, E.H.: Career Anchors: Discovering Your Real Values. University Associates San Diego, San Diego (1990)
26. Super, D.E.: A life-span, life-space approach to career development (1990)
27. Thompson, M.N., Subich, L.M.: The relation of social status to the career decision-making process. J. Vocat. Behav. **69**(2), 289–301 (2006)
28. Varshney, K.R., Chenthamarakshan, V., Fancher, S.W., Wang, J., Fang, D., Mojsilović, A.: Predicting employee expertise for talent management in the enterprise. In: Proceedings of KDD 2014, pp. 1729–1738. ACM (2014)
29. Varshney, K.R., Wang, J., Mojsilovic, A., Fang, D., Bauer, J.H.: Predicting and recommending skills in the social enterprise. In: Proceedings of AAAI ICWSM Workshop Social Computing for Workforce, vol. 2, pp. 20–23 (2013)
30. Wang, J., Zhang, Y., Posse, C., Bhasin, A.: Is it time for a career switch? In: Proceedings of WWW 2013, pp. 1377–1388. International World Wide Web Conferences Steering Committee (2013)
31. Wilson, T.D., Dunn, E.W.: Self-knowledge: its limits, value, and potential for improvement. Annu. Rev. Psychol. **55**, 493–518 (2004)
32. Xu, H., Yu, Z., Xiong, H., Guo, B., Zhu, H.: Learning career mobility and human activity patterns for job change analysis. In: Proceedings of ICDM 2015, pp. 1057–1062. IEEE (2015)
33. Xu, Y., Li, Z., Gupta, A., Bugdayci, A., Bhasin, A.: Modeling professional similarity by mining professional career trajectories. In: Proceedings of KDD 2014, pp. 1945–1954. ACM (2014)
34. Yang, Y.: An evaluation of statistical approaches to text categorization. Inf. Retrieval **1**(1–2), 69–90 (1999)

Online Prediction for Forex with an Optimized Experts Selection Model

Jia Zhu[1]([⊠]), Jing Yang[2], Jing Xiao[1], Changqin Huang[1], Gansen Zhao[1], and Yong Tang[1]

[1] School of Computer Science, South China Normal University, Guangzhou, China
{jzhu,xiaojing,cqhuang,zhaogansen,ytang}@m.scnu.edu.cn
[2] Institute of Computer Technology, Chinese Academy of Sciences, Beijing, China
jyang@ict.ac.cn

Abstract. Online prediction is a process to repeatedly predict the next element from a sequence of given previous elements. It has a broad range of applications on various areas, such as medical and finance. The biggest challenge of online prediction is sequence data does not have explicit features, which means it is difficult to remain good predictions. One of popular solution is to make prediction with expert advice, and the challenge is to pick the right experts with minimum cumulative loss. In this article, we use forex prediction as a case study, and propose a model that can select a good set of forex experts by learning a set of previous observed sequences. To achieve better performance, our model not only considers the average mistakes made by experts but also takes the average profit earn by experts into account. We demonstrate the merits of our model on a real major currency pairs data set.

1 Introduction

Online prediction is a process to repeatedly predict the next element from a sequence of given previous elements. It has a broad range of applications on various areas, such as medical and finance [1–4]. In general, there are two methods for the problems of online prediction, with and without prior knowledge. The method without prior knowledge called universal sequence prediction method [5–7], which guarantees the prediction model can achieve optimal loss only after seeing long sequences because no prior knowledge associated with the sequence is used. However, it is difficult for this kind of methods to achieve good performance with limited information even with feature selection techniques [8].

The other method is with prior knowledge by predicting using expert advice. Assume we have a set of x experts, with certain algorithms, e.g., the Weighted Majority (WM) algorithm [9], is guaranteed to perform almost as good as the best expert. As we do not know which one is the best expert in the next sequence, thus, the WM algorithm is a good choice as it can handle various lengths of sequence with expert advice as long as there is an expert existed.

Therefore, it is clear that the key task is to choose a set of experts that can fit the sequence and provide the best performance based on their combination. For instance, as shown in Fig. 1, this is the history record of five forex experts to give their advice for

© Springer International Publishing Switzerland 2016
F. Li et al. (Eds.): APWeb 2016, Part I, LNCS 9931, pp. 371–382, 2016.
DOI: 10.1007/978-3-319-45814-4_30

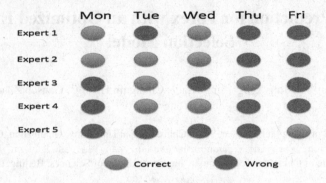

Fig. 1. Sample of expert advice

the daily move of a currency pair. Though Expert 3 and Expert 5 are the only experts have correct prediction on Thursday and Friday respectively, but they may not achieve good results with Expert 2 because their performance are not good on other weekdays.

In this paper, our contribution is to propose a model based on the WM algorithm by adapting empirical risk minimization, which can select a set of experts from training sequences. Our model not only treats the hindsight loss is a function to optimize but also can overcome the overfitting problem, which means the pool of experts our model picked can always perform well on various lengths of sequence. In addition, our model takes the average profit earn by experts into account because profit is also an important factor to judge if an expert is a good expert in forex prediction. Our theoretical analysis provides generalization bounds that show no overfitting for longer histories, and guarantee the advantage of learning in online prediction. The aggregated decision can achieve similar precision to the best expert when the expert set is not large, even the best expert is unknown. We apply our model to forex prediction, and the experimental results demonstrate the merits of our model.

The rest of this paper is organized as follows. In Sect. 2, we discuss related works in online prediction. In Sect. 3, we describe the details of our method. In Sect. 4, we show our experiments, evaluation metrics, and results. Lastly, we conclude this study in Sect. 5.

2   Related Work

The previous studies of online prediction differ significantly with respect to two scenarios: learning from examples and predicting from expert advice. The following review is presented in thematic order, and focuses on important works in these two scenarios.

Littlestone [10] first introduced the Winnow algorithm with Mistake-Bound Learning model in 1988. The Winnow algorithm is designed for learning from examples with few mistakes when the number of relevant variables is much less than the total number of variables. The way of update parameters is similar to the Perceptron algorithm introduced by Rosenblatt [11], and its Mistake Bound model is harder than the PAC learning model of Valiant [12] which is a model required to select examples from a fixed distribution [13].

Model based prediction has also widely been studied. For example, the Naive Bayes sequence classifier which is one of the simplest generative model was proposed by Lewis [14]. It assumes the features in the sequences are independent of each other, and the conditional probabilities of the features are learned in the training stage. However, the independence assumption required by Naive Bayes is often violated in practice because dependence often exists among elements in sequences. Yakhnenko et al. [15] further applied a k-order Markov model that can model the dependence among data. The model is trained in a discriminative setting rather than the generative setting so that the performance of generative model based methods can be improved.

Above classical algorithms of learning from examples can not obtain sound results without long sequences of data [5, 16]. Littlestone and Warmuth [9] proposed the simpler version of WM algorithm to study the problem of predicting from experts, and later a number of refinements are introduced by Cesa-Bianchia et al. [17]. The problem can be specify as a loss function optimal problem according to different settings, e.g., the square loss introduced by Cesa-Bianchia et al. [17] and the log loss function introduced by Foster and Vohra [18].

The other part of work is multitask prediction problem which seeks a common feature space and receives training instances for each class [19]. However, their approach is designed for the data from multiple sources, which is not suitable for the data that only has a set of individual sequences, e.g., the daily changes of foreign exchange rate, which is the kind of data we are going to evaluate in this article.

Yu et al. [21] proposed another approach to map each sequence to an variable that corresponds to the best expert, and then generate a probabilistic suffix tree for prediction based on the previous histories. But this approach is likely to cause overfitting as it strongly depends on the chosen expert but not a set of collaborative experts that is the goal we want to achieve.

Eban et al. [16] described an algorithm that learns a good set of experts using prediction suffix trees (PST) [22]. They used the empirical risk minimization for learning experts, which will also be used in our model with modifications for forex data. Though the idea of their approach is to find the best number of experts which is as small as possible, but the data they used to train is a very long sequence so that the generated experts are not meaningful for short sequence data.

Recently Zhao et al. [23] proposed a framework for online learning with expert advice, which tries to extend two regular forecasters. Though they proved their algorithm under some assumptions, but their work is designed for reducing the query which is not suitable for forex prediction.

3   Randomized Weighted Majority Algorithm

Before proposing our model for experts selection, we introduce the randomized weighted majority algorithm (RWMA) to formulate the problem we intend to resolve because it can achieve a better mistake bound according to [9].

Because we use forex prediction as a case study, therefore we define each forex expert is a function f, and the set of previous history is $X = x_1, ..., x_t$, the set of predictions by the experts is $R = r_1, ..., r_n$, then the output is $\hat{x}$. The algorithm can be formally

Algorithm 1. Randomized Weighted Majority Algorithm

---

  input : A set of predictions $r_1, ..., r_k$ by the experts
1 Initialize the weights $\omega_1, ..., \omega_n$ and probability $p_1, ..., p_n$ of all experts to 1 and $1/n$ respectively.
2 **foreach** *time t* **do**
      output: $\hat{x} = f_i(x_{1:t-1})$ with probability p_t^i
3   induces loss vector l_t and update all the weights.
4   **if** $l_t^i = 1$ **then** $\omega_{t+1}^i = (1 - \beta)\omega_t^i$;
5   where β is the penalized parameter.
6   ;
7   **else** $\omega_{t+1}^i = \omega_t^i$.
8   ;
9      $p_{t+1}^i = \omega_{t+1}^i / W_{t+1}$, where $W_{t+1} = \sum_{i \in R} \omega_{t+1}^i$.
10 **end**

---

described below, where β is the penalized parameter to adjust the weight each time when the expert makes a mistake. In our definition, *mistake* means a wrong prediction.

The following theorem and corollary of the algorithm are provided:

Theorem 1. *On any sequence of trials, the expected number of mistakes M made by the Randomized Weighted Majority algorithm satisfies:*

$$M \leq \frac{m \ln(1/\beta) + \ln n}{1 - \beta},$$

where m is the number of mistakes made by the best expert so far.

Corollary 1. *On any sequence of trials, the expected number of mistakes M made by a modified version of the Randomized Weighted Majority algorithm satisfies:*

$$M \leq m + \ln n + O(\sqrt{m \ln n}),$$

where m is the number of mistakes made by the best expert so far.

Proof of Theorem 1. Define F_j to be the fraction of the weight on the wrong answers at the jth trial, and we have seen t sample, then $M = \sum_{j=1}^{t} F_j$, where M is the expected number of mistakes so far.

Hence the final weight is:

$$W = n \prod_{j=1}^{t} (1 - (1 - \beta))F_j.$$

Let m be the number of mistakes made by the best expert so far, then the total weight must not be smaller than the weight on the best expert:

$$n \prod_{j=1}^{t} (1 - (1 - \beta))F_j \geq \beta^m.$$

If we take the natural log of both sides, we have:

$$\ln n + \sum_{j=1}^{t} \ln(1-(1-\beta))F_j \ge m\ln\beta,$$

$$\ln n - \sum_{j=1}^{t}(1-\beta)F_j \ge m\ln\beta,$$

$$M \le \frac{m\ln 1/\beta + \ln n}{1-\beta}.$$

The problem now is clear, to achieve the smallest M mistakes, we need to select a set of experts and keep the number of experts as small as possible so that the $\ln n$ will be small. However, since none of experts can work well on different sequences, we will also need to find out a set of experts that can cover as many sequences as possible because can handle various lengths of sequence with expert advice unlike most of existing approaches, which means the value of n will be increased. In the next section, we will introduce our experts selection model that can learn a good set of experts based on sequences.

4   Experts Selection

As we mentioned earlier, we apply the empirical risk minimization (ERM) principle [16,24] to find the best set of experts on the training data. This set of experts should achieve a small hindsight generalization loss/mistake. We will prove this corollary later in this section. Firstly, given an forex expert f, and the set of currency pairs price move history $X = x_1, ..., x_t$, then the average number of mistakes of f on X is:

$$L(f, X) = \frac{1}{t}\sum_{i=1}^{t} l(f(x_{1:i-1}), x_i), \tag{1}$$

where function l calculates the mistake of f on each x. The goal is to learn a set of experts F than can minimize the average number of mistakes of RWMA, which is M_F. If the number of mistakes made by the best expert is $m = min_{f \in F} L(f, X)$, according to the Theorem 1, we know for a sequence X:

$$M_F \le \frac{min_{f \in F} L(f, X)\ln 1/\beta + \ln n_F}{1-\beta}, \tag{2}$$

where n_F is the size of F. As the average number of mistakes of F is $L(F, X)$, we have $min_{f \in F} L(f, X) \le L(F, X)$, so the Eq. (2) can be changed to:

$$M_F \le \frac{L(F, X)\ln 1/\beta + \ln n_F}{1-\beta}. \tag{3}$$

According to the Eq. (3), given a set of experts H, we have the best set of experts $F, F \subset H$, with minimum mistakes:

$$L(F, X) = \min_{F \subset H} \frac{1}{t}\sum_{i=1}^{t} L(F, x_i), \tag{4}$$

and can be rewritten as:

$$L(F,X) = \min_{F \subset H} \frac{1}{t} \sum_{i=1}^{t} \sum_{j=1}^{|F|} L(f_j,x_i). \tag{5}$$

Theorem 2. *On any sequence of trials, the expected number of mistakes M made by the Randomized Weighted Majority algorithm should not be smaller than the expected number of mistakes M_F made by the same algorithm if M_F is the average mistakes from the best set of experts.*

Proof of Theorem 2. According to Theorem 1, we have:

$$M \leq \frac{m \ln 1/\beta + \ln n}{1-\beta},$$

and

$$M_F \leq \frac{L(F,X) \ln 1/\beta + \ln n_F}{1-\beta},$$

as $n_F < n$, and in the best case $m = L(F,X)$, therefore $M_F \leq M$. From the proof, we know that the set of experts will achieve smaller hindsight generalization loss in the next sequence if the average number of mistakes $L(F,X)$ is small.

To achieve the best $L(F,X)$, we use Eq. (5) to do the calculation derived from ERM. However, since we intend to use the expert advice for forex prediction, it is also necessary to consider the average profit of prediction in addition to the prediction mistakes. If an expert has made more mistakes than others but obtained bigger profit, we also consider this expert is a good expert. Similarly, we have the average profit of f on X:

$$P(f,X) = \frac{1}{t} \sum_{i=1}^{t} p(f(x_{1:i-1}),x_i), \tag{6}$$

where function p calculates the profit of f on each x. If we assume the average number of mistakes of RWMA is calculated based on profit, then the Eq. (3) can be changed to:

$$M_F \leq \frac{P(F,X) \ln 1/\beta + \ln n_F}{1-\beta}, \tag{7}$$

we then have the best set of experts with maximum profit:

$$P(F,X) = \max_{F \subset H} \frac{1}{t} \sum_{i=1}^{t} \sum_{j=1}^{|F|} P(f_j,x_i). \tag{8}$$

We now have the set of experts $F_{mistake}$ calculated using the Eq. (5), and the set of experts F_{profit} calculated using the Eq. (8). To achieve better performance, we use two fundamental mathematics operations to combine these two set of good candidates:

$$F_{union} = F_{mistake} \cup F_{profit},$$

$$F_{intersection} = F_{mistake} \cap F_{profit}.$$

From the Eqs. (3) and (7) we know that we can minimize the M_F as long as we get the best set of experts on the training instances. In other words, there is no overfitting for longer histories because the outcome does not depend on the length.

5   Experiments

5.1   Data Preparation and Evaluation Metrics

In our experiment, we adopted a real major currency pairs data set, EUR/USD. We manually collected the data from eight websites as shown in Table 1. These eight websites has experts to provide daily prediction for EUR/USD. The prediction has three results, "UP", "DOWN" and "NEUTRAL". The data of the currency pairs price change is retrieved from Yahoo Finance[1]. The period we picked is from 29-Apr-2013 to 25-Apr-2014 consists of 260 daily predictions. Each prediction may have three outcomes, "UP" means the price of the currency pairs will go up next day, "DOWN" means the price of the currency pairs will go down next day, and "NEUTRAL" means the price of the currency pairs will not have big change next day. If the change is less than 10 pips, e.g., the price of EUR/USD drops from 1.4132 to 1.4130, then we think the "NEUTRAL" prediction is correct. Pips is the measure of change in currency pairs in the forex market.

We split the data to four equal size based on time stamp. The data of the first three quarters will be used for learning to experts. Specifically, the third quarter data will be used for 1 quarter learning evaluation, which means we use 1 quarter data to learn. The fourth quarter data will be used for testing including three parts, 1 quarter, 40 days and 20 days. Our goal is to generate prediction based on the advices of a set of experts and select a set of high quality experts.

Table 1. List of forex experts

Expert no.	Expert name	Website
0	ActionForex	www.actionforex.com
1	RoboForex	www.roboforex.com
2	Fxopen	www.fxopen.com
3	Fxempire	www.fxempire.com
4	dailyForex	www.dailyfx.com
5	Jingsong	www.17forex.com
6	Feedo	www.feedoo.com
7	Bank of China	www.boc.cn

Regards to evaluation metrics, we used two indicators, one is prediction accuracy, the other is the net profit. The prediction accuracy is calculated as $Accuracy = \frac{NC}{NC+NW}$, where NC is number of times the prediction is correct, and NW is number of times the prediction is wrong. The calculation of net profit is simple, we just used the total gain minus the total loss. We do not consider other cost in our experiments, e.g., trading fee. In addition, the proposed algorithm can be completed in seconds on the data set, therefore, we are not going to discuss the complexity in this article.

---

[1] http://finance.yahoo.com/.

5.2   Evaluation Results

There are two methods need to be evaluated, namely, $RWMA_{mistake}$ and $RWMA_{profit}$. They are two set of experts based on mistake and profit respectively selected by RWMA algorithm. The union and interaction of these two set of experts were also being tested, namely, $RWMA_{union}$ and $RWMA_{intersection}$. For the parameters of RWMA, we set the value of β is 0.5, and the standard number of selected experts is 3. As a comparison, we used all experts, namely, $RWMA_{all}$ as the baseline method. Table 2 shows the set of experts based on different criteria.

Table 2. Outcomes of learning experts

Criteria	No. of experts
3Q Learning using Mistake	034
3Q Learning using Profit	134
3Q Learning Union	0134
3Q Learning Intersection	34
1Q Learning using Mistake	457
1Q Learning using Profit	134
1Q Learning Union	13457
1Q Learning Intersection	4

Comparisons of Prediction Accuracy. We first compared the EUR/USD prediction accuracy of experts learning from 1 and 3 quarters data as shown in Figs. 2 and 3. From these two figures, we can see that all methods generated by our model are better than the $RWMA_{all}$ method. It is quite impressive that $RWMA_{intersection}$ method achieves better accuracy in the 20 days prediction based on 3 quarters learning compared to other methods, nearly 30 % higher than the baseline method for example. This result prove

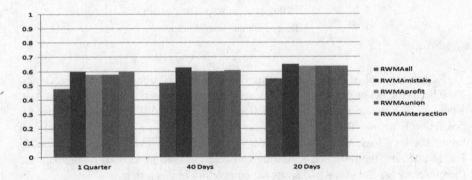

Fig. 2. EUR/USD accuracy of 1Q learning

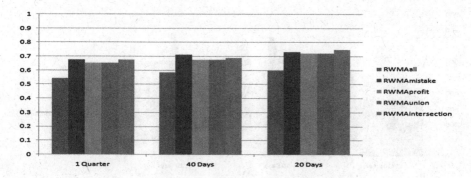

Fig. 3. EUR/USD accuracy of 3Q learning

that the profit factor is useful at some stages maybe because some experts' investment strategy is to maximize profit in each trade rather than trade accuracy.

Comparisons of Prediction Profit. In forex prediction, many people including famous investment funds focus more on prediction profit rather than prediction accuracy. We compared the EUR/USD prediction profit of experts learning from 1 and 3 quarters data as shown in Figs. 4 and 5. The numbers in the figure are price interest point (pips), e.g., if the EUR/USD price rises from 1.3508 to 1.3608 and the prediction is "UP", then the profit is 100 pips. From these two figures, $RWMA_{profit}$ method performs well in the selection period in terms of profit, which achieves nearly 800 pips in the 3 quarters learning. We discover that the results of all methods generated based on 3 quarters learning are better than 1 quarter learning. In addition, $RWMA_{intersection}$ again performs quite well in all three prediction periods if there is sufficient data to learn. The results show that $RWMA_{all}$ can lose a lot of money. For example, in the 1 quarter prediction based on 1 quarter learning, $RWMA_{all}$ earn nearly 40 % less than others.

Comparisons with Other Methods. In our experiments, we adopt two existing methods, namely, Ensemble Tracking and Active Forecaster, which have been proposed by

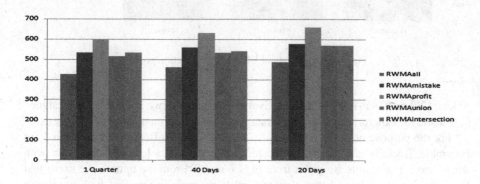

Fig. 4. EUR/USD profit of 1Q learning

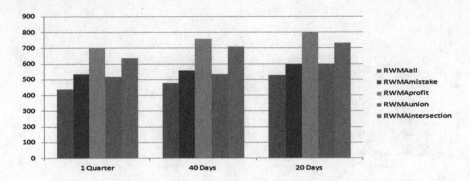

Fig. 5. EUR/USD profit of 3Q learning

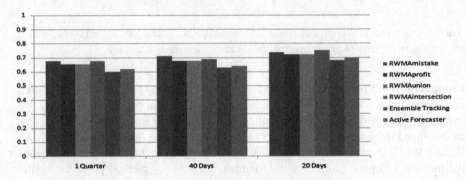

Fig. 6. Comparisons of EUR/USD prediction accuracy

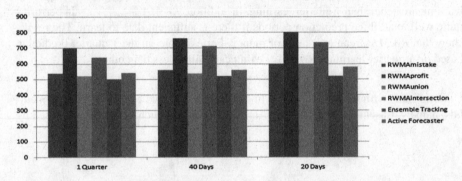

Fig. 7. Comparisons of EUR/USD prediction profit

Avidan [20] and Zhao et al. [23] respectively because these two methods were designed to suitable for experts selection rather than mixed the decision output for all experts.

For the purpose of evaluation, we also set the number of selected experts to 3 for Ensemble Tracking and Active Forecaster using 3 quarters learning consistent with our method. The results are given from Figs. 6 and 7. From the figures, we know that Ensemble Tracking is no better than any other methods in terms of both prediction accuracy and profit because this method was designed for long sequence video tracking data. Though Active Forecaster has similar performance to our $RWMA_{mistake}$ and $RWMA_{profit}$ methods, but it still does not exceed our methods.

6 Conclusions

In this paper, we addressed the problem of online prediction using expert advice. We analyzed the limitations of prediction using experts with the WM algorithm, and found out the key challenge is to choose a set of experts that can fit the sequence. To tackle this problem, we novelly proposed a model to learn these experts from a set of training sequences. Our model not only treats the hindsight loss is a function to optimize but also can overcome the overfitting problem, which means the pool of experts can always perform well using various lengths of sequence. The advantages of our proposed model were proved by theoretical analysis, and supported by extensive experimental evaluation on a real major currency pairs data set with comparisons to other methods. In the future, we will focus on investigating the improvement of the model so that it can effectively update the hypothesis for next prediction after each iteration.

Acknowledgments. This work was supported by the Youth Teacher Startup Fund of South China Normal University (No. 14KJ18), the Natural Science Foundation of Guangdong Province, China (No. 2015A030310509), the National Science Foundation of China (61370229, 61272067, 61303049), and the S&T Projects of Guangdong Province (No. 2013B090800024, No. 2014B010103004, No. 2014B010117007, No. 2016A030303055, No. 2016B030305004, 2016B010109008).

References

1. Duskin, O., Feitelson, D.G.: Distinguishing humans from robots in web search logs: preliminary results using query rates and intervals. In: Proceedings of the 2009 Workshop on Web Search Click Data, WSCD 2009, pp. 15–19 (2009)
2. Kadous, M.W., Sammut, C.: Classification of multivariate time series and structured data using constructive induction. In: Proceedings of the IEEE ISI PAISI, PACCF, and SOCO International Workshops on Intelligence and Security Informatics, PAISI, PACCF and SOCO 2008, pp. 179–261 (2006)
3. Liu, X., Zhang, P., Zeng, D.: Sequence matching for suspicious activity detection in anti-money laundering. In: Yang, C.C., Chen, H., Chau, M., Chang, K., Lang, S.-D., Chen, P.S., Hsieh, R., Zeng, D., Wang, F.-Y., Carley, K.M., Mao, W., Zhan, J. (eds.) ISI Workshops 2008. LNCS, vol. 5075, pp. 50–61. Springer, Heidelberg (2008)
4. Tan, P.N., Kumar, V.: Discovery of web robot sessions based on their navigational patterns. Data Min. Knowl. Discov. **6**, 9–35 (2002)
5. Hutter, M.: On the foundations of universal sequence prediction. In: Cai, J.-Y., Cooper, S.B., Li, A. (eds.) TAMC 2006. LNCS, vol. 3959, pp. 408–420. Springer, Heidelberg (2006)
6. Kajan, L., Kertesz-Farkas, A., Franklin, D., Ivanova, N., Kocsor, A., Pongor, S.: Application of a simple likelihood ratio approximant to protein sequence classification. Bioinformatic 2865–2869 (2006)
7. Wang, J., Zhao, P., Hoi, S.C.H.: Cost-sensitive online classification. In: ICDM, pp. 1140–1145 (2012)

8. Wang, J., Zhao, P., Hoi, S.C.H., Jin, R.: Online feature selection and its applications. IEEE Trans. Knowl. Data Eng., 1–14 (2013)
9. Littlestone, N., Warmuth, M.K.: The weighted majority algorithm. Inf. Comput. **108**, 212–261 (1994)
10. Littlestone, N.: Learning quickly when irrelevant attributes abound: a new linear threshold algorithm. Mach. Learn. **2**, 285–318 (1988)
11. Rosenblatt, F.: The perceptron: a probabilistic model for information storage and organization in the brain. Psychol. Rev. **65**, 386–408 (1958)
12. Valiant, L.G.: A theory of the learnable. Comm. ACM **27**, 1134–1142 (1984)
13. Blum, A.: Separating distribution-free and mistake-bound learning models over the boolean domain. SIAM J. Comput. **23**(5), 990–1000 (1994)
14. Lewis, D.D.: Naive (bayes) at forty: the independence assumption in information retrieval. In: The 10th European Conference on Machine Learning, ECML 1998, pp. 4–15 (1998)
15. Yakhnenko, O., Silvescu, A., Honavar, V.: Discriminatively trained markov model for sequence classification. In: Proceedings of the Fifth IEEE International Conference on Data Mining, ICDM 2005, pp. 498–505 (2005)
16. Eban, E., Globerson, A., Shalev-Shwartz, S., Birnbaum, A.: Learning the exerts for online sequence prediction. In: ICML, pp. 1–8 (2012)
17. Cesa-Bianchia, N., Freund, Y., Helmbold, D.P., Haussler, D., Schapire, R.E., Warmuth, M.K.: How to use expert advice. In: Annual ACM Symposium on Theory of Computeing, pp. 382–391 (1993)
18. Foster, D.P., Vohra, R.V.: A randomeization rule for selecting forecasts. Oper. Res. **41**, 704–709 (1993)
19. Abernethy, J., Bartlett, P.L., Rakhlin, A.: Multitask learning with expert advice. In: Bshouty, N.H., Gentile, C. (eds.) COLT. LNCS (LNAI), vol. 4539, pp. 484–498. Springer, Heidelberg (2007)
20. Avidan, S.: Ensemble tracking. IEEE Trans. Pattern Anal. Mach. Intell. **29**(2), 261–270 (2007)
21. Yu, C.N., Joachims, T.: Learning structural svms with latent variables. In: ICML, 1169–1176 (2009)
22. Ron, D., Singer, Y., Tishby, N.: The power of amnesia: learning probabilistic automata with variable memory length. Mach. Learn. **25**, 117–149 (1996)
23. Zhao, P., Hoi, S., Zhuang, J.: Active learning with expert advice. In: Proceedings of the Twenty-Ninth Conference on Uncertainty in Artificial Intelligence, pp. 1–10 (2013)
24. Vapnik, V.: The Nature of Statistical Learning Theory. Information Science and Statistics. Springer, New York (2000)

Fast Rare Category Detection Using Nearest Centroid Neighborhood

Song Wang[1], Hao Huang[1(✉)], Yunjun Gao[2], Tieyun Qian[1], Liang Hong[3],
and Zhiyong Peng[1(✉)]

[1] State Key Laboratory of Software Engineering, Wuhan University, Wuhan, China
qty@whu.edu.cn, {xavierwang,haohuang,peng}@whu.edu.cn
[2] College of Computer Science, Zhejiang University, Hangzhou, China
gaoyj@zju.edu.cn
[3] School of Information Management, Wuhan University, Wuhan, China
hong@whu.edu.cn

Abstract. Rare category detection is an open challenge in data mining. The existing approaches to this problem often have some flaws, such as inappropriate investigation scopes, high time complexity, and limited applicable conditions, which will degrade their performance and reduce their usability. In this paper, we present FRANC an effective and efficient solution for rare category detection. It adopts an investigation scope based on k-nearest centroid neighbors with an automatically selected k, which helps the algorithm capture the real changes on local densities and data distribution caused by the presence of rare categories. By using our proposed pruning method, the identification of k-nearest centroid neighbors, which is the most computationally expensive step in FRANC, will be much faster for each data example. Extensive experimental results on real data sets demonstrate the effectiveness and efficiency of FRANC.

1 Introduction

Rare category detection (abbreviated as RCD henceforth) helps discover rare categories (a.k.a. rare classes) in an unlabeled data set by proposing their candidate data examples to human experts for labeling [8]. It has various applications, especially in the field like financial fraud detection [4] and network intrusion detection [10], where the rare categories (e.g., fraud transactions and intrusion activities) are of key importance but often hidden in massive normal data.

There are two challenges in RCD, namely, how to efficiently analyse the data set to identify the candidate data examples of rare categories, and how to minimize the number of labeling requests needed to discover all rare categories. To address either or both of the two challenges, so far four types of approaches have been proposed. (1) The clustering or Gaussian mixture model based approaches [6,11,13] prefer data examples that are isolated from clusters or Gaussians as the candidates such that they are not ideal for identifying rare categories hidden in majority categories; (2) the density based approaches [4] are able to find out all rare categories by exploring the abrupt change in local density caused by the

© Springer International Publishing Switzerland 2016
F. Li et al. (Eds.): APWeb 2016, Part I, LNCS 9931, pp. 383–394, 2016.
DOI: 10.1007/978-3-319-45814-4_31

presence of rare categories with a compensation of quadratic time complexity; (3) the wavelet analysis based approaches [10] reduce computational cost by tabulating the data set into bins and calculating the change of local density between adjacent bins via wavelet analysis, although a fixed bin size may degrade their detection performance since rare categories often have various sizes and densities; (4) the nearest neighbor based approaches [3,5,8,9] investigate the density differences between each candidate and its nearest neighbors instead of within a fixed-sized local region, while this kind of neighborhood would make the investigation scope shift to high-density local regions around the candidates and miss the low-density ones, resulting in bias in the investigation.

In order to avoid the flaws of the existing approaches, we present FRANC (Fast Rare cAtegory detection using Nearest Centroid neighborhood) algorithm, a novel and more effective nearest neighbor based approach to RCD. In FRANC, we investigate each candidate and its k-nearest centroid neighbors (abbreviated as kNCN henceforth) to involve both the high-density and low-density local regions around the candidates into the investigation scope. To be more efficient, we exploit a clustering based pruning method to reduce the search space of the kNCN for each data example. Furthermore, to optimize the size of investigation scope for the data examples of rare categories, we propose a k selection method for kNCN by exploring the characteristics of rare categories. Extensive experimental results on real data sets demonstrate that compared with existing four types of approaches, our FRANC algorithm can achieve a reasonably better performance on addressing the two challenges in RCD.

The remaining sections are organized as follow. In Sect. 2, we review the related work. In Sect. 3, we give the problem statement of RCD, a discussion on investigation scope, and a pruning method for kNCN search, following which we present our kNCN based RCD approach in Sect. 4, and report the experimental results in Sect. 5 before concluding the paper in Sect. 6.

2 Related Work

The existing RCD algorithms can be classified into four types, i.e., (1) the clustering or Gaussian mixture model based, (2) the density based, (3) the wavelet analysis based, and (4) the nearest neighbor based approaches.

Clustering or Gaussian mixture model based algorithms [6,11,13] decompose the data set into clusters and Gaussians, and assume that data examples that are isolated from the clusters and Gaussians have the characteristics of rare categories. However, in practice, rare categories are often hidden in majority categories. For example, fraud financial transactions are usually disguised as legal transactions. For these "hidden" rare categories, this type of RCD algorithms requires much more labeling requests to find them out.

Density based algorithms [4] assume that abrupt changes in local density indicate the presence of rare categories. Hence, they perform semi-parametric density estimation over the given data set, and calculate the gradient of the estimated density, which measures the change rate of local density. Nevertheless, this type of RCD algorithms suffers from a quadratic time complexity.

Wavelet analysis based algorithms [10] tabulate data examples into bins and estimate the local density of each bin by Histogram Density Estimation. They measure the change rate of local density by using wavelet analysis, and select candidate data examples from the bins with the greatest change rates. When the bins are large enough, this framework is much more efficient than semi-parametric density estimation for each data example. Nonetheless, if the bin sizes are inappropriate (e.g., oversize or undersize), the performance of this type of RCD algorithms will degrade.

Nearest neighbor based algorithms [3,5,8,9] investigate the maximal change or the variance of local density around each candidate data example. The investigation scope is the k-nearest neighbors (abbreviated as kNN henceforth) of the candidate. The size of investigation scope can be dynamically adjusted according to the local density of the investigated candidate. However, this kind of investigation scope may bring bias for investigation since these nearest neighbors may locate at one side of the candidate rather than really "around" it.

3   Preliminaries

In this section, we present the RCD problem statement and a kNCN based RCD investigation scope, followed by introducing a pruning method for kNCN search.

3.1   Problem Statement

A general problem statement of RCD can be formulated as follows.

Given: (1) a set of N unlabeled data examples $D = \{x_1, \ldots, x_N\}$, $x_i \in \mathbb{R}^d$, which come from m distinct categories; (2) a human expert, who is able to give users the category label of each $x_i \in D$.

Goal: find at least one data example from each category especially each rare category with as less label queries to the human expert as possible.

To minimize the number of label queries, a RCD approach should select the candidates by investigating which data examples are most likely from rare categories. As each rare category has only a few data examples that often appear at a small local region [4,8,10,13], for each candidate, the investigation scope is often a relatively small area around the candidate. Compared with a fixed-sized bin [10] or a fixed bandwidth [4], nearest neighbors would be a more flexible choice for investigation scope, but are still not good enough.

In what follows, we analyse the flaws of investigation scope based on nearest neighbors, and introduce a kNCN based investigation scope for RCD.

3.2   Investigation Scope

Since the data examples of a rare category form a cluster in a small local region, their local densities are often higher than that of the nearby data examples. Hence, for a data example at the boundary of a rare category, such as point x_i in Fig. 1, there will be a high-density local region and a low-density local region

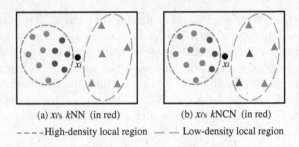

(a) x_i's kNN (in red)   (b) x_i's kNCN (in red)

- - - -High-density local region — — Low-density local region

Fig. 1. Examples of kNN and kNCN. (Color figure online)

around it. In order to utilize this density differential to identify rare categories, an appropriate investigation scope should include both of the two local regions.

Nonetheless, using the nearest neighbors as the investigation scope would miss the low-density local region around each candidate. Take Fig. 1(a) for example, x_i finds its kNN ($k = 5$) only in the high-density local region. This is because most data examples in high-density local region are close enough to x_i as they are from the same rare category of x_i.

To avoid the above flaw, we propose to use kNCN as the investigation scope. kNCN is a common tool for classification [2]. A standard kNCN search for a data example x_i keeps finding a new nearest centroid neighbor x_j ($i \neq j$) in the given data set D such that the mean (a.k.a. centroid) of x_j and current nearest centroid neighbors is closest to x_i, until the total number of nearest centroid neighbors achieves k. Take Fig. 1(b) for example, in order to have a mean close to x_1, kNCN has to involve data examples in low-density local region to prevent the mean from shifting to high-density local region. Given this property, kNCN is a more appropriate investigation scope for RCD.

3.3 Pruning for kNCN Search

A standard kNCN search requires $O(kN^2)$ time. To be more efficient, we propose a clustering based pruning method to reduce the search space from the given data set D to a subset of D with much less data examples inside. The pruning is carried out as follows. We first decompose D into smaller groups via K-means with $K = \lceil N^{\frac{1}{2}} \rceil$, which takes $O(N\lceil N^{\frac{1}{2}} \rceil)$ time, and then perform a standard kNCN search on the set of group centers, which takes $O(k\lceil N^{\frac{1}{2}} \rceil^2)$ time. For each data example in the jth group ($j \in \{1, \ldots, \lceil N^{\frac{1}{2}} \rceil\}$), the search space for its kNCN is the other data examples in this group as well as data examples in groups of which the group centers are the kNCN of the jth group center. As the average number of data examples in each group is $N/\lceil N^{\frac{1}{2}} \rceil \gg k$, there will be enough neighbors for the data example in the jth group for kNCN search. Since $N/\lceil N^{\frac{1}{2}} \rceil \leq \lceil N^{\frac{1}{2}} \rceil$, the expectation of time required for kNCN search in the pruned search space is $O((k+1)N\lceil N^{\frac{1}{2}} \rceil)$. In summary, with pruning, the overall time complexity of kNCN search is expected to be reduced to $O(N\lceil N^{\frac{1}{2}} \rceil)$.

4 kNCN Based Rare Category Detection

In this section, we first propose a kNCN based rare category criterion to help identify the candidate data examples of rare categories, and then introduce how to select an appropriate k value, followed by presenting our FRANC algorithm.

4.1   Rare Category Criterion

For a small local region inside a majority category, if there appears a rare category, the local density and local data distribution will change. These changes will be more obvious at the boundary. To better capture these changes, we propose a kNCN based investigation scope. To quantify these changes, in what follows, we present two measures based on kNCN, i.e., *density variation coefficient* and *distribution variation coefficient*, and combine them as a new rare category criterion called *rare category candidacy*.

Definition 1. Given a data example x_i, let $NCN(x_i, k)$ be the kNCN of x_i, the density variation coefficient around x_i, denoted by $\phi(\cdot)$, is defined as

$$\phi(x_i, k) = \frac{dens(x_i, k)}{\min_{x_j \in NCN(x_i, k)} dens(x_j, k)}$$

where $dens(x_i, k) = 1/\max_{x_\ell \in NCN(x_i, k)} \|x_i - x_\ell\|$.

In this measure, $dens(x_i, k)$ reflects the local density around x_i, $\phi(x_i, k)$ refers to the maximal descending rate between the local densities of x_i and its kNCN. When x_i is a outlier, it will have a low local density $dens(x_i, k)$ and need to find kNCN from nearby clusters, so that its $\phi(x_i, k)$ will be less than 1; when x_i is a majority-category data example, its $\phi(x_i, k)$ will be close to 1 since the data distribution in a small local region of the majority category is usually locally smooth [4]; when x_i is a rare-category data example and the k is large enough to make the kNCN of x_i contain a few of data examples from the nearby majority category, the corresponding $\phi(x_i, k)$ will be greater than 1 since the local density of a rare category is often much higher than that of the nearby majority category.

Definition 2. Given a data example x_i, let $S(x_i, k)$ be the set of the distances between x_i and its kNCN, the distribution variation coefficient around x_i, denoted by $\psi(\cdot)$, is defined as

$$\psi(x_i, k) = \frac{var\big(S(x_i, k)\big)}{mean\big(S(x_i, k)\big)}$$

where $var(\cdot)$ and $mean(\cdot)$ return the variance and mean of a data set respectively.

When x_i is a outlier, the majority of its kNCN will be from nearby clusters, resulting in a great $mean\big(S(x_i, k)\big)$; when x_i is a majority-category data example, its $var\big(S(x_i, k)\big)$ will be small due to the locally smooth data distribution inside the majority category; when x_i is a rare-category data example and the k is large enough to make the kNCN of x_i contain a few data examples from the nearby majority category and the majority of data examples from the rare category, the $var\big(S(x_i, k)\big)$ will be great and $mean\big(S(x_i, k)\big)$ will be small, resulting in a greater $\psi(x_i, k)$.

Definition 3. Given a data example x_i, the rare category candidacy of x_i, denoted by $RCC(\cdot)$, is defined as

$$RCC(x_i, k) = \phi(x_i, k) * \psi(x_i, k). \tag{1}$$

With an appropriate k value, the data examples from rare categories will have greater $\phi(x_i, k)$ and $\psi(x_i, k)$, and then higher scores on rare category candidacy (abbreviated as RCC) than the other types of data examples.

4.2 The Selection of k

According to the discussions below the Definitions 1 and 2, an appropriate k value should be large enough to make the kNCN of a rare-category data example contain a few data examples from the nearby majority category and the majority of data examples from the rare category. In order words, if the selected k value is close to or slightly larger than the number of data example of the objective rare category, it will be an appropriate choice for our RCC criterion. Given the conclusion above, in the next we propose an HDE-based selection for k value.

For the sake of simplicity, we start from one-dimensional data sets. For example, Fig. 2(a) illustrates the underlying data distribution of an one-dimensional data set containing N data examples, in which the small abrupt probability peak around $x = 28$ indicates there is a rare category. To select a k value for this data set, we first perform HDE (Histogram Density Estimation) to tabulate the data examples into w bins. Let h denotes the bandwidth of each bin, and v_t be the bin count (i.e., the number of data examples) of the tth bin B_t ($t \in \{1, \ldots, w\}$). An optimal bandwidth h^* will balance the smoothness and precision of density estimation, and can be calculated as [12]:

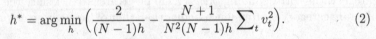

$$h^* = \arg\min_h \left(\frac{2}{(N-1)h} - \frac{N+1}{N^2(N-1)h} \sum_t v_t^2 \right). \tag{2}$$

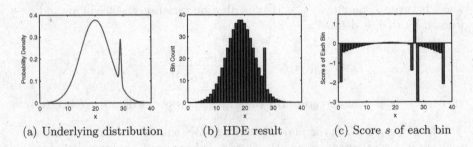

(a) Underlying distribution   (b) HDE result   (c) Score s of each bin

Fig. 2. Example of k selection on one-dimensional data sets.

In other words, each bin with the optimal bandwidth will contain as many as possible data examples that are similar enough in locations and local densities [7]. Furthermore, as a rare category often forms a cluster in a small local region, it will appear in a bin with the optimal bandwidth (see Fig. 2(b) for example). For this bin, its bin count will be an appropriate choice for k selection, since it contains all (or most of) data example of the rare category as well as a few data examples of the majority category. We can find this bin by exploring the density differentials between itself and its adjacent bins. Formally, for each bin B_t, we calculate a score $s(B_t) = \sum_{j=t-1}^{t+1} \frac{v_t - v_j}{v_t}$ to quantify the density differentials. Figure 2(c) illustrates the score s of each bin in Fig. 2(b). After finding out the bin with the maximal score s, we set the bin count of this bin as the k value.

For a multi-dimensional data set, the data distribution is not stretched and distinct enough on each dimension. Therefore, we do not need to carry out HDE on all dimensions. Instead, we utilize PCA (Principal Component Analysis) to extract the principal components, on which the data distribution is most stretched and distinct, and then tabulate each of the principal components by HDE and find the bin with maximal score s for our k selection. In PCA, the eigenvalues of the covariance matrix of the data set indicate how principal each component is. Hence, we can determine which components are the most principal ones by identifying the largest eigenvalues from the remaining ones. To this end, we look for a threshold τ to partition the eigenvalues into two groups by maximizing the variance of the group centers while minimizing the variance of the eigenvalues of each group separately. Formally, the optimal τ should be $\max_{\tau} \frac{(\mu_1 - \mu_2)^2}{\sigma_1 + \sigma_1}$, where μ_1 and μ_2 are the group centers with current threshold τ, σ_1 and σ_2 are the corresponding variances of eigenvalues in the two groups.

Given the discussion above, we present our k selection method as follows.

Step 1. (a) If data set D is multi-dimensional, reduce its dimensionality d to c_p (i.e., the number of most principal components determined above) by PCA, and store the dimensionality-reduced data set as D'. (b) If D is one-dimensional, set $c_p = 1$, $D' = D$.

Step 2. Perform HDE on each dimension i of D' ($i \in \{1, \ldots, c_p\}$) with an optimal bandwidth h_i determined by Eq. (2), and tabulate D' into $\prod_{i=1}^{c_p} \lceil e_i/h_i \rceil$ bins, where e_i refers to the extent of data distribution on dimension i of D'.

Step 3. Find the bin with maximal score $s(B_t) = \sum_{B_j \in A(B_t)} \frac{v_t - v_j}{v_t}$, where $t \in \{1, \ldots, \prod_{i=1}^{c_p} \lceil e_i/h_i \rceil\}$, $A(B_t)$ refers to the set of bins adjacent to B_t, and then set the bin count of this bin as the k value.

4.3   The FRANC Algorithm

With the RCC criterion and an automatically selected k, we present our FRANC algorithm, of which the intuition is to select data examples with the maximal RCC scores as the candidates of rare-category data examples for labeling.

Algorithm 1 outlines the pseudo-code of FRANC. It takes inputs a given data set D containing N data examples, and a k value which is automatically selected from D that maximizes the RCC scores of the data examples from

---

Algorithm 1. The FRANC Algorithm

Input : Unlabeled data set D, automatically selected k.
Output: The set Q of queried data examples, and the set L of their labels.
1 Initialize $Q = \varnothing$, $L = \varnothing$;
2 $\forall x_i \in D$, calculate $RCC(x_i, k)$ by Eq. (1) with pruning method for kNCN;
3 **while** the number of discovered categories is less than m **do**
4   Query $q = \arg\max_{x_i \in D} \left(RCC(x_i, k) \right)$ for its category label ℓ_q;
5     $Q = Q \bigcup q$, $L = L \bigcup \ell_q$;
6     $\forall x_i \in q \bigcup NCN(q, k)$, set $RCC(x_j, k) = -\infty$; $//NCN(q, k)$ is the kNCN of q
7 **end while**

---

rare categories and helps FRANC to detect rare categories more accurately. The algorithm starts from initializing two empty sets (line 1), namely the set Q, which is used to record the selected data examples for labeling, and the set L, which is used to record the corresponding labeling results, following which it finds the kNCN of each data example in D with the help of our proposed pruning method for kNCN search, and calculates the RCC score for each data example (line 2). Then, FRANC keeps selecting data example $q \in D$ with the maximal RCC score, and querying its category label ℓ_q until all m categories in D are discovered (line 4). After each query, q and ℓ_q will be added into sets Q and L respectively (line 5); the RCC scores of q and its kNCN will be set as minus infinity (line 6) to avoid the algorithm repeatedly querying data examples around q and save labeling requests.

In FRANC, the most time-consuming step is kNCN search. With our proposed pruning method, the dominant item in the time complexity of FRANC is expected to be reduced to a level of $O(N\lceil N^{\frac{1}{2}}\rceil)$, which is lower than most of the existing RCD algorithms. Moreover, extensive experimental results demonstrate that compared with the existing approaches, FRANC has a reasonably better performance on efficiency, and a significantly better performance on RCD.

5   Experimental Evaluation

In this section, we first introduce the experimental setup, and then evaluate the RCD performance and efficiency performance of FRANC, followed by discussing the effect of our k selection method.

5.1   Experimental Setup

We evaluate our algorithm on nine commonly used real data sets from the UCI data reposity [1], namely the Glass, Vertebral, Ecoli, Statlog, Pen Digits, Wine Quality, Shuttle, Page Block, and Letters data sets. Following the experimental setup in [8,10,13], the Verterbal, Statlog, Pen Digits and Letters data sets are subsampled to create imbalanced data sets that suit the RCD scenario, since in their original data sets, each category has almost the same number of data examples.

Table 1. Description of UCI data sets

Data set	N	d	m	Largest category	Smallest category
Glass	214	9	6	35.51 %	4.21 %
Vertebral	310	6	3	48.39 %	19.35 %
Ecoli	336	7	6	42.56 %	2.68 %
Statlog	512	19	7	50.00 %	1.56 %
Pen digits	1040	16	10	49.23 %	0.77 %
Wine quality	1589	11	5	42.86 %	1.13 %
Shuttle	4515	9	7	75.53 %	0.13 %
Page block	5473	20	5	89.77 %	0.51 %
Letters	19500	16	26	4.17 %	1.20 %

The property of the data sets are listed in Table 1, where N denotes the number of data examples, d the dimensionality, and m the number of categories in each data set, the values in the "Largest Category" and "Smallest Category" columns refer to the proportions of data examples from the largest and smallest categories, respectively. Moreover, to prevent the kNCN search being dominated by only a few dimensions of the data examples, we pre-process each data sets by min-max normalization to map the data distribution on each dimension to the range $[0, 1]$. All algorithms in the experiments are implemented in Java, running on a PC with Intel Core i7 at 2.0 GHz and 8.0 GB RAM.

5.2   RCD Performance

Since the goal of RCD is to find at least one data example from each category in an unlabeled data set by as less label queries as possible, the performance of a RCD algorithm is usually evaluated by the number of queries needed by the algorithm to discover all the categories in the data set.

In this experiment, we compare the performance of our FRANC algorithm with that of the HMS [13], SEDER [4], vFRED [10], and CLOVER [8] algorithms, which are the high-performance representatives of the existing four types of RCD approaches. Figure 3 illustrates each algorithm's performance curve, in which the y-axis is the percentage of discovered categories, and the x-axis is the corresponding number of required queries. From the figure, we can observe that FRANC outperforms the existing approaches in terms of the number of queries required for discovering all categories in each tested data set.

5.3   Efficiency Performance

In this experiment, we compare the runtime across the tested algorithms, namely HMS, vFRED, SEDER, CLOVER, and FRANC. Table 2 shows the comparison result, from which we have the following observations. (1) the runtime of FRANC

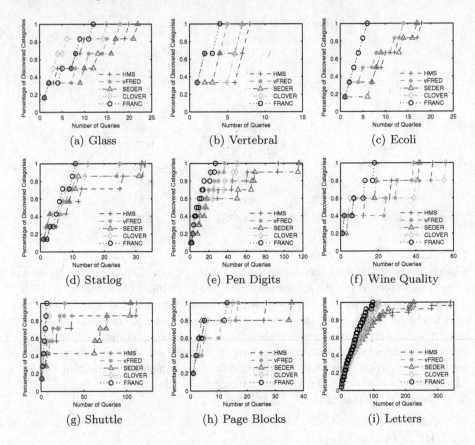

Fig. 3. Performance comparisons on real data sets.

is significantly less than that of HMS, SEDER, and CLOVER, of which the time complexities are nearly quadratic or even cubic; (2) the runtime of FRANC is close to (in most cases, is slightly better than) that of vFRED, which is the one of the fastest RCD algorithm in existing work. But note that the RCD performance of FRANC is significantly better than that of vFRED (see Fig. 3).

5.4   Effect of Selected k Value

The k value is of importance for the RCD performance of FRANC, since it affects the size of the investigation scope and the RCC score of each candidate data example. To verify the effectiveness of our selection of the k value, we vary the value of k used in FRANC and report the corresponding total number of queries required by FRANC in Fig. 4, in which the FRANC's performance with the selected k value is marked with a red asterisk. From the figure, we can observe that although our selected k value may not be the optimum k value for FRANC, it can still bring our algorithm a reasonably improved RCD performance.

Table 2. Runtime of each algorithm (seconds)

Data set	HMS	vFRED	SEDER	CLOVER	FRANC
Glass	0.68	0.30	2.54	0.29	0.28
Vertebral	1.07	0.32	3.14	0.38	0.30
Ecoli	1.11	0.40	3.54	0.41	0.31
Statlog	24.87	1.18	44.07	3.16	0.86
Pen digits	148.36	17.35	206.49	13.11	11.41
Wine quality	289.31	21.59	302.58	13.55	12.22
Shuttle	>1000.00	76.93	>1000.00	41.06	37.13
Page blocks	>1000.00	87.58	>1000.00	43.14	39.10
Letters	>1000.00	>1000.00	>1000.00	>1000.00	852.33

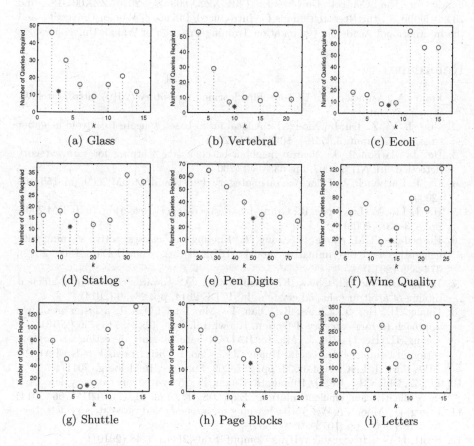

(a) Glass   (b) Vertebral   (c) Ecoli

(d) Statlog   (e) Pen Digits   (f) Wine Quality

(g) Shuttle   (h) Page Blocks   (i) Letters

Fig. 4. FRANC's performance with various k values.

6   Conclusion

In this paper, we have introduced an effective and efficient approach called FRANC for RCD. To better capture the changes of local density and data distribution around a data example, FRANC uses the kNCN of the data example as the investigation scope. The candidate data examples of rare categories are selected by FRANC based on a kNCN based rare category criterion called rare category candidacy (RCC). To improve the RCD performance of FRANC, we have proposed a k selection method to maximize the RCC scores of rare-category data examples. To be more efficient, we have presented a pruning method for kNCN search, which is the most computationally expensive step in FRANC. Extensive experiments on real data sets have been conducted, and the experimental results have verified the effectiveness and efficiency of our algorithm.

Acknowledgements. This work was supported in part by NSFC Grants (61502347, 61522208, 61572376, 61303025, 61379033, and 61232002), the Fundamental Research Funds for the Central Universities (2015XZZX005-07, 2015XZZX004-18, and 2042015kf0038), the Research Funds for Introduced Talents of Wuhan University, and the International Academic Cooperation Training Program of Wuhan University.

References

1. Frank, A., Asuncion, A.: UCI machine learning repository (2010). http://archive.ics.uci.edu/ml/
2. Gou, J., Yi, Z., Du, L., Xiong, T.: A local mean-based k-nearest centroid neighbor classifier. Comput. J. **55**(9), 1058–1071 (2012)
3. He, J., Carbonell, J.: Nearest-neighbor-based active learning for rare category detection. In: NIPS 2007, pp. 633–640 (2007)
4. He, J., Carbonell, J.: Prior-free rare category detection. In: SDM 2009, pp. 155–163 (2009)
5. He, J., Liu, Y., Lawrence, R.: Graph-based rare category detection. In: ICDM 2008, pp. 833–838 (2008)
6. Hospedales, T.M., Gong, S., Xiang, T.: Finding rare classes: active learning with generative and discriminative models. IEEE Trans. Knowl. Data Eng. **25**(2), 374–386 (2013)
7. Huang, H., Gao, Y., Chiew, K., Chen, L., He, Q.: Towards effective and efficient mining of arbitrary shaped clusters. In: ICDE 2014, pp. 28–39 (2014)
8. Huang, H., He, Q., Chiew, K., Qian, F., Ma, L.: CLOVER: a faster prior-free approach to rare-category detection. Knowl. Inf. Syst. **35**(3), 713–736 (2013)
9. Huang, H., He, Q., He, J., Ma, L.: RADAR: rare category detection via computation of boundary degree. In: Huang, J.Z., Cao, L., Srivastava, J. (eds.) PAKDD 2011, Part II. LNCS, vol. 6635, pp. 258–269. Springer, Heidelberg (2011)
10. Liu, Z., Chiew, K., He, Q., Huang, H., Huang, B.: Prior-free rare category detection: more effective and efficient solutions. Expert Syst. Appl. **41**(17), 7691–7706 (2014)
11. Pelleg, D., Moore, A.W.: Active learning for anomaly and rare-category detection. In: NIPS 2004, pp. 1073–1080 (2004)
12. Scott, D.W.: Histogram. WIREs Comput. Stat. **2**(1), 44–48 (2010)
13. Vatturi, P., Wong, W.: Category detection using hierarchical mean shift. In: KDD 2009, pp. 847–856 (2009)

Latent Semantic Diagnosis in Traditional Chinese Medicine

Wendi Ji[1], Ying Zhang[1], Xiaoling Wang[1(✉)], and Yiping Zhou[2]

[1] Shanghai Key Laboratory of Trustworthy Computing,
Institute for Data Science and Engineering,
East China Normal University, Shanghai, China
{wendygee,ying.zhang}@ecnu.cn, xlwang@sei.ecnu.edu.cn
[2] Basic Medical College, Shanghai University of Traditional Chinese Medicine,
Pudong, China
sunrising318@163.com

Abstract. Traditional Chinese Medicine (TCM) is the main route of disease control for ancient Chinese. Through thousands of years' development and inheriting, TCM is the most influential traditional medical system which lasts the longest time and used by the largest population. However, the theory of TCM lacks objective and quantitative standards. In this paper, we propose a statistical diagnosis approach to find out the pathogenesises based on the latent semantic analysis of symptoms and the corresponding herbs. We assume that the latent pathogenesis is the inherent connection between symptoms and herbs within a medical case. Previous topic models mostly focus on single content documents, but medical cases have two different contents: symptoms and herbs. We therefore develop a novel muti-content model based on LDA. We used the proposed model to analysis two TCM domains *amenorrhea* and *lung cancer*. Experiment results illustrate that the pathogenesises found by our model correspond well with the theory of TCM and it provides a theoretical data-driven approach to establish diagnosis standards.

Keywords: Latent semantic model · Traditional Chinese Medicine · Clustering

1 Introduction

As a system of methodology and approach for disease diagnosis and treatment, Traditional Chinese Medicine has a long story in Chinese society. Actually, TCM has been successfully applied to the treatment of various complex diseases such as cancer, rheumatoid arthritis, irregular menstruation. TCM diagnosis consists of two steps: in the first step, a doctor collects symptoms through observation, auscultation and olfaction, inquiry and pulse diagnosis; and in the second step, the doctor gives diagnosis conclusions and herb prescriptions by analyzing patient information based on TCM theories and own experience.

Matching symptoms and herbs is the way that forms TCM theory and is the main manifestation of TCM theory in early literatures. Many distinguished

F. Li et al. (Eds.): APWeb 2016, Part I, LNCS 9931, pp. 395–407, 2016.
DOI: 10.1007/978-3-319-45814-4_32

classical books in Chinese have been written recording clinical cases or knowledge about diagnosises. However, most of the diagnosises were given empirically and subjectively. Furthermore, the inheriting way of TCM is based on the analysis and summarization of previous cases, so that the theoretical system of TCM is structured subjectively. This is the reason why, in practice, people raise doubts about TCM. Therefore, exposing the relationships of the traditional Chinese medical entities (e.g. symptoms, herbs, environment), and finding scientific knowledge or hypotheses among these entities are of great importance.

Data mining method has been applied to researches of TCM in recent years. Pointwise mutual information (PMI)[5], which quantifies the discrepancy between a pair of data, can be used in discovering the relationship between single symptom and single herb. However, we demand to find the relationship between several symptoms and several herbs. Zhang et al. developed a new clustering method in the form of hierarchical latent class model and applied this model to discover the latent pathogenesis in TCM diagnosis [7]. This is a significant work for syndrome differentiation but the pathogenesis is only based on symptoms, without considering the herbs corresponding to the symptoms. To solve this problem, we treat pathogenesis as the latent factor associated with symptoms and herbs. To the best of our knowledge, our work is the first exploration to find the relationship between symptoms and herbs using latent semantic analysis.

In this paper, we propose a theoretical data-driven model to find the latent factor associating symptoms with herbs. A Multi-Content LDA Model (MCLDA) is proposed based on Latent Dirichlet Allocation(LDA). With the additional assumption that symptom terms are exchangeable among them and it is also true for the herb terms, we can model them with a bag-of-words model. In the model, symptom terms and herb terms share the same topic, and each term is associated with two Dirichlet-Multinomial distributions. The topic here is the latent variable, pathogenesis, that we aim to find. Then we use Gibbs Sampling algorithm to estimate the parameters. To test our model, we use medical cases in ancient Chinese medicine books and modern medicine medical cases as our experimental datasets. The results demonstrated the effectiveness of the proposed model in finding the latent relationship between symptoms and herbs.

The main contributions of this paper are:

1. A new research problem has been proposed to find the relationship between symptoms and herbs in TCM diagnosis using latent semantic analysis.
2. A Multi-Content LDA Model has been proposed to explore the latent pathogenesis of different patients, which associates symptoms with herbs in medical cases.
3. The proposed model has been evaluated. Experimental results confirmed the latent pathogenesises found by the data-driven model are mostly consistent with the syndrome factors in TCM theories.

The rest of the paper is organized as follows. Section 2 gives a review of the related work. We definite the problem in Sect. 3. In Sect. 4, we detail the Multi-Content LDA Model. Then in Sect. 5, we report the experimental results and its evaluation. Finally, we conclude our work in Sect. 6.

2   Related Work

Text mining plays an important role in the research of Traditional Chinese Medicine. In Chinese medical formula (CMF) research, Zhou et al. performed experiments to discover combination rules of herbs based on frequent itemset mining [4]. In TCM clinical diagnosis, Yao et al. discovered characteristic features for tongue manifestations, and these features were treated by K-means clustering to predict the diagnostic result [8]. K-means is a popular clustering algorithm which have been applied widely on CMF and clinical diagnosis. However, in TCM theory, one symptom may be caused by various latent pathogeneses, and one herbs may have effect on various disease. Therefore, the hard clustering that assign TCM entities to only one group of latent pathogeneses is not suitable.

Latent semantic models, such as matrix factorization, PLSA and LDA, have been proved to perform well in finding latent variable. Probabilistic topic modeling has achieved great success in text mining. Lei et al. applied PLSA in herbal prescription development, aiming to calculate some new herbal combinations [9]. But the major drawback of PLSA is the parameter space is proportional to dataset size, and there is no natural way to handle unseen data point. Another latent semantic analysis in TCM was conducted by Zhang et al., they proposed hierarchical clustering model, latent tree model, to discover latent structures of symptoms and applied this model to establish objective standards for syndrome differentiation in TCM diagnosis of kidney deficiency syndromes [7]. Their analysis confirmed that the nature clusters found by latent tree model correspond well to the TCM syndrome types.

Since Blei et al. proposed LDA for document classification [2], it was applied in various domains. On one hand, LDA is a soft clustering method that assign TCM entities to various topics. On the other hand, it is a latent semantic model that regards each topic as a distribution on words and each document as a distribution on topics, so the parameter space doesn't increase when dataset increases. Therefore, we develop the Multi-Content LDA Model (MCLDA) on the basis of traditional LDA for modeling relationship between the symptoms and herbs. This model can simultaneously learn more than one type of entities sharing the same topic.

3   Problem Definition

In this paper, we aim to find the relationship between symptoms and herbs of TCM cases and propose an objective and statistical analysis method to identify pathogenesises that associate symptoms and herbs. A medical case is defined as a document that contains a group of symptoms and a group of corresponding herbs. Determining the pathogenesis is the premises of TCM diagnosis and treatment. In our model, a pathogenesis is the latent topic that leads to a list of symptoms and is the basis for doctors to prescribe a list of herbs.

In TCM theories, doctors understand human body from an energetic and functional perspective. It believes that disease is caused by some imbalance of the body, such as "kidney deficiency". Let's track back to the development of TCM. Ancient doctors first found that some herbs are able to treat a disease, where the disease just refers to some symptoms and the actual disease was not defined at that time. In the early period, the correspondence between herbs and symptoms was rough and empirical. Then, doctors started to build TCM theories to tackle multiple and complex symptoms. They defined pathogenesises as types of imbalance within body through the analysis of herbs and symptoms. For example, if a herb has effects on a symptom, they may belong to the same pathogenesis. As shown in Fig. 1, the pathogenesises are the linkage between symptoms and herbs. However, the analysis and inference of TCM theory is subjective and empirical, which results in doubts by modern medicine.

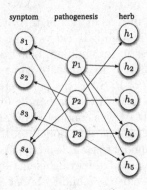

Fig. 1. Tripartite graph among symptoms, herbs and pathogenesises.

Inspired by the traditional thoughts of TCM, in this paper, we use latent pathogenesises to connect symptoms and herbs. We assume that symptoms and herbs of a medical case are independent given the latent pathogenesises of the patient, which means that the symptoms are the appearance of inherent pathogenesises and herbs are treatments that target at the inherent pathogenesises. To find out the latent pathogenesises, we regard each medical case as a document, symptoms and herbs as two kinds of words. We propose a multi-content model based on LDA to classify the two types of words into several topics, each topic contains several symptoms and several herbs. The topic here is the latent pathogenesis that we found in medical cases, which is the essential connection between symptoms and herbs.

4   Multi-content LDA Model

In this section, we propose the Multi-Content LDA Model to find latent pathogenesises. We first illustrate the proposed model and its corresponding inference algorithm. Then we introduce the parameter estimation method. The notations that will be used in the paper are presented in Table 1.

Table 1. Notations

Notation	Description
K	the number of pathogenesis topics
M_d, N_d	the number of symptom words, herb words
S, W	the number of terms in symptom dictionary, in herb dictionary
d, D	a document, a document set
z, z	pathogenesis topics, a topic list
s, s	a symptom term, the symptom term list representation of the corpus
w, w	a herb term, the herb term list representation of the corpus
l	an indicator representing the term's type.
θ	multinomial distribution over pathogenesis topics
ϕ	multinomial distribution over symptom terms
ψ	multinomial distribution over herb terms
α	Dirichlet prior vector for θ
β	Dirchlet prior vector for ϕ
γ	Dirchlet prior vector for ψ
i_s	$i_s = (d_s, m)$ is a two-dimensional coordinates, which is the mth word of the dth document's symptom part
i_w	$i_w = (d_w, m)$ is a two-dimensional coordinates, which is the mth word of the dth document's herb part
$N_{sm}^{(k)}$	the number of symptoms that are assigned topic k in symptom part of document m
$N_{wm}^{(k)}$	the number of herbs that are assigned topic k in medicine part of document m
$N_{sk}^{(p)}$	the number of symptoms with the value p that are assigned topic k
$N_{wk}^{(q)}$	the number of herbs with the value q that are assigned topic k
$N_{*m,j}$	the number of term j in document m

4.1 Multi-content LDA Model

Our topic modeling approach is based on Latent Dirichlet Allocation(LDA), a hierarchical Bayesian model proven successful in learning a set of latent topics used widely in document classification and recommendation systems. In LDA, each document is assumed to be characterized by a particular set of topics. Each topic is identified on the basis of supervised labeling on the basis of their likelihood of co-occurrence. A word may occur in several topics with a different probability, however, with a different typical set of neighboring words in each topic. Each topic in LDA is a multinomial distribution over only one kind of words. However, in our Traditional Chinese Medincine(TCM) problem, one pathogenesis, which is the latent topic in our model, is associated with both symptoms and herbs. We therefore extend the LDA model and propose the Multi-Content LDA Model (MCLDA) for TCM.

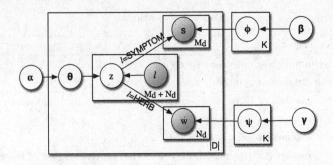

Fig. 2. Graphical model for MCLDA

The proposed Multi-content LDA Model is a generative probabilistic model extending LDA, which can be presented by the graphical model in Fig. 2. A medical case as a document contains two kinds of items: symptom terms and herb terms. These two items are independent given the latent pathogenesises of the patient, which are the topics of the case. Indicator l is a binary variable, whose value is either $SYMPTOM$ or $HERB$. When $l = SYMPTOM$, a symptom word will be generated, and when $l = HERB$, a herb word will be generated.

Each document is generated by drawing a document-specific mixture factor θ over topic $1, ..., K$ and the mix θ is drawn from a Dirichlet prior α. We constrain that given the distribution of K topics, the symptoms and herbs are independent. Therefore, each symptom has an individual symptom-specific topic z and each herb has an individual herb-specific topic z. Both topic z is drawn from multinomial distribution θ. Symptom terms s are drawn from multinomial distribution ϕ based on the corresponding topic z and herb terms w are drawn from multinomial distribution ψ based on the corresponding topic z. Dirichlet prior β and γ are the prior distributions for term distribution ϕ and ψ.

The generative process of MCLDA proceeds in Algorithm 1.

Algorithm 1. Generative process of MCLDA

1: **for** each topic $k \in K$ **do**
2:    Draw $\phi_k \sim Dirchlet(\beta)$;
3:    Draw $\psi_k \sim Dirchlet(\gamma)$;
4: **end for**
5: **for** each document $d_m \in D$ **do**
6:    Draw $\theta_m \sim Dirchlet(\alpha)$;
7:    **for** each term in d_m **do**
8:       Draw topic $z_{m,i} \sim Disc(\theta_m)$;
9:       **if** l=SYMPTOM **then**
10:          Draw symptom term $s_{m,i} \sim Disc(\phi_{z_{m,i}})$;
11:       **end if**
12:       **if** l=HERB **then**
13:          Draw herb term $w_{m,i} \sim Disc(\psi_{z_{m,i}})$;
14:       **end if**
15:    **end for**
16: **end for**

4.2   Parameter Estimation

We aim to compute of the joint likelihood of the observed symptoms and herbs of different medical cases. Given the hyperparameters α, β and γ, the joint distribution of document-topic distribution θ, topic-word distribution ϕ and ψ, topic assignments z, indicator l, symptoms s and herbs w is given by:

$$
\begin{aligned}
p(s, w, z, l, \theta, \phi, \psi | \alpha, \beta, \gamma) &= p(\phi|\beta)p(\psi|\gamma)p(\theta|\alpha)p(s, w, z, l|\phi, \psi, \theta) \\
&= p(\phi|\beta)p(\psi|\gamma)p(\theta|\alpha)p(s, w|z, l, \phi, \psi)p(z|l, \theta)p(l)
\end{aligned}
\tag{1}
$$

The topic distribution $p(z|\theta)$ is the same as standard LDA. The symptom terms s and the herb terms w eventually make similar contribution to the full likelihood. The sampling formulas update the individual topic assignments $z_{d,m}$ or $z_{d,n}$. According to the independence assumption of symptoms and herbs, we consider the probability of the observations of the model.

When $l = SYMPTOM$, the probability of the symptom terms is given as follows:

$$
p(s|z, \beta) = \int \prod_{m=1}^{|D|} \prod_{j=1}^{S} p(s_{m,j}|\phi_{z_m})^{n_{s_{m,j}}} \prod_{z=1}^{K} p(\phi_{z_m}|\beta) d\Phi
\tag{2}
$$

When $l = HERB$, the probability of the herb terms is given as follows:

$$
p(w|z, \gamma) = \int \prod_{m=1}^{|D|} \prod_{j=1}^{W} p(w_{m,j}|\psi_{z_m})^{n_{w_{m,j}}} \prod_{z=1}^{K} p(\psi_{z_m}|\gamma) d\Psi
\tag{3}
$$

To estimate the parameters of our model, we use Gibbs sampling algorithm as the learning algorithm. Gibbs sampling is the most widely used version of the Hastings-Metropolis algorithm for Markov Chain Monte Carlo (MCMC) simulation. It simulates a sequence of random variables whose joint distribution always converges [10]. In Gibbs sampling method, to sample the observed document containing $s = (s_1, s_2, ..., s_S)$ and $w = (w_1, w_2, ..., w_W)$ from joint distribution $p(s_1, s_2, ..., s_S, w_1, w_2, ..., w_W)$, we sample symptom $p(s_i|s_{\neg i})$ and each herb $p(w_i|w_{\neg i})$ for all the values of latent variable $z = \{1, ..., K\}$, and update the topic assignment with its newest value after it has been sampled, where $s_{\neg i}$ is defined as symptoms excluding s_i and $w_{\neg i}$ is defined as herbs excluding w_i.

When $l = SYMPTOM$, given conditional distribution for a symptom with index $i_s = (m, n)$, the sampling sampler draws the hidden topic z_{i_s} as follows:

$$
p(z_{i_s} = k | l = SYMPTOM, z_{\neg i_s}, s, w) \propto \frac{N_{s_m, \neg i_s}^{(k)} + N_{w_m}^{(k)} + \alpha_k}{\sum_{k=1}^{K}(N_{s_m, \neg i_s}^{(k)} + N_{w_m}^{(k)} + \alpha_k)} \cdot \frac{N_{s_k, \neg i_s}^{(p)} + \beta_p}{\sum_{p=1}^{S} N_{s_k, \neg i_s}^{(p)} + \beta_p}.
\tag{4}
$$

When $l = HERB$, given conditional distribution for a herb with index $i_w = (m, n)$, the sampling sampler draws the hidden topic z_{i_w} as follows:

$$
p(z_{i_w} = k | l = HERB, z_{\neg i_w}, s, w) \propto \frac{N_{w_m, \neg i_w}^{(k)} + N_{s_m}^{(k)} + \alpha_k}{\sum_{k=1}^{K}(N_{w_m, \neg i_w}^{(k)} + N_{s_m}^{(k)} + \alpha_k)} \cdot \frac{N_{w_k, \neg i_w}^{(q)} + \gamma_q}{\sum_{q=1}^{W} N_{w_k, \neg i_w}^{(q)} + \gamma_q}.
\tag{5}
$$

In Formulas (4) and (5), z_{i_*} is the topics of all excluding item i_*, the count $N_{*_*, \neg i_*}^{(*)}$ indicates that item i is excluded, α_k is the Dirchlet prior of topic k, β_s is the Dirchlet prior of symptom term s, γ_w is the Dirchlet prior of herb term w.

5 Experiments

5.1 Data Sets

To verify the correctness and feasibility of our method, we use Multi-Content LDA Model to study two subdomain of TCM, namely *amenorrhea* and *lung cancer*. One of the dataset we used is the medical cases from ancient Chinese medicine books. The dataset contains 106 medical cases, which focus on the same disease *amenorrhea*. The statistics of this dataset shows that these cases contain 152 types of symptoms and 248 types of herbs. The other dataset is a collection of modern medicine clinical cases which mainly focus on *lung cancer*, provided by a famous TCM hospital in China. This dataset contains 952 medical cases, including 77 types of symptoms and 356 types of herbs. Each medical case of both datasets contains a list of symptoms and a list of corresponding herbs. For simplification, we don't consider the dose of each herb.

5.2 Pathogenesises Diagnosis

In this section, we setup a serious experiments to illustrate how MCLDA works to find the relationship between symptoms and herbs, and why single content model, such as LDA, fails in this task.

The proposed MCLDA is designed to model the relationship among symptoms, herbs and pathogenesises from probability generation aspect. The MCLDA not only provides the clustering of symptoms and herbs, but also gives the one-to-one correspondences of symptom clusters and herb clusters.

With the help of specialists of TCM, we tested the accuracy of the result of the MCLDA model by manual rating. We ran the training algorithm for multiple times to test the general performance. Table 2 shows the statistical results when the number of topics is 3 and 5, and both of them were run 20 times. Row 1 represents the reasonable clusters number k. Row 2 and row 4 show among these 20 groups of results, how many groups have k reasonable clusters separately. Row 3 and row 4 show the corresponding ratios. For example, in *amenorrhea* dataset, when clustered into 3 topics, there is 13 of 20 groups of results who have 3 reasonable clusters. The results show that most latent pathogenesises found by MCLDA are reasonable based on the theory of TCM.

Table 2. The correspondence of symptoms and herbs based on MCLDA

(a) *amenorrhea*

	5	4	3	2	1	0
K=3			13	5	2	0
			65%	25%	10%	0
K=5	6	7	4	2	1	0
	30%	35%	20%	10%	5%	0

(b) *lung cancer*

	5	4	3	2	1	0
K=3			15	4	1	0
			75%	20%	5%	0
K=5	9	6	4	1	0	0
	45%	30%	20%	5%	0	0

Table 3. The correspondence of symptoms and herbs based on LDA (making no distinction between symptoms and herbs)

(a) *amenorrhea*

	5	4	3	2	1	0
K=3			0	5	9	6
			0	25%	45%	30%
K=5	0	0	1	4	12	3
	0	0	5%	20%	60%	15%

(b) *lung cancer*

	5	4	3	2	1	0
K=3			0	4	7	9
			0	20%	35%	45%
K=5	0	0	0	6	11	3
	0	0	0	30%	55%	15%

Table 4. The correspondence of symptoms and herbs based on LDA (treating symptoms and herbs separately)

(a) *amenorrhea*

	5	4	3	2	1	0
K=3			6	8	2	0
			37.5%	50%	12.5%	0
K=5	4	6	4	2	0	0
	25%	37.5%	35%	12.5%	0	0

(b) *lung cancer*

	5	4	3	2	1	0
K=3			4	9	3	0
			25%	56.25%	18.75%	0
K=5	3	4	6	2	1	0
	18.75%	25%	37.5%	12.5%	6.25%	0

Next, we discuss whether LDA could be used to find the correlation between symptoms and herbs. First, we make no distinction between symptoms and herbs. We trained 20 groups of pathogenesises for both 3 topics and 5 topics and asked specialists to check the accuracy of each cluster. The results are presented in Table 3. It is clear that the relationship between symptoms and herbs obtained by this method is scarcely explained by the theory of TCM, where symptoms and herbs within a cluster make no sense to belong to the same pathogenesis.

Another straight-forward optional method is clustering symptoms and herbs by LDA separately and asking experts to check the correspondences between symptom clusters and herb clusters manually. The first problem of this method is the complexity of manual comparison. The number of comparisons among K clusters is K^2. If we train the model N times for both symptoms and herbs, the total number of comparisons is N^2K^2. It is almost impossible to train symptoms and herbs separately 20 times for $K = 5$, since that the comparisons between symptom clusters and herb clusters are as much as 10,000 times. Therefore, we only ran LDA on each type 4 times and obtained 16 pairs of symptom clusters and herb clusters for doctors to rate. The results for both datasets are shown in Table 4. Comparing Tables 2 and 4, the purposed MCLDA performs better in finding the latent pathogenesises to connect symptoms and herbs. Furthermore, the complexity of matching symptom clusters and herb clusters manually makes LDA fail in this work.

5.3   Effectiveness of MCLDA Model

First, we adopt perplexity to measure the overall performance of MCLDA, which is used to measure how well the model fits the test data [1].

It is the most common metrics in language models [1] and topic models [2]. It decreases monotonically when the likelihood of the test data set increases, which means that a lower perplexity score indicates better overall performance. The perplexity of M medical cases can be defined to [3]:

$$perplexity\,(d_{test}) = exp\left[-\frac{\sum_m^M log\,p(s_m, w_m)}{\sum_m^M (N_{sm} + N_{wm})}\right], \tag{6}$$

As discussed in last section, clustering symptoms and herbs separately fails in finding latent pathogenesises. Therefore, we only compare the propose MCLDA with LDA which does not distinguish symptom terms and herb terms. Figure 3 (c) and (d) present the perplexity for each model on both *amenorrhea* and *lung cancer* for different values of K. Regardless of the meaning of pathogenesises in TCM, the MCLDA still performs better than the LDA in modeling medical cases, which indicates that the two contents of a case should be modeled separately. And the descent of perplexity decreases when K increases validates the existing of the latent pathogenesises. Once K is large enough to cover the pathogenesises of the disease, the performance is not improved by increasing topics. Figure 3 (a) and (b) present the perplexity of symptoms, herbs and the whole medical cases in *amenorrhea* and *lung cancer* of the MCLDA model.

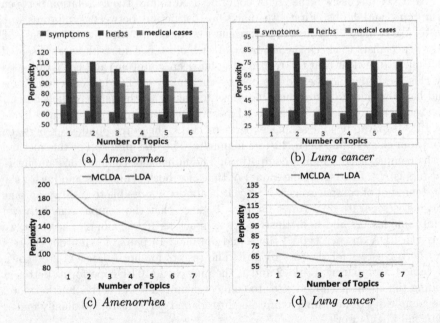

(a) *Amenorrhea*   (b) *Lung cancer*

(c) *Amenorrhea*   (d) *Lung cancer*

Fig. 3. The perplexity of symptoms and herbs in *amenorrhea* and *lung cancer*.

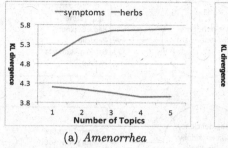
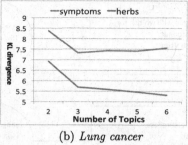

(a) *Amenorrhea*   (b) *Lung cancer*

Fig. 4. The KL-divergence of symptoms and herbs in *amenorrhea* and *lung cancer*.

Then, we use the Kullback-Leibler (KL) divergence to show the distinctiveness of the latent pathogenesises [11]. KL-divergence is a non-symmetric measure of the difference between two distributions. The higher the KL-divergence is, the more distinct the clusters are. We use the symptom distribution $p(s|z)$ and the herb distribution $p(w|z)$ to calculate the KL-divergence of pathogenesises. We should note that a higher KL-divergence does not mean a better model because the multiple-multiple correspondence is the inherent characteristic for both symptom-pathogenesis and herb-pathogenesis.

Figure 4 shows the KL divergence of symptoms and herbs for different number of topics. The diversity of pathogenesises for different values of K varies a lot. In *amenorrhea* dataset, the KL-divergence of symptoms increases with K and the KL-divergence of herbs decreases with K, which indicates that one herb tends to have effect on multiple *amenorrhea* pathogenesises and one symptom unlikely belongs to various pathogenesises. In *lung cancer*, the diversity of symptoms and herbs between pathogenesises is different from *amenorrhea*. Both symptoms and herbs in *lung cancer* could attribute to the multiple pathogenesises.

5.4   Evaluation According to Ancient Books

TCM is based on several theories. One of the theories is called Yin-Yang theory [12]. TCM hold the opinion that yin and yang always exist in the human body. The illness is caused by the disharmony of yin and yang, such as excess of yin or yang, or deficiency of yin or yang. The other theory of TCM is called five meridians. Five meridians refer to the meridians of heart, liver, spleen, lung and kidney. It is considered that the organs of human body cooperate with each other to complete a variety of normal physiological function and they are not exist in isolation. In conclusion, only five meridians are harmonious and yin and yang are in balance can the human body healthy. Different symptoms may be caused by different abnormal meridians of organs and the excess or deficiency of yin or yang or some other latent factors.

To evaluate the correlation of symptoms and herbs, we consult specialists and look up in ancient books. According to the yin-yang, five meridians and other

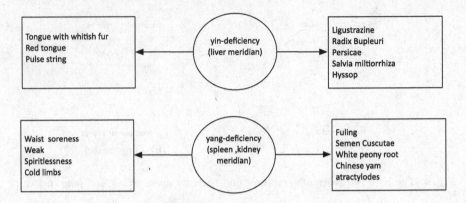

Fig. 5. Two topics result

theories, both dataset are clustered into one to six clusters. Each cluster represents a topic which in our problem represent a pathogenesis. We will interpret the rationality of the result by TCM theory.

The experiment result which clustered the data into two topics are shown in Fig. 5. We pick up some of the symptoms and herbs. As we can see, the main symptoms of topic one are tongue with whitish fur, red tongue and pulse string. This usually means an excess of fire (or Yang) in the body, which can be called yin-deficiency. And these symptoms also mean liver-qi stagnation so the main task is to dredges channel of liver and nourish yin. The representative herbs given by our model is Ligustrazine, Radix Bupleuri, Persicae, Salvia miltiorrhiza and Hyssop. It's proved that these herbs all have effect on healing amenorrhoea caused by liver-qi stagnation. For example, according to "Compendium of Materia Medica", Ligustrazine can remove the liver-qi stagnation, moisten liver, tonify deficiency and so on [6]. Other herbs have such effects as well according to medical book.

Other clustering results also have rationality according to specialists. Especially the result which clustered the data into five topics, the specialists said that these five clusters our model gives can be regard as kidney-yang deficiency, kidney-yin deficiency, spleen deficiency, yin deficiency, liver-qi exuberant, which demonstrate the good clustering effect. And the specialists also proved that the corresponding herbs have reasonable effects on healing the diseases. Because of space limitations, we do not interpret in detail here.

6   Conclusion

This research has focused on finding the latent relationship between symptoms and herbs in tradition Chinese medicine. We assume that the pathogenesises are the inherent connection between symptoms and herbs. For a patient, the symptoms are caused by pathogenesises and herbs given by doctors treat the corresponding pathogenesises. We proposed a novel latent semantic model based LDA to model pathogenesises as the hidden topics of two contents. We have collected two datasets about *amenorrhea* and *lung cancer* by working with TCM specialists. The experiment results show that the pathogenesises found by MCLDA matches TCM theory well.

Acknowledgements. This work was supported by NSFC grants (No. 61532021, 61472141 and 61021004), Shanghai Knowledge Service Platform Project (No. ZF1213)and Shanghai Leading Academic Discipline Project(Project NumberB412).

References

1. Brown, P.F., et al.: An estimate of an upper bound for the entropy of English. Comput. Linguist. **18**(1), 31–40 (1992)
2. Blei, D.M., Ng, A.Y., Jordan, M.I.: Latent dirichlet allocation. J. Mach. Learn. Res. **3**, 993–1022 (2003)
3. Murphy, K.P.: Machine Learning: A Probabilistic Perspective. MIT Press, Cambridge (2012)
4. Zhou, X., Liu, B., Wu, Z.: Text mining for clinical chinese herbal medical knowledge discovery. In: Hoffmann, A., Motoda, H., Scheffer, T. (eds.) DS 2005. LNCS (LNAI), vol. 3735, pp. 396–398. Springer, Heidelberg (2005)
5. Church, K.W., Hanks, P.: Word association norms, mutual information, and lexicography. Comput. Linguist. **16**(1), 22–29 (1990)
6. Li, S.: Compendium of Materia Medica: (Bencao Gangmu). Foreign Languages Press, Beijing (2003)
7. Zhang, N.L., et al.: Latent tree models and diagnosis in traditional Chinese medicine. Artif. Intell. Med. **42**(3), 229–245 (2008)
8. Ying, J., et al.: Collection and analysis of characteristics of tongue manifestations in patients with cerebrovascular diseases. J. Beijing Univ. Tradit. Chin. Med. **28**(4), 62 (2005)
9. Lei, L., et al.: Study on application of probability latent semantic analysis (PLSA) in herbal prescription development. World Sci. Technol. (Modernization Tradit. Chin. Med. Mater. Med.) **5**, 1976–1980 (2012)
10. Heinrich, G.: Parameter estimation for text analysis. Technical report (2005)
11. Hofmann, T.: Probabilistic latent semantic indexing. In: Proceedings of the 22nd Annual International ACM SIGIR Conference on Research and Development in Information Retrieval. ACM (1999)
12. Zhou, X., Peng, Y., Liu, B.: Text mining for traditional Chinese medical knowledge discovery: a survey. J. Biomed. Inform. **43**(4), 650–660 (2010)

Correlation-Based Weighted K-Labelsets for Multi-label Classification

Jingyang Xu and Jun Ma$^{(\boxtimes)}$

School of Computer Science and Technology, Shandong University,
Jinan 250101, China
xujingyang1991@gmail.com, majun@sdu.edu.cn

Abstract. RAkEL(RAndom k-labELsets) is an effective ensemble multi-label classification method where each sub classifier is trained on a small randomly-selected subset of k labels, called k-labelset. However, random combination of labels may lead to the poor performance of sub classifiers and the method can not make full use of the label correlations. In this paper, we propose a novel ensemble multi-label classification method named LCWkEL(Label Correlations-based Weighted k-labELsets). Instead of randomly choosing subsets, we select a number of k-labelsets based on a label correlation matrix. Furthermore, considering the label correlations in different k-labelsets may have different influence on an instance, we construct a weight coefficient vector for an instance. Each dimension of the vector represents the weight coefficient for each sub classifier. For the multi-label classification of an unlabeled instance, LCWkEL calculates the weighted sum of all sub classifiers' predictions, which can improve the classification performance effectively. Experimental results on three areas of data sets show that the method proposed in this paper can obtain competitive performance compared with the RAkEL method and other high-performing multi-label classification methods.

Keywords: Multi-label classification · Label correlation · Ensemble method · k-labelsets

1 Introduction

In traditional supervised learning tasks, each instance is associated with only one single label λ from a set of disjoint labels $\mathcal{L}(|\mathcal{L}| > 1)$. However, there are many cases where each instance is associated with a set of labels $Y \subseteq \mathcal{L}$ simultaneously. For example, in text classification [1], a document may be associated with several pre-defined topics such as "basketball" and "sports". In audio sentiment classification [2], an audio text can have multiple types of emotions. Hence multi-label classification has attracted much attention during the past years [3,4], and widely been used in the fields such as text classification, music classification, image classification, etc.

At present, multi-label classification methods are mainly divided into two main categories: algorithm adaptation and problem transformation [5,6]. Algorithm adaptation methods extend common learning algorithms to handle multi-label classification tasks, such as MLkNN [7] algorithm which is derived from

© Springer International Publishing Switzerland 2016
F. Li et al. (Eds.): APWeb 2016, Part I, LNCS 9931, pp. 408–419, 2016.
DOI: 10.1007/978-3-319-45814-4_33

the kNN [8]. Problem transformation methods transform a multi-label classification problem into several single-label classification problems. Representative methods such as Label Powerset(LP) [9], Calibrated Label Ranking(CLR) [10], and Random k-Labelsets(RAkEL) [11].

In reality, there are often some correlations between labels, and the label correlations may provide helpful information. A large number of researches show that exploiting label correlations in algorithms can enhance the prediction performance [12,13]. Different multi-label classification methods have tried to exploit different orders(first-order,second-order and high-order) of label correlations. The RAkEL method mentioned above has the advantage of taking label correlations into consideration, the insufficient is that it considers the label correlations in the way of random combination, and it ignores the different influence of different label correlations on an instance.

In this paper, we propose the LCWkEL(Label Correlations-based Weighted k-labELsets) method. We use the k-labelsets definition from RAkEL and employ LP to train sub classifiers as well. The contribution of this paper is summarized as follows: (1) We select a number of k-labelsets based on a label correlation matrix then train corresponding sub classifiers, which can avoid the performance loss of sub classifiers. (2) In order to exploit different influence of different label correlations on an instance, we construct a weight coefficient vector for an instance. Each dimension of the vector represents a weight coefficient for each sub classifier. (3) The multi-label classification of a new instance is achieved by a weighted sum of sub classifiers' results, which can improve the classification performance effectively.

The rest of this paper is organized as follows: In Sect. 2, we briefly review the related works about LP method and RAkEL method. In Sect. 3, we describe the LCWkEL method in detail. In Sect. 4, we discuss the corresponding experiments and some metrics to evaluate our method and other multi-label classification methods. Section 5 concludes our work and introduces future work.

2   Related Work

In this section, we first introduce the formal definition of multi-label classification, then briefly review the LP method and RAkEL method. The following notion is used throughout this paper. Let $\mathcal{X} = \mathcal{R}^d$ be the d-dimensional input space, and $\mathcal{Y} = \{1,0\}^L$ be the output label space of L possible labels denoted by $\mathcal{L} = \{\lambda_1, \lambda_2, \cdots, \lambda_L\}$. Given a training set $(x_i, y_i)_{i=1}^N$ that consists of N instances, where each instance $x_i = [x_{i1}, x_{i2}, \cdots, x_{id}]$ is a d-dimensional vector and $y_i = [y_{i1}, y_{i2}, \cdots, y_{iL}]$ is the label vector of x_i, y_{il} is 1 if x_i has the ℓ-th label λ_ℓ of $\mathcal{L}$ and 0 otherwise. The goal of multi-label classification is to learn a function $H : \mathcal{X} \to \mathcal{Y}$ which can predict the label vector for a test instance x.

2.1   The LP Method

Label Powerset(LP) [9] is a simple problem transformation method which directly transforms multi-label classification problems into multi-class

Table 1. An example of the classification process of RAkEL. $k=3$ and $M=6$

Predictions							
Classifier	Labelset	λ_1	λ_2	λ_3	λ_4	λ_5	λ_6
h_1	$\{\lambda_1,\lambda_2,\lambda_6\}$	1	0	-	-	-	1
h_2	$\{\lambda_2,\lambda_3,\lambda_4\}$	-	1	1	0	-	-
h_3	$\{\lambda_3,\lambda_5,\lambda_6\}$	-	-	0	-	0	1
h_4	$\{\lambda_2,\lambda_4,\lambda_5\}$	-	0	-	0	0	-
h_5	$\{\lambda_1,\lambda_4,\lambda_5\}$	1	-	-	0	1	-
h_6	$\{\lambda_1,\lambda_2,\lambda_3\}$	1	0	1	-	-	-
Average votes		3/3	1/4	2/3	0/3	1/3	2/2
Final prediction		1	0	1	0	0	1

classification problems. LP considers each distinct subset of $\mathcal{L}$, called *labelset*, that exists in the training set as a different class value of a single-label classification task. However LP method is challenged by large number of labelsets appearing in the training set.

2.2   The RAkEL Method

This paper focuses on the Random k-Labelsets(RAkEL) [11] method. RAkEL constructs an ensemble of LP classifiers. In order to avoid the disadvantage of LP method, RAkEL randomly breaks the original set of labels $\mathcal{L}$ into M small overlapping k-labels subsets, called *k-labelsets*, and employs LP to train a corresponding muliti-label classifier, where $k(k < L)$ is a parameter that specifiers the size of a subset, $M(M > 1)$ is the number of k-labelsets. The classification of a new instance is achieved by voting of the LP classifiers in the ensemble. An example of the classification process of RAkEL is shown in Table 1. Each row represents the classification result of a LP classifier denoted by h_m trained on a randomly-selected k-labelset. RAkEL calculates the mean of these predictions for each label $\lambda_\ell \in \mathcal{L}$ and outputs 1 if it is greater than a 0.5 threshold, 0 otherwise. RAkEL uses LP as the sub classifiers to transform multi-label classification problems into integrated multi-class classification problems.

3   The LCWkEL Method

The basic idea of this method is inspired by the RAkEL method. However, in RAkEL labels are randomly combined in a k-labelset, labels with no relationship may be put into the same subset, and converted to a single label to be trained. This will cause irrelevant labels appearing in the classification results of sub classifiers. In order to avoid the performance loss caused by randomly selecting k-labelsets, for a given desired number M of k-labelsets, we select M k-labelsets that have the most correlation information based on a label correlation matrix and employ

LP to train corresponding sub classifiers as well. Further considering the different influences of different label correlations(correlated labels in a k-labelset) on an instance, we introduce a weight coefficient vector $w = [w_1, w_2, \cdots, w_M]$ for each instance x where each dimension represents the weight coefficient of a corresponding sub classifier on x. Then the classification of a new instance x is a weighted sum of the outputs of the LP classifiers.

The model proposed in this paper for multi-label classification can be written as the following form:

$$H(x) = \sum_{m=1}^{M} w_m h_m(x) \qquad (1)$$

where each basic function h_m is a LP classifier trained on a k-labelset consisting of k correlated labels. The weight coefficient w_m represents the m-th dimension of weight vector w of x. For an instance x, the coefficient w_m reflects the contribution of a corresponding classifier h_m which is also used to measure the influence of a k-labelset on x.

3.1 k-Labelsets Selection Based on Label Correlations

RAkEL method randomly selects k-labelsets, then trains the sub classifiers. The relationship between labels in each subset is random, which may lead to the poor prediction accuracy of sub classifiers. In this paper, we use correlated labels to construct the k-labelsets. By taking the relationship between labels into account, sub classifiers prediction accuracy can be improved. The correlations between labels can be described as the form of network structure, we call it a label correlation network. The label correlation network is defined as a weighted network $G = (V, E, W)$, where V is the node collection(label collection), E is the edge collection, and W is the weight collection, the weight of an edge shows the correlation between two labels.

We use the label correlation matrix as the adjacency matrix of the label correlation network, calculated as following:

Firstly, we define a $L \times N$ label matrix $T = (t_{ij})$, where L is the number of labels and N is the number of instances in training data set, $t_{ij} = 1$ expresses that instance j has label i, $t_{ij} = 0$ otherwise.

In order to enhance the representation of each label's information, we define a $L \times N$ incidence matrix $A = (a_{ij})$, where a_{ij} represents the incidence of label i and instance j. Referencing the idea of TF-IDF, we use Eq. 2 to denote the incidence of a label and an instance:

$$a_{ij} = t_{ij} \times \log \frac{N}{I_i} \qquad (2)$$

I_i represents the number of instances labeled by i, N is the number of instances.

Therefore we can obtain the $L \times L$ label correlation matrix $C = (c_{ij})$. Each label's information is represented as a row vector A_i in the incidence matrix A. As the correlation between labels can be calculated by the label's row vector distance in space, we use cosine similarity to calculate the correlation between the two vectors. The label correlation matrix is:

$$c_{ij} = sim(A_i, A_j) = \cos(\theta_{ij}) = \frac{\sum_{n=1}^{N} a_{in} a_{jn}}{\sqrt{\sum_{n=1}^{N} a_{in}^2 \sum_{n=1}^{N} a_{jn}^2}} \tag{3}$$

For a given k, we define I_C to represent the correlation information that a k-labelset has. I_C can be calculated as the sum of the label correlation between each pair of labels appearing in the k-labelset. For one k-labelset, we have a total of p such pairs denoted by (i, j), p is given by the binomial coefficient: $p = \binom{k}{2}$. $I_C = \sum_{n=1}^{p} c_{ij}$. Then we can get a k-labelsets list Q, where all items are in a descending order according to the value of I_C.

Referring to RAkEL, in order to reduce the complexity of training process but cover all labels, we suggest a suitable value $2L$ for our desired number of k-labelsets M. So we select top $2L$ k-labelsets in Q to train corresponding sub classifiers.

3.2   Weight Coefficient Matrix Calculation for Training Set

As introduced in last subsection, we need to train M sub classifiers in total. Considering the different influence of label correlations on different instances, one instance may receive different influence of the M k-labelsets. We introduce a M dimension weight coefficient vector $w_i = [w_{i1}, w_{i2}, \cdots, w_{iM}]$ for each instance x_i in training set. Each dimension of w_i represents the weight coefficient of each sub classifier trained on a corresponding k-labelset. Therefore, for the training set, we will get a $N \times M$ weight coefficient matrix $W = (w_{ij})$. Each instance's weight vector is a row vector w_i in W, N is the number of instances in training set. The bigger the value of one dimension in w_i, the more contribution the sub classifier has on x_i. In training stage, the weight coefficient matrix W is learned by solving a minimization problem formulated as follows:

$$\min_{W} \frac{1}{2} \sum_{i=1}^{N} \|Y_i - H(x_i)\|^2 + \frac{\beta}{2} \|W\|^2$$

$$s.t. \sum_{m=1}^{M} W_{im} = 1, \forall i \in \{1, \cdots, N\} \tag{4}$$

$$0 \leq W_{im} \leq 1, \forall i \in \{1, \cdots, N\}, m \in \{1, \cdots, M\}$$

The first term in the object function aims to minimize the global error between the prediction $H(x)$ of LCWkEL and the multi-label ground truth Y. The second term is a two-norm regularization term of the weight coefficient matrix. The parameter β is used to trade off the relationship between the two terms.

Algorithm 1. (The training process of LCWkEL)

Input:

size of labelsets k, number of k-labelsets M, learning parameter β, set of labels $\mathcal{L}$, and the training set $D = (x_i, y_i)_{i=1}^{N}$.

Output:

an ensemble of LP classifiers h_m, and the corresponding k-labelsets $\mathcal{R}_m$, the weight coefficient matrix W of training set,regression models R_m for W.

1: initialize $S \leftarrow M$ k-labelsets selected via the method proposed in 3.1
2: **for** $m \leftarrow 1$ to M **do**
3:     select $\mathcal{R}_m$, a k-labelset in S
4:     train the LP classifier h_m based on D and $\mathcal{R}_m$
5:     $S \leftarrow S \backslash \mathcal{R}_m$
6: **end for**
7: learn W by minimize (4)
8: **for** $m \leftarrow 1$ to M **do**
9:     train a regression model R_m for the m-th column of weight matrix W
10: **end for**

Obviously, Eq. 4 can be decomposed into N optimization problems, and the i-th one is:

$$\min_{w_i} \frac{1}{2} \|Y_i - \sum_{m=1}^{M} w_{im} h_m(x_i)\|^2 + \frac{\beta}{2} \|w_i\|^2$$

$$s.t. \sum_{m=1}^{M} w_{im} = 1 \tag{5}$$

$$0 \le w_{im} \le 1, m \in \{1, \cdots, M\}$$

where h_m is the prediction of a LP classifier for a k-labelset.

The weight coefficient matrix W is obtained through the training process of the LCWkEL method.

3.3   Predict the Labels of a Test Instance

We can use the weight coefficient matrix W and the LP classifiers ensemble obtained in training stage to predict the labels of a test instance.

Given a test instance x, its weight coefficient vector w is unknown. To solve this problem, we train M regression models on the training instances and the weight coefficient matrix W. Each regression model is trained for predicting a dimension of the weight coefficient vector. Then we use the regression models to calculate the weight coefficient vector w of a test instance x.

Our output label space has L possible labels denoted by $\mathcal{L} = \{\lambda_1, \lambda_2, \cdots, \lambda_L\}$, but each LP classifier $h_m(m \in \{1, \cdots, M\})$ only provides k binary predictions for the λ_ℓ in the corresponding k-labelset. Therefore, We extend the prediction of a LP classifier h_m to a L dimension vector represented by $V_m \in \{1, 0\}^L$.

Table 2. An example of the classification process of LCWkEL. $k=3$ and $M=6$

Predictions								
Classifier	Labelset	Weight	λ_1	λ_2	λ_3	λ_4	λ_5	λ_6
h_1	$\{\lambda_1, \lambda_2, \lambda_6\}$	0.1	1	0	0	0	0	1
h_2	$\{\lambda_2, \lambda_3, \lambda_4\}$	0.2	0	1	1	0	0	0
h_3	$\{\lambda_3, \lambda_5, \lambda_6\}$	0.05	0	0	0	0	1	1
h_4	$\{\lambda_2, \lambda_4, \lambda_5\}$	0.2	0	1	0	0	0	0
h_5	$\{\lambda_1, \lambda_4, \lambda_5\}$	0.2	1	0	0	1	1	0
h_6	$\{\lambda_1, \lambda_2, \lambda_3\}$	0.25	1	1	0	0	0	0
Weighted sum			0.55	0.65	0.2	0.2	0.25	0.15
Final prediction			1	1	0	0	0	0

The ℓ-th element v_ℓ in V_m is calculated as:

$$v_\ell = \begin{cases} 1, & \text{if } \lambda_\ell \in \mathcal{R}_m \text{ and } \lambda_\ell \text{ is 1 in } h_m(x) \\ 0, & \text{if } \lambda_\ell \in \mathcal{R}_m \text{ and } \lambda_\ell \text{ is 0 in } h_m(x) \text{ or } \lambda_\ell \notin \mathcal{R}_m \end{cases} \qquad (6)$$

where $\mathcal{R}_m$ is a k-labelset associated with a LP classifier h_m.

For example, for the labelset $\{\lambda_2, \lambda_3, \lambda_5\}$, if the prediction of h_m on a test instance x is[1,1,0] ,we will get V_m ,that is, $[0, 1, 1, 0, 0...]$.

LCWkEL uses the weight coefficient vector w of x just obtained to calculate the weighted sum of V_m for each label $\lambda_j \in \mathcal{L}$, then outputs 1 if it is greater than a 0.5 threshold, 0 otherwise. Table 2 shows an example of the classification process of LCWkEL. Algorithms 1 and 2 describe the training and classification processes of the proposed LCWkEL, respectively.

4 Experiment and Analysis

In this section, we describe our experiment to evaluate the performance of LCWkEL method. Our method is realized based on the platform of Mulan [14]. Mulan is a Weka [15]-based open source project, and contains some common multi-label classification algorithms. The compared methods used in our experiment are: RAkEL [11], BR [16], CLR [10] and ECC [17]. We use three data sets from three different domains. The evaluation is based on five measures, estimated via 3-fold cross-validation ten times, each using a random 66 % of data set for training and the rest for evaluation. We also try to explore the influence of parameter k.

4.1 Experimental Datasets

We use three benchmark data sets belonging to different domains. *Emotions* is a data set in music domain, there are 593 instances and 6 possible labels.

Algorithm 2. (The classification process of LCWkEL)

Input:

a test instance x,number of k-labelsets M,number of possible labels L.an ensemble of LP classifiers h_m and corresponding k-labelsets $\mathcal{R}_m$,regression models R_m for weight matrix W.

Output:

the multi-label classification vector $Y = (y_1, y_2, \cdots, y_L)$ be assigned to x.

1: initialize the L dimension label vector, $Y = [0, 0, \cdots, 0]$
2: **for** $m \leftarrow 1$ to M **do**
3:    calculate the m-th dimension weight vector w_m for x using regression model R_m
4:    obtain $h_m(x)$ via LP classifier h_m
5:    compute the vector representation V_m for $h_m(x)$ using (6)
6:    $Y = Y + w_m \cdot V_m$
7: **end for**
8: **for** $j \leftarrow 1$ to L **do**
9:    **if** $y_j \geq 0.5$ **then**
10:      $y_j = 1$;
11:    **else**
12:      $y_j = 0$;
13:    **end if**
14: **end for**

Table 3. The basic statistic information of data sets

Data	Data information				
	Domain	Instances	Labels	Cardinality	Density
Emotions	music	593	6	1.869	0.311
Scene	image	2407	6	1.074	0.179
Medical	text	978	45	1.245	0.028

Scene is an image data set which has 2407 images and 6 possible labels. *Medical* is a text data set which has 978 instances and 45 possible labels. The basic statistic information is shown in Table 3. *Label cardinality* represents the average label number of each instance, *label density* is the average ratio of label number of each instance. More details on these data sets are available at the Mulan library website.[1]

4.2 Evaluation Measures

In multi-label classification tasks, each instance may belong to multiple categories at the same time. Traditional evaluation measures of the single label classification are not suitable for multi-label classification. In this paper, we select five common multi-label classification evaluation measures summarized in [5]: *hamming loss, coverage, example-based recall,example-based precision,* and

---

[1] http://mulan.sourceforge.net/datasets.html.

example-based F-measure to evaluate the performance of the compared methods, details are defined as follows:

Hamming Loss: measures the mismatch of real label set and prediction label set.

$$hamming\ loss = \frac{1}{t}\sum_{i=1}^{t}\frac{1}{q}|Z_i \Delta Y_i| \tag{7}$$

Z_i is the prediction label set of instance x_i, Y_i is the real label set of instance x_i. Δ stands for the symmetric difference of two sets. The $\frac{1}{q}$ factor is used to obtain a normalized value in $[0, 1]$. t is the number of instances in test set.

Coverage: measures the average depth in the ranking in order to cover all the labels associated with an instance.

$$coverage = \frac{1}{t}\sum_{i=1}^{t}\max_{\lambda \in Y_i} \iota_i(\lambda) - 1 \tag{8}$$

$\iota_i(\lambda)$ represents the position of label λ in the ranking.

It is also common to use a group of example-based metrics from the information retrieval:

$$example\text{-}based\ recall = \frac{1}{t}\sum_{i=1}^{t}\frac{|Z_i \cap Y_i|}{|Y_i|} \tag{9}$$

$$example\text{-}based\ precision = \frac{1}{t}\sum_{i=1}^{t}\frac{|Z_i \cap Y_i|}{|Z_i|} \tag{10}$$

$$example\text{-}based\ F\text{-}measure = \frac{1}{t}\sum_{i=1}^{t}\frac{2|Z_i \cap Y_i|}{|Y_i| + |Z_i|} \tag{11}$$

4.3   Experiment Results and Analysis

We use 3-fold cross-validation ten times and calculate the mean and standard deviation of the results on the data sets mentioned above. Our method is firstly compared with RAkEL [11] method, where each LP classifier learns a randomly-combined k-labelset. Our method is also compared with three other state-of-the-art multi-label classification methods: BR [16] which considers first-order label correlations(ignores label correlations), CLR [10] which considers second-order label correlations(the relationship between a pair of labels), ECC [17] which considers high-order label corrections(the label corrections among all the possible labels). All compared methods use the parameters recommended in the corresponding literatures. The C4.5 decision tree learning algorithm is used as the base-level single-label classification algorithm of BR, the LP classifiers of RAkEL and LCWkEL, and the binary classifiers of CLR. For LCWkEL method, the parameter k is set to 3, parameter β is set to 1, parameter M is set to 2L.

Table 4. Results(means ± std.) on music data set.

Data	Criteria	LCWkEL	RAkEL	BR	CLR	ECC
Emotions	hloss	0.229 ± 0.002(3)	0.224 ± 0.029	0.263 ± 0.022	0.249 ± 0.009	0.201 ± 0.004
	coverage	1.922 ± 0.114(2)	1.975 ± 0.167±	2.584 ± 0.234	1.922 ± 0.086	1.790 0.048
	recall	0.731 ± 0.021(1)	0.573 ± 0.064	0.566 ± 0.026	0.640 ± 0.019	0.581 ± 0.032
	precision	0.619 ± 0.010(2)	0.617 ± 0.062	0.544 ± 0.024	0.564 ± 0.005	0.646 ± 0.031
	F-measure	0.631 ± 0.004(1)	0.565 ± 0.059	0.525 ± 0.022	0.560 ± 0.008	0.581 ± 0.026

Table 5. Results(means ± std.) on image data set.

Data	Criteria	LCWkEL	RAkEL	BR	CLR	ECC
scene	hloss	0.103 ± 0.011(2)	0.105 ± 0.008	0.135 ± 0.004	0.138 ± 0.009	0.096 ± 0.001
	coverage	0.611 ± 0.064(3)	0.615 ± 0.064	1.335 ± 0.150	0.591 ± 0.065	0.551 ± 0.055
	recall	0.673 ± 0.026(1)	0.653 ± 0.032	0.625 ± 0.022	0.651 ± 0.034	0.625 ± 0.011
	precision	0.643 ± 0.038(1)	0.634 ± 0.034	0.544 ± 0.015	0.534 ± 0.032	0.634 ± 0.017
	F-measure	0.647 ± 0.032(1)	0.633 ± 0.032	0.566 ± 0.017	0.565 ± 0.032	0.624 ± 0.014

Table 6. Results(means ± std.) on text data set.

Data	Criteria	LCWkEL	RAkEL	BR	CLR	ECC
Medical	hloss	0.010 ± 0.004(1)	0.011 ± 0.001	0.011 ± 0.001	0.018 ± 0.004	0.010 ± 0.001
	coverage	4.206 ± 0.870(3)	4.305 ± 0.455	3.918 ± 0.257	4.670 ± 2.198	2.102 ± 0.333
	recall	0.796 ± 0.013(1)	0.796 ± 0.033	0.794 ± 0.033	0.460 ± 0.239	0.796 ± 0.024
	precision	0.820 ± 0.037(1)	0.773 ± 0.026	0.772 ± 0.025	0.494 ± 0.197	0.784 ± 0.023
	F-measure	0.777 ± 0.027(2)	0.770 ± 0.028	0.769 ± 0.028	0.464 ± 0.216	0.778 ± 0.023

We first perform the experiment on music data set *emotions*, Table 4 summarizes the performance on five evaluation measures. Tables 5 and 6 show the results on image and text data sets, respectively.

The numbers in parentheses represent the ranking of LCWkEL among five methods. For the first two evaluation measures, the smaller the value, the better the performance. For the rest three evaluation measures, the bigger the value, the better the performance. Notice that our method outperforms RAkEL on all the three data sets. Compared with other three multi-label classification methods which consider different orders of label correlations, LCWkEL still achieves the best performance on example-based metrics from the information retrieval in most cases. These metrics can measure the accuracy of a classification method. But it is less effective on *coverage* which is a ranking-based metric. ECC can achieve the best performance on *coverage*, the reason may be that ECC is a chaining method that can model label correlations while maintaining acceptable ranking loss. Our method is more concerned about the accuracy of the classification.

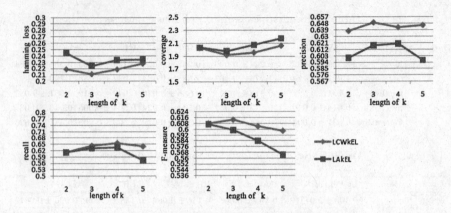

Fig. 1. Influence of parameter k

4.4 Influence of Parameter k

In our method, there are two important user-specified parameters M(number of classifiers) and k(size of labelsets). Based on the experience of RAkEL, we set the number of classifiers $M = 2L$. In order to examine the influence of parameter k, we run LCWkEL and RAkEL with k varying from 2 to 5. Due to the page limit, we only report the results on the *emotions* data set shown in Fig. 1, the results are similar in other data sets, so we set $k = 3$ in our experiment.

5   Conclusion and Future Work

This paper proposes a new ensemble multi-label classification method, named LCWkEL. The basic idea of this method is inspired by the RAkEL which creates an ensemble of multi-label classifiers by manipulating the label space using randomization. However, random combination of k-labelsets may lead to the performance loss in RAkEL, and RAkEL can not make full use of label correlations. LCWkEL has the advantage of combining label correlations and ensemble method into a unified framework. Experimental results show that our method has competitive performance against RAkEL and other state-of-the-art multi-label classification methods. In the future, it is possible to develop other approaches to construct the k-labelsets based on label correlations and determine the appropriate number of sub classifiers. We may also try to incorporate semantic knowledge between labels into our framework. In addition, we will also look for new data sets from different areas to validate our method.

Acknowledgment. This work was supported by Natural Science Foundation of China (61272240, 71402083).

References

1. Katakis, I., Tsoumakas, G., Vlahavas, I.: Multilabel text classification for automated tag suggestion. In: Proceedings of the ECML/PKDD 2008 Discovery Challenge, Antwerp, Belgium (2008)
2. Trohidis, K., Tsoumakas, G., Kalliris, G., Vlahavas, I.: Multilabel classification of music into emotions. In: Proceedings of 2008 International Conference on Music Information Retrieval (ISMIR 2008), Philadelphia, PA, USA, pp. 325–330 (2008)
3. Bucak, S.S., Jin, R., Jain, A.K.: Multi-label learning with incomplete class assignments. In: Proceedings of the IEEE Computer Society Conference on Computer Vision and Pattern Recognition, Colorado Springs, CO, pp. 2801–2808 (2011)
4. Hariharan, B., Zelnik-Manor, L., Vishwanathan, S.V.N., Varma, M.: Large scale max-margin multi-label classification with priors. In: Proceedings of the 27th International Conference on Machine Learning, Haifa, Israel, pp. 423–430 (2010)
5. Gibaja, E., Ventura, S.: A tutorial on multilabel learning. ACM Comput. Surv. (CSUR) **47**(3), 52 (2015)
6. Tsoumakas, G., Katakis, I., Vlahavas, I.: Mining multi-label data. In: Maimon, O., Rokach, L. (eds.) Data mining and knowledge discovery handbook, pp. 667–685. Springer, New York (2009)
7. Zhang, M.L., Zhou, Z.H.: ML-KNN: a lazy learning approach to multi-label learning. Pattern Recogn. **40**(7), 2038–2048 (2007)
8. Borrajo, D., Veloso, M.: Lazy incremental learning of control knowledge for efficiently obtaining quality plans. AI Rev. J. **11**(1–5), 371–405 (1997). Special Issue on Lazy Learning
9. Zhang, M.L., Zhou, Z.H.: A review on multi-label learning algorithms. IEEE Trans. Knowl. Data Eng. **26**(8), 1819–1837 (2014)
10. Fürnkranz, J., Hüllermeier, E., Mencía, E.L., Brinker, K.: Multilabel classification via calibrated label ranking. Mach. Learn. **73**(2), 133–153 (2008)
11. Tsoumakas, G., Katakis, I., Vlahavas, I.: Random k-labelsets for multilabel classification. IEEE Trans. Knowl. Data Eng. **23**(7), 1079–1089 (2011)
12. Tsoumakas, G., Dimou, A., Spyromitros, E., Mezaris, V.: Correlation-based pruning of stacked binary relevance models for multi-label learning. In: Proceedings of the 1st International Workshop on Learning from Multi-Label Data, Bled, Slovenia, pp. 101–116 (2009)
13. Jin, B., Muller, B., Zhai, C., Lu, X.: Multi-label literature classification based on the gene ontology graph. BMC Bioinform. **9**(1), 525 (2008)
14. Tsoumakas, G., Spyromitros-Xioufis, E., Vilcek, J., Vlahavas, I.: Mulan: a java library for multi-label learning. J. Mach. Learn. Res. **12**, 2411–2414 (2011). JMLR. org
15. Bouckaert, R.R., Frank, E., Hall, M.A., Holmes, G., Pfahringer, B., Reutemann, P., Witten, I.H.: WEKA-experiences with a java open-source project. J. Mach. Learn. Res. **11**, 2533–2541 (2010). JMLR. org
16. Boutell, M.R., Luo, J., Shen, X., Brown, C.M.: Learning multi-label scene classification. Pattern Recogn. **37**(9), 1757–1771 (2004)
17. Read, J., Pfahringer, B., Holmes, G., Frank, E.: Classifier chains for multi-label classification. Mach. Learn. **85**(3), 333–359 (2011)

Classifying Relation via Bidirectional Recurrent Neural Network Based on Local Information

Xiaoyun Hou, Zhe Zhao, Tao Liu$^{(\boxtimes)}$, and Xiaoyong Du

School of Information, Renmin University of China, Beijing, China
{xiaoyunhou,helloworld,tliu,duyong}@ruc.edu.cn

Abstract. Relation classification is an important research task in the field of natural language processing (NLP). In this paper, we apply a bidirectional recurrent neural network upon local windows of entities for relation classification. In contrast to previous approaches, only word tokens around entities are taken into consideration in our model. Upon word tokens, a bidirectional recurrent neural network is used to extract local context features of entities. To retain the important features for classification , we propose to use a novel weighted pooling layer upon hidden layers of RNN. Experiments on the SemEval-2010 dataset show that our proposed method achieves competitive results without introducing any external resources.

Keywords: Relation extraction · Bidirectional recurrent neural network · Weighted pooling

1 Introduction

Relation classification is an important task in NLP and normally acts as an intermediate step in various applications such as information extraction [1,14] and question answering [16]. The goal of the relation classification is to predict semantic relations between pairs of marked nominals (entities) in given texts and can be formally defined as follows [7]: given a sentence S with the annotated pairs of nominals e_1 and e_2, we aim to recognise the semantic relation between e_1 and e_2.

Recently, state-of-the-art results in this task are achieved by neural models, such as convolutional neural networks (CNN), recurrent neural networks (RNN) and recursive neural networks (RecNN). Complex compositionality can be learned by these models to determine the relations between entities. Usually, external resources such as dependency trees are required for these models to achieve competitive results. In this paper, a bidirectional recurrent neural network (BRNN) is proposed to extract local context features for relation classification. Compared to previous approaches, our proposed model has the following distinct features:

(1) Only the local window around entities are considered in our model. We discover that information beyond local contexts of entities has little value for

© Springer International Publishing Switzerland 2016
F. Li et al. (Eds.): APWeb 2016, Part I, LNCS 9931, pp. 420–430, 2016.
DOI: 10.1007/978-3-319-45814-4_34

"The beautiful hydrothermal features in the park (geysers, hot springs, mud pots, etc.), the uplift and subsidence, and many of the <e1>*earthquakes*</e1> are caused by the <e2>*movements*</e2> of hydrothermal or magmatic fluids."

Fig. 1. A sample in SemEval-2010 dataset

predicting entities' relations. Words around entities are sufficient to identify the relations between entities. The example shown in Fig. 1 illustrates the effectiveness of local information. The word 'caused' provides crucial clue, while words beyond local contexts of entities contain little information for determining the relation. By discarding irrelevant information, our model not only achieves better results, but also saves a large amount of computational resources.

(2) A weighted pooling layer is proposed to introduce the position information into our model. Intuitively, words play different roles according to their relative positions to entities. Position features have been demonstrated to be important for relation classsification in many researches [10,19]. Though RNN can capture the ordering information of the word sequences, we discover that explicitly introducing position information by weighting pooling is still helpful.

Experiment results show that significant improvements are achieved when these two techniques are introduced into bidirectional recurrent neural network (BRNN). As a result, competitive results are achieved by our model without exploiting any external resources.

The rest of the paper is organized as follows: we firstly review previous work in Sect. 2, and then present detailed descriptions about the proposed model in Sect. 3. Section 4 presents the experimental results and conclusions are provided by Sect. 5.

2 Related Work

Relation classification is an important topic in NLP. Various approaches have been explored to detect and identify the entities' relations. These models fall into one of the two classes, unsupervised and supervised approaches, according to whether the data is labeled or not. Most of the unsupervised approaches are based on distributional hypothesis, which states that the words that occur in the similar contexts tend to have similar meanings. Accordingly, it is reasonable to assume that the pairs of nominals that appear in similar contexts tend to have similar relations. [5] adopts hierarchical clustering method to cluster the contexts of nominals, and the most frequent words in the contexts are chosen to represent the relation between the nominals.

In the supervised approaches, relation classification is usually considered as a multi-classification problem. Depending on the input to classifier, these approaches can be further divided into feature-based, kernel-based and neural network-based models.

In feature-based approaches, different feature sets such as lexical features, syntactic features, semantic features are exploited and fed to a chosen classifier [4,8]. These features need to be defined subjectively, and it is difficult to decide which kinds of features is more useful.

Kernel-based approaches compute the similarity between two data samples in a more natural way, without explicit feature representation. Zelenko et al. [18] designed a tree kernel to compute the similarity between two parse trees. Bunescu and Mooney [2] proposed a new kernel which work on the shortest dependency path for relation classification. In addition to considering structural information, semantic information [9] is introduced into kernel methods. For this class of models, designing an effective kernel is of vital importance for the performance of the models since all information is measured by the kernel function. However, this process requires a lot human labours.

Recently, deep neural networks (DNNs) are developed for relation extraction. DNNs can learn underlying features automatically and have attracted growing interests. Socher et al. [12] proposed a recursive neural network to extract the compositional semantics using parse trees; Zeng et al. [19] propose to use combination of Convolutional neural network (CNN), lexical-level features and position features for predicting relation categories. Xu et al. [15] utilized bidirectional long short term memory to pick up the information on the shortest path between entities in dependency tree. Hashimoto et al. [6] explicitly weighted phrases' importance in RNNs to improve performance. Ebrahimi et al. [3] rebuilt an RNN on the dependency path between two marked entities. Yu et al. [17] proposed the Factor-based Compositional Embedding Model (FCM) which extracted features from the substructures of a sentence and combined them through a sum-pooling layer.

3   The Proposed Model

3.1   Overview

Figure 2 illustrates the overall architecture of our model. Firstly, local information centered around two entities is extracted from the original sentence as the input for the model. For instance, when the window size is set to be 3, the input of the model shown in Fig. 1 is 'many of the earthquakes are caused by' and 'are caused by the movement of hydrothermal or'. And then, the word tokens are projected to vectors by looking up word embeddings matrix, upon which the bidirectional recurrent neural network is built to extract local information around entities. The forward RNN is responsible for concluding history information of the entity, accordingly the backward RNN is for representing the future information of the entity. A pooling layer is used to gather information from all hidden layers of BRNN in different weights to generate the context features for the entity. Each entity is represented as the concatenation of its embedding and

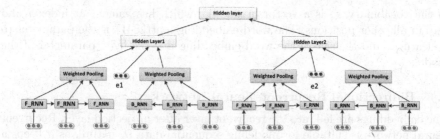

Fig. 2. The framework of BRNN model

its context features. Finally, the feature vector is fed into a softmax classifier to predict the relationship between entities.

3.2 Local Information Around Entities

It is crucial for relation classification task to capture the distinguished features from texts. Although the deep neural networks have the ability to learn under-lying features automatically, we still wish to put less irrelevant information to the networks. We discover that contents beyond local contexts of entities contain little information for relation classification. As a result, only local information around entities is used as the inputs of our model. To be more specific, the model is constructed upon entities' local context for the following reasons:

- The local information around entities is informative. Generally, the key fea-tures that are crucial for determining the relationships between entities fall into the local contexts of entities.
- RNN is used in this paper. Limiting inputs to local contexts avoids the problem of gradient vanishing or exploding while processing the long-distance text.
- Since the inputs are short, only a fraction of time and computational resources are required in our model.

3.3 Word Embedding

Word embedding layer is used to project discrete word symbols to low-dimensional dense word vectors through lookup table operation. Given a sentence x consisting of N words $x = \{w_1, w_2, ..., w_N\}$. We extract the local input I $= \{w_{l1-k}, ..., w_{l1}, e_1, w_{r1}, ..., w_{r1+k} \ w_{l2-k}, ..., w_{l2}, e_{l2}, w_{r2}, ..., w_{r2+k}\}$, then every word w_i is converted into a real-valued vector r_i, thus the input to the next layer is a sequence of real-valued vectors r $= \{r_{l1-k}, ..., r_{l1}, r_{e1}, r_{r1}, ..., r_{r1+k} \ r_{l2-k}, ..., r_{l2}, r_{e2}, r_{r2}, ..., r_{r2+k}\}$. The project operation is defined as Eq.(1)

$$r_i = W_v v_i \tag{1}$$

$W_v \in \mathbb{R}^{d \times |V|}$ is the word embedding matrix, where $|V|$ is the size of vocabulary. Each column $W_v^i \in \mathbb{R}^{d \times 1}$ corresponds to the word embedding of the i-th word in the vocabulary, v_i is a vector of size $|V|$ which has value 1 at index i and zero in all other positions. The word embedding matrix W_v is the parameter to be learned, and the size of the word embedding d is the hyperparameter of the model.

3.4   Bidirectional Recurrent Neural Network

The embeddings are fed into the recurrent layer after projection layer. Recurrent neural network is suitable for modeling sequential data by nature, as it keeps a hidden state vector h, which changes with input data at each step accordingly. We use the bidirectional network to gather information along both side of entity to get the left and right context features respectively. Taking entity e_1 for example. First, we introduce the forward RNN for the history (left) context of e_1, the embeddings are fed into the recurrent layers one by one. For each step, the network learn hidden layer h_t by combining the current input r_t and the previous hidden layer h_{t-1}. Specifically, h_{t-1} and r_t are transformed by matrix W_1^f and W_2^f and are added together with bias b_f, the result of which is then nonlinearly squashed by an activation function. Formally, we have

$$h_t^f = f(W_1^f r_t + W_2^f h_{t-1}^f + b^f) \tag{2}$$

where W_1^f and W_2^f are weight matrices for the input and recurrent connections respectively, b_f is a bias term for the hidden state vector, and f is a tanh activation function. The backward RNN computes in the similar way, formally denoted as follows:

$$h_t^b = f(W_1^b r_t + W_2^b h_{t-1}^b + b^b) \tag{3}$$

3.5   Weighted Pooling

Many researches have shown that word's relative distances to the entities play an important role for relation classification. In CDNN [19], a position feature vector which indicates the relative distance from the current word to the marked entities is appended to each word embedding; in [10], four position indicators which specify the start and end of the entities are used to replace the position vector. In this paper, we propose to use weighted pooling for introducing position information into the model, which is more concise and efficient compared to previous approaches. Different weights are assigned to hidden layers according to their relative distances to the entities. See Eq. 4 below.

$$h = \alpha_1 h_1 + \alpha_2 h_2 + ...\alpha_t h_t + ...\alpha_k h_k \tag{4}$$

where h_t is the hidden layer of t-th word, α_t is the weight from h_t to the final context hidden layer h. The weight α_t is computed as the following equation.

$$\alpha_t = \frac{l - d_t}{l} \tag{5}$$

l is the length of entity's left or right context, d_t is the distance from t-th word to the entity. The word which is closer to the entity obtains the larger weight, and thus provides more information to the context hidden layer.

3.6   Output

The context features learned from recurrent layer are concatenated with entity embedding to form the entity's representation. And the two entity representations in the sentence are concatenated into a single vector $f = [f_1, f_2]$, which can be regarded as the final representation of the sample. Finally, the feature vector f is fed into a softmax classifier to predict the relationship between entities.

$$o = W_{cat}f \tag{6}$$

W_{cat} is the transformation matrix and o is the output of the network. Each dimension in output o can be interpreted as the confidence score of the corresponding relation.

3.7   Model Training

The model in Fig. 2 involves the parameters $\theta = \{W_1^f, W_2^f, b^f, W_1^b, W_2^b, b^b, W_v, W_{cat}\}$. Given an input example x, the network with parameters θ outputs the vector o, where the i-th component o_i represents the score of the relation i given the input. To obtain the conditional probability $p(i|x, \theta)$, a softmax operation is employed on the output. The equation is as follows:

$$p(i|x, \theta) = \frac{e^{o_i}}{\sum_{k=1}^{K} e^{o_k}} \tag{7}$$

where K is the number of relation classes. Given all training example $(x^{(i)}, y^{(i)})$, the training objective is cross-entropy error with L2 penalization term, and formally defined as follows:

$$J = - \sum_{m=1}^{M} \sum_{k=1}^{K} y_k^{(i)} log p_k^{(i)} + \lambda (\sum_{i=1}^{w} \|W_i\|^2) \tag{8}$$

where $y^{(i)} \in \mathbb{R}^{K \times 1}$ is the one-hot represented ground truth and $p^{(i)} \in \mathbb{R}^{K \times 1}$ is the estimated probability for each class by softmax (K is the number of target classes). M is the number of samples. w is the number of weight matrix, λ is a hyperparameter that specifies the magnitude of penalty on weights. We use the pretrained word embedding provided by Turian [13], other parameters are initialized randomly. Stochastic gradient descent is applied for optimization.

4   Experiments

4.1   Dataset

We test our model on the SemEval-2010 Task 8 dataset, which contains 8,000 sentences for training and 2,717 for testing with 9 directed natural relations and an artificial relation *Other*. *Other* indicates that the relation does not belong

Table 1. Statistics of the data sets

Relation	Train	Test
Other	1,410(17.63 %)	454(16.71 %)
Cause-effect	1,003(12.54 %)	328(12.07 %)
Component-whole	941(11.76 %)	312(11.48 %)
Entity-destination	845(10.56 %)	292(10.75 %)
Product-producer	717(8.96 %)	231(8.5 %)
Entity-origin	716(8.95 %)	258(9.5 %)
Member-collection	690(8.63 %)	233(8.58 %)
Message-Topic	634(7.92 %)	261(9.61 %)
Content-container	540(6.75 %)	192(7.07 %)
Instrument-agency	504(6.3 %)	156(5.74 %)
Total	8000(100 %)	2717(100 %)

to any natural classes. Table 1 shows the statistics of the annotated relation types of this dataset. We can see that the distribution of relation types in the test set is similar to that of the training set. Each sentence is marked with two nominals e_1 and e_2. The official evaluation framework takes directionality into consideration. A pair is counted as correct only if the order of nominals in the relationship is correct. Take the sentence shown in Fig. 1 for example, the corresponding relationship is *Cause-Effect(e_2,e_1)*, not *Cause-Effect(e_1,e_2)*. The official evaluation metric is the macro-averaged F1-score (excluding *Other*)

4.2 Parameter Settings

In this section, we present the experimental setup. Word-embedding is set to be 50-dimensional. Since there is no official development dataset, we tuned the hyperparameters by using 5-fold cross-validation. After the hyperparameters are determined, the model is trained with the best configuration. We get the best performance when the window size is set to be 3. The best dimension of recurrent hidden layer is 400. We use mini-batch training and the batch size is set to be 1000. The learning rate $\lambda = 10^{-5}$. Table 2 reports all hyperparameters used in the experiments.

4.3 MLP vs. BRNN

In this section, we discuss the effectiveness of BRNN over multi-layer perceptron (MLP) baselines. The inputs are both entities and their local contexts' embeddings (as shown in Table 3). [19] use concatenation of entities and their contexts' embeddings as inputs of the multi-layer perceptron. In our model, BRNN is used for extracting information from local contexts. Window size is set to be 2. The MLP gets a F_1 of 66.4 % and the BRNN obtains a F_1 of 71.4 %. The results suggest that the BRNN has a strong ability for semantic modeling.

Table 2. Hyperparameters used in experiments

Parameter	Name	Value
d	Word Embedding Size	50
h	Recurrent Hidden Layer Size	400
k	Window Size	3
bs	Batch Size	1000
λ	Learning Rate	10^{-5}

Table 3. Lexical level features

Features	Remark
L1	Noun 1
L2	Noun 2
L3	Left and right tokens of noun 1
L4	Left and right tokens of noun 2

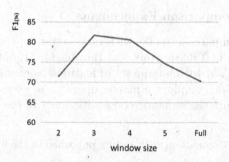

Fig. 3. Comparison of different window size

4.4 Local Context Vs. Global Context

Based on the analysis in Sect. 3.2, we perform the experiments to demonstrate the advantage of local context. In Fig. 3, We compare the results of different window sizes. The tendency shows that the proposed model performs best when the window size is 3, which means the words whose distance to the entity less than three is the more useful than the distant words. Moreover, too large or too small window size does not bring well consequences. Too large window size means more noises are introduced into the model. While small window limits the information the models can exploit, which is inadequate to identify the relation between the entities.

4.5 Different Pooling Methods

This subsection analyzes the impact of pooling layer which is responsible for gathering information from all hidden layers. We use the last hidden layer as a

Table 4. Comparison of different pooling methods

Pooling method	F1(%)
No pooling	79.8
Mean pooling	80.2
Max pooling	80.7
Weighted pooling	81.7

baseline, which is called 'no pooling' in this paper. From Table 4, we can observe that 'no pooling' yields a remarkable performance of 79.8 %, which outperforms CNN 78.9 % and MVRNN 79.1 %. Mean pooling treats all hidden layers equally, getting a F_1 of 80.2 % and max-pooling expects to find the most distinct feature at each dimension, boosting the F_1 by 0.5 %. The weight-pooling pays different attention to words in different positions, obtaining a F_1 of 81.7 %. The results suggest that the position feature used in pooling layer is helpful.

4.6   Results of Comparison Experiments

In this section, we compare our model with various state-of-the-art baselines. The results are presented in Table 5. Rink and Harabagiu [11] obtained the highest F1 of 82.2 % using SVM with a large set of features. Socher et al. [12] proposed a MVRNN model which employs a matrix operator to every node in the parse tree in order to capture the semantic compositionality of the sentence. MVRNN

Table 5. Comparison with results published in the literature

Classifier	Feature set	F_1
SVM	POS, WordNet, prefixes and other morphological features, dependency parse, Levin classes, PropBank, FanmeNet, NomLex-Plux, Google n-gram, paraphrases, TextRunner	82.2
MVRNN	Word embeddings	79.1
	Word embeddings, POS, NER, WordNet	82.4
CDNN	Word embeddings	69.7
	Word embeddings, word position embeddings	78.9
	Word embeddings, word position embeddings, WordNet	82.7
FCM	Word embeddings, dependency parsing	80.6
	Word embeddings, dependency parsing, NER	83.0
SDP-LSTM	Word embeddings, dependency parsing	82.4
	Word embeddings, POS embeddings, WordNet embeddings, grammar relation embeddings	83.7
BRNN	Word embeddings	79.8
	Word embeddings, Position Feature	81.7

achieves a *F1* of 78.9 %. CDNN proposed by Zeng got the *F1* of 79.1 %. When only the embedding is used, our model obtain F_1 of 79.8 %, outperforming both MVRNN and CNN model. This indicates that BRNN has a strong ability for semantic modeling without any explicit linguistic knowledge. While the position feature is added, we get 1.9 % improvement, which shows the weighted pooling is effective. Besides that, our model can even rival the models that exploit rich external resources, such as FCM and SDP-LSTM. In the future work, we will further explore the method for introducing external resources into our model.

5 Conclusion

In this paper, we propose a novel neural network for relation classification. We discover that the local information around entities is sufficient to achieve decent results for predicting entities' relations. Besides that, A weighted pooling method is proposed to leverage the position feature explicitly. We demonstrate the effectiveness of our model on SemEval-2010 relation classification task, and achieve a *F1*-score of 81.7 %.

Acknowledgments. This work is supported by National Natural Science Foundation of China (61472428, 61003204), the Fundamental Research Funds for the Central Universities, the Research Funds of Renmin University of China No. 14XNLQ06 and Tencent company.

References

1. Banko, M., Cafarella, M.J., Soderland, S., Broadhead, M., Etzioni, O.: Open information extraction from the web. In: Proceedings of the 20th International Joint Conference on Artificial Intelligence, IJCAI 2007, Hyderabad, India, 6–12 January 2007, pp. 2670–2676 (2007)
2. Bunescu, R.C., Mooney, R.J.: A shortest path dependency kernel for relation extraction. In: Proceedings of the Conference on Human Language Technology and Empirical Methods in Natural Language Processing, pp. 724–731. Association for Computational Linguistics (2005)
3. Ebrahimi, J., Dou, D.: Chain based RNN for relation classification. In: NAACL HLT 2015, The 2015 Conference of the North American Chapter of the Association for Computational Linguistics: Human Language Technologies, Denver, Colorado, USA, 31 May–5 June 2015, pp. 1244–1249 (2015)
4. GuoDong, Z., Jian, S., Jie, Z., Min, Z.: Exploring various knowledge in relation extraction. In: Proceedings of the 43rd Annual Meeting on Association for Computational Linguistics, pp. 427–434. Association for Computational Linguistics (2005)
5. Hasegawa, T., Sekine, S., Grishman, R.: Discovering relations among named entities from large corpora. In: Proceedings of the 42nd Annual Meeting on Association for Computational Linguistics, p. 415. Association for Computational Linguistics (2004)

6. Hashimoto, K., Miwa, M., Tsuruoka, Y., Chikayama, T.: Simple customization ofrecursive neural networks for semantic relation classification. In: Proceedings of the 2013 Conference on Empirical Methods in Natural Language Processing, EMNLP 2013, 18–21 October 2013, Grand Hyatt Seattle, Seattle, Washington, USA, A meeting of SIGDAT, a Special Interest Group of the ACL, pp. 1372–1376 (2013)

7. Hendrickx, I., Kim, S.N., Kozareva, Z., Nakov, P., Ó Séaghdha, D.,Padó, S., Pennacchiotti, M., Romano, L., Szpakowicz, S.: Semeval-2010task 8: multi-way classification of semantic relations between pairs of nominals. In: Proceedings of the Workshop on Semantic Evaluations: Recent Achievements and Future Directions, pp. 94–99. Association for Computational Linguistics (2009)

8. Kambhatla, N.: Combining lexical, syntactic, and semantic features with maximum entropy models for extracting relations. In: Proceedings of the ACL 2004 on Interactive Poster and Demonstration Sessions, p. 22. Association for Computational Linguistics (2004)

9. Plank, B., Moschitti, A.: Embedding semantic similarity in tree kernels for domain adaptation of relation extraction. In: ACL, vol. 1, pp. 1498–1507 (2013)

10. Qin, P., Xu, W., Guo, J.: An empirical convolutional neural network approach for semantic relation classification. Neurocomputing **190**, 1–9 (2016)

11. Rink, B., Harabagiu, S.: Utd: classifying semantic relations by combining lexical and semantic resources. In: Proceedings of the 5th International Workshop on Semantic Evaluation, pp. 256–259. Association for Computational Linguistics (2010)

12. Socher, R., Huval, B., Manning, C.D., Ng, A.Y.: Semantic compositionality through recursive matrix-vector spaces. In: Proceedings of the 2012 Joint Conference on Empirical Methods in Natural Language Processing and Computational Natural Language Learning, EMNLP-CoNLL 2012, 12–14 July 2012, Jeju Island, Korea, pp. 1201–1211 (2012)

13. Turian, J.P., Ratinov, L., Bengio, Y.: Word representations: a simple and general method for semi-supervised learning. In: ACL 2010, Proceedings of the 48th Annual Meeting of the Association for Computational Linguistics, 11–16 July 2010, Uppsala, Sweden, pp. 384–394 (2010)

14. Wu, F., Weld, D.S.: Open information extraction using wikipedia. In: ACL 2010, Proceedings of the 48th Annual Meeting of the Association for Computational Linguistics, 11–16 July 2010, Uppsala, Sweden, pp. 118–127 (2010)

15. Xu, Y., Mou, L., Li, G., Chen, Y., Peng, H., Jin, Z.: Classifying relations via long short term memory networks along shortest dependency paths. In: Proceedings of Conference on Empirical Methods in Natural Language Processing (2015) (to appear)

16. Yao, X., Durme, B.V.: Information extraction over structured data: question answering with freebase. In: Proceedings of the 52nd Annual Meeting of the Association for Computational Linguistics, ACL 2014, 22–27 June 2014, Baltimore, MD, USA, vol. 1, Long Papers, pp. 956–966 (2014)

17. Yu, M., Gormley, M., Dredze, M.: Factor-based compositional embedding models. In: NIPS Workshop on Learning Semantics (2014)

18. Zelenko, D., Aone, C., Richardella, A.: Kernel methods for relation extraction. J. Mach. Learn. Res. **3**, 1083–1106 (2003)

19. Zeng, D., Liu, K., Lai, S., Zhou, G., Zhao, J.: Relation classification via convolutional deep neural network. In: COLING 2014, 25th International Conference on Computational Linguistics, Proceedings of the Conference: Technical Papers, 23–29 August 2014, Dublin, Ireland, pp. 2335–2344 (2014)

Psychological Stress Detection from Online Shopping

Liang Zhao[1]([✉]), Hao Wang[2], Yuanyuan Xue[1], Qi Li[1], and Ling Feng[1]

[1] Tsinghua National Laboratory for Information Science and Technology,
Department of Computer Science and Technology, Centre for Computational
Mental Healthcare Research, Institute of Data Science,
Tsinghua University, Beijing 100084, China
zhaoliang0415@gmail.com, {xyy12,qili13}@mails.tsinghua.edu.cn,
fengling@tsinghua.edu.cn
[2] School of Computer Science and Technology,
North China University of Technology, Beijing 100144, China
wanghaomails@gmail.com

Abstract. The increasingly faster life pace in modern society makes people always feel stressful and it is of great significance to discover a users suffering stress in time. According to psychological study, shopping is chosen as an effective way for stress relief, especially for females. Compared with non-stress cases, a user may perform different shopping patterns when under stress. An interesting issue then arises: can we detect one's psychological stress from online shopping data? By investigating stress-related outlier features from both content and behavior of online purchase orders, we learn a users stress status by classification. A real user study of 20 experienced female online customers aged 23–30 verifies the effectiveness of shopping based stress detection, achieving an F1-measure of more than 80 % with J48 classifier. None of the features negatively affect the detection result. Feature combinations bring dramatic improvements than single feature. In total, shopping content features are proved to be more significant than behavior features.

1 Introduction

With the increasingly developed information technology as well as the global popularization of e-commerce, tens of thousands of users tend to online shopping. Conventionally, online shopping data is widely used in personalized user profile analysis, advertisement targeting, and also commodity recommendation. However, what shopping data tells us is far more than these. [1] defines six broad categories of hedonic shopping motivations. Shopping for stress relief and shopping to alleviate a negative mood is one main category. "When the going gets tough, the tough go shopping"[1]. The notion of "retail therapy" is proved to be a strategic effort to improve mood [2,17]. And shopping has been acknowledged as a form of emotion-focused coping in response to stressful events or simply to get

---

[1] http://www.imdb.com/title/tt0502439/, a television series in 1991.

© Springer International Publishing Switzerland 2016
F. Li et al. (Eds.): APWeb 2016, Part I, LNCS 9931, pp. 431–443, 2016.
DOI: 10.1007/978-3-319-45814-4_35

one's mind off a problem [10]. A recent online poll of over 1,000 U.S. adults commissioned by HuffPost shows that nearly one in three Americans uses shopping as a stress-reliever, and women are twice as likely as men to use retail therapy as a way to cope with stress (40 % vs. 19 %) [3]. An interesting issue then arises: *can we detect a user's psychological stress from his/her online shopping data?* As psychological stress is a hot and significant health problem especially in this faster life-pace society, by addressing this issue can we get a new convenient and low-cost channel for stress detection compared to traditional consultant with psychologists. Besides, analyzing the difference of shopping patterns with/without stress will further facilitate conventional shopping issues mentioned above, i.e., offering more details of user profile analysis and providing more accurate advertisement and commodity recommendation according to user's stress status.

In daily experience, individual shopping is highly random and triggered by various factors, such as personal shopping interests, popularity, or external events like birthday or festivals, not merely by stress. Thus, the challenge to detect stress from online shopping lies in how to capture stress-related outlier from individual shopping data. With preliminary rules, we first filter out those non-stress related purchase orders, such as orders of daily necessities (e.g., utility bills, telephone bills, etc.), or containing keywords like birthday, marriage or festival gifts. According to psychological studies, impulsive behaviors can be a strategic effort to repair a bad mood [11,18,19], and adopting retail therapy, individuals will treat themselves to unplanned indulgences to improve mood [2]. Focusing on such shopping indulgences and impulse, we then define and extract stress-related features from the perspective of both abnormal shopping content and impulsive shopping behavior (e.g., infrequent commodity category purchased, the indulgently expensive cost, impulsive order frequency, etc.). A real user study with 20 experienced female online customers (aged 23–30) was conducted to verify the effectiveness of our shopping-based stress detection. Decision tree (J48) classifier is the most suitable method upon our real data set, achieving an F1-measure of more than 80 %. In more detail, none of the stress-related outlier features extracted negatively affect the detection result. Shopping content features contribute almost three times more than behavior features in our data set. The F1-measure improves by 39.7 % when involving content features. It is also interesting that feature combination brings dramatically bigger performance improvement than single feature alone.

Our contributions lie in the following two aspects.

- To the best of our knowledge, this is the first exploration detecting psychological stress from online shopping data.
- We novelly define and extract various stress-related features of both shopping content and behavior from the purchase data, and also investigate the impact of both single feature and feature combinations to the detection performance.

2   Related Work

2.1   Stress Analysis from Psychophysiological Signals

Traditional stress analysis and detection techniques in the field of psychology use subjective questionnaires, such as the most widely used *Perceived Stress Scale* (PSS) [6]. Various sensors are also leveraged to monitor the changes of psychophysiological signals (e.g., galvanic skin response GSR, heart rate variability HRV, electroencephalogram EEG, electrocardiogram ECG, blood pressure, electromyogram, and respiration) and physical signals (e.g., voice, gesture and interaction, facial expressions, eye gaze, pupil dilation and blink rates) for people under stress [7,8,16]. The limitations of these methods are the invasive user experience caused by the body contact and the deviation induced by different physical or environmental conditions.

2.2   Stress-Related Analysis on Social Network

Computer-aided stress analysis and applications in social network has drawn much attention in recent years [12,13,21,22]. [4,5] analyzed users' twitting behaviors to measure their depression risks, and found out that social media contains useful cues in predicting one's depression tendency. [20] built a depression detection model based on a sentiment analysis method on micro-blog. Looking into the contents and temporal features of users' BBS posts, [15] tried to detect depressed users through a supervised learning approach. Focusing on the four main kinds of adolescent stress, [21] extracted different features from teenagers' tweets and leverage classifiers to learn the potential stress category and corresponding stress level behind the tweet. [13] presented a deep sparse neural network to detect tweet-level stress for arbitrary micro-blog users. [14] detected user-level stress from cross-media micro-blog via a deep convolution network on sequential tweeting time series in a certain time period. [23] further improved the tweet-based adolescent stress detection by involving the details of social interactions (linguistic content as well as interacting behaviors) under the tweet. [9] proposed a co-training-based semi-supervised learning approach integrating both GPS trajectories and micro-blog for adolescent stress detection.

To the best of our knowledge, state-of-the-art stress detection focus on micro-blog platform, and there are no literature published exploring the issue upon online shopping data set.

3   Problem Definition

We focus on individual purchase data of online shopping as our data source. To be brief, we use purchase data and shopping data exchangeably in the rest of the paper. Figure 1 illustrates a purchase order of Taobao[2]. A purchase order consists of six parts, detailed as following:

- **Order time** is the time when the customer decides to buy the chosen commodities and submits the purchase order (not paid yet);

---

[2] http://www.taobao.com, the biggest C2C e-commerce site in China.

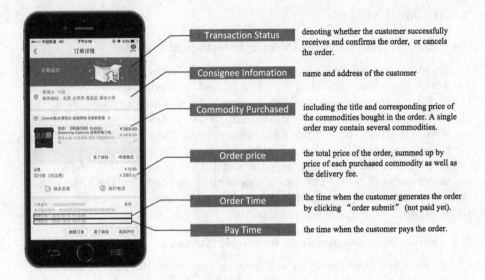

Fig. 1. An example of an online purchase order.

- **Pay time** is the time when the customer pays for the order.
- **Consignee info** is the basic information of the consignee of the order, including name, address and phone number. In this study, we simply consider consignee name;
- **Commodity** represents the commodities the customer buys in the order, denoted as the title of the commodities. A single order may contain several different commodities;
- **Item price** is the price of a certain commodity purchased in the order;
- **Order price** is the price of the whole order, summed up by the item price of each commodity as well as the delivery fee;
- **Transaction status** specifies the status of the order. The transaction status is *success*, if and only if the customer receives and is satisfied with the commodities, and then confirms the order. Otherwise, if the customer changes his/her mind before payment or is dissatisfied with the commodities then returns them, then the transaction status changes to *cancel*, and we call such orders *invalid orders*.

As described, a purchase order p_order is represented as a six-dimensional vector, denoted as $p\_order = (t_o, t_p, cons, \{< commo_1, price_1 >, \ldots, < commo_n, price_n >\}, s\_price, trans)$, where t_o, t_p correspond to order time and pay time, respectively. $cons$ is denotes the name of consignee, sum is the sum price of the order and $trans$ specifies the transaction status. Particularly, $< commo_i, price_i > (1 \leq i \leq n)$ illustrates the i-th commodity (by title) in the order as well as its corresponding price, and n is the number of commodities bought in the order. Given a set of online purchase orders of user u, we aggregate the orders of day d into a set $Orders(u, d)$. We then extract stress-related outlier features from $Orders(u, d)$ represented as an outlier feature vector $\mathbf{fv}_u^d$. Classifiers are exploited to learn the user's stress status. Let $Stress = \{true, false\}$ be the two stress status. The detection result $res(u, d) = true$ suggests that the user is stressful in day d, otherwise the user is non-stressed.

4 Outlier Feature Extraction from Purchase Orders

Psychological studies show that impulsive behaviors can be a strategic effort to repair a bad mood [11,18,19]. And adopting retail therapy, individuals will treat themselves to unplanned indulgences to improve mood [2]. Such shopping indulgences and impulse outlier are powerful cues for stress detection, for example, spending more money or shopping more frequently than usual.

4.1 Non-Stress-Related Orders Filtering

However, not all shopping indulgences are triggered by stress. A user may treat herself a luxury handbag as birthday gift which is much expensive than those she ever buys, or she purchases much more frequently just for festival celebration. Before analyzing stress-related outlier features of purchase orders, we first need to filter out such non-stress-related orders. Four kinds of non-stress-related cases are considered in this paper.

- Orders whose consignee name are inconsistent with the customer himself/herself. With this strategy, all the orders maintained accurately describe the shopping requirements of the customer.
- Orders of daily necessities, such as utility bills, telephone bills, etc.
- Orders whose order time or pay time fall in major festivals and holidays. It is highly possible that in the 17 major festivals[3] (e.g., Spring Festival, Valentines' Day, etc.) customers tend to buy something expensive treating themselves or make a great many orders for celebration instead of abreaction.
- Orders whose commodity titles specifying birthday, pregnancy, marriage, or festival (the same with the festivals in the third case) gift. The reason of ignoring such orders is the same with the third case.

Given a user u, a day d and the set of orders in the day denoted as $Orders(u, d)$ after the above preprocessing, we novelly interpret stress-related shopping indulgence from the two perspectives, *shopping content* and *shopping behavior*.

---

[3] 1. New Year's Day 2. Spring Festival 3. Valentines' Day 4. Women's Day 5. Easter Day 6. May Day 7. Mother's Day 8. Father's Day 9. Dragon Boat Festival 10. Children's Day 11. National Day 12. Teachers' Day 13. Mid-Autumn Day 14. Halloween 15. Thanksgiving Day 16. Christmas Day 17. Lantern's Day.

4.2 Shopping Content Features

• **Commodity Category Outlier.** Usually, one's shopping preference is relatively stable and focus on several commodity categories. Once s/he buys something of his/her infrequent categories, it is a signal of outlier. For example, a fat girl who prefers buying snacks suddenly orders diet pills one day due to self-cognition stress (agonized with the her fat shape). We define such commodity category outlier as a boolean value with *true* representing the user buys commodities of infrequent categories in this day.

Hierarchical commodity catalogs are crawled from *Taobao* (Fig. 2(a)) covering seven main categories with totally 1587 sub-category keywords. Figure 2(b) illustrates the hierarchy of the *women clothes* catalog, of which the nodes represent sub-category keywords from general (e.g., tops) to specific (e.g., coat). For each history purchase order of the user, we extract the commodity category/sub-category words from the commodity title and map it to the seven coarse-grained categories according to the hierarchical catalogs. Assume $C = \{women\ clothes, shoes\ and\ bags, accessories, cosmetics, pets, electronics, sports\ and\ outdoors\}$ denote the set of commodity categories. For $c_i \in C$ ($1 \leq i \leq 7$), the distribution of c_i is calculated as $dr_u^i = \frac{num\_item_u^i}{total\_item_u}$, where $num\_item_u^i$ is the number of commodities which user u buys and belong to c_i, and $total\_item_u$ is the total number of commodities u buys. We take c_i as the frequent category of user u if $dr_u^i \geq \theta$. Here we set $\theta = 0.3$. If there exist commodities of $Orders(u,d)$ beyond the frequent category set, then the commodity category outlier is set *true*; otherwise the feature value is set *false*.

• **Commodity Price Outlier.** Buying something much more expensive than those similar commodities one ever buys is also a kind of indulgence. We consider such commodity-level price outlier as a stress-related feature. The feature is also boolean valued, with *true* denoting that the outlier exists. Particularly, such price outlier is category-dependent. Here we distinguish the commodity price outlier by the fine-grained sub-categories (e.g., skirt, coat) in the hierarchies illustrated as

Commodity Category	# Hierarchy Levels	# Sub-category Keywords
Women clothes	5	42
Shoes and bags	4	37
Accessories	5	106
Cosmetics	7	230
Pets	6	470
Electronics	7	653
Sports and outdoors	4	49
Total	/	1587

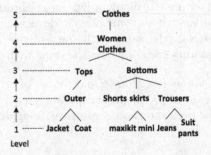

(a) Seven commodity catalogs crawled from *Taobao* (b) An illustration of a hierarchical commodity category

Fig. 2. Hierarchical commodity catalogs crawled from *Taobao*

Fig. 2(b) instead of the seven coarse-grained categories (e.g., women clothes). For each commodity purchased in history orders of a user, we extract the commodity sub-category words by title, and compute the price expectation as well as the standard deviation of that sub-category.

$$exp(u, sub\_c) = \frac{1}{N} \sum_{commo_i \ of \ sub\_c}^{N} price_i$$

$$dev(u, sub\_c) = \sqrt{\frac{1}{N} \sum_{commo_i \ of \ sub\_c}^{N} (price_i - exp(u, sub\_c))^2}$$

where sub_c is the sub-category extracted from u's purchase history, $commo_i$ is a commodity the user purchased, $price_i$ denotes the price of $commo_i$, and N is the number of commodities of sub_c the user ever buys. Assume the commodity price of a given sub-category a user buys observe a random distribution with the expectation $exp(u, sub\_c)$ and the standard deviation $dev(u, sub\_c)$. For the commodities in $Orders(u, t)$, if there exists a commodity of sub_c which is expensive than $exp(u, sub\_c)$ by more than $dev(u, sub\_c)$, then we think commodity price outlier exists in day d. Particularly, if a sub-category sub_c does not yet appear in user's purchase history, we use the expectation and standard deviation of its hierarchically parent to estimate the outlier. And if the coarse-grained category of sub_c does not even appear in the user's purchase history, then we simply set the expectation and standard deviation as 0.

• **Total cost outlier.** There is one possible case that all the commodities in the orders are with regular price but the user buys too many, which may also lead to a stress-related outlier. Thus, beside commodity-level price outlier, we also investigate the outlier of total price the user spends in a certain day. For each order p in $Orders(u, d)$, we sum up the order price of p and obtain the total cost in day d, denoted as $cost(u, d)$. Similarly, assume a user's everyday total cost observe a random distribution of an expectation of exp_c and a standard deviation dev_c. The total cost outlier is extracted when $cost(u, d) - exp(u, d) > dev(u, d)$ and the feature is set *true*.

4.3 Shopping Behavior Features

From the perspective of time, purchase frequency and transaction status, five features are defined to describe a user's shopping behavior under stress.

Time-Related Behavior Features

• **Ratio of orders with abnormal order/pay time.** We live with regular circadian rhythm (i.e., work at daylight and sleep in the night). As investigation of micro-blog based stress detection [13,21,23], abnormal posting time of micro-blog suggests potential stress of a user. Similarly, we check the orders with abnormal order/pay time and take the ratio of such orders in total $Orders(u, d)$ as a stress-related feature, valued in [0, 1], where the value 0 represents no orders are submitted or paid in abnormal time, and value 1 suggests the opposite case. To be simple, we take the time segment from 10:00 p.m. to 6:00 a.m. next morning as the abnormal interval. If the order/pay time of an order is in this abnormal interval, then the corresponding purchase order is considered with time abnormality. The bigger the ratio of orders with time abnormality, the higher probability that the user is suffering stress.

• **Average time gap between order and pay time.** The time gap between the order time and pay time of a purchase order (short for *O-P time gap*) always shows the user's purchase determination. Generally, a user pays immediately after submitting the order if s/he decides to buy it. A user under stress may be impulsive and submit some purchase orders for hedonic abreaction, but when coming to payment, s/he may hesitate. A longer O-P time gap just reveals such anomaly behavior, providing a stress-related cue. Here we calculate the O-P time gap by hour. For each order $p\_order \in Orders(u, d)$, the average O-P time gap is calculated as

$$T\_gap(u, d) = \frac{1}{n} \sum_{p\_order \in Orders(u,d)} (p\_order.t_p - p\_order.t_o)$$

where $T\_gap(u, d)$ is the average O-P time gap of user u in day d, and n is the number of orders purchased in the day.

Density-Related Behavior Features

• **Number of Orders in a Single Day.** Number of orders the user purchases in a single day directly reflects his/her indulgent shopping behavior. Obviously, more orders reveal a bigger purchase impulse and consequently more possibly corresponds with stress shopping.

• **Recent Purchase Frequency.** Considering the continuity of stress, we study the recent purchase frequency before day d and check whether day d is involved in an abnormally frequent purchase interval. Shopping during the recent m days before day d is investigated. The recent purchase frequency is a ratio valued in $[0, 1]$ which is represented as the proportion of days with at least one order in the recent m days. A bigger recent purchase frequency suggests a bigger probability that day d is covered in the indulgent purchase interval and the user is still buried in stress this day. In this study, we set $m = 3$.

Transaction Status Related Features

• **Ratio of Invalid Orders.** Orders are *canceled* (invalid orders) in two cases: (1) the user changes his/her mind before payment; (2) the user returns the commodity due to dissatisfaction. The former case corresponds to one's hesitation under stress shopping as mentioned in the description of O-P time gap feature, while the latter case also reveals the user's impulsive shopping. We record the ratio of invalid orders in $Orders(u, d)$, to evaluate the degree of a user's impulsive shopping, valued in $[0, 1]$. A bigger ratio of such orders suggests a more unstable shopping behavior and the user risks a bigger stress potential.

5   Experimental Study

5.1   Setup

Psychological stress detection is a highly personalized issue and there are no available benchmark specially for the problem. In this work, we collect data ourselves and conduct a user study on a real online shopping data set. 20 experienced female online customers aged between 23 and 30 participated in the user study. 8060 purchase orders from 2013/1/1 to 2016/4/1 were collected from their accounts in Taobao, averagely 403 purchase orders per user. They are all experienced online customers since in average they purchase 134 orders per year, and almost one order every three days. Each participants were asked to scan her own purchase history, recall the real situations and annotate whether she is stressful on a certain day. We take such annotation as the ground truth to evaluate the shopping based stress detection. For each user, chronologically, we use the early 80 % purchase orders as the training data, and the rest 20 % as the testing data. The training data is also used to estimate the distribution of the user's shopping habit (i.e., the expectation and standard deviation of category-dependent item price as well as total cost every day) and help compute corresponding outlier.

5.2   General Performance of Shopping Based Stress Detection

Performance Comparison of Different Classifiers. We apply four different classifiers, i.e., J48, Naive Bayes, Logistic Regression, and SVM, to detect stress from shopping features. Assume the stressful days annotated by users be positive cases, TP, FP, TN, FN respectively represent true positive, false positive, true negative and false negative cases detected by classifiers. Precision (measuring *how many detected stressful shopping days match the participants' annotations*) and recall (measuring *how many stressful shopping days are not detected while the participants annotate as stressful*) are used to evaluate the performance, where $precision = TP/(TP + FP)$, $recall = TP/(TP + FN)$.

Figure 3(a) compares the performance of different classifiers. Obviously, decision tree based classifier is more suitable for shopping based stress detection. Averagely, J48 works the best, improving the F-measure by 7.1 %, 17.8 %, 10.9 % compared with Naive Bayes, Logistic Regression and SVM, respectively. The average F-measure by J48 achieves more than 80 %, which proves that for experienced online customers it is available to detect their stress from online shopping data. In the rest of the experiments, we select J48 as the classifier.

Performance of Data Preprocessing. As mentioned, before extracting stress-related features, we first filter out non-stress related orders. In our data set, 18.7 % orders in average are filtered out as non-stress related orders. Figure 3(b) illustrates the impact of such preprocessing. Our preprocessing is proved to be effective, dramatically improving the precision by 23.5 % and merely damaging the recall by 3.7 %, compared to that without preprocessing.

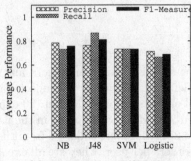
(a) Comparison of classifiers

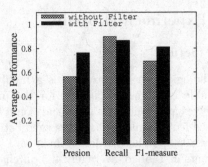
(b) Performance of data preprocessing

Fig. 3. General performance of shopping based stress detection

Single feature	Impact
Category outlier (c)	0
Item price outlier (pr)	0.6%
Total cost outlier (tc)	**3.2%**
Abnormal time (abt)	0.3%
O-P time gap (OPtg)	0
Order number (num)	0.1%
Recent frequency (rf)	**10.9%**
Invalid orders (inv)	0.2%

(a) single feature

Feature combination	Impact
Pr+tc	**13.9%**
c+pr	1.5%
tc+num	8.4%
tc+freq	**21.9%**
tc+OPtg	6.0%

(b) two-feature combinations

Feature combination	Impact
Shopping content	**37.1%**
Shopping behavior	14.5%

(c) different kinds of features

Fig. 4. Investigation of feature impact

5.3   Feature Impact

To evaluate the impact of a feature or a feature combination, we define the impact of a feature (combination) as the F-measure improvement

$$Imp(\mathbf{F}) = \frac{fmeasure\_total - fmeasure\_no\_\mathbf{F}}{fmeasure\_no\_\mathbf{F}} * 100\,\%$$

where $\mathbf{F}$ is the set of features, $fmeasure\_no\_\mathbf{F}$, $fmeasure\_total$ denotes the f-measure without $\mathbf{F}$ and with all the features, respectively. $Imp(\mathbf{F})$ values in $[-1, 1]$, where a bigger $|Imp(\mathbf{F})|$ corresponds to a bigger feature impact. When the value is positive, the feature positively affect the detection result; otherwise, the opposite. Tables in Fig. 4 demonstrate the impact of each feature alone as well as feature combinations. None of the features negatively affect the detection result (Fig. 4(a)). Total cost outlier and recent purchase frequency are proved as powerful single features in our data set. It is interesting that impact of some features are tiny when considered alone, but when combined with other features, they contain much more significance. For example, item price outlier only contributes 0.6 % to the total performance, while combining with total cost outlier, the impact dramatically improves to 13.9 %, which is also much bigger than that of total cost outlier alone (Fig. 4(b)). Commodity category outlier itself almost contributes 0 to the detection, but when joining with item price and total cost (the shopping content features), the impact achieves unbelievably 37.1 %. The result coincides with the fact that shopping under stress contains various cases and the features need to be considered simultaneously. In addition, shown in Fig. 4(c), shopping content features contribute almost three times more than shopping behavior features in our experiment.

6   Conclusions and Future Work

Inspired by the psychological studies that shopping is acknowledged as an effective stress-reliever, we present a novel exploration detecting one's psychological stress from online shopping data in this paper. Filtering out non-stress-related purchase orders, we novelly define and extract various stress-related outlier features of both shopping content and shopping behaviors. A real user study of 20 experienced female online customers verifies the effectiveness of our shopping-based stress detection. Decision tree (J48) classifier is the most suitable approach and achieves an F1-measure of more than 80 %. Shopping content features contributes almost three times more than behavior features in our data set. It is also interesting that feature combination brings dramatically bigger contributions than single feature alone.

In this paper, we aggregate the shopping data simply by natural date. We further plan to locate stressful intervals with flexible time length with clustering algorithms. On the other hand, compared to the existing stress detection from micro-blog, shopping data lacks of linguistic hints and is not able to provide more semantics such as stress categories and stress source events. The other interesting issue in future work is to combine the two data source together, exploit the semantic hints of micro-blog, overcome data sparsity to the greatest extent, and present a more accurate stress detection method.

Acknowledgement. The work is supported by National Natural Science Foundation of China (61373022, 61532015, 71473146) and Chinese Major State Basic Research Development 973 Program (2015CB352301).

References

1. Arnold, M.J., Reynolds, K.E.: Hedonic shopping motivations. J. Retail. **79**(2), 77–95 (2003)
2. Atalay, A.S., Meloy, M.G.: Retail therapy: a strategic effort to improve mood. Psychol. Market. **28**(6), 638–659 (2011)

3. Gregoire, C.: Retail therapy: one in three recently stressed americans shops to deal with anxiety (2013). http://www.huffingtonpost.com/2013/05/23/retail-therapy-shopping_n_3324972.html

4. Choudhury, M., Counts, S., Horvitz, E.: Social media as a measurement tool of depression in populations. In: ACM Web Science, pp. 47–56 (2013)

5. Choudhury, M., Gamon, M., Counts, S., Horvitz, E.: Prediction depression via social media. In: ICWSM, pp. 128–137 (2013)

6. Cohen, S., Kamarck, T., Mermelstein, R.: A global measure of perceived stress. J. Health Soc. Behav. **4**, 385–396 (1983)

7. Hamid, N.H.A., Sulaiman, N., Aris, S., Murat, Z., Taib, M.: Evaluation of human stress using eeg power spectrum. In: CSPA, pp. 1–4 (2010)

8. Hosseini, S., Khalilzadeh, M.: Emotional stress recognition system using EEG and psychophysiological signals: using new labelling process of EEG signals in emotional stress state. In: ICBECS, pp. 1–6 (2010)

9. Jin, L., Xue, Y., Li, Q., Feng, L.: Integrating human mobility and social media for adolescent psychological stress detection. In: Navathe, S.B., et al. (eds.) DASFAA 2016. LNCS, vol. 9643, pp. 367–382. Springer, Heidelberg (2016). doi:10.1007/978-3-319-32049-6_23

10. Lee, E., Moschis, G.P., Mathur, A.: A study of life events and changes in patronage preferences. J. Bus. Res. **54**(1), 25–38 (2001)

11. Leith, K.P., Baumeister, R.F.: Why do bad moods increase self-defeating behavior? emotion, risk tasking, and self-regulation. J. Person. Soc. Psychol. **71**(6), 1250–1267 (1996)

12. Li, Q., Xue, Y., Jia, J., Feng, L.: Helping teenagers relieve psychological pressures: a micro-blog based system. In: EDBT, pp. 660–663 (2014)

13. Lin, H., Jia, J., Guo, Q., Xue, Y., Huang, J., Cai, L., Feng, L.: Psychological stress detection from cross-media microblog data using deep sparse neural nework. In: ICME, pp. 1–6 (2014)

14. Lin, H., Jia, J., Guo, Q., Xue, Y., Li, Q., Huang, J., Cai, L., Feng, L.: User-level psychological stress detection from social media using deep neural network. In: MM, pp. 507–516 (2014)

15. Shen, Y.-C., Kuo, T.-T., Yeh, I.-N., Chen, T.-T., Lin, S.-D.: Exploiting temporal information in a two-stage classification framework for content-based depression detection. In: Pei, J., Tseng, V.S., Cao, L., Motoda, H., Xu, G. (eds.) PAKDD 2013, Part I. LNCS, vol. 7818, pp. 276–288. Springer, Heidelberg (2013)

16. Shi, Y., Ruiz, N., Taib, R., Choi, E., Chen, F.: Galvanic skin response (GSR) as an index of cognitive load. In: CHI, pp. 2651–2656 (2007)

17. Thayer, R.E., Newman, J.R., McClain, T.M.: Self-regulation of mood: strategies for changing a bad mood, raising energy, and reducing tension. J. Person. Soc. Psychol. **67**(5), 910–925 (1994)

18. Tice, D.M., Bratslavsky, E.: Giving in to feel good: the place of emotion regulation in the context of general self-control. Psychol. Inq. **11**(3), 149–159 (2000)

19. Tice, D.M., Bratslavsky, E., Baumeister, R.F.: Emotional distress regulation takes precedence over impulse control: if you feel bad, do it!. J. Person. Soc. Psychol. **80**(1), 53–67 (2001)

20. Wang, X., Zhang, C., Ji, Y., Sun, L., Wu, L., Bao, Z.: A depression detection model based on sentiment analysis in micro-blog social network. In: Li, J., Cao, L., Wang, C., Tan, K.C., Liu, B., Pei, J., Tseng, V.S. (eds.) PAKDD 2013 Workshops. LNCS, vol. 7867, pp. 201–213. Springer, Heidelberg (2013)

21. Xue, Y., Li, Q., Jin, L., Feng, L., Clifton, D.A., Clifford, G.D.: Detecting adolescent psychological pressures from micro-blog. In: Zhang, Y., Yao, G., He, J., Wang, L., Smalheiser, N.R., Yin, X. (eds.) HIS 2014. LNCS, vol. 8423, pp. 83–94. Springer, Heidelberg (2014)
22. Zhang, Y., Tang, J., Sun, J., Chen, Y., Rao, J.: Moodcast: emotion prediction via dynamic continuous factor graph model. In: ICDM, pp. 1193–1198 (2010)
23. Zhao, L., Jia, J., Feng, L.: Teenagers stress detection based on time-sensitive micro-blog comment/response actions. In: Dillon, T. (ed.) IFIP AI 2015. IFIP AICT, vol. 465, pp. 26–36. Springer, Heidelberg (2015)

Confidence-Learning Based Collaborative Filtering with Heterogeneous Implicit Feedbacks

Jing Wang$^{(\boxtimes)}$, Lanfen Lin, Heng Zhang, and Jiaqi Tu

College of Computer Science, Zhejiang University, Hangzhou, China
{cswangjing,llf,hengzhang,shellytu}@zju.edu.cn

Abstract. Implicit feedbacks, which indirectly reflect opinions through observing user behaviors, have recently received more and more attention in recommendation communities due to their accessibility and richness in real-world applications. Most of the existing implicit-feedback-based recommendation algorithms only exploit one type of implicit feedback. In real-world applications, there is usually more than one type of implicit feedback. Considering the sparsity problem of recommender systems, it is significant to leveraging more available data. In this paper, we study the **heterogeneous implicit feedbacks** problem, where more than one type of implicit feedback is available. We study the characteristics of different types of implicit feedbacks, and propose a unified approach to infer the confidence that we can believe a user prefers an item. Then we apply the inferred confidence to both point-wise and pair-wise matrix factorization models, and propose a more generic strategy to select training samples for pair-wise methods. Experiments on real-world e-commerce data show that our methods outperform the state-of-art approaches, considering several commonly used ranking oriented evaluation criterions.

Keywords: Recommender systems · Heterogeneous implicit feedbacks · Confidence · Collaborative filtering · E-commerce

1 Introduction

E-commerce has been growing rapidly in recent years and this has resulted in a huge volume of products and services. Users are provided with more options, making it more difficult for them to find the right choice. This emphasizes the prominence of recommender systems, which aim at helping users to find products and services that best meet their needs and interests [1].

Recommender systems rely on different types of input. The most convenient is the high quality **explicit feedback**, such as 5-star ratings, or thumbs-up/down, like and dislike. Collaborative filtering (CF), which is one of the most successful recommendation techniques, has been well studied to exploit explicit feedback [2–4]. However, high-quality explicit feedback is not always available in real-world applications. **Implicit feedback**, which indirectly reflects opinion through observing user behavior, such as purchase and click, are easy to gather, without incurring into any overhead on users. Recently, implicit feedbacks have received more and more attention in

© Springer International Publishing Switzerland 2016
F. Li et al. (Eds.): APWeb 2016, Part I, LNCS 9931, pp. 444–455, 2016.
DOI: 10.1007/978-3-319-45814-4_36

recommendation communities due to their availability and richness. Most of the existing implicit-feedback-based algorithms only consider one type of implicit feedback, such as ImplicitALS [5] and BPR [6]. ImplicitALS [5] thinks that more frequent interactions indicate stronger confidence that the user prefers the item. However, when more than one type of implicit feedback exists, only considering the frequency of interactions is unreasonable, since different implicit feedbacks indicate different confidence levels. BPR [6] is based on the assumption that a user prefers a consumed item to an unconsumed item. However, BPR cannot compare two interacted items when more than one type of implicit feedback exists, which will lose a lot of valuable information. A typical example to explain the limitations of ImplicitALS and BPR is: an item that is clicked five times vs. an item in the shopping cart, which is more preferred? Considering the sparsity problem of recommender systems, it is significant to leveraging more available data effectively. Pan et al. [7] are the first to study the **heterogeneous implicit feedback** problem, where more than one type of implicit feedback is available. They take users' transaction records as certain data and examination records as uncertain data, and propose ABPR. ABPR [7] treats all uncertain data equally, however, different kinds of implicit feedbacks represent different levels of uncertainty. For example, adding an item into shopping cart indicates higher confidence of preference than clicking an item. What's more, the number of times an event has occurred and the combinations of different events also indicate different confidence levels. For example, a frequently clicked item may be more preferred than once clicked item, but it is hard to compare a frequently clicked item with an item in the favorites that is clicked only once. It is a challenge to characterize the confidence of user preference from heterogeneous implicit feedbacks, and it is also meaningful to solve the sparsity problem by incorporating more information.

We try to find a unified way to deal with heterogeneous implicit feedbacks, and take full advantage of the information contained in various implicit feedbacks. We first study the characteristics of different types of implicit feedbacks, and propose a unified approach to infer the confidence of preference from heterogeneous implicit feedbacks. Then we apply the inferred confidence to both point-wise and pair-wise matrix factorization model, and propose a more generic strategy to select training samples for pair-wise methods. Experiments on real-world e-commerce data show that our methods outperform the state-of-art approaches, considering several commonly used ranking oriented evaluation criterions. The contributions of this paper are summarized as follows:

- We study the heterogeneous implicit feedbacks problem where both certain and multiple types of uncertain feedbacks are available.
- We analyze the characteristics of different types of implicit feedbacks, and propose a unified approach to infer the confidence of preference from heterogeneous implicit feedbacks.
- We apply the inferred confidence to both point-wise and pair-wise matrix factorization model. We propose a more generic strategy to select training samples for pair-wise methods, which brings more effective comparable pairs and further relieve the sparsity problem.

- We conduct extensive experiments on real-world e-commerce data from Tmall.-com. The results show that our approach can greatly improve the original point-wise and pair-wise methods, considering several commonly used ranking oriented evaluation criterions

The rest of this paper is organized as follows: in the next section, we summarize the related work; in Sect. 3, we describe our proposed methods; in Sect. 4, we introduce the datasets, baselines, evaluation metrics and experimental results; finally, we summarize our method and discuss our future work.

2 Related Work

One of the earliest solutions for handling implicit feedback is Hu et al.'s ImplicitALS [5] that consists of a matrix factorization model, and a point-wise (each user-item instance) based objective function which is weighted by the confidence derived from the observations of behaviors. Pan et al. [8] proposed two point-wise based frameworks for handling one-class collaborative filtering (OCCF) problem, in which data usually consist of binary data reflecting a user's action or inaction. The first one is wALS, which gives different weights to the error terms of positive examples and inaction examples in the objective function. The second one is sampling-based ALS, which selects a part of the inaction samples as negative samples, together with all positive samples to generate the training data.

Pair-wise methods directly minimize a ranking objective function, such as BPR [6], which is based on the assumption that a user prefers a consumed item to an unconsumed item. The task is to provide the user with a personalized total ranking of all items. BPR with confidence (BPRC) extends BPR, by adding a given confidence for each sample, and optimizes a confidence-weighted objective function. BPRC works well when the confidence for each implicit feedback that is obtained from external context information. However, in most applications the confidence cannot be easily obtained, which motivates us to learn the confidence.

All of the above methods only consider one type of implicit feedback, and unable to deal with heterogeneous implicit feedbacks. Pan et al. [7] is the first to study heterogeneous-implicit-feedback-based recommendation algorithms. They take users' transaction records as certain feedback and users' examination records as uncertain feedback, and propose ABPR to extend BPR on heterogeneous implicit feedbacks. ABPR learns a confidence weight for each examination record, which denotes a probability that the corresponding user likes the examined item. Although ABPR classifies the implicit feedbacks into certain and uncertain types, it treats all uncertain feedbacks equally. However, on most of websites, there is always more than one type of uncertain feedback, and they indicate different confidence levels. It is obvious that an item in shopping cart is more likely to be preferred than a clicked item (without other actions). We propose a unified method to characterize the confidence of user preference from heterogeneous implicit feedbacks.

3 Confidence-Learning Based Collaborative Filtering

3.1 Preliminaries

Let U be the set of all users and I the set of all items. We reserve special indexing letters for distinguishing users from items: for users u, v and for items i, j. Let $R = \{r_{ui}\}$ be the ratings to indicate the preference of user u on item i. In implicit feedback problems, we set $r_{ui} = 1$ if implicit feedback exists, or $r_{ui} = 0$ otherwise, and use additional $C = \{c_{ui}\}$ to characterize the confidence that we believe the user likes the item. In this paper, we infer C from heterogeneous implicit feedbacks.

3.2 Confidence Learning

Different kinds of implicit feedbacks represent different levels of confidence, and it is inflexible for man to design a function to calculate the confidence for point-wise methods, such as ImplicitALS [5], or design a set of rules to decide which item is more preferred for pair-wise methods, such as ABPR [7]. Our aim is to transform heterogeneous implicit feedbacks into a comparable space.

According to the ability to indicate confidence of preference, implicit feedbacks can be classified into **certain feedback** and **uncertain feedback**. Certain feedback is the implicit feedback that indicates full confidence, such as purchase, while uncertain feedback shows uncertain confidence, such as click. We have certain feedback and multiple types of uncertain feedbacks. Certain feedback is denoted as $T = \{(u, i)\}$, and uncertain feedback is denoted as $E = \{(u, i)\}$. For example, purchase can be regarded as certain feedback, because a user votes for it by paying for it. Conversely, users do not need to pay for other behaviors, such as click, adding things into the collection or shopping cart. When $(u, i) \in T$, we can fully believe that user u likes item i. If we can find the relation between certain feedback and uncertain feedbacks and quantize the closeness of relation, then we can transform heterogeneous implicit feedbacks into the same space. In our problem settings, (u, i) pairs in the system can be in the following cases, as shown in Fig. 1:

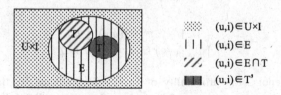

⣿	$(u,i) \in U \times I$
|||	$(u,i) \in E$
///	$(u,i) \in E \cap T$
▉	$(u,i) \in T'$

Fig. 1. Certain feedback and uncertain feedbacks

- $(u, i) \in U \times I$, includes all (u, i) pairs in the system;
- $(u, i) \in E$, means user u has uncertain implicit feedbacks on item i;
- $(u, i) \in T$, means user u has certain feedbacks on item i. Note that $(u, i) \in E$ is a precondition of $(u, i) \in T$, i.e., if user u has certain feedbacks on item i, there must

also be uncertain feedbacks. For example, there are always click records before a user purchases something. For these (u, i) pairs, we believe user u prefers item i, and set confidence $c_{ui} = 1$;

- If $(u, i) \in E$, but $(u, i) \notin T$, there are two possible situations for this (u, i): user u will have certain feedbacks on item i in the future, or not. This is easy to understand in e-commerce settings. For example, the clicked but not purchased items have small probabilities to be purchased later. So maybe $(u, i) \in T$ will be true in the future, denoted as $(u, i) \in T'$. Both T and T' mean the certain feedback, but T can be observed in the training data, while T' is the future data, which cannot be observed in the training data. We are not sure whether $(u, i) \in T'$ is true or not before the user actually do it, so we use dashed circle to represent this part.

We try to predict the probability of $(u, i) \in T'$ for $(u, i) \in E$, $(u, i) \notin T$, then we can use it as the confidence c_{ui} that we believe user u prefers item i. Thus the task turns into modeling the internal relations between certain feedback and uncertain feedbacks. This problem is actually similar to OOCF [8]. If $(u, i) \in E$ and $(u, i) \in T$, we can label this (u, i) as label "1", But if $(u, i) \in E$ and $(u, i) \notin T$, we cannot know in advance whether $(u, i) \in T'$ or not, so their labels are unknown. We only have one kind of label. The unlabeled instances contain a great proportion of negative samples, and a small proportion of positive samples. We borrow ideas from OOCF [8], and sample a portion of $\{(u, i)|(u, i) \in E, (u, i) \notin T\}$ as negative examples to balance the extent of treating unlabeled instances as negative examples ("0"). Thus the task turns into a binary classification problem. The features used for classification model can be derived from: (1) statistics of uncertain feedbacks, such as click times, reading time; (2) user profiles, such as age, gender and behavior bias; (3) item profiles, such as category, brand, price, popularity. The specific features are related to the application, so we do not go in depth in this section. By using feature engineering, we can easily incorporate complex factors that influence the confidence of preference. The machine learning methods for binary classification problem have been well studied [9–12], so we just need to choose the most suitable one in a real-world application. Once the classification model is learned, we can predict the probability that (u, i) belongs to class "1", i.e., $(u, i) \in T'$, for the unlabeled user-item pairs in $\{(u, i)|(u, i) \in E, (u, i) \notin T\}$.

Finally, we obtain the confidence of preference from heterogeneous implicit feedback by:

$$c_{ui} = \begin{cases} 1 & (u, i) \in T \\ p_{(u,i)\in T'} & (u, i) \in E, (u, i) \notin T \end{cases} \tag{1}$$

where $p_{(u,i)\in T'}$ denotes the probability of $(u, i) \in T'$ predicted by the classification model. We call this process as **confidence learning**. We will briefly introduce the features and classification models in the experimental part, and explore the relation between the accuracy of classification and the final accuracy of recommendation.

3.3   CL-ImplicitALS

Now we have c_{ui} to quantize the confidence that we can believe a user prefers an item. Next, we apply c_{ui} to matrix factorization models. Matrix factorization models map both users and items to a joint latent factor space of dimensionality f, such that user-item interactions are modeled as inner products in that space. Accordingly, each item i is associated with a vector $y_i \in R^f$, and the elements of y_i measure the extent to which the item possesses those latent factors. Similarly, each user u is associated with a vector $x_u \in R^f$. The resulting dot product, $x_u^T y_i$, captures the interaction between user u and item i—the user's overall interest in the item's characteristics. The major challenge is computing the mapping $x_u, y_i \in R^f$ of each item and user to factor vectors. We apply the inferred confidence c_{ui} to a point-wise matrix factorization model. The objective function is shown in Eq. (2):

$$\min \sum_{(u,i) \in U \times I} c_{ui}(r_{ui} - x_u^T y_i)^2 + \lambda(\sum_u ||x_u||^2 + \sum_i ||y_i||^2) \tag{2}$$

where $c_{ui}(r_{ui} - x_u^T y_i)^2$ is the point-wise loss function designed to compare the difference between the real user preference r_{ui} and the predicted preference $x_u^T y_i$, and in the implicit feedback setting, r_{ui} is binary user preference ($r_{ui} = 1$ if implicit feedback exists, and $r_{ui} = 0$ otherwise); c_{ui} indicates the confidence that user u likes item i, and reflects the contribution of minimizing the error term to the overall objective function; $\lambda(\sum_u ||x_u||^2 + \sum_i ||y_i||^2)$ is the regularization function to avoid overfitting. We name this method as confidence-learning-based ImplicitALS (**CL-ImplicitALS**). The difference between this method and ImplicitALS [5] is: ImplicitALS only considers one type of implicit feedback, and calculate c_{ui} based on pre-defined function, considering the counts of observed behaviors, which cannot deal with heterogeneous implicit feedbacks; our method can deal with heterogeneous implicit feedbacks and differentiate various kinds of user behaviors, by using the previously learned confidence.

3.4   CL-BPR

We also apply the inferred confidence c_{ui} to a pair-wise matrix factorization model. Pair-wise methods directly minimize a ranking objective function. If user u prefers item i than item j, it is denoted as $i \succ_u j$. The goal is to correctly order such item pairs. Equation (3) is an extended form of pair-wise matrix factorization model, based on the previously inferred confidence c_{ui}:

$$\min \sum_{(u,i,j):i \succ_u j} -\ln(1/(1 + \exp(-c_{uij}(x_u^T y_i - x_u^T y_j)))) + \lambda(\sum_u ||x_u||^2 + \sum_i ||y_i||^2) \tag{3}$$

where $-\ln(1/(1 + \exp(-c_{uij}(x_u^T y_i - x_u^T y_j))))$ is the loss function designed to encourage pairwise comparison; $c_{uij} = c_{ui} - c_{uj}$ indicates how much we trust that user u prefers item i than item j, $\hat{r}_{uij} = \hat{r}_{ui} - \hat{r}_{uj} = x_u^T y_i - x_u^T y_j$ is the difference of predicted preference values between item i and item j; $\lambda(\sum_u \|x_u\|^2 + \sum_i \|y_i\|^2)$ is the regularization term used to avoid overfitting.

The challenge for pair-wise method is to choose suitable (u, i, j) triples that satisfy $i \succ_u j$. BPR [6] assumes that a user prefers a consumed item to an unconsumed item. ABPR [7] thinks that a user prefers an item with certain feedback than that with uncertain feedback, and prefers an item with uncertain feedback than that with no feedback, but ABPR cannot differentiate the items both with uncertain feedbacks. We already introduced the method to learn the confidence c_{ui}, and then we can also compare the preference order between item i and item j that both have uncertain feedbacks. In our approach, (u, i, j) is a triple from $\{(u, i, j)|(u, i) \in T \cup E, c_{ui} > c_{uj}\}$, which means if we are more confident that user u likes item i than item j, then we guess that they may satisfy the $i \succ_u j$ relation. However, we are not one hundred percent sure that user u prefers item i than j even if $c_{ui} > c_{uj}$, so we also use $c_{uij} = c_{ui} - c_{uj}$ to describe the confidence of $i \succ_u j$ relation. Note that certain feedback indicates full confidence $(c_{ui} = 1)$, uncertain feedbacks indicate smaller nonnegative value $(0 < c_{ui} < 1)$, and if user u has no implicit feedback on item i, we set $c_{ui} = 0$, so choosing from $c_{ui} > c_{uj}$ includes the following situation: (1) $(u, i) \in T$, $(u, j) \notin T$; (2) $(u, i) \in E$, $(u, j) \notin E$; (3) $(u, i) \in E$, $(u, j) \in E$ and $c_{ui} > c_{uj}$. BPR [6] contains situation (1), ABPR [7] contains situation (1) and (2), and our method contains all of those situations. We have a unified way to choose training samples for pair-wise methods, which can bring more effective comparable pairs and further relieve the sparsity problem. We name this method as confidence-learning-based **BPR (CL-BPR)**, because it is an improved version of BPR and ABPR, and based on a confidence-learning step. We can see that by introducing a confidence-learning step, heterogeneous implicit feedbacks can be processed in a unified form, and the inferred confidence is also meaningful (the probability of $(u, i) \in T'$), unlike the confidence learned in ABPR [7].

4   Experiments

4.1   Dataset and Statistics

We conduct extensive experiments on two real-world e-commerce datasets provided by Tmall.com. The first one is released in the first stage of *Tmall Recommendation Prize 2014,*[1] and we name it *Tmall-Rec* in the rest of paper. This dataset is focused on brand recommendation, that is, an item in this dataset means a brand. The second one is the dataset released in *IJCAI-15 Competition,*[2] which contains the interactions between users and products. Both datasets contain data fields listed in Table 1.

Those implicit feedbacks are very common on online retail platforms: (1) a user clicks an item when he wants to know more details of that item, but we do not know whether he is satisfied with this item or not, so this is uncertain feedback; (2) a user collects an item (adds an item into collection) indicates that he might review this item

---

[1] https://102.alibaba.com/competition/addDiscovery/index.htm.

[2] http://tianchi.aliyun.com/datalab/dataSet.htm?spm=5176.100073.888.13.nt1XTA&id=1.

Table 1. Data fields

Column	Description	Instructions
User Id	The unique identifier of a user	Sampling and encryption
Product/Brand Id	The unique identifier of a product/brand	Sampling and encryption
Time	The time when the interaction occurred	Precision level to the specific day
Action type	The type of action	Buy, click, collect and cart

later. Although this behavior shows higher preference than clicks, it is also uncertain feedback; (3) a user adds an item into his shopping cart indicates he may want to buy this item. But sometimes a user puts competitive items into shopping cart but only purchases one of them, or finds it not interesting enough later, so it is also uncertain feedback; (4) a user buys an item and pays for it. Since the ultimate goal of recommender systems on an online retail platform is to prompt users to make a purchase decision, it's natural to think that purchase behaviors show certain preferences. Even if the user is unsatisfied with the received product, it also shows his preference for such kind of products. So in this paper, we regard purchase behavior as certain feedback. Certain feedback is very precious but sparse, while uncertain feedback is more abundant. Statistics about two datasets are shown in Table 2. We first split each dataset into two parts (*Dataset1* and *Dataset2*) by users evenly. *Dataset1* is used for training and evaluating the classification model, and the classification model is used to predict the confidence values for the heterogeneous implicit feedbacks in *Dataset2*, and then *Dataset2* is used for training and testing the collaborative filtering model. Specifically, we split *Dataset1* into two parts (*Dataset1-train* and *Dataset1-test*) by users evenly. *Dataset1-train* is used to train the classification model, and *Dataset1-test* is used to evaluate the classification model. We split the *Dataset2* according to the time, since recommenders normally predict users' future preference exploiting historical data. On the *Tmall-Rec* dataset, we use 0–90 days as *Dataset2-train*, and 91–122 days as *Dataset2-test*; on the *IJCAI-15* dataset, we use 0–110 days as *Dataset2-train*, and 111–160 days as *Dataset2-test*.

Table 2. Statistics about *Tmall-Rec* and *IJCAI-15*

Dataset	#User	#Item	#Click	#Buy	#Collect	#Cart	Sparsity (T)	Sparsity (T∪E)
Tmall-Rec	884	9531	174 539	6984	1204	153	0.08 %	0.684 %
IJCAI-15	424170	1090390	48550713	3292144	3005723	76750	0.0007 %	0.0069 %

4.2 Features and Classification Methods

In the confidence-learning step, features are needed for the classification model. Since feature engineering is not the focus of this paper, and considering the space limit, we only introduce the features that are easy to understand. Noting that we have 3 types of uncertain feedbacks (click, collect and cart), so each group has 3 features. We list the following groups of features:

- User u's behavior count of each uncertain feedback on item i.
- User u's average behavior count of each uncertain feedback over items.
- Item i's average behavior count over users.
- User u's total behavior count.
- Item i's total behavior count.

After feature extraction, we use logistic regression [9], random forest [10] and GBDT [11, 12] in the classification step. We evaluate the classification accuracy use area under the curve (AUC), which is equal to the probability that a classifier will rank a randomly chosen positive instance higher than a randomly chosen negative one. We adjust the features and classification methods to obtain different versions of confidence-learning models, and get different AUCs, and then we observe the accuracy of the CF step. Thus we can observe: when the accuracy of classification improves, whether the accuracy of recommendation will improve or not.

4.3 Baselines and Evaluation Metrics

Noting that although we use the datasets of *Tmall-Rec* and *IJCAI-15*, the problem settings are different from ours. What's more, the competitions have additional test sets while we downloaded the original training sets and then split them into our own training and test sets, so the released leaderboard scores are not comparable to our results. We run several state-of-art algorithms on the same datasets as ours, and compare their performances.

The ultimate goal of recommender systems on an online retail platform is to prompt users to make a purchase decision. Unlike explicit ratings, the purchase either happens or not, thus evaluation metrics such as MAE, RMSE are not appropriate. Ranking-oriented evaluation metrics, including Precision@5, Precision@10, Recall@5, Recall@10, AUC, MAP, NDCG and MRR are used to evaluate the algorithms. A purchased item in the test set is treated as a positive instance. We compare our approaches, CL-ImplicitALS and CL-BPR, with the following algorithms:

- ImplicitALS [5]: including ImplicitALS (T) for certain feedback (purchase) only, and ImplicitALS (T∪E) for the combination of certain feedback and uncertain feedback. In ImplicitALS (T), the purchase count is used to compute c_{ui}. In ImplicitALS (T∪E), the total count of interactions is used to compute c_{ui}.
- BPR [6]: including BPR (T) for certain feedback only, and BPR (T∪E) for the combination of certain feedback and uncertain feedback. In BPR (T), user u prefers item i than item j if $(u, i) \in T$ and $(u, j) \notin T$. In BPR (T∪E), user u prefers item i than item j if $(u, i) \in T \cup E$ and $(u, j) \notin T \cup E$.
- ABPR: user u prefers item i than item j if $(u, i) \in T$ and $(u, j) \notin T$, and user u prefers item i than item j if $(u, i) \in T \cup E$ and $(u, j) \notin T \cup E$. Other settings exactly follow the paper of Pan et al. [7].

All of the parameters for baselines are well tuned to ensure fairness of the comparison.

4.4 Results and Analysis

The recommendation performance of CL-ImplicitALS, CL-BPR and baselines are shown in Tables 3, 4, 5 and 6, from which we can have the following observations:

Table 3. Recommendation performance of point-wise methods on Tmall-Rec dataset

Algorithms	P@5	P@10	R@5	R@10	AUC	MAP	NDCG	MRR
ImplicitALS (T)	0.0158	0.0112	0.0664	0.0857	0.5511	0.0336	0.0483	0.0416
ImplicitALS (T∪E)	0.0158	0.0125	0.0664	0.0903	0.5572	0.0339	0.0499	0.0428
CL-ImplicitALS	**0.0211**	**0.0158**	**0.0725**	**0.1091**	**0.5763**	**0.0472**	**0.0676**	**0.0681**

Table 4. Recommendation performance of point-wise methods on IJCAI-15 dataset

Algorithms	P@5	P@10	R@5	R@10	AUC	MAP	NDCG	MRR
ImplicitALS (T)	0.0131	0.0094	0.0097	0.0140	0.5364	0.0063	0.0139	0.0317
ImplicitALS (T∪E)	0.0103	0.0082	0.0076	0.0122	0.5295	0.0054	0.0118	0.0256
CL-ImplicitALS	**0.0212**	**0.0156**	**0.0155**	**0.0231**	**0.5576**	**0.0110**	**0.0233**	**0.0522**

Table 5. Recommendation performance of pair-wise methods on Tmall-Rec dataset

Algorithms	P@5	P@10	R@5	R@10	AUC	MAP	NDCG	MRR
BPR (T)	0.0172	0.0152	0.0634	0.1098	0.5603	0.0350	0.0554	0.0439
BPR (T∪E)	0.0198	0.0165	0.0634	0.0867	0.5721	0.0352	0.0538	0.0541
ABPR	0.0211	0.0165	0.0662	0.1041	0.5759	0.0372	0.0592	0.0612
CL-BPR	**0.0238**	**0.0185**	**0.0725**	**0.1135**	**0.5821**	**0.0459**	**0.0683**	**0.0688**

Table 6. Recommendation performance of pair-wise methods on IJCAI-15 dataset

Algorithms	P@5	P@10	R@5	R@10	AUC	MAP	NDCG	MRR
BPR (T)	0.0156	0.0129	0.0116	0.0191	0.5401	0.0090	0.0184	0.0376
BPR (T∪E)	0.0177	0.0133	0.0138	0.0208	0.5498	0.0091	0.0196	0.0408
ABPR	0.0162	0.0094	0.0115	0.0182	0.5429	0.0056	0.0139	0.0350
CL-BPR	**0.0203**	**0.0157**	**0.0149**	**0.0233**	**0.5535**	**0.0116**	**0.0240**	**0.0529**

- ImplicitALS (T∪E) is slightly better than ImplicitALS (T) in *Tmall-Rec* dataset, but worse than ImplicitALS (T) in *IJCAI-15* dataset. We think that's because the original ImplicitALS treats all types of implicit feedbacks equally, which is not reasonable for heterogeneous implicit feedbacks, so incorporating more data does not help to improve the performance of recommendation.
- BPR (T∪E) outperforms BPR (T), because incorporating uncertain feedbacks can bring more comparable item pairs. However, ABPR does not outperform BPR

($T \cup E$) in *IJCAI-15* dataset. We analyzed the automatically learned confidence in ABPR, and find these values vary very little.

- On both datasets, our approach, either CL-ImplicitALS or CL-BPR, achieves the best results. It seems that our approaches can better leverage various types of implicit feedbacks.

We then study the relation between the accuracy of confidence-learning step (classification model to predict the probability of $(u, i) \in T'$), and the accuracy of collaborative filtering step, on *Tmall-Rec* dataset and *IJCAI-15* dataset, evaluated by AUC, as shown in Fig. 2. We can see the coherence of their changing trends: better performance in confidence-learning step can generally result in better performance in recommendation. That means, using the probability of $(u, i) \in T'$ as the confidence of preference is effective.

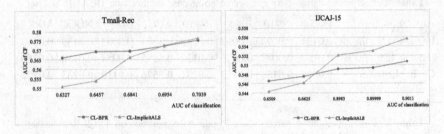

Fig. 2. AUC of classification and AUC of collaborative filtering

5 Conclusions and Future Work

In this paper, we propose a unified method to transform heterogeneous implicit feedbacks to a comparable space, which indicates the confidence that we can believe a user prefers an item, by studying the internal relations from two categories of implicit feedbacks: certain feedback and uncertain feedback. Our method can deal with certain feedback and multiple types of uncertain feedbacks in a unified way, while existing methods either deal with only one type of implicit feedback or treat all uncertain feedbacks equally. We apply the inferred confidence to two mainstream recommender models, point-wise (ImplicitALS) and pair-wise (BPR) models, named CL-ImplicitALS and CL-BPR, and propose a more generic strategy to select training samples for pair-wise methods. Experiments on two real-world e-commerce datasets, one for brands and the other for products, show that CL-ImplicitALS and CL-BPR outperform the baseline approaches. We also validate the effectiveness of using the probability of $(u, i) \in T'$ as the confidence.

In the future work, more features can be added, and more advanced classification algorithms can be used to improve the accuracy of the confidence-learning, since our experiments show that better performance of confidence-learning can generally results in better performance of recommendation. We can apply the inferred confidence to more comprehensive collaborative filtering methods, such as context-aware approaches.

Acknowledgment. This work is supported by grants from the Doctoral Program of the Ministry of Education of China (Grant No. 20110101110065), the National Key Technology R&D Program of China (Grant No. 2012BAD35B01-3), and the National Natural Science Foundation of China (Grant No. U1536118).

References

1. Ricci, F., Rokach, L., Shapira, B., et al.: Recommender Systems Handbook. Springer, US (2011)
2. Adomavicius, G., Tuzhilin, A.: Toward the next generation of recommender systems: a survey of the state-of-the-art and possible extensions. IEEE Trans. Knowl. Data Eng. **17**(6), 734–749 (2005)
3. Bobadilla, J., Ortega, F., Hernando, A., et al.: Recommender systems survey. Knowl. Based Syst. **46**, 109–132 (2013)
4. Park, D.H., Kim, H.K., Choi, I.Y., et al.: A literature review and classification of recommender systems research. Expert Syst. Appl. **39**(11), 10059–10072 (2012)
5. Hu, Y., Koren, Y., Volinsky, C.: Collaborative filtering for implicit feedback datasets. IEEE Computer Society, Pisa (2008)
6. Rendle, S., Freudenthaler, C., Gantner, Z., et al.: BPR: Bayesian personalized ranking from implicit feedback. AUAI Press, Montreal (2009)
7. Pan, W., Zhong, H., Xu, C., et al.: Adaptive Bayesian personalized ranking for heterogeneous implicit feedbacks. Knowl. Based Syst. **73**, 173–180 (2015)
8. Pan, R., Zhou, Y., Cao, B., et al.: One-class collaborative filtering. In: Gunopulos, D., Turini, F., Zaniolo, C. et al., (eds.) Proceedings of the IEEE International Conference on Data Mining, pp. 502–511 (2008)
9. Freedman, D.A.: Statistical Models: Theory and Practice. Cambridge University Press, New York (2009)
10. Liaw, A., Wiener, M.: Classification and regression by randomForest. R news **2**(3), 18–22 (2002)
11. Friedman, J.H.: Greedy function approximation: a gradient boosting machine. Ann. Stat. **29**, 1189–1232 (2000)
12. Friedman, J.H.: Stochastic gradient boosting. Comput. Stat. Data Anal. **38**(4), 367–378 (2002)

Star-Scan: A Stable Clustering by Statistically Finding Centers and Noises

Nan Yang, Qing Liu$^{(\boxtimes)}$, Yaping Li, Lin Xiao, and Xiaoqing Liu

Information School, Renmin University of China, No. 59, Zhongguancun Street,
Haidian District, Beijing, China
{yangnan,qliu,ypli}@ruc.edu.cn, {netruc,blueapple1}@126.com
http://info.ruc.edu.cn

Abstract. In this paper, we present a new clustering algorithm, called A **St**able Clustering by Statistically Finding **C**enters and **N**oises (Star-Scan). Star-Scan is a density-based clustering algorithm that can find arbitrary shape clusters and resists to the noise in a dataset. It borrows the idea from Rodriguez's Clustering by Fast Search and Find of Density Peaks (CFSFDP) that the cluster centers are characterized by the points with both higher density and farther distance to other centers than their neighbors. Different from CFSFDP, instead of manual operation, Star-Scan uses a statistical method, box plot, to select cluster centers automatically. Furthermore, due to inadequate selection of cluster centers in CFSFDP, we apply a merging post-process to the produced clusters to get stable and correct results. Finally, we also use box plot to filter out noises on each of final clusters to solve the problem of over-filtering in CFSFDP. We have demonstrated the good performance of Star-Scan algorithm on several synthetic datasets.

Keywords: Density-based clustering · Box plot · Statistics

1 Introduction

Clustering is a discovery process, which separates a dataset into several partitions based on similarity in which intra cluster similarity is maximized and inter cluster is minimized. Applications of clustering are broad which include grouping sequence of genes in molecular biology [6], segmentation of images and information retrieval [7]. Among them, since the density-based clustering algorithm can find arbitrary shape clusters and resists to the noise of datasets, it attracts many researcher's attention. The famous one is Density-Based Spatial Clustering of Applications with Noise (DBSCAN) [13]. It has two parameters, radius ε and density threshold $MinPt$. But the determination of appropriate value of density threshold is not an easy task. For this reason, Rodriguez et al. had proposed a new density-based clustering algorithm, Clustering by Fast Search and Find of Density Peaks (CFSFDP) [14], which uses decision graph to select the points with high local density as cluster centers. CFSFDP has following advantages: (1) the cluster centers, corresponding to the outliers in decision graph, are

© Springer International Publishing Switzerland 2016
F. Li et al. (Eds.): APWeb 2016, Part I, LNCS 9931, pp. 456–467, 2016.
DOI: 10.1007/978-3-319-45814-4_37

visualized and can be identified from the data points easily. (2) the clustering process is performed in a single step, in contrast with other clustering algorithms where an objective function is optimized iteratively. In their experiments, the authors have demonstrated that the algorithm has very excellent performance under different datasets. Even though CFSFDP has several advantages, there are still many problems remained. First, the selection of cluster centers is performed manually. It is not adaptive to large scale clustering analysis. Second, the number of centers selected by decision graph would not always be same as that of true clusters. Third, the way of filtering out noises in CFSFDP is too strict, which causes many data points filtered out as noises. To solve these problems, we propose a new clustering algorithm, called A **Sta**ble Clustering by **S**tatistically Finding **C**enters and **N**oises (Star-Scan). In Star-Scan algorithm, we employ a statistical method, box plot [15], instead of manual operation, to select outliers as cluster centers automatically. Furthermore, a merging post-process is applied to the produced clusters to get stable and correct clusters. We also employ box plot to filter out noises on each of final clusters and gain more accurate result. The rest of the paper is organized as follows. In Sect. 2, we briefly review previous research works of density-based clustering. Section 3 introduces the clustering method of CFSFDP and describes its problems. In Sect. 4, we present a new clustering method, Star-Scan, which is two-phases algorithm. Box plot is introduced and the principle of selecting centers and noises is discussed. After that, the merging strategy and the definition of measurement are described. In Sect. 5, the experiments are set up to evaluate Star-Scan and the results are analyzed. We give conclusions and future works in Sect. 6.

2 Related Works

Clustering methods can be classified into five types [8]: partitioned, hierarchical, density-based, grid-based and model-based method. In comparison with partitioned and hierarchical approaches density-based clustering is less constrained and robust to noise, its clusters need not be bound in the convex shape and can be arbitrarily shaped in the data space [9]. As density-based clustering is the basis of our method, now we give a brief survey of related works on density-based clustering. DBSCAN [13] forms the practical framework of density-based clustering. It bases on two assumptions: (1) high density areas are separated by low density areas. (2) the density of any point in the clusters is higher than that of noises in global data space. DBSCAN adopts the two input parameters ε and $MinPt$ as global variables and uses the initial value in whole clustering process. However, a constant density level $MinPt$ cannot completely describe the structure of dataset [9] and the real dataset contains clusters with varying densities. OPTICS [10] algorithm solves the constant density level problem by extending the DBSCAN such that several parameters are processed at the same time. Definition of cluster density is essential, DENCLUSE [11,12] uses a kernel density estimator to define clusters and DBSCAN can be seen as special case of DENCLUE using a uniform spherical kernel. To decrease the high computational complexity of DBSCAN, Act-DBSCAN [5] exploits the pairwise lower-bounding (LB) similarities to initialize the cluster structure and obtain the results same as in DBSCAN with a small cost. SSDBSCAN [4] is semi-supervised method, uses labeled objects to help finding density parameters for each natural cluster in a dataset. Currently, density-based clustering has been successfully used in many new applications, such as analyzing of uncertain data [3], data streams [2], high dimensional RNA-sequencing data [1] and big data [16]. Recently, CFSFDP [14] is proposed by Rodriguez et al. In contrast with above clustering algorithms in which an objective function is optimized iteratively, CFSFDP assigns the clusters in a single step.

3   CFSFDP and Problems

Among density-based algorithms, CFSFDP is excellent for its good performance. The basis of CFSFDP is on two assumptions: (1) cluster centers with higher density are surround by their neighbors with lower density. (2) the distances between centers are far from each other comparing to their neighbors. CFSFDP computes two quantities for each data point i: its local density ρ_i and distance δ_i.

Definition 1. ρ_i *is the local density of a data point* i, *which is the Gaussian kernel function of the point* i *over all data points with a parameter* ε.

$$\rho_i = \sum_{j \neq i, j=1}^{n} e^{-(\frac{d_{ij}}{\varepsilon})^2}. \tag{1}$$

where ε is given radius, d_{ij} is Euclidean distance between point i and j, and n is the number of data points.

Definition 2. δ_i *is the minimum distance between* i *and any other points with higher density than* i.

$$\delta_i = \min_{j:\rho_j > \rho_i} (d_{ij}) \tag{2}$$

The value of ε is determined by a parameter μ, which is a percent. Let *sortlist* represent the list of sorted distances of point pairs. ε is the distance value indexed by μ in *sortlist*. According to [14], the empiric value μ is 2.0. A simple example of CFSFDP is illustrated in Fig. 1. It shows that 28 points embedded in two-dimension space. Figure 1(a) shows data points distribution, point 1 and 10 are cluster centers. Figure 1(b) shows the decision graph in which x coordinates is ρ and y coordinates is δ, point 1 and 10 have very large δ value. From the decision graph, cluster centers can be found easily. After centers are determined and next step is to assign rest points to different centers. The assignment process is only single step, that makes CFSFDP very fast.

The final process is noises filtering. Let $C_l, l = 1, 2, ..., k$, denote the produced k clusters. $th_{noise}(l)$ is noise threshold to filter noises in C_l.

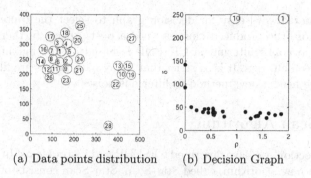

(a) Data points distribution   (b) Decision Graph

Fig. 1. A dataset with two clusters

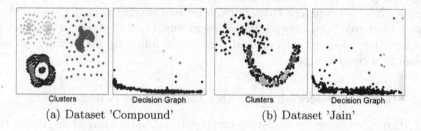

(a) Dataset 'Compound'   (b) Dataset 'Jain'

Fig. 2. Two test cases under 'Compound' and 'Jain'

Definition 3. $th_{noise}(l)$ *is the maximum average density of point pairs between cluster C_l and other clusters. For any clusters C_m, $m \neq l$, if the point pairs $u \in C_l$ and $v \in C_m$ and the distance between two points is below ε, the value of $th_{noise}(l)$ is the maximum average of ρ_u and ρ_v.*

$$th_{noise}(l) = \max_{u \in C_l, v \in C_m, m \neq l, d_{uv} \leq \varepsilon} \{\frac{\rho_u + \rho_v}{2}\} \qquad (3)$$

Although CFSFDP has several significant advantages, there are still some drawbacks. First, the cluster centers are selected manually, it can be referenced to their demo matlab program [14]. This makes CFSFDP not adaptive for large scale clustering analysis. It is necessary to explore a automatical method to select cluster centers from decision graph instead of manual operation. Second, two assumptions of CFSFDP are not always hold. The number of cluster centers selected by decision graph would not always be same as that of the true clusters. Moreover, the distances between these centers are not farther than their neighbors under all conditions. We have tested CFSFDP under four 2-dimensional 'Shape sets' from a synthetic datasets [17]. Two of experimental results are illustrated in Fig. 2. By analyzing the graphs, we can see from the results that the decision graph doesn't work well. Let us assume a situation where a user who will select centers by decision graph. Whether or not the user knows the ground truth classes, he could not select correct centers under above datasets. As a result, it is not always easy for decision graph to select the centers correctly. For this reason, the produced clusters need a post-process further. Third, the method of the noises filtering in CFSFDP is based on the maximum average density between clusters. It is so strict that many data points are filtered out as noises and we need a new method to filter out noises.

4   Star-Scan Algorithm

In previous section, we have described CFSFDP and its problems. In this section, we propose a new algorithm, called Star-Scan. Star-Scan consists of two phases. In phase one, we use box plot to select cluster centers from decision graph and assign rest points to the centers. After the assignment, the clusters are produced. In phase two, we compare clusters pairwise and merge them if there exists a density reachable path between two cluster centers and then we use box plot to filter out the noises on each of final clusters. In following subsections, we will explain Star-Scan in detail.

4.1   Statistical Selection of Cluster Centers and Noises

Box Plot. In descriptive statistics, box plot is a standard way of displaying the distribution of data set based on five number summary: minimum, first quartile, second quartile, third quartile, and maximum. The first quartile ($q1$) is defined as the middle number between the minimum (smallest number) and the median of the data set. The second quartile ($q2$) is the median of the data. The third quartile ($q3$) is the middle value between the median and the maximum (highest value) of the data set. In tradition outliers was defined as those observations which lie outside the following interval: $[q1 - 1.5 \times IR, q3 + 1.5 \times IR]$, IR is the interquartile range ($IR = q3 - q1$). As is well-known, if the data come from the normal distribution, the former interval contains 99.3 % of the values. The outliers in our research are corresponding to the cluster centers and noises, which are characterized by a higher quantity than their neighbors. In large scale data clustering the number of the cluster centers and noises is less than $(1 - 99.3\%)$ percent. Then we narrow the range of the outliers which lie in the interval: $[q3 + 3 \times IR, \infty]$ and use value $u = q3 + 3 \times IR$ as a threshold to detect outliers.

Centers Selection. Now we describe how box plot select centers in detail. After two quantities ρ and δ of all points are calculated, the distribution of both values is same as the graph depicted in Fig. 1(b). We can observe from the decision graph that a point i with both values extreme great is considered as an outlier. Hence, we define a factor α_i for point i, $\alpha_i = \sqrt{\rho_i \times \delta_i}$. If the α_i value of any data point i is above u, i is selected as a cluster center (outlier).

We set up experiments to verify box plot. The results have demonstrated that box plot works as well as manual selection. But if the cluster centers in data distribution are not easily separated from others by decision graph visually, box plot also fails. We have set up the experiments under 4 test cases to find

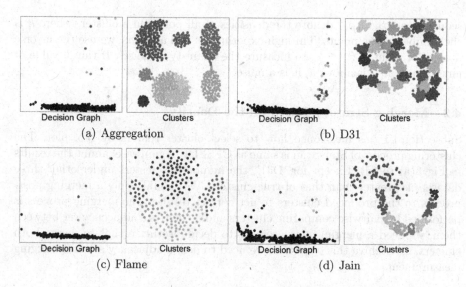

(a) Aggregation   (b) D31

(c) Flame   (d) Jain

Fig. 3. Centers selected by box plot

the reasons. The results are illustrated in Fig. 3(a) to (d). The color points are centers selected by box plot in part (a) and corresponding cluster assignments in part (b). The clusters are correct in the test case of 'D31' and not correct in the other cases. The reason is that the correct selection of cluster centers by decision graph is not always easy. Through the analysis of the results, we observe that the number of centers selected by decision graph in generally is greater or equal to that of true clusters. If both numbers are different, neither user nor box plot can produce correct clusters. In this case, the clusters produced by CFSFDP are heavily relied on user's selection. Different from user, box plot can statistically discover centers (outliers) objectively, not affected by human judgement. To address the problem in this case, we apply a merging post-process to the produced clusters to get more correct and stable results.

Noises Selection. In DBSCAN, noises are filtered out over whole dataset before clustering process. It is a pre-process and global method. Different from DBSCAN, CFSFDP dose clustering process first and then filters out noises on each cluster. It is a post-process and local method. We think the way of CFSFDP is more reasonable. There are two reasons: (1) before clustering, some points of a cluster may be mistaken as noises. (2) a global density threshold may lead to the situation in which some clusters with lower density level are filtered out as noises. Although CFSFDP uses a suitable noise-filtering strategy, their measurement does not works perfectly. The experimental results in Sect. 5 have proven this. In Star-Scan, we use post-process and local method. But different from CFSFDP, we employ statistical method, box plot, to filter out the noises. In fact, the noises can also be treated as outliers and identified by box plot. Let a sequence$\langle \rho_1, \rho_2, ..., \rho_k \rangle$ denote the densities of all points in a cluster, where k is the number of the points. Through experiments and analysis, we use the inverse density of point i, $\beta_i = \frac{1}{\rho_i}$, to measure the quantity of noises. If the β_i value of any data point i is above u, it is a noise.

4.2 Merging Strategy Based on the Density

Subsection 4.1 has mentioned how to select cluster centers by box plot. The clustering process of Star-Scan is same as CFSFDP. Again we examine the results in Fig. 3(a) to (d). Except for 'D31', the number of clusters under other three datasets is greater than that of true clusters. We should apply a merging post-process to the produced clusters to get correct results. The merging process is performed by pairwise comparing clusters and if there is a adjacency set between them, we need a merging measurement to decide wether or not to merge both clusters. To archive this objective, we need to define adjacency set and merging measurement.

Adjacency Set. Adjacency set is the set of point pairs$\langle u, v \rangle$ such that u and v satisfy two criterions: (1) u and v belong to different clusters; (2) the distance between two points is below ε. After phase one, assume there are k clusters produced, they are denoted as $C_l, l = 1, 2, ..., k$. Then, we use $AS_{l,m}$ to denote the adjacency set between C_l and C_m.

Definition 4. $AS_{l,m}$ *is a set of point pairs$\langle u, v \rangle$, where both points belong to different clusters and the distance of between two points is below ε, formulated as following.*

$$AS_{l,m} = \{ \langle u, v \rangle | u \in C_l, v \in C_m, d_{u,v} \leq \varepsilon \} \tag{4}$$

Merging Measurement. In this subsection, we will explain how to merge two clusters. After assigning process, along with the clusters produced, a span tree is constructed for each cluster. The center is the root and the each of rest points has a path from itself to the center. Because the density of center is the greatest among the points in the cluster, the densities of the points along the paths from themselves to the center are increased. Our merging strategy is based on the intuition: if the maximum densities of the points in adjacency set is greater than the minimum of average densities of two clusters, C_l and C_m can be merged. Therefore, for each pair C_l and C_m, we can derive a adjacency set $AS_{l,m}$. We only take $AS_{l,m} \neq \Phi$ into consideration. Let $th_{merg}(l, m)$ denote merging factor of C_l and C_m, avg_l and avg_m denote average densities of the C_l and C_m, the merging formula is as follow.

$$th_{merg}(l, m) = max\{(\rho_u, \rho_v)| <u, v> \in AS_{i,j}\} - min(avg_l, avg_m) \tag{5}$$

So if the value $th_{merg}(l, m)$ of two cluster C_l and C_m is above zero, both clusters can be merged.

5   Experiments and Results Analysis

5.1   Datasets

Two groups of datasets are used to evaluate proposed algorithm. Group one includes 'Shape sets' synthetic datasets [17], which is used to compare proposed algorithm with others. We choose four datasets from it: 'Aggregation', 'Compound', 'D31' and 'Jain' as benchmarks. Group two includes the Chameleon datasets [18], which is used to evaluate the capability of the algorithms to correctly identify clusters in noisy environments. The first dataset, 't4.8k', has six clusters of different size, shape, and orientation, as well as random noise points and special artifacts such as streaks running across clusters. The second dataset, 't5.8k', has eight clusters of different shape, size, and orientation, some of which are inside the space enclosed by other clusters. Moreover, the third dataset, 't7.10k' also contains random noises and special artifacts, such as a collection of points forming vertical streaks. Finally, the fourth dataset, 't8.8k', has eight clusters of different shape, size, density, and orientation, as well as random noise.

5.2   Results and Analysis

We run DBSCAN, CFSFDP and Star-Scan algorithms on the datasets in group one and two respectively. In the test on each of all datasets, we adjust parameters manually until optimal result is reached by visual inspection. In the test of DBSCAN, we adjust two parameters ε and $MinPt$. In the test of CFSDSP, we run the program in [14], adjust the parameter μ and select the centers manually. In the test of Star-Scan algorithm, we adjust the parameter ε.

Clusters Analysis. In the test cases of group one, the compared results of the benchmark, DBSCAN, CFSFDP and Star-Scan are illustrated in Figs. 4, 5, 6 and 7. The colored dots represent clusters and black dots represent noises. Under datasets 'Aggregation' and 'D31', Star-Scan performs as well as DBSCAN and CFSFDP, the clusters produced same as benchmark. Under dataset 'Compound', Star-Scan and DBSCAN can find correct clusters, CFSFDP fails. Under dataset 'Jain', Star-Scan can find correct clusters, DBSCAN and CFSFDP failed. In the test cases of group two, DBSCAN performs well under datsets 't4.8k', 't5.8k' and 't7.10k', but fails under 't8.8k'. CFSFDP fails under all four datasets. Under four datasets, Star-Scan performs well, the results are illustrated Figs. 8, 9, 10 and 11. The left parts are clusters after phase one and the right parts are clusters after phase two.

Noises Filtering Analysis. Under datasets 'Aggregation', 'D31' and 'Jain', the noises found by CFSFDP is worse than others, Star-Scan perform a little better than DBSCAN. Under dataset 'Compound', DBSCAN filter out many points belong to a cluster due to its pre-process and global method, Star-Scan performs well than CFSFDP. From the analysis of both clusters and noises, the Star-Scan performs effectively and produces more stable results.

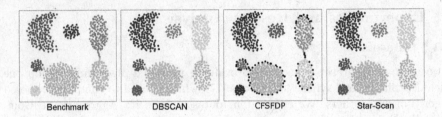

Fig. 4. Clustering results on dataset 'Aggregation'

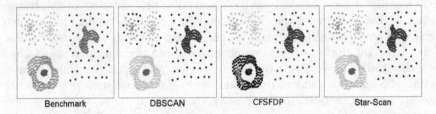

Fig. 5. Clustering results on dataset 'Compound'

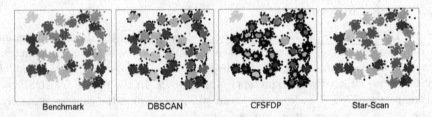

Fig. 6. Clustering results on dataset 'D31'

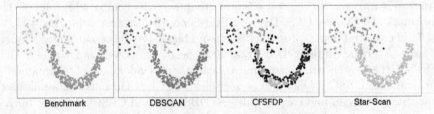

Fig. 7. Clustering results on dataset 'Jain'

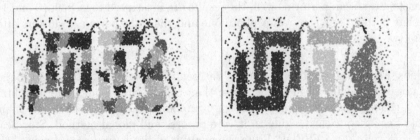

Fig. 8. Star-Scan clustering results on dataset 't4.8k'

Fig. 9. Star-Scan clustering results on dataset 't5.8k'

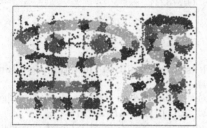
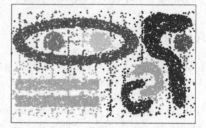

Fig. 10. Star-Scan clustering results on dataset 't7.10k'

Fig. 11. Star-Scan clustering results on dataset 't8.8k'

6 Conclusions and Future Works

We propose a new clustering algorithm, called Star-Scan. It borrows the idea from Rodriguez's Clustering by Fast Search and Find of Density Peaks(CFSFDP). CFSFDP uses decision graph to select the cluster centers manually and the clustering process is to assign the rest points to the centers. CFSFDP only need one parameter ε and the clustering process is fulfilled in one step. This makes it excellent in performance. Even though CFSFDP has these advantages, there are still some problems. To overcome the shortcoming of manually selecting the cluster centers in CFSFDP, we use a statistical method, box plot, to select cluster centers automatically. Furthermore, to deal with the inadequate selection of cluster centers by decision graph in many cases, we apply a merging post-process to the produced clusters to get stable and correct results.

Finally, due to noises over-filtering method in CFSFDP leads to many data points treated as noises, we also use box plot to filter out noises and gain more accurate result. To evaluate Star-Scan, we set up experiments on two group of synthetic datasets [17,18]. The results demonstrate that Star-Scan performs effectively. The future works include three aspects: (1) we will further study the method on computing δ value to make outliers more effectively. (2) the way of selecting cluster centers should be investigated intensively. (3) we will explore the impact on the result under other density kernel function.

Acknowledgments. This work was supported by National Natural Science Foundation of China under Grant No. 60773216, No. 60773217.

References

1. Inuk, J., Jong, C.P., Sun, K.: piClust: a density based piRNA clustering algorithm. J. Comput. Biol. Chem. **50**, 60–67 (2014)
2. Amineh, A., Ying, W.T., Hadi, S.: On density-based data streams clustering algorithms: a survey. J. Comput. Sci. Technol. **29**(1), 116–141 (2014)
3. Xianchao, Z., Han, L., Xiaotong, Z., Xinyue, L.: Novel density-based clustering algorithms for uncertain data. In: The Twenty-Eighth AAAI Conference on Artificial Intelligence, pp. 2191–2197. AAAI Press (2014)
4. Levi, L., Jörg, S.: Semi-supervised density-based clustering. In: The Ninth IEEE International Conference on Data Mining, pp. 842–847. IEEE Computer Society (2009)
5. Son, T., Xiao, H., Nina, H., Claudia, P., Christian, B.: Active density-based clustering. In: IEEE 13th International Conference on Data Mining, pp. 508–517. IEEE Computer Society (2013)
6. Michael, B.E., Paul, T.S., Patrick, O.B., David, B.: Cluster analysis and display of genome-wide expression patterns. In: National Academy of Sciences of the United States of America (PNAS), pp. 14863–14868. HighWire Press (1998)
7. Jain, A.K., Murty, M.N., Flynn, P.J.: Data clustering: a review. ACM Comput. Surv. CSUR **31**, 264–323 (1999)
8. Jiawei, H., Micheline, K.: Data Mining: Concepts and Techniques. Morgan Kaufmann, San Francisco (2000)
9. Hans-Peter, K., Peer, K., Jörg, S., Arthur, Z.: Density-based clustering. WIREs Data Min. Knowl. Discov. **1**, 231–240 (2011)
10. Mihael, A., Markus, M.B., Hans-Peter, K., Jörg, S.: OPTICS: ordering points to identify the clustering structure. In: ACM SIGMOD International Conference on Management of Data, pp. 49–60. ACM Press, Philadelphia (1999)
11. Alexander, H., Daniel, K.: An efficient approach to clustering in large multimedia databases with noise. In: The Fourth International Conference on Knowledge Discovery and Data Mining (KDD-98), pp. 58–65. AAAI Press, New York (1998)
12. Hinneburg, A., Gabriel, H.-H.: DENCLUE 2.0: fast clustering based on kernel density estimation. In: Berthold, M., Shawe-Taylor, J., Lavrač, N. (eds.) IDA 2007. LNCS, vol. 4723, pp. 70–80. Springer, Heidelberg (2007)
13. Martin, E., Hans-Peter, K., Jörg, S., Xiaowei, X.: A density-based algorithm for discovering clusters in large spatial databases with noise. In: The Second International Conference on Knowledge Discovery and Data Mining (KDD-1996), pp. 226–231. AAAI Press, Portland, Oregon, USA (1996)

14. Alex, R., Alessabdro, L.: Clustering by fast search and find of density peaks. Science **344**, 1492–1496 (2014)
15. Robert, D.M., Douglas, A.L., William, G.M.: Statistics: An Introduction. Duxbury Press, London (1994)
16. Junhao, G., Yufei, T.: DBSCAN revisited: mis-claim, un-fixability, and approximation. In: ACM SIGMOD International Conference on Management of Data (SIGMOD 2015), pp. 519–530. ACM Press, Melbourne, Victoria, Australia (2015)
17. Clustering Datasets. http://cs.joensuu.fi/sipu/datasets/
18. Chameleon Datasets. http://glaros.dtc.umn.edu/gkhome/cluto/cluto/download

Aggregating Crowd Wisdom with Instance Grouping Methods

Li'ang Yin[✉], Zhengbo Li, Jianhua Han, and Yong Yu

Computer Science Department, Shanghai Jiao Tong University,
No. 800 Dongchuan Road, Shanghai, China
{yinla,zhengbo,hanjianhua44,yyu}@apex.sjtu.edu.cn

Abstract. With the blooming of crowdsourcing platforms, utilizing crowd wisdom becomes popular. Label aggregation is one of the key topics in crowdsourcing research. The goal is to infer true labels from multiple labels provided by different users. Most researchers make their efforts in modeling user ability and instance difficulty. However, these methods may suffer from sparsity of labels in practice. In this paper, we consider label aggregation from the view of grouping instances. We assume instances are sampled from latent groups and instances in the same group share the same true label. A probabilistic graphical model named InGroup (Instance Grouping model) is constructed to infer latent group assignments as well as true labels. Further, we combine user ability and group difficulty into InGroup to achieve a better model called InGroup+ (InGroup Plus). The experiments conducted on a real-world dataset show the advantages of instance grouping methods compared with other methods.

Keywords: Crowd wisdom · Label aggregation · Instance grouping · Probabilistic graphical model

1 Introduction

With the blooming of online crowdsourcing platforms such as Amazon Mechanical Turk and CrowdFlower, crowdsourcing has become more and more popular. Crowdsourcing is useful for different types of tasks. We are particularly interested in labeling tasks which exploit platform users to assign labels to specific instances instead of experts where employing online users is affordable but experts are usually costly. In many labeling scenarios, one instance may receive multiple labels from different users. Since labels from one single online user may not be very accurate, multiple labels help reduce this noise. For example, in order to identify the main object in an image, the image is shown to several users and each user gives his/her label. Having multiple labels, we want to know the most proper label (true label) of the image. Inferring true label from multiple labels is called label aggregation and it is one of the key topics in the research of utilizing crowdsourcing data [1,9,12].

© Springer International Publishing Switzerland 2016
F. Li et al. (Eds.): APWeb 2016, Part I, LNCS 9931, pp. 468–479, 2016.
DOI: 10.1007/978-3-319-45814-4_38

One simple and useful method for label aggregation is majority voting, which is used for election, trials, opinion polls, and so on. Recent research works have proposed more sophisticated models [1–7,9,10,12,13]. These models adopt following assumptions: (1) User ability: users with high abilities may give labels more correctly; (2) Instance difficulty: difficult instances may cause more labeling mistakes. There are roughly two kinds of modeling frameworks: trust propagation and generative models. Details are discussed in the section of related work.

Real-world datasets are always sparse, in other words, the number of user labels is very small compared to the number of users. In the CUB-200-2010 dataset [8], each instance only receives 5 labels though there are more than 1500 users. This fact makes it difficult to model user ability or instance difficulty. More recent works consider this problem and try to group users to overcome label sparsity [5,7].

We consider label aggregation from the view of grouping instances. A real-world example (from CUB-200-2010 [8]) is used to illustrate the idea. In the CUB-200-2010 dataset, users are asked to label some local attributes (such as bill shape) of bird images. Images are clawed from websites and the key challenge besides sparsity of labels is that some images do not show the corresponding attribute. For example, one user can hardly label bill shape when the bird's bill is covered by a tree branch or the image only shows the backside of bird's head. Though it seems impossible to label that attribute, we observe that birds belong to the same category have the same bill shape. For example, black footed albatross (category) has hooked bill, then we can infer that an image (instance) of black footed albatross contains the attribute of hooked bill even the bird's bill is not shown to users.

We introduce 'latent group' to represent the concept of category since the prior knowledge about category is usually unknown. The concept of group also alleviates the sparsity problem because one group contains several instances. Say n instances in a group, then the group has n times user labels than a single instance, which makes it easier to infer the group label. We construct a probabilistic graphical model named InGroup (**In**stance **Group**ing model) to infer latent group assignments as well as true labels. Furthermore, we combine user ability and group difficulty into InGroup to achieve a better model InGroup+ (InGroup Plus).

Experiments conducted on the CUB-200-2010 dataset show the advantages of instance grouping methods compared with other methods. We illustrate the effect of user ability in InGroup+ for different tasks. We also discuss the influence of number of latent groups through experiments.

2   Related Work

Label Aggregation is also known as truth discovery [12] or wisdom of the crowds [9] which aims to infer true labels from multiple labels given by online users. There are roughly two kinds of modeling frameworks: trust propagation and generative models.

TruthFinder is the first trust propagation model in crowdsourcing [12]. It assumes facts provided by trustworthy sources are more reliable and sources providing reliable facts are more trustworthy. The model iteratively updates source trustworthiness and fact reliability until convergence. Similar assumptions and modeling methods are adopted by Investment and 3-Estimate models [2,4]. CATD not only estimates source reliability, but also considers the confidence interval of the estimation [3].

More recent works use generative models. Most works have the assumptions of user ability and instance difficulty. GLAD uses a bilinear function to model these factors [10]. DARE adopts a probabilistic model to generate user labels from true labels, user ability, and instance difficulty [1]. TEM models the existence of truth besides user ability and instance difficulty [13]. CBCC considers the sparsity of labels and tries to group users to alleviate the problem of sparsity [7].

Other works [5,6,9] find that instance feature is important in label aggregation. Therefore they add a linear classifier into the generative process. However, a linear classifier is usually too simple to capture the underlying structure of instance feature. That is one of the motivations that we propose to group instances in our previous work [11].

3   Methods

Given M users and N instances, let l_{ui} denote the label user u assigns to instance i. We use binary labels in this paper, i.e. $l_{ui} \in \{0,1\}$. Furthermore, let x_i be a column vector to present the feature of instance i. Our goal is to infer a true label t_i for each instance i. We first construct InGroup model which implements the idea of instance grouping, then combine user ability and group difficulty into it to achieve a better model named InGroup+.

3.1   Instance Grouping Model

Assume there are K latent groups. Each instance i is assigned to group $k \in \{1,...,K\}$ with probability $p(z_i = k)$, where z_i is the partition index. To simplify our model and avoid overfitting, we use hard partition as in classical k-means algorithm. That is

$$p(z_i = k) = \begin{cases} 1, & \text{instance } i \text{ belongs to group } k \\ 0, & \text{otherwise.} \end{cases}$$

We use $z_i = k$ to denote instance i is assigned to group k in the rest of the paper.

The generative process of InGroup is described as follows:

For each group k, its center c_k has a prior multivariate normal distribution

$$p(c_k) \sim \mathcal{N}(\mu_0, I),$$

where I is the identity matrix and μ_0 can be set to the average value of all x_i. q_k denotes the probability of group k's label being '1'. We call q_k group label in the rest of the paper. q_k has a prior beta distribution

$$p(q_k) \sim \text{Beta}(\alpha, \beta).$$

where α and β are pseudo counts in beta function.

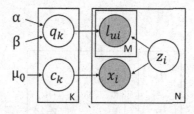

Fig. 1. Probabilistic graphical model of InGroup.

For each instance i, its feature x_i is sampled from group k where $z_i = k$ with the probability

$$p(x_i) \sim \mathcal{N}(c_k, \sigma^2 I).$$

where σ^2 is the variance of instances in a group. Here we assume groups have the same variance and elements of instance feature are mutually independent. Because of data sparsity, this assumption reduces the risk of overfitting. The group variance σ^2 can be estimated by averaging the variances of all groups after instance assignment.

Each label l_{ui} is sampled from group k where $z_i = k$ with the probability

$$p(l_{ui}) \sim \text{Bernoulli}(q_k).$$

Figure 1 illustrates the generative process of InGroup.

Let X, L, C, Q, Z denote sets $\{x_i\}, \{l_{ui}\}, \{c_k\}, \{q_k\}, \{z_i\}$ respectively. Our goal is to find proper C, Q, and Z which maximize the posterior probability

$$p(C, Q \mid X, L; Z) \propto p(X, L \mid C, Q; Z) p(C) p(Q). \tag{1}$$

By adopting the EM algorithm, we can maximize formula (1) in an iterative approach. When iteration converges, we obtain q_k for each group. By applying the assumption that instances belong to the same group share the same true label, we can predict instance label in a simple way by rounding q_k:

$$t_i = \text{round}(q_k), \text{if } z_i = k. \tag{2}$$

3.2 InGroup Plus

Further, we consider combining user ability and group difficulty into the proposed model above. Assume each user u has ability a_u and each latent group k has difficulty b_k. We construct InGroup+ (InGroup Plus) also by a generative process. Figure 2 gives the probabilistic graphical model.

There are four plates in the generative process, namely group, instance, user, and label plates. We describe them respectively in the following.

Group plate contains variables of each latent group. The group center c_k and group label q_k are generated in the same way as in InGroup model. Group difficulty b_k has a prior normal distribution

$$p(b_k) \sim \mathcal{N}(\mu_b, \sigma_b^2)$$

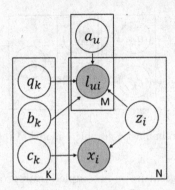

Fig. 2. Probabilistic graphical model of InGroup+. Corresponding priors are omitted for a clean view. There are four plates, namely group(K), instance(N), user(M), and label plates($M \times N$).

where μ_b is the mean and σ_b^2 is the variance of group difficulty. Small variance indicates that group difficulty tends to be close to the mean while high variance indicates that group difficulty is very spread out.

Instance plate contains instance feature x_i and partition index z_i. They are the same as corresponding variables of InGroup.

User plate contains user ability a_u, which has a prior normal distribution

$$p(a_u) \sim \mathcal{N}(\mu_a, \sigma_a^2)$$

where μ_a is the mean and σ_a^2 is the variance of user ability.

Now we look into the label plate. Assume that users with high abilities may give labels more correctly while difficult groups may cause more labeling mistakes for instances in the group. We use following equations to generate each user label. A user u assigns label l_{ui} to an instance i in group k with probability

$$p(l_{ui} = 1 | a_u, b_{z_i=k}, q_{z_i=k}) = (q_k - 0.5)\delta(\lambda(a_u - b_k)) + 0.5$$
$$p(l_{ui} = 0 | a_u, b_{z_i=k}, q_{z_i=k}) = (0.5 - q_k)\delta(\lambda(a_u - b_k)) + 0.5$$

where $\delta(\lambda x)$ denotes the sigmoid function $1/(1+e^{-\lambda x})$ and λ controls its smoothness. In our case of binary label, $p(l_{ui} = 1) + p(l_{ui} = 0) = 1$. Here $(a_u - b_k)$ denotes the relative ability of user u in classifying instance i in group k. We can regard equations above as interpolation between q_k and 0.5: if q_k is close to 1, and a user u is good at classifying instance i in group k, which leads to a large positive value of $(a_u - b_k)$, then the probability of $p(l_{ui} = 1)$ is high; Similarly, if q_k is close to 0 and a user has relatively high ability, we have the probability of $p(l_{ui} = 0)$ being high; If a user cannot classify an instance at all, which leads to a large negative value of $(a_u - b_k)$, then he/she randomly picks a label with probability 0.5.

Since the relative user ability $(a_u - b_k)$ is used in equations above, we can set the mean of group difficulty $\mu_b = 0$ and adjust the mean of user ability μ_a.

Usually, an optimistic prior is set, which is $\mu_a > 0$ since users give more correct labels than incorrect ones.

Putting all plates together, we have the posterior probability

$$p(C, Q, B, A|X, L; Z) \propto p(X, L|C, Q, B, A; Z)p(C, Q, B, A)$$
$$= p(X|C; Z)p(L|A, B, Q; Z)p(C)p(A)p(B)p(Q) \quad (3)$$

where X, L, C, Q, B, A, Z denote sets $\{x_i\}, \{l_{ui}\}, \{c_k\}, \{q_k\}, \{b_k\}, \{a_u\}, \{z_i\}$ respectively. After we obtain proper model parameters C, Q, B, A, Z by maximizing the formula above, the true label t_i is inferred by Eq. (2). In the following section we describe how to maximize the posterior probability.

4 Parameter Estimation

We need to maximize posteriors to obtain proper model parameters for InGroup and InGroup+ respectively. In both formulas, partition indexes $\{z_i\}$ acts as the latent variables, therefore the Expectation-Maximization(EM) algorithm is utilized to iteratively update model parameters. We only illustrate how to maximize formula (3) because of page limit. Formula (1) can be maximized in a similar and simpler way which benefits from the property of conjugate distributions.

Before applying EM to formula (3), we take the logarithm of it to obtain

$$\log p(X|C; Z) + \log p(L|A, B, Q; Z) + \log p(C) + \log p(A) + \log p(B) + \log p(Q) \quad (4)$$

which is convenient in the following calculation.

In E-step, we estimate each partition index z_i by fixing other parameters. From (4), we keep terms related to z_i and substitute corresponding probability functions, that yields

$$z_i^* = \arg\max_{z_i} \log p(x_i|c_{z_i}) + \sum_u \log p(l_{ui}|a_u, b_{z_i}, q_{z_i})$$

$$= \arg\max_{z_i} -\frac{1}{2\sigma^2}(x_i - c_{z_i})^2$$

$$+ \sum_u \mathcal{I}(l_{ui} = 1) \log \left((q_{z_i} - 0.5)\delta(\lambda(a_u - b_{z_i})) + 0.5 \right)$$

$$+ \sum_u \mathcal{I}(l_{ui} = 0) \log \left((0.5 - q_{z_i})\delta(\lambda(a_u - b_{z_i})) + 0.5 \right) \quad (5)$$

where $\mathcal{I}(x)$ is an indicator function. $\mathcal{I}(x) = 1$ if x is true and $\mathcal{I}(x) = 0$ otherwise.

In M-step, we maximize other model parameters by fixing $\{z_i\}$ obtained above. Specifically we want to estimate group centers $\{c_k\}$, user ability $\{a_u\}$, group difficulty $\{b_k\}$, and group labels $\{q_k\}$. Similar with (5), we keep terms related to each parameter and obtain their maximizations respectively:

$$c_k^* = \arg\max_{c_k} \sum_{z_i=k} \log p(x_i|c_{z_i}) + \log p(c_k) \quad (6)$$

$$a_u^\star = \arg\max_{a_u} \sum_i \log p(l_{ui}|a_u, b_{z_i}, q_{z_i}) + \log p(a_u) \tag{7}$$

$$b_k^\star = \arg\max_{b_k} \sum_{u,z_i=k} \log p(l_{ui}|a_u, b_k, q_k) + \log p(b_k) \tag{8}$$

$$q_k^\star = \arg\max_{q_k} \sum_{u,z_i=k} \log p(l_{ui}|a_u, b_k, q_k) + \log p(q_k) \tag{9}$$

where $\sum_{z_i=k}$ is the summation over all instances in group k.

Benefiting from the property of conjugate distributions, we have closed form solution for $c_k^\star$

$$c_k^\star = \frac{\mu_0 + \frac{1}{\sigma^2}\sum_{z_i=k} x_i}{1 + \frac{1}{\sigma^2}\|i : z_i = k\|}$$

where $\|i : z_i = k\|$ denotes the number of instances in group k.

$a_u^\star$, $b_k^\star$ and $q_k^\star$ do not have closed form solutions, therefore we use gradient ascent to find their maximums respectively. Let $\delta_{u,k}$ denote $\delta(\lambda(a_u - b_k))$ for short. After some calculation, we have gradient

$$\Delta a_u = \frac{\partial \sum_i \log p(l_{ui}|a_u, b_{z_i}, q_{z_i}) + \log p(a_u)}{\partial a_u}$$

$$= \sum_i \mathcal{I}(l_{ui} = 1)\frac{\lambda}{(q_{z_i} - 0.5)\delta_{u,z_i} + 0.5}(q_{z_i} - 0.5)\delta_{u,z_i}(1 - \delta_{u,z_i})$$

$$\sum_i \mathcal{I}(l_{ui} = 0)\frac{\lambda}{(0.5 - q_{z_i})\delta_{u,z_i} + 0.5}(0.5 - q_{z_i})\delta_{u,z_i}(1 - \delta_{u,z_i}) - \frac{1}{\sigma_a^2}(a_u - \mu_a)$$

Similarly, we have

$$\Delta b_k = -\sum_{u,z_i=k} \mathcal{I}(l_{ui} = 1)\frac{\lambda}{(q_k - 0.5)\delta_{u,k} + 0.5}(q_k - 0.5)\delta_{u,k}(1 - \delta_{u,k})$$

$$- \sum_{u,z_i=k} \mathcal{I}(l_{ui} = 0)\frac{\lambda}{(0.5 - q_k)\delta_{u,k} + 0.5}(0.5 - q_k)\delta_{u,k}(1 - \delta_{u,k}) - \frac{1}{\sigma_b^2}(b_k - \mu_b)$$

and

$$\Delta q_k = \sum_{u,z_i=k} \mathcal{I}(l_{ui} = 1)\frac{1}{(q_k - 0.5)\delta_{u,k} + 0.5}\delta_{u,k}$$

$$- \sum_{u,z_i=k} \mathcal{I}(l_{ui} = 0)\frac{1}{(0.5 - q_k)\delta_{u,k} + 0.5}\delta_{u,k} + \frac{\alpha - 1}{q_k} - \frac{\beta - 1}{1 - q_k}$$

By using gradient ascent, we update

$$a_u' = a_u + \eta\Delta a_u$$
$$b_k' = b_k + \eta\Delta b_k$$
$$q_k' = q_k + \eta\Delta q_k$$

for several iterations to achieve convergence. η is the learning rate. A small η like 0.05 is preferred in the experiments.

Algorithm 1 illustrates the process of parameter estimation for InGroup+.

Algorithm 1. InGroup+

Input:
    M users, N instances
    $\{l_{ui}\}$, set of user labels
    $\{x_i\}$, set of instance features
    $\alpha, \beta, \mu_0, \mu_a, \sigma_a, \mu_b, \sigma_b$, prior parameters
    K, number of clusters
Output:
    $\{z_i\}, \{c_k\}, \{a_u\}, \{b_k\}, \{q_k\}$, sets of model parameters
    $\{t_i\}$, set of inferred labels
 1: **while** not convergent **do**
 2:   **if** not initialized **then**
 3:      Initialize $\{z_i\}$ with k-means
 4:      Initialize other model parameters
 5:   **else**
 6:      **for** $i = 1$ to N **do**
 7:         Update z_i with (5)
 8:      **end for**
 9:   **end if**
10:   **for** $k = 1$ to K **do**
11:      Update c_k with (6)
12:      Update b_k with (8)
13:      Update q_k with (9)
14:   **end for**
15:   **for** $u = 1$ to M **do**
16:      Update a_u with (7)
17:   **end for**
18: **end while**
19: **for** $i = 1$ to N **do**
20:   $t_i = \text{round}(q_{z_i})$
21: **end for**

---

5   Experiments

We conduct experiments on the CUB-200-2010 dataset [8]. In CUB-200-2010, users are required to label several local attributes for given bird images. The dataset contains 6033 images and 287 local attributes for each image. There are about 1500 users giving their labels to each local attribute and each attribute receives 5 user labels. All user labels are binary which judge the existence of corresponding attributes. We regard inferring true labels on a specific attribute as a label aggregation task. 8 attributes are used, namely **bill** (bill shape is all-purpose or not), **head** (head pattern is plain or not), **shape** (shape is perching-like or not), **forehead** (forehead is black or not), **throat_b** (throat is black or not), **throat_w** (throat is white or not), **underparts** (underparts is yellow or not), and **breast** (breast pattern is solid or not). These attributes are challenging (user labels for an instance often disagree with each other) and their ground truth can be found from whatbird.com and other bird websites.

Feature vectors are constructed from user labels. Specifically, for an image, labels on each attribute are averaged as one element in feature vector, then a 287-dimensional vector is obtained since there are 287 attributes for each image. We discard some attributes because they are zeros for almost all instances, and yield a feature vector of length 200 for each instance which contains less noise. We have tried image features (like SIFT features and features extracted from a deep neural network), but they seem do not work on the data. This indicates features of local attributes are not easy to capture by these feature extractors.

Several methods are compared with our instance grouping methods (InGroup and InGroup+), namely MV (majority voting), TruthFinder [12], Investment [4], 3-Estimate [2], CATD [3], LC (Learning from Crowds [6]), GLAD [10], DARE [1], and CBCC [7] models.

For both InGroup and InGroup+, we set prior parameters of beta function $\alpha = \beta = 3$. For InGroup+, we set user ability prior parameters $\mu_a = 0.5$ and $\sigma_a = 1.0$, and group difficulty prior parameters $\mu_b = 0.0$ and $\sigma_b = 1.0$. Smoothness parameter λ is set to 0.1.

The results of model accuracy is illustrated in Table 1. The bold font indicates the best or comparable results of each column. Trust propagation methods (TruthFinder and Investment) are no better than majority voting and 3-Estimate is even worse. This is mainly because of the sparsity of labels. Trust propagation methods cannot find sufficient evidence to model trustworthiness of sources or reliability of facts. CATD works better since it considers the confidence interval of the estimation. LC adopts a linear classifier to model instance feature but it does not show any advantages, from which we can see that instances are not linearly classifiable in the dataset. Generative models are slightly better because prior distributions keep them from severe overfitting, though they still suffer from the sparsity of labels. On the other hand, proposed instance grouping methods achieve significant improvement compared with other methods. This is because instance grouping methods can capture relationships between instances and latent groups. The experimental results verify our assumptions about grouping instances: Instances form several groups and in each group instances agree on the same group label.

Table 1. Accuracy comparison

	Bill	Head	Shape	Forehead	Throat_b	Throat_w	Underparts	Breast
MV	0.8168	0.8760	0.8754	0.8651	0.8934	0.7814	0.9385	0.7608
TruthFinder	0.8170	0.8760	0.8797	0.8651	0.8934	0.7814	0.9385	0.7608
Investment	0.7616	0.8294	0.8873	0.8233	0.8457	0.7590	0.9388	0.7253
3-estimate	0.7239	0.7268	0.6743	0.7686	0.8130	0.7321	0.9121	0.6905
CATD	0.8170	0.8710	0.9113	0.8613	0.8919	0.7774	0.9373	0.7636
LC	0.8187	0.8768	0.8865	0.8646	0.8928	0.7882	**0.9433**	0.7606
DARE	0.8145	0.8652	0.9105	0.8619	0.8913	0.7946	**0.9430**	0.7630
CBCC	0.8140	0.8661	0.9010	0.8647	0.8881	**0.8185**	0.9453	0.7580
InGroup	**0.8586**	**0.8938**	0.9161	**0.8947**	**0.9098**	0.8071	**0.9475**	**0.7968**
InGroup+	**0.8604**	**0.8949**	**0.9252**	**0.8952**	**0.9115**	0.8087	**0.9473**	**0.7931**

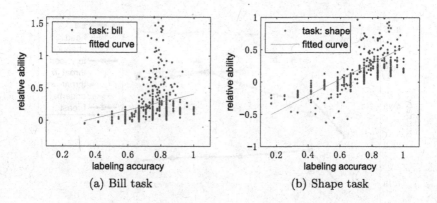

(a) Bill task   (b) Shape task

Fig. 3. Comparison of ability and accuracy for two tasks.

InGroup+ has a slight improvement compared with InGroup for most tasks. Results of shape task show the advantages of InGroup+. Figure 3 compares labeling accuracy and user ability on bill and shape tasks. We use labeling accuracy as values of x-axis and relative user ability as values of y-axis for both sub-figures. Here labeling accuracy is the ratio of correct labels to total labels given by a user. Relative user ability is calculated as $a_u - (\sum_k b_k)/K$ where $(\sum_k b_k)/K$ denotes the average group difficulty. If the model captures user ability, we shall have strong positive correlation between labeling accuracy and relative user ability. We use a red line to illustrate the linear fit of data for each task. On shape task (the right sub-figure), we can observe the red line has a steep slope which means InGroup+ captures user ability successfully. However, the slope of red line of bill task (left sub-figure) is much flatter which means there is no significant difference between user abilities. Therefore the improvement of modeling user ability is small on bill task.

The latent group number K is a key parameter in proposed methods. Figure 4 shows how the accuracy changes along different values of K for each task. Here we illustrate curves of InGroup+ only. InGroup has similar curves. From the figure, we can see InGroup+ achieves high accuracy when K ranges from 200 to 800 for most tasks. This observation shows our model is relatively robust and can adjust to different group size. The best accuracy is usually achieved with latent group number ranging from 200 to 400 (about 15–30 instances in one group). In fact, the CUB-200-2010 dataset contains images from 200 bird categories which is in accord with our experimental results. When K is too small, one group contains too many instances that cannot be distinguished from each other. When K is too large, there are not enough instances in each group to infer reliable group label. Note that setting K to the number of all the instances makes each instance a group, then InGroup becomes majority voting and InGroup+ becomes a generative model similar to DARE.

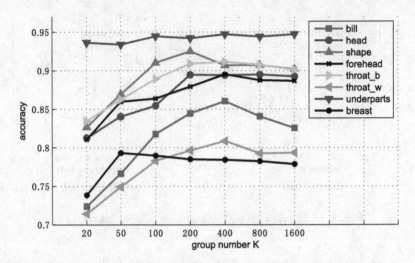

Fig. 4. Accuracy changes along latent group number K of InGroup+.

6   Conclusion

In this paper, we consider label aggregation from the view of grouping instances. We construct a probabilistic graphical model named InGroup to infer latent group assignments as well as true labels. Furthermore, we combine user ability and group difficulty into InGroup to achieve a better model InGroup+. The EM algorithm is adopted to estimate model parameters.

Experiments conducted on the CUB-200-2010 dataset show the advantages of instance grouping methods compared with other methods. We illustrate the effect of user ability in InGroup+ for different tasks. We also discuss the influence of number of latent groups through experiments.

This work considers difficulty on the latent group level. However, instances in the same group may have different instance difficulty which is on the instance level. In the CUB-200-2010 dataset we are confronted with the problem of label sparsity, which makes it likely to overfit to model instance level difficulty. We defer the generalization of instance grouping methods and the experiments on other datasets to future work.

References

1. Bachrach, Y., Graepel, T., Minka, T., Guiver, J.: How to grade a test without knowing the answers–a bayesian graphical model for adaptive crowdsourcing and aptitude testing. In: Proceedings of the 29th International Conference on Machine Learning (ICML 2012), pp. 1183–1190 (2012)
2. Galland, A., Abiteboul, S., Marian, A., Senellart, P.: Corroborating information from disagreeing views. In: Proceedings of the Third ACM International Conference on Web Search and Data Mining, pp. 131–140. ACM (2010)

3. Li, Q., Li, Y., Gao, J., Su, L., Zhao, B., Demirbas, M., Fan, W., Han, J.: A confidence-aware approach for truth discovery on long-tail data. Proc. VLDB Endow. **8**(4), 425–436 (2014)
4. Pasternack, J., Roth, D.: Knowing what to believe (when you already know something). In: Proceedings of the 23rd International Conference on Computational Linguistics, pp. 877–885. Association for Computational Linguistics (2010)
5. Qi, G.-J., Aggarwal, C.C., Han, J., Huang, T.: Mining collective intelligence in diverse groups. In: Proceedings of the 22nd International Conference on World Wide Web, pp. 1041–1052. International World Wide Web Conferences Steering Committee (2013)
6. Raykar, V.C., Yu, S., Zhao, L.H., Valadez, G.H., Florin, C., Bogoni, L., Moy, L.: Learning from crowds. J. Mach. Learn. Res. **11**, 1297–1322 (2010)
7. Venanzi, M., Guiver, J., Kazai, G., Kohli, P., Shokouhi, M.: Community-based bayesian aggregation models for crowdsourcing. In: Proceedings of the 23rd International Conference on World Wide Web, pp. 155–164. International World Wide Web Conferences Steering Committee (2014)
8. Welinder, P., Branson, S., Mita, T., Wah, C., Schroff, F., Belongie, S., Perona, P.: Caltech-UCSD birds 200
9. Welinder, P., Branson, S., Perona, P., Belongie, S.J.: The multidimensional wisdom of crowds. In: Advances in Neural Information Processing Systems, pp. 2424–2432 (2010)
10. Whitehill, J., Wu, T.-F., Bergsma, J., Movellan, J.R., Ruvolo, P.L.: Whose vote should count more: optimal integration of labels from labelers of unknown expertise. In: Advances in Neural Information Processing Systems, pp. 2035–2043 (2009)
11. Yin, L., Han, J., Yu, Y.: Label aggregation with instance grouping model. In: Proceedings of the 25th International Conference Companion on World Wide Web, pp. 135–136. International World Wide Web Conferences Steering Committee (2016)
12. Yin, X., Han, J., Yu, P.S.: Truth discovery with multiple conflicting information providers on the web. IEEE Trans. Knowl. Data Eng. **20**(6), 796–808 (2008)
13. Zhi, S., Zhao, B., Tong, W., Gao, J., Yu, D., Ji, H., Han, J.: Modeling truth existence in truth discovery. In: Proceedings of the 21th ACM SIGKDD International Conference on Knowledge Discovery and Data Mining, pp. 1543–1552. ACM (2015)

CoDS: Co-training with Domain Similarity for Cross-Domain Image Sentiment Classification

Linlin Zhang[1], Meng Chen[1], Xiaohui Yu[1,2($\boxtimes$)], and Yang Liu[1]

[1] School of Computer Science and Technology, Shandong University, Jinan, China
zhanglinlinsd@163.com, chenmeng114@hotmail.com, {xyu,yliu}@sdu.edu.cn
[2] School of Information Technology, York University, Toronto, Canada

Abstract. Classifying images according to the sentiments expressed therein has a wide range of applications, such as sentiment-based search or recommendation. Most existing methods for image sentiment classification approach this problem by training general classifiers based on certain visual features, ignoring the discrepancies across domains. In this paper, we propose a novel **co**-training method with **d**omain **s**imilarity (CoDS) for cross-domain image sentiment classification in social applications. The key idea underlying our approach is to use both the images and the corresponding textual comments when training classifiers, and to use the labeled data of one domain to make sentiment classification for the images of another domain through co-training. We compute image/text similarity between the source domain and the target domain and set the weighting of the corresponding classifiers to improve performance. We perform extensive experiments on a real dataset collected from Flickr. The experimental results show that our proposed method significantly outperforms the baseline methods.

1 Introduction

In recent years, social applications (e.g., Flickr) have been growing rapidly and it is increasingly common for users to share pictures and interact with others using comments. The availability of such data in large volumes makes it possible to classify image sentiments, that is, to understand the positive or negative emotions embedded in the images. Some existing methods [1, 2] to solve this problem tend to build a general classifier with some user-defined features (e.g., saturation, brightness), which may perform well only for the images in the domain the classifier is trained for. However, there are usually numerous domains in real applications (e.g., Flickr has 1,187 domains [3]), and a general classifier for all domains may be problematic, as similar features may express diverse emotions in separate domains. For example, as shown in Table 1, the color histograms of the negative image in the face domain and the positive image in the car domain are similar, but the emotions are opposite. Further, it is ususally infeasible to label the emotions of images manually for the vast number of domains, and we thus cannot obtain sufficient training data in each domain to construct a robust classifier.

F. Li et al. (Eds.): APWeb 2016, Part I, LNCS 9931, pp. 480–492, 2016.
DOI: 10.1007/978-3-319-45814-4_39

Table 1. The positive and negative images with comments and their color histograms (Hue (H), Saturation (S), and Brightness (B)) in face domain and car domain.

Domain	Face		Car	
	Positive	Negative	Positive	Negative
Image				
Histogram				
Comment	*nice girl!*	*why are you so upset?*	*it looks awesome, I like it!*	*such a pity it is abandoned!*

In this paper, we propose a novel method for cross-domain image sentiment classification. Given a large volume of labeled images from one source domain, we make sentiment classification for the images from a new target domain. The idea of our methods comes from observations on real social datasets. First, we observe that about half of the images posted on Flickr have comments and the sentiment polarity of comments is usually consistent with that of the corresponding image. For example, people comment on a positive smiling face with *"nice girl!"* and a negative sad face with *"why are you so upset?"* (see Table 1). Second, despite the differences across domains, people are willing to use some common sentiment words to express their emotions in comments. For example, "great", "perfect", and "wonderful" are usually used to comment on the positive images while "terrible", "bad", and "sad" often occur in negative ones. Finally, the labeled image data are relatively rare, leaving a large volume of images unlabeled.

We therefore develop a method that could take advantage of the first two facts mentioned above by bringing textual data into the task of image classification. Also, this method will be able to use the classifier built for a known domain to help build a new classifier for a new unknown domain. We call this method the **co**-training method with **d**omain **s**imilarity (CoDS), which means to iteratively enlarge the labeled dataset by adding high-confidence predicted unlabeled data. First, with the source domain data, we train an image classifier with the images and a text classifier with the comments. Second, we compute the similarity between the images/comments of the source domain and the ones of the target domain, and set it as the weighting of the corresponding classifier. Finally, we use the co-training method to improve the classification accuracy.

We perform thorough experiments using data collected from various domains in Flickr, and compare CoDS with several baseline methods. The experimental results show that CoDS outperforms existing methods in terms of standard classification measures.

Specifically, the main contributions of this paper are summarized as follows.

- We propose CoDS for cross-domain image sentiment classification. To the best of our knowledge, we are the first to use the labeled data of one domain to make sentiment classification for another domain.
- CoDS takes images and comments as two views and attempts to improve the co-training algorithm by setting the image/text similarity between the source domain and the target domain as the weighting of the corresponding classifier.
- We compare CoDS with some traditional baselines, and demonstrate significant improvement over standard classification metrics.

The rest of this paper is organized as follows. We review the related work in Sect. 2, and give the preliminaries of our work in Sect. 3. Section 4 introduces the proposed method CoDS. We present the experimental results and performance analysis in Sect. 5, and conclude the paper in Sect. 6.

2   Related Work

2.1   Sentiment Classification

There is considerable body of study on sentiment classification, which can be roughly divided into three groups according to data source, namely classification based on texts [4,5], images [1,3,6], and both [7].

Many researchers devote to make sentiment classification on text content. The main methods consist of: (1) exploiting natural language processing (NLP) techniques [4], and (2) adopting machine learning methods [5]. Park and Paroubek [4] develop a multinomial Naïve Bayes (NB) classifier by adopting n-grams features extracted from textual reviews in Twitter. Pang and Lee [5] perform document-level sentiment classification with multiple machine learning methods, e.g., NB, Maximum Entropy and SVM, and conclude that SVM achieves the best performance. Despite its great success, there are still several broad challenges in making text sentiment classification. For example, (1) the huge amount of available text data makes it difficult to filter noises; (2) a significant number of users prefer to use images or videos to express their sentiments rather than text, as the old saying quotes *"A picture is worth a thousand words"*.

Image sentiment classification is prevalent along with the openness of social networks and video websites. Current methods of sentiment classification for social media images include low-level visual feature based [1], mid-level visual feature based [3] and deep learning based approaches [6]. Siersdorfer et al. [1] propose a machine learning algorithm to predict image sentiment by using pixel-level features. Borth et al. [3] infer image sentiment using adjective and noun pairs (ANPs) by understanding visual concepts to overcome the affective gap between the low-level features and the high-level sentiments. You et al. [6] train a Convolutional Neural Networks (CNN) image sentiment classification model with the ANPs from Flickr and transfer it to Twitter data. In addition, Zhang et al. propose to combine textual and visual content to make sentiment classification [7]. They train cross-media public sentiment classifiers by extracting useful features from both texts and images. However, all of the researchers train a general classifier for the images of all domains, which are ill-considered since image features of different domains vary greatly.

2.2 Cross-Domain Classification

In cross-domain classification, the labeled and unlabeled data come from different domains, and their underlying distributions vary a lot from each other, which violates the basic assumption of traditional classification learning. Traditional research along this problem mainly performs domain adaptation or transfer learning algorithms [8,9]. Dai et al. [8] investigate a translated learning method for different feature space across domains. Wang et al. [9] extract cross-domain image knowledge for the vision classification task. However, these methods cannot exploit useful comment information.

Co-training is another methodology for using a large unlabeled dataset to boost the performance of a learning algorithm. It performs well in many applications such as web page classification [10] and cross-lingual sentiment classification [11]. Blum and Mitchell [10] propose using two independent and redundant views of web pages together to allow inexpensive unlabeled data to augment a much smaller set of labeled examples. Wang [11] overcomes the problem of a lack of Chinese corpus by co-training method through the abundant English corpus in machine translations. In this paper, we propose a new method based on co-training strategy to better solve our problem.

3 Preliminaries

In this section, we explain a few definitions required for the subsequent discussion, and define the problem addressed in this paper.

Definition 1 (Domain). *A domain dataset crawled from Flickr contains numerous images and their corresponding comments. Images from one domain often describe the same object, such as "face", "car" or "flower".*

Definition 2 (Instance). *We describe an instance x containing image and comments data as (x^i, x^c).*

Definition 3 (Source domain). *A source domain, which contains numerous labeled instances, is the initial training dataset.*

Definition 4 (Target domain). *A target domain, which contains a large volume of unlabeled instances, is a domain that we will train classifiers for.*

With the aforementioned definitions, the problem of cross-domain image sentiment classification is formally defined as:

Given some labeled instances from a source domain, we assign positive(1) or negative(0) sentiment labels to the images from a target domain.

4   CoDS Method

To train a cross-domain image sentiment classifier, we first build an image classifier and a text classifier with instances from the source domain. Second, we compute the image similarity and text similarity between the source domain and the target domain, and set the result as the weighting of the corresponding classifier. Finally, we use the co-training method to improve the classification accuracy of polarity identification.

4.1   Image and Text Classifiers

Image Classifier. There are two main methodologies for image sentiment classification. One uses some traditional classifiers (e.g., SVM) to make polarity identification with representative visual features such as saturation and brightness. The other employs neural networks to train sentiment classifiers, and CNN is the most popular one. It has been shown that CNN performs better than the feature-based ones due to its strong reflection of human feelings [6,12], and we choose CNN to train the image classifier in this paper.

Typically, CNN contains several convolutional layers and full connected layers. Pooling layers and normalization layers exist between the convolutional layers. CNN is a supervised learning algorithm, in which the parameters of different layers are learned through back-propagation. Inspired by the satisfying performance achieved in [6], we use their CNN architecture for image sentiment classification. In order to overcome the lack of enormous labeled datasets and save training time, we decide to explore the possibility of fine-tuning an already existing model learned on the 1,000,000 ImageNet images [13].

Text Classifier. Text sentiment classification is heavily dependent on the sentiment lexicon. Two common approaches for lexicon generation are dictionary-based and corpus-based. The former is based on bootstrapping a seed of sentiment words from dictionaries like WordNet or SentiWordNet; the latter is based on the corpus and, thus, is inherently domain dependent. Considering the sparsity of emotional words in short comments, we apply the corpus-based lexicon generation approach.

Indeed, a careful analysis of image comments shows that the volume of positive comments is significantly larger than the negatives. The reason is that people often use positive sentences like *"Great photo."*, or *"Nice picture."* to express their opinion on the quality of the photography instead of sentiments towards the content of the image. To solve this problem, as mentioned in [14], we first extract a list of terms that describe the photo attributes or components, e.g., "shot", "photo", and "composition"; then we extract two types of comments (opinions on the photo quality and sentiments on the photo content) for each image according to whether the comment contains the extracted terms. Based on the sentiment comments, we propose using two types of features (meta-features and syntax features) to carry out the text sentiment classification.

Meta-features. We extract the nouns, verbs and adjectives from the sentiment comments, and sort them according to occurrence frequency in descending order. All the words whose frequencies exceed the threshold (we set five in this paper) are considered meta-features. Specifically, for a comment, if it contains the meta-feature word, we set the corresponding element of the vector as 1, and 0 otherwise. Here *tf-idf* can also be used as the term weighting according to the characteristic of the text data.

Syntax Features. As meta-features are very sparse, we choose two discriminating features for comment representation. The first is, the number of positive and negative sentiment words in the comments of an image. Here we use SentiWordNet as the sentiment dictionary. The second feature is the number of words in positive or negative word lists. The positive/negative word list is generated by the words in positive/negative comments.

Having determined all the features, we use SVM as the classifier, as SVM is proved to be robust [5].

4.2   Domain Similarities

To determine whether image or text classifier contributes more for the prediction, we compute the similarity between the images/comments of the source domain and the images/comments of the target domain respectively.

Image Similarity. Image similarity computing is mainly based on the various visual presentations employed to describe images. The most common types of visual information are color, texture and shape, and they are widely studied and applied for many applications, especially measuring the similarity between images in retrieval systems [15]. However, existing image similarity computing methods usually just consider pairwise image similarities, i.e., image-to-image similarities. In this paper, we propose a domain-to-domain method to measure the domain similarity based on images. Denote $u^i \in \mathbb{R}^{D \times 1}$ and $u^j \in \mathbb{R}^{D \times 1}$ are the visual image feature vectors of domain i and domain j respectively, where D represents the number of visual features. Then we compute the cosine similarity between domain i and domain j as their image similarity, which can be formulated as follows:

$$ImageSim(i,j) = \frac{u^i \cdot u^j}{\| u^i \|_2 \cdot \| u^j \|_2}, \tag{1}$$

where u^i is the average vector for M images from domain i, which is defined as:

$$u^i = \frac{1}{M} \cdot \sum_{l=1}^{M} u_l^i, \tag{2}$$

and u_l^i is the lth image visual feature of domain i.

Here we use color and texture features [7] since both of them have been proven to perform well for image sentiment classification.

Text Similarity. We use Jensen-Shannon divergence to measure the text similarity of two domains based on their sentiment word distributions [16]. $v^i \in \mathbb{R}^{P \times 1}$ and $v^j \in \mathbb{R}^{P \times 1}$ are the textual sentiment word distribution vectors of domain i and domain j respectively, where P is the sentiment lexicon size, and $v_w^i \in [0, 1]$ stands for the frequency of word w occurring in domain i. The text similarity between domain i and domain j can be described as follows:

$$
\begin{aligned}
TextSim(i, j) &= 1 - D_{JS}(v^i \parallel v^j) \\
&= 1 - \frac{1}{2}(D_{KL}(v^i \parallel \overline{v}) + D_{KL}(v^j \parallel \overline{v})),
\end{aligned}
\tag{3}
$$

where $\overline{v} = \frac{1}{2}(v^i + v^j)$ is the average distribution, and $D_{KL}(\cdot)$ is the Kullback-Leibler divergence which is defined as:

$$
D_{KL}(p \parallel q) = \sum_{w \in D} p(w) \log \frac{p(w)}{q(w)}.
\tag{4}
$$

4.3 Co-training Algorithm

Co-training takes images and comments as two views and increases the amount of annotated data S using some amounts of unlabeled data U in an incremental way. The added data are selected according to the prediction results which are computed below.

For the lth instance (x_l^i, x_l^c) of U from the target domain, the predicted label is:

$$
y^l = \begin{cases} 0 & p_{neg}^l > p_{pos}^l \\ 1 & p_{neg}^l \le p_{pos}^l \end{cases},
\tag{5}
$$

where p_{neg}^l and p_{pos}^l are the probabilities of the lth instance labeled as negative or positive, which can be computed as follows:

$$
\begin{aligned}
p_{neg}^l &= \frac{p_{neg}^l(I) \times ImageSim(S, U) + p_{neg}^l(C) \times TextSim(S, U)}{ImageSim(S, U) + TextSim(S, U)}, \\
p_{pos}^l &= \frac{p_{pos}^l(I) \times ImageSim(S, U) + p_{pos}^l(C) \times TextSim(S, U)}{ImageSim(S, U) + TextSim(S, U)}.
\end{aligned}
\tag{6}
$$

Here $p_{neg}^l(I)$, $p_{pos}^l(I)$, $p_{neg}^l(C)$ and $p_{pos}^l(C)$ are probabilities generated by image classifier and text classifier, and $ImageSim$ and $TextSim$ are image similarity and text similarity between S and U.

We select the top k (the growth size) instances with pseudo labels as candidate set K according to the descending order of $| p_{neg}^l - p_{pos}^l |$. The input dataset for the next iteration is: $S = S \cup K$ as the training set and $U = U - K$ as the unlabeled set. The process carries out iteratively until U is empty. Finally, we use the image classifier trained on pseudo labeled image dataset S to classify the test image dataset T from the target domain. The algorithm of CoDS is described in Algorithm 1.

Algorithm 1. The algorithm of CoDS.

Input:

    -Training data S (image data S_I and comment data S_C), corresponding labels Y;

    -Unlabeled data U (image data U_I and comment data U_C);

    -Test image data T_I;

    -Growth size k;

Output:

    Labels of T_I;

1: **while** U is not empty **do**
2:    train image classifier C_I on S_I;
3:    use C_I to predict probabilities of U_I belonging to each class;
4:    train textual classifier C_C on S_C;
5:    use C_C to predict probabilities of U_C belonging to each class;
6:    compute $ImageSim(S_I, U_I)$ and $TextSim(S_C, U_C)$ according to Eqs. (1) and (3);
7:    compute the final probabilities according to Eq. (6) and assign pseudo labels according to Eq. (5);
8:    select top k instances sorted by $\mid p_{neg}^l - p_{pos}^l \mid$ as set K;
9:    $S = S \cup K$;
10:    $U = U - K$;
11: **end while**
12: train image classifier C on S;
13: use C to assign sentiment labels for T_I;

---

5   Experiments

We have conducted extensive experiments to evaluate the performance of CoDS method on a real Flickr dataset. In this section, we first describe the dataset, followed by the baseline methods used for comparison. Then we show the performances evaluated with several parameters.

5.1   Dataset

The dataset used in this paper is crawled from Flickr by Flickr API according to the ANPs released in [3] and the sentiment labels are set according to the sentiment scores of ANPs with manually revised. Here we choose *"face"*, *"flower"* and *"car"* domains as our experimental dataset, because they are not only frequently tagged in social multimedia, but also associated with diverse adjectives to form a large dataset. The numbers of instances in each domain are listed in Table 2. When used as a target domain, the dataset is divided in half: 50 % as unlabeled set U and 50 % as test set T. Otherwise, the whole dataset is used as a training set S for the source domain.

5.2   Baselines

In our experiments, the proposed CoDS approach is compared with the following four baselines.

Table 2. Data scales of the three experimental domains

Face		Flower		Car	
Positive	Negative	Positive	Negative	Positive	Negative
3317	3947	3830	3456	4443	5240
7264		7286		9683	

SVM-I: This method applies the inductive SVM classifier trained on the images of a source domain directly to the images of the target domain. The features used in this method are extracted according to [7].

CNN-I: This method trains CNN model with the images of a source domain by fine-tuning the existing model in [13] and applies CNN model directly to the images of the target domain.

SVM-(T+I): This method is proposed in [7]. It trains a SVM classifier by integrating text and image features.

Co-Train: This method performs traditional co-training algorithm, which takes image and textual data as two views without considering domain similarity.

5.3   Performance Evaluation

In our experiments, we use standard precision, recall and F-measure to evaluate the performance of positive and negative classes, and use accuracy metrics to measure the overall performance of the algorithm. We choose "face" as the source domain and "car" as the target domain ($face \rightarrow car$) as an example and discuss the comparison results with several parameters (e.g., iteration number I and growth size k).

Method Comparison. In our experiments, we first compare our proposed CoDS ($I = 30$, $k = 100$) approach with the four baselines. Table 3 shows the comparison results. The parameters ($I = 30$, $k = 100$) are set by considering the size of unlabeled car dataset U.

Table 3. Comparison results of $face \rightarrow car$.

Method	Positive			Negative			Total
	Precision	Recall	F-measure	Precision	Recall	F-measure	Accuracy
SVM-I	0.57	0.66	0.61	0.61	0.53	0.57	0.59
CNN-I	0.61	0.70	0.65	0.67	0.55	0.60	0.64
SVM-(T+I)	0.66	0.69	0.67	0.78	0.74	0.76	0.72
Co-Train	0.74	0.79	0.76	0.80	0.75	0.77	0.76
CoDS	**0.79**	**0.86**	**0.82**	**0.87**	**0.77**	**0.81**	**0.82**

Seen from the table, CoDS outperforms all four baselines over all the metrics. Due to the low generalizability of classifiers trained on one specific domain, both SVM-I and CNN-I have poor performances. SVM-(T+I), which takes comments into account, performs 13 % and 8 % better than SVM-I and CNN-I on accuracy metrics, providing evidence for our observation that the associated textual information contains important cues for image sentiments. Co-Train performs better than SVM-(T+I), as it considers the unlabeled data of the target domain. CoDS performs the best since we set different weightings for image and text classifiers according to corresponding domain similarity.

Influences of Iteration Number (I) and Growth Size (k). To explore whether the iteration number (I) and growth size (k) would affect the performance of the proposed method, we make more experiments and show the results in Figs. 1 and 2.

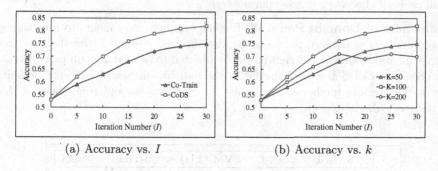

(a) Accuracy vs. I (b) Accuracy vs. k

Fig. 1. Accuracy vs. Parameters

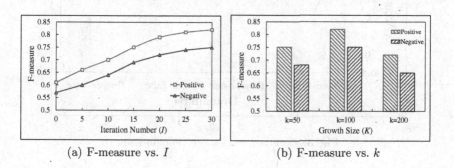

(a) F-measure vs. I (b) F-measure vs. k

Fig. 2. F-measure vs. Parameters

Figure 1(a) shows the accuracy curves of Co-Train and CoDS with iteration numbers I vary from 0 to 30 ($k = 100$). When I is set to 0, both Co-Train and CoDS are the same with CNN-I. Clearly, as we increase the iteration number, the accuracy of both the methods improves. We can see the proposed CoDS method outperforms the Co-Train method in every iteration result. Both curves become smooth after dozens of iterations, because both methods have gradually reached upper limits.

We then explore the effects of growth size by setting k to 50, 100 and 200. Figure 1(b) shows the accuracy curves of CoDS on different growth sizes. Seen from the comparison results, the method evaluated on the moderate parameter ($k = 100$) performs better than the two extreme ones. While the smaller growth size ($k = 50$) brings slower performance improvements, the bigger one ($k = 200$) causes unstable results. The reason behind this is that the growth size is too large, adding wrong instances of prediction, which will act as noisy training data.

Figure 2(a) shows the F-measure curves of positive class and negative class over iteration number I, and Fig. 2(b) shows the F-measure metric over growth size k on the two classes. Both the results show that our method CoDS performs well on each class over various parameters.

Influences of Domain Pairs. We further conduct experiments to investigate the influences of domain pairs on the classification results. Figure 3 shows the experimental results of different methods adapted to several domain pairs. It can be seen that CoDS always outperforms the four baselines with different domain pairs. The results further demonstrate the effectiveness and robustness of the proposed CoDS approach.

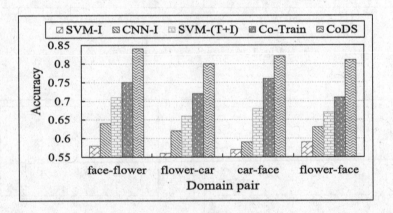

Fig. 3. Accuracy vs. Various domain pairs

6   Conclusions

In this paper, we propose a novel co-training approach (CoDS) to address the problem of cross-domain image sentiment classification. CoDS has two advantages: (1) it makes full use of the image and text data of the source domain; (2) it considers the unlabeled data of the target domain to optimize the classifier.

We crawl a real dataset from Flickr and use it to validate our proposed method. Several comparison experiments show that CoDS significantly performs better than the state-of-the-art methods.

Acknowledgment. This work was supported in part by the National Basic Research 973 Program of China under Grant No. 2015CB352502, the National Natural Science Foundation of China under Grant Nos. 61272092 and 61572289, the Natural Science Foundation of Shandong Province of China under Grant Nos. ZR2012FZ004 and ZR2015FM002, the Science and Technology Development Program of Shandong Province of China under Grant No. 2014GGE27178, and the NSERC Discovery Grants.

References

1. Siersdorfer, S., Minack, E., Deng, F., Hare, J.: Analyzing and predicting sentiment of images on the social web. In: Proceedings of MM, pp. 715–718. ACM (2010)
2. Machajdik, J., Hanbury, A.: Affective image classification using features inspired by psychology and art theory. In: Proceedings of MM, pp. 83–92. ACM (2010)
3. Borth, D., Ji, R., Chen, T., Breuel, T., Chang, S.F.: Large-scale visual sentiment ontology and detectors using adjective noun pairs. In: Proceedings of MM, pp. 223–232. ACM (2013)
4. Pak, A., Paroubek, P.: Twitter as a corpus for sentiment analysis and opinion mining. LREC **10**, 1320–1326 (2010)
5. Pang, B., Lee, L., Vaithyanathan, S.: Thumbs up? Sentiment classification using machine learning techniques. In: Proceedings of EMNLP, pp. 79–86. Association for Computational Linguistics (2002)
6. You, Q., Luo, J., Jin, H., Yang, J.: Robust image sentiment analysis using progressively trained and domain transferred deep networks (2015). arXiv:1509.06041
7. Zhang, Y., Shang, L., Jia, X.: Sentiment analysis on microblogging by integrating text and image features. In: Cao, T., Lim, E.-P., Zhou, Z.-H., Ho, T.-B., Cheung, D., Motoda, H. (eds.) PAKDD 2015. LNCS, vol. 9078, pp. 52–63. Springer, Heidelberg (2015)
8. Dai, W., Chen, Y., Xue, G.R., Yang, Q., Yu, Y.: Translated learning: transfer learning across different feature spaces. In: Proceedings of NIPS, pp. 353–360 (2008)
9. Wang, H., Nie, F., Huang, H., Ding, C.: Dyadic transfer learning for cross-domain image classification. In: Proceedings of ICCV, pp. 551–556. IEEE (2011)
10. Blum, A., Mitchell, T.: Combining labeled and unlabeled data with co-training. In: Proceedings of COLT, pp. 92–100. ACM (1998)
11. Wan, X.: Co-training for cross-lingual sentiment classification. In: Proceedings of ACL, pp. 235–243. Association for Computational Linguistics (2009)
12. Chen, T., Borth, D., Darrell, T., Chang, S.F.: Deepsentibank: Visual sentiment concept classification with deep convolutional neural networks. Comput. Sci. (2014)
13. Krizhevsky, A., Sutskever, I., Hinton, G.E.: Imagenet classification with deep convolutional neural networks. In: Proceedings of NIPS, pp. 1097–1105 (2012)
14. Kisilevich, S., Rohrdant, C., Keim, D.: "Beautiful picture of an ugly place". Exploring photo collections using opinion and sentiment analysis of user comments. In: Proceedings of IMCSIT, pp. 419–428. IEEE (2010)

15. Kontschieder, P., Donoser, M., Bischof, H.: Beyond pairwise shape similarity analysis. In: Zha, H., Taniguchi, R., Maybank, S. (eds.) ACCV 2009, Part III. LNCS, vol. 5996, pp. 655–666. Springer, Heidelberg (2010)
16. Remus, R.: Domain adaptation using domain similarity-and domain complexity-based instance selection for cross-domain sentiment analysis. In: Proceedings of ICDMW, pp. 717–723. IEEE (2012)

Feature Selection via Vectorizing Feature's Discriminative Information

Jun Wang, Hengpeng Xu, and Jinmao Wei[✉]

College of Computer and Control Engineering,
Nankai University, Tianjin 300350, China
{junwang,xuhengpeng}@mail.nankai.edu.cn, weijm@nankai.edu.cn

Abstract. Feature selection is a popular technology for reducing dimensionality. Commonly features are evaluated with univariate scores according to their classification abilities, and the high-score ones are preferred and selected. However, there are two flaws for this strategy. First, feature complementarity is ignored. A subspace constructed by the partially predominant but complementary features is suitable for recognition task, whereas this feature subset cannot be selected by this strategy. Second, feature redundancy for classification cannot be measured accurately. This redundancy weakens the subset's discriminative performance, but it cannot be reduced by this strategy. In this paper, a new feature selection method is proposed. It assesses feature's discriminative information for each class and vectorizes this information. Then, features are represented by their corresponding discriminative information vectors, and the most distinct ones are selected. Both feature complementarity and classification redundancy can be easily measured by comparing the differences between these new vectors. Experimental results on both low-dimensional and high-dimensional data testify the new method's effectiveness.

Keywords: Feature selection · Discriminative information · Feature complementarity · Feature redundancy

1 Introduction

For pattern recognition problems, feature selection is important to speed up learning and improve concept quality. It aims at choosing a subset of features for decreasing the size of the structure and improving prediction accuracy [1]. Feature selection methods can be categorized into supervised, semi-supervised, and unsupervised one, according to whether the instances are labeled, partially labeled or not [2]. For supervised cases, class labels are employed for measuring features' discriminative abilities. Many popular and efficient feature selection methods belong to this group. So supervised methods are further divided into three well-known models, i.e., filter, wrapper and embedded [3,4].

A filter model evaluates features by considering their intrinsic characteristics [5]. It ranks features in terms of a univariate scoring metric, and the features

© Springer International Publishing Switzerland 2016
F. Li et al. (Eds.): APWeb 2016, Part I, LNCS 9931, pp. 493–505, 2016.
DOI: 10.1007/978-3-319-45814-4_40

superior in recognizing a majority of classes will be highly scored and selected. However, two critical issues, i.e., feature complementarity and feature redundancy, are both not well considered. First, partially predominant features as well as their complementary subset receive scant attention. These features denote those discriminative for a certain class but not all the classes. They are evaluated as inferior under the univariate scoring metric. Yet the performance of their complementary combination is excellent. This subset is suitable for constructing the recognition space, which attributes to its features' complementary abilities in recognizing different classes. Unfortunately, a univariate score will submerge these features' superior abilities.

Second, feature redundancy is ignored or hardly to be measured accurately. Redundant features can't provide any new information for learning the target concept. Common filter methods that prefer to high-score features can't reduce this redundancy. Actually there are some methods taking redundancy reduction into consideration, like some information theory based methods [6,7]. However, it is still complicated for these methods to measure multi-feature redundancy specific to classification. This redundancy information is shared by more than two features for recognizing classes. It involves high-order probability estimation and is intractable for the common methods to tackle.

A new feature selection method is proposed in this paper for addressing the above problems. It evaluates feature's discriminative ability for each class, and extracts this information for constructing a new discriminative information vector. Different from the common feature selection methods, the new method quantifies feature's recognitive performance specific to a certain class, rather than averaging its performance to all the classes and assessing with a global score. This vectorization representation facilitates finding those partially predominant features and obtaining their complementary subset. Also multi-feature redundancy for classification can be easily measured through these discriminative vectors.

2   Related Work

Dozens of feature evaluation criteria are used in the filter selection methods. They can be roughly categorized into the dependency measure, information measure, distance measure, and consistency measure. As to the dependency measure, some statistical criteria are included, like Pearson Correlation Coefficient, Chi-square Statistic, T Statistic, F Statistic, Gini Index, etc. [8]. Besides, Fisher Score [9], Laplacian Score [10], Hilbert-Schmidt Independence Criterion [11] and SPEC [12] are also used for measuring feature's dependency with classes. They can be subsumed into the group of preserving instance similarity [13]. They enhance the instance similarities in the same class and weaken those in the different classes. So they commonly achieve superior separability performance.

As to the information measure, it derives from the information theory, including some popular evaluation criteria like Mutual Information [6,7], Information Gain [14], Gain Ratio [14], Symmetrical Uncertainty [15], Representation

Entropy [16], etc. In this group, mutual information is representative and widely used in feature selection. It quantifies how much information is shared by two variables via measuring the divergence of their distributions. Although mutual information is efficient in measuring two variables' relevance, its computation is sensitive to even small probability estimation error.

The above evaluation criteria are univariate and commonly used in top-k models. This is beneficial for those discriminative but also redundant features winning out. So amount of redundant information exists in the selected features. This exactly explains why the performance of these features is often below expectation. In fact, some efforts are made to reduce redundancy, such as $mRMR$ [6], $FCBF$ [15], CFS [17], etc. But the reduced redundancy is not that for recognizing classes. Detailedly speaking, this information incorporates two parts. One part is two features' shared information for classification, and the other part is unrelated with classification. Clearly, only reducing the first part conduces to enhance features' joint discriminative performance. Yet the above criteria cannot measure this part. Note that the conditional mutual information based methods [7,18] actually endeavor to reduce this classification redundancy. However, when more than two features are assessed, it's still tough for these methods to accomplish the high-order probability estimation involved in redundancy reduction.

On the other hand, feature complementarity is less discussed by the common evaluation criteria. This issue involves the relationship of more than two variables, so its measurement is out of the reach of these criteria. Furthermore, the partially predominant but complementary features are scarcely paid attention to. The common criteria assign features with global evaluation scores, so partially predominant features superior in discriminating some or even only one class are poorly rated and discarded. The optimal subset constituted by these features is also hard to be built.

3   Discriminative Information Vectorization

3.1   Discriminative Information

Suppose instances $\mathbf{X} = \{\mathbf{x}_i\}_{i=1}^{n}$ characterized by d features $\mathbf{F} = \{\mathbf{f}_j\}_{j=1}^{d}$ and classified to c target classes $\mathbf{C} = \{\mathbf{c}_l\}_{l=1}^{c}$. For class $\mathbf{c}_l$, its class center at the jth dimension is denoted as $\mu_{lj} = \frac{1}{N_l} \sum\limits_{\mathbf{x}_i \in \mathbf{c}_l} x_{ij}$, and its class-conditional dispersion at the jth dimension is $\Sigma_{lj} = \frac{1}{N_l - 1} \sum\limits_{\mathbf{x}_i \in \mathbf{c}_l} (x_{ij} - \mu_{lj})^2$.

Feature's discriminative ability rests with the class separability at this observation dimension. Take Fig. 1 for explanation. In the figure, instances belong to 3 classes, and the class centers are represented by stars. It is shown that when projecting instances to different features, recognizing a certain class is also different. Feature $\mathbf{f}_1$ well scatters class 1 from class 2 and class 3, but it cannot discriminate the instances belonging to the latter two classes. The same situation happens to feature $\mathbf{f}_2$ with class 2 and feature $\mathbf{f}_3$ with class 3. These features are excellent in recognizing a certain class rather than all the classes. They can be

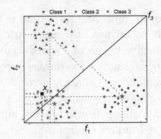

Fig. 1. Instances belonging to 3 classes are projected on features $\mathbf{f}_1$, $\mathbf{f}_2$ and $\mathbf{f}_3$.

deemed as the partially predominant ones. When examining a given instance $\mathbf{x}_i$, it is nearer to the center of class 3 than to the other two classes when projected on $\mathbf{f}_3$. So it will be correctly predicted at this observation dimension. But it may be incorrectly classified along the directions of the other two features because of its farther distance to its true class center. According to this distance measure, feature's discriminative information for a certain class is defined as follows:

Definition 1 (Discriminative Information). *Suppose* $\beta_j = (\beta_{1j}, ..., \beta_{ij}, ...,$ $\beta_{nj})^T \in \mathbb{R}^n, \beta_{ij} = \exp\left[-\frac{(x_{ij} - \mu_{lj})^2}{2\Sigma_{lj}}\right]$, *and* $\lambda = (\lambda_1, ..., \lambda_i, ..., \lambda_n)^T \in \mathbb{R}^n, \lambda_i = \begin{cases} 0, & \mathbf{x}_i \text{ belongs to } \mathbf{c}_l \\ 1, & \text{otherwise} \end{cases}$. *Then, feature* $\mathbf{f}_j$ *'s discriminative information for class* $\mathbf{c}_l$ *is defined as:*

$$DI_j(\mathbf{c}_l) = p(\mathbf{c}_l) \|\beta_j - \lambda\|^2. \tag{1}$$

Feature's discriminative information in Eq. (1) has two terms. The first term $p(\mathbf{c}_l)$ is the prior probability of class $\mathbf{c}_l$. Here class imbalance is taken into consideration [19]. When evaluating feature's discriminative information for different classes, the class containing a majority number of inner-class instances is prior considered. This term is also important for assessing feature's general classification ability to all the classes.

The second term measures the difference of two vectors β_j and λ through the ℓ_2 norm quadratic function. λ represents the false classification information. β_j describes the prediction information of feature $\mathbf{f}_j$. It derives from the distance between each instance $\mathbf{x}_i$ and class center μ_l. Closer it is, more probably $\mathbf{x}_i$ is predicted to class $\mathbf{c}_l$. β_{ij} is accordingly assigned with a larger value. A perfect case is that β_j is completely inconsistent with λ. It means that for the inner-class instances β_j is maximized, and for the outer-class ones it is minimized. In this case, the difference of β_j and λ is maximal, and $DI_j(\mathbf{c}_l)$ is maximized. This can also be explained by Fig. 1. When evaluating $\mathbf{f}_1$'s discriminative information for class 1, its inner-class instances can be correctly predicted according to their nearer distances to μ_1. The outer-class instances are most probably not classified to it. So $\mathbf{f}_1$ is good at recognizing class 1, and $DI_1(\mathbf{c}_1)$ is higher. If β_j is consistent with λ, which indicates that most of the instances are mistakenly classified and the observed feature is inferior in scattering the corresponding class from the other ones, the discriminative information is lower.

The second term can also be interpreted via the quadratic loss function [20]. This function is widely used for estimating the prediction error of a learning algorithm. A high score is achieved if the algorithm classifies most of the instances correctly. Feature $\mathbf{f}_j$'s quadratic loss for recognizing class $\mathbf{c}_l$ is defined as:

$$QL_j(\mathbf{c}_l) = \sum_{i=1}^{n} [p_j(\mathbf{x}_i|\mathbf{c}_l) - \alpha_i]^2, \alpha_i = \begin{cases} 1, & \mathbf{x}_i \, belongs \, to \, \mathbf{c}_l \\ 0, & otherwise \end{cases}, \tag{2}$$

where $p_j(\mathbf{x}_i|\mathbf{c}_l)$ is the prediction probability of $\mathbf{x}_j$ classified to $\mathbf{c}_l$ at the jth dimension. Actually, the second term in Eq. (1) is a variant of feature's quadratic loss for a class. Theorem 1 further elaborates this issue as follows:

Theorem 1. *Suppose* $\mathbf{X}$ *follows a normal distribution. Then* $DI_j(\mathbf{c}_l)$ *is negatively correlated to* $QL_j(\mathbf{c}_l)$.

Proof. According to the assumption of $\mathbf{X}$'s distribution, $p_j(\mathbf{x}_i|\mathbf{c}_l)$ can be represented as $p_j(\mathbf{x}_i|\mathbf{c}_l) = \frac{1}{\sqrt{2\pi\Sigma_{lj}}} \exp\left[-\frac{(x_{ij}-\mu_{lj})^2}{2\Sigma_{lj}^2}\right]$. It is clear that $p_j(\mathbf{x}_i|\mathbf{c}_l)$ is positively correlated to β_{ij}.

According to the definitions of α_i and λ_i, it can be obtained that $\alpha_i + \lambda_i = 1$. Thus, $QL_j(\mathbf{c}_l)$ can be denoted as $QL_j(\mathbf{c}_l) = \sum_{j=1}^{n} (p_j(\mathbf{x}_i|\mathbf{c}_l) + \lambda_i - 1)^2$.

Because $0 \leq p_j(\mathbf{x}_i|\mathbf{c}_l) \leq 1$, $0 \leq \beta_{ij} \leq 1$, and $p_j(\mathbf{x}_i|\mathbf{c}_l)$ is positively correlated to β_{ij}, it is easy to conclude that $\|\beta_j - \lambda\|^2$ is negatively correlated to $QL_j(\mathbf{c}_l)$. Hence, $DI_j(\mathbf{c}_l)$ is negatively correlated to $QL_j(\mathbf{c}_l)$.   □

Theorem 1 demonstrates that more discriminative information is provided, less prediction error is made. A feature which can not only scatter the inner-class instances from the outer-class ones but also maximize this separation will be highly scored by this measurement. It entirely depends on this feature's inconsistent decisions for the instances with their false classification information.

3.2   Feature Selection via Discriminative Information Vector

Since discriminative information describes feature's recognitive ability for a class, feature $\mathbf{f}_j$ can be represented by a discriminative information vector, defined as:

$$\varepsilon_j \triangleq [DI_j(\mathbf{c}_1), ..., DI_j(\mathbf{c}_l), ..., DI_j(\mathbf{c}_c)]^T, \tag{3}$$

where c is the number of classes, and $DI_j(\mathbf{c}_l)$ is the discriminative information of feature $\mathbf{f}_j$ for class $\mathbf{c}_l$. Note that ε_j is actually a vectorization representation of feature's dependency with classes rather than a univariate evaluation score. Each element of this vector corresponds to feature's classification ability to a class. Thus, it is easy to find those partially predominant features whose some elements of discriminative information vectors are larger. Furthermore, multi-feature classification redundancy is also easy to estimate by comparing the differences of the new vectors. So the two complicated problems for the common feature evaluation criteria, finding the partially predominant features and measuring multi-feature redundancy, are both solved by constructing these new vectors.

Based on the discriminative information vector, a new filter feature selection method is proposed, named DivF. It searches the optimal feature subset in the discriminative information vector space. Suppose the candidate feature subset $\mathbf{S} = \{\mathbf{f}_1, \mathbf{f}_2, ..., \mathbf{f}_k\}$. Then DivF selects k^* features from $\mathbf{S}$ for constructing the optimal subset $\mathbf{S}^*$ by maximizing the following objective function:

$$J_{DivF}(\mathbf{S}^*) = \sum_{\mathbf{f}_i, \mathbf{f}_j \in \mathbf{S}^*, |\mathbf{S}^*|=k^*} \|\varepsilon_i - \varepsilon_j\|^2. \tag{4}$$

It's shown from Eq. (4) that DivF exploits a direct way for comparing the difference of the discriminative information vectors, i.e., the ℓ_2 norm distance. A farther distance represents features' significant difference in classification abilities. In other words, these features are excellent in recognizing different classes, and their discriminative abilities are exactly very complementary. A farther distance also represents less redundant classification information shared by features. So DivF is actually a multi-feature redundancy reduction method. Moreover, its computation is easily accomplished and will not be troubled by any probability estimation problem. Thus, the dilemma of reducing multi-feature redundancy for the common feature selection methods can be solved by DivF.

Algorithm 1. DivF algorithm

---

input: F,C,k,k^*
output: $\mathbf{S}^*$
 1: **begin**
 2: **for** $j = 1$ to $|\mathbf{F}|$ **do**
 3:     calculate ε_j via Eqs. (1) and (3);
 4: **end for**;
 5: **for** $l = 1$ to $|\mathbf{C}|$ **do**
 6:     calculate n_l via Eq. (5);
 7:     select top-n_l features $\mathbf{f}_1', \mathbf{f}_2', ..., \mathbf{f}_{nl}'$ according to ε_{jl}';
 8:     $\mathbf{S} = \mathbf{S} \bigcup \{\mathbf{f}_1', \mathbf{f}_2', ..., \mathbf{f}_{nl}'\}$;
 9: **end for**;
10: select k^* features $\mathbf{f}_1^*, \mathbf{f}_2^*, ..., \mathbf{f}_{k^*}^*$ from $\mathbf{S}$ using a special searching scheme by maximizing Eq. (4); $\cdots\cdots\cdots\cdots\cdots\cdots\cdots\cdots\cdots\cdots\cdots\cdots\cdots\cdots\cdots\cdots\cdots\cdots\cdots\cdots$ ($\star$)
11: $\mathbf{S}^* = \{\mathbf{f}_1^*, \mathbf{f}_2^*, ..., \mathbf{f}_{k^*}^*\}$;
12: **end;**

The procedure of DivF is summarized in Algorithm 1. Note that a candidate feature subset $\mathbf{S}$ is bulit in Lines 3–7. This is mainly for promising the partially predominant features selected. $\mathbf{S}$ is constituted of the features excellent in recognizing different classes. These features are top-ranked due to their higher discriminative information to a certain class. The number of the features selected according to class c_l depends on c_l's prior probability, defined as:

$$n_l = [k \times p(c_l)], \tag{5}$$

where $[\cdot]$ is the rounding operation, and k is the cardinality of $\mathbf{S}$.

The special searching scheme signed by $(\star)$ in Algorithm 1 exploits two methods for testing in the following experiment: (1) Sequential Forward Selection (SFS): a greedy search strategy. It starts from an empty feature subset and includes one feature at a time. First, the feature with maximal $\|\varepsilon_j\|_1$ is selected. Then, $\mathbf{S}$ is expanded with features maximizing Eq. (4); and (2) Genetic Algorithm Selection (GAS): a random search strategy. It randomly searches features maximizing Eq. (4). The candidate solutions are represented by 0/1 strings. Then crossover and mutation operations on populations are executed according to the crossover rate and mutation rate.

4   Experiment and Analysis

4.1   Experimental Setting

Real-world data sets are tested in this section for comparing the performances of DivF based methods and other feature selection methods. Both low-dimensional and high-dimensional data in Table 1 are tested. They are UCI data[1], microarray expression data and face recognition data [13,21]. The number of classes ranges from 2 to 50. So both binary-class and multi-class cases are involved.

Five popular feature selection methods are tested with DivF-SFS and DivF-GAS, i.e., Minimal Redundancy Maximal Relevance (mRMR) [6], Fisher Score [9], Correlation-Based Feature Selection (CBFS) [17,22], ReliefF [23] and Spectral Feature Selection (SPEC) [12]. mRMR is an efficient mutual information based selection method. It takes redundancy reduction into consideration, and is proved more efficient than the max-relevance method. Both within-class and between-class scatters are measured in Fisher Score. It prefers the feature subset in which the within-class relationship is tight and between-class relationship is loose. CBFS employs a correlation measure and balances the feature-class correlation and the feature-feature correlation in its evaluation. So it is also capable of reducing redundancy. ReliefF is a well-known feature weighting method extending Relief. It randomly samples a certain number of instances and weights features based on the differences between each sampled instance and its nearest instances in the same and different classes. SPEC preserves the original spectral information for obtaining a high recognitive performance. It suits for both the supervised and the unsupervised selection tasks.

In the experiments, CBFS employs Pearson Correlation Coefficient as the correlation measure. For ReliefF, the number of the nearest neighbor and sampled instances are set as 10 and n (all instances sampled). Similarity matrix in

---

[1] http://archive.ics.uci.edu/ml.

Table 1. Benchmark data sets

Data sets	♯ Features	♯ Instances	♯ Classes	Source
Qsar (Qsa)	41	1055	2	UCI
Libras (Lib)	90	360	15	UCI
UrbanLand (Urb)	147	675	9	UCI
Semeion (Sem)	256	1593	10	UCI
UJIIndoorLoc (Uji)	528	21048	3	UCI
Isolet (Iso)	617	7797	26	UCI
Cnae (Cna)	856	1080	9	UCI
Srbct (Srb)	2308	88	5	Micro
Pie10P (Pie)	2420	210	10	Face
Dlbcl (Dlb)	4026	88	6	Micro
DBWorld (Dbw)	4702	64	2	UCI
Breast (Bre)	9216	84	5	Micro
Amazon (Ama)	10000	1500	50	UCI
Pix10P (Pix)	10000	100	10	Face
Orl10P (Orl)	10304	100	10	Face
Cancers (Can)	12533	174	11	Micro

SPEC is calculated as $K_{ij} = \begin{cases} a_{ij}, & class(\mathbf{x}_i) = class(\mathbf{x}_j) \\ 0, & otherwise \end{cases}$, where a_{ij} is the similarity between instance $\mathbf{x}_i$ and $\mathbf{x}_j$ and determined by the RBF kernel function: $a_{ij} = \exp\left(-\frac{\|\mathbf{x}_i - \mathbf{x}_j\|^2}{2\delta^2}\right)$, $\delta^2 = mean(\|\mathbf{x}_i - \mathbf{x}_j\|^2)$. Besides, $\hat{\varphi}_1(\cdot)$ ranking function is adopted according to the literature [12]. The candidate feature number k in DivF based method is 100. For DivF-GAS, the initial population, crossover rate, mutation rate and generations are respectively set to 100, 0.8, 0.01 and 100.

4.2   Classification Performance

Each method sequentially selects optimal $1, 2, ..., 50$ features, i.e., k^* increases from 1 to 50. Note that the number of the features for Qsar is less than 50, so k^* is up to m. Classification performance of the selected features is tested in the weka environment [24] by three classifiers, i.e., Naive Bayes (NB) classifier, Support Vector Machine (SVM) classifier and 1-Nearest Neighbor (1NN) classifier. Ten-fold cross-validation is repeated for 10 runs, and the maximal classification precisions are recorded in Table 2. They represent the method's performance in the best case. Also pairwise t-tests at 5 % significance level are conducted, and the best results and those not significantly worse than them are highlighted in bold. The average results on all the data sets are counted in the last row.

Generally speaking, the performances of DivF-SFS and DivF-GAS are comparable or superior to the other feature selection methods. The computation

Table 2. Classification precisions (mean±std., in percentage) of different classifiers built by the selected features (pairwise t-test at 5 % significance level)($\uparrow$).

(a) NB classification precisions

Data	Methods						
	mRMR	Fisher Score	CBFS	ReliefF	SPEC	DivF-SFS	DivF-GAS
Qsa	76.90±0.50	76.75±0.23	76.96±0.60	76.39±0.46	76.75±0.23	77.09±0.40	**77.32±0.32**
Lib	61.30±1.29	53.91±1.60	52.86±1.29	56.77±1.16	56.29±1.54	63.21±0.90	**64.61±0.90**
Urb	68.11±0.43	72.41±0.59	79.01±0.42	72.24±0.94	72.41±0.71	**83.47±0.51**	**83.40±0.53**
Sem	74.86±0.62	71.59±0.83	66.01±0.48	75.89±0.49	72.71±0.57	78.93±0.51	**79.74±0.49**
Uji	98.72±0.02	**99.03±0.03**	**99.03±0.03**	**99.03±0.03**	**99.03±0.03**	**99.03±0.03**	99.03±0.03
Iso	67.11±0.11	56.36±0.13	66.72±0.16	58.96±0.19	55.76±0.20	**78.88±0.18**	76.53±0.16
Cna	86.61±0.26	85.84±0.20	80.14±0.28	85.99±0.17	81.44±0.36	86.37±0.26	**86.68±0.25**
Srb	96.65±0.54	96.17±0.77	95.69±0.73	97.30±0.58	96.75±0.65	**97.53±0.57**	97.38±0.61
Pie	79.80±0.78	92.18±0.69	69.39±1.29	87.82±0.90	**92.91±0.93**	86.63±0.97	91.53±0.86
Dlb	98.66±1.13	**98.84±1.17**	94.04±1.08	92.34±1.58	93.09±1.82	**98.84±0.33**	98.86±0.28
Dbw	94.29±0.81	93.66±0.77	86.19±1.92	93.74±0.76	92.23±1.01	95.18±0.70	**96.88±0.70**
Bre	**100.00±0.00**	67.62±1.82	69.31±1.57	88.11±0.97	79.53±1.42	98.64±0.43	**100.00±0.00**
Ama	44.49±0.63	38.05±0.26	38.34±0.49	37.98±0.42	40.83±0.32	47.43±0.83	**47.79±0.63**
Pix	**100.00±0.00**	96.87±0.50	95.78±1.04	96.05±0.60	97.25±0.53	99.78±0.17	**100.00±0.00**
Orl	82.02±1.14	91.03±1.34	85.90±0.99	73.62±2.01	90.82±1.31	88.64±0.63	**95.28±0.44**
Can	77.58±1.18	83.20±1.90	77.83±1.93	84.55±0.99	77.37±1.96	85.77±1.46	**86.52±1.06**
AVG.	*81.69*	*79.59*	*77.08*	*79.80*	*79.70*	*85.34*	***86.35***

(b) SVM classification precisions

Data	Methods						
	mRMR	Fisher Score	CBFS	ReliefF	SPEC	DivF-SFS	DivF-GAS
Qsa	85.71±0.23	85.76±0.22	**85.91±0.25**	85.82±0.23	85.85±0.26	85.75±0.23	85.76±0.25
Lib	**71.55±0.80**	68.71±0.94	63.52±0.58	67.85±0.80	66.23±0.92	70.20±0.86	70.58±0.60
Urb	74.78±0.56	77.23±1.72	81.91±0.42	78.19±1.88	78.90±1.03	84.98±0.31	**85.36±0.51**
Sem	79.63±0.59	75.77±0.46	73.39±0.39	79.53±0.48	77.19±0.31	83.20±0.52	**83.32±0.48**
Uji	99.50±0.02	99.90±0.02	99.46±0.01	99.87±0.02	98.86±0.01	99.92±0.01	**99.93±0.00**
Iso	83.25±0.10	77.07±0.15	82.78±0.12	78.19±0.16	77.16±0.11	88.16±0.17	**89.56±0.14**
Cna	85.72±0.16	85.17±0.41	80.88±0.35	85.74±0.33	82.25±0.50	85.91±0.30	**86.65±0.29**
Srb	96.84±0.63	96.80±0.65	95.04±0.74	96.63±0.68	96.75±0.65	97.75±0.65	**98.61±0.53**
Pie	94.98±0.72	96.91±0.92	92.91±1.02	96.46±0.33	96.73±0.52	96.52±0.51	**97.66±0.69**
Dlb	98.66±1.13	97.65±1.19	97.19±1.24	96.08±1.23	93.54±1.28	**100.00±0.00**	100.00±0.00
Dbw	91.79±1.37	90.21±1.39	86.19±1.92	90.49±1.23	92.36±0.99	88.81±1.42	**93.26±1.04**
Bre	98.04±0.49	77.15±1.33	61.62±2.01	90.18±1.22	86.53±2.30	96.54±0.58	**98.27±0.21**
Ama	53.35±0.36	33.81±0.29	38.34±0.29	41.84±0.28	37.02±0.25	**56.69±0.42**	55.18±0.59
Pix	96.36±0.90	95.14±1.01	97.01±0.55	93.77±1.21	94.27±0.85	97.84±0.61	**100.00±0.00**
Orl	95.60±0.64	97.09±0.51	92.10±1.03	77.75±2.00	94.96±0.65	96.37±0.70	**100.00±0.00**
Can	81.70±0.68	64.27±1.09	78.30±1.23	80.30±1.26	63.74±1.02	83.65±1.23	**87.12±1.32**
AVG.	*86.72*	*82.42*	*81.66*	*83.67*	*82.65*	*88.27*	***89.45***

(c) 1NN classification precisions

Data	Methods						
	mRMR	Fisher Score	CBFS	ReliefF	SPEC	DivF-SFS	DivF-GAS
Qsa	83.93±0.41	84.02±0.41	84.03±0.47	84.19±0.39	84.31±0.51	84.48±0.44	**84.51±0.35**
Lib	83.31±0.74	80.10±0.56	81.99±0.82	83.79±0.96	76.99±0.88	87.38±0.61	**87.90±0.65**
Urb	68.98±0.81	79.53±0.37	75.28±0.40	78.73±0.48	80.04±0.46	80.12±0.45	**81.71±0.56**
Sem	78.64±0.41	73.92±0.46	69.56±0.38	79.84±0.74	76.53±0.43	**82.74±0.32**	82.55±0.48
Uji	99.60±0.02	99.98±0.00	99.90±0.01	**99.99±0.00**	99.80±0.02	99.92±0.01	99.98±0.00
Iso	78.78±0.13	73.97±0.29	70.04±0.09	72.50±0.27	73.65±0.31	81.72±0.11	**83.57±0.15**
Cna	83.76±0.47	85.26±0.35	81.39±0.58	**85.54±0.39**	82.05±0.59	83.40±0.26	82.93±0.28
Srb	98.23±0.58	98.16±0.39	85.48±0.53	97.54±0.52	98.03±0.54	98.17±0.63	**98.31±0.57**
Pie	95.61±0.54	97.82±0.52	96.45±0.45	97.10±0.46	97.46±0.50	96.53±0.38	**98.30±0.49**
Dlb	98.66±1.13	97.52±1.05	97.82±1.84	97.48±1.15	94.14±1.24	**100.00±0.00**	100.00±0.00
Dbw	**92.64±1.29**	90.75±1.19	87.36±1.87	90.59±1.22	89.53±1.41	88.81±1.42	89.82±1.14
Bre	98.13±0.43	71.92±1.56	62.45±2.44	87.42±0.93	85.25±1.71	96.23±0.55	**98.81±0.24**
Ama	35.72±0.58	35.51±0.32	39.73±0.39	35.46±0.28	38.05±0.57	**42.54±0.76**	40.60±0.51
Pix	**99.11±0.30**	97.89±0.46	96.11±0.77	99.07±0.27	97.89±0.46	97.43±0.66	99.15±0.24
Orl	95.15±0.78	96.01±0.68	89.38±1.31	86.46±2.17	96.01±0.68	97.23±0.68	**100.00±0.00**
Can	76.80±0.82	79.46±1.00	64.55±1.04	78.16±1.45	67.43±1.17	81.70±1.13	**83.42±1.08**
AVG.	*85.44*	*83.86*	*80.10*	*84.62*	*83.57*	*87.40*	*88.22*

cost of DivF-SFS is much lower than DivF-GAS due to adopting greedy search strategy. It just selects k^* best features from the k candidates, and meanwhile the both parameters are assigned with relatively smaller values. This means that the selections of the candidate features and best features are both limited within small ranges. The effects of k and k^* will be further discussed in Sect. 4.3. Unlike DivF-SFS, the optimal solution of DivF-GAS in each iteration is unrelated with the former search result. This is beneficial for avoiding local optimal solutions. A large initial populations, high crossover and mutation rate and plenty of generations conduce to DivF-GAS's superior results. Correspondingly, DivF-GAS is more time-consuming than DivF-SFS.

In Table 3, other metrics except classification precision are conducted for testing the features' performance. The number of the selected features k^* increases from 5 to 50 in interval of 5. Average balanced error rate (BER) and area under ROC curve (AUC) of all the test data on the three classifiers are counted. BER concerns the unbalanced distributions of classes and counts the average classification error of each class. AUC uses a trapezoidal method to estimate the area under the ROC curve, which is widely used for judging the classifier's discriminative ability. Both quantitative metrics evaluate features' general predictive power from different aspects. From the table it indicates that the DivF based methods are superior to the other methods. Excluding DivF-SFS and DivF-GAS, mRMR also performs excellent when comparing to the other methods.

Table 3. Average BER ($\downarrow$) and AUC ($\uparrow$) measures (in percentage) for the classification performance of the selected features at each searching step.

k^*	mRMR		Fisher Score		CBFS		ReliefF		SPEC		DivF-SFS		DivF-GAS	
	BER	AUC	BER	AUC	BER	AUC	BER	AUC	BER	AUC	BER	AUC	BER	AUC
5	**34.49**	**86.44**	45.50	79.73	50.18	76.60	43.29	81.56	49.91	76.33	40.07	84.49	41.01	85.19
10	28.01	89.15	38.61	82.97	43.30	81.24	35.37	86.38	41.59	80.67	28.35	89.07	**27.69**	**89.44**
15	24.90	90.65	34.79	85.13	39.28	82.81	31.08	87.70	38.11	82.80	24.03	90.54	**22.43**	**91.05**
20	22.48	91.67	31.99	86.37	37.00	84.24	27.63	88.99	30.79	86.90	22.46	**92.91**	22.40	92.89
25	21.83	91.74	29.53	87.27	35.90	84.30	25.21	89.89	28.77	87.71	21.40	92.33	**20.34**	**92.68**
30	20.46	**92.16**	26.88	88.61	34.96	84.70	24.16	90.43	26.01	89.45	19.83	91.79	**18.97**	92.16
35	19.69	92.23	24.79	89.99	33.91	85.13	22.69	91.08	24.70	92.46	18.37	92.27	**17.94**	**92.52**
40	19.10	92.50	23.46	90.35	32.70	85.48	21.77	91.55	23.82	90.29	17.72	92.48	**16.85**	**92.71**
45	17.30	94.89	21.39	92.70	32.82	86.86	19.76	93.83	22.17	92.35	15.71	**95.82**	**15.33**	95.36
50	17.16	94.76	19.84	93.44	32.94	86.59	19.20	94.05	21.07	93.02	15.10	94.16	**14.78**	**95.09**

4.3   Effects of Parameters k and k^*

The parameter k ascertains DivF's search range. A larger value leads to more candidate features included. k^* ascertains the size of the feature subset. More best features selected may improve the classification performance. Correspondingly, computation cost will increase evidently. Figure 2 illustrates the effects of k

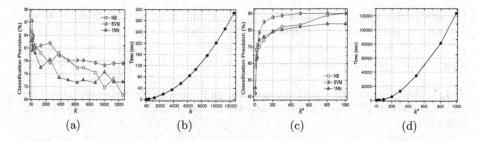

(a)   (b)   (c)   (d)

Fig. 2. Variation of classification precision and computation time with different parameters k and k^*: (a) and (b) reflect the variations as k increases, and (c) and (d) show the situations when k^* increases.

and k^* on both classification precision and computation cost. DivF-SFS is tested on Cancers data in Table 1. The classification precisions of NB, SVM and 1NN classifiers are shown in the figure. Computation cost merely records the feature selection time, excluding the classification cost. All the tests are implemented in Matlab on an Intel Core i7-3770 CPU (@3.4 GHz) with 32 GB memory.

In Fig. 2(a) and (b), k is set sequentially to 50, 100, 200, 500, $\left[\frac{1}{10}m\right]$, $\left[\frac{2}{10}m\right]$,..., and m, while k^* constantly equals to 50. In Fig. 2(c) and (d), k^* varies as 10, 20, 30, 50, 100, 200, 300, 500, 800 and 1000, while k is fixed to m.

It is clear that the classification precisions on the three classifiers present declining trends as k increases. It indicates that more candidate features may probably not help to enhance feature subset's performance. DivF is a redundancy reduction based method. It prefers the features maximally irrelevant with each other. So a feature with low relevance to class is perhaps selected if it is irrelevant with the other selected features.

In contrast, classification performance is enhanced as more best features are selected. For filter models, largening k^* may increase the probability of finding the optimal subset, although more time consumption is needed. Certainly, monotonic increase of k^* is unnecessary in some cases. It means that the top classification precisions of some data sets will be achieved when k^* adopts a relatively small value. Take SVM and 1NN classifiers in the figure for another example. When k^* is sequentially set as 800 and 1000, there are almost no variations for the precisions of the both classifiers, and relatively stable ranges are achieved. In this situation, continually increasing k^* makes little sense to enhance subset's performance. When $k = k^* = m$, an exhaustive search is performed.

5   Conclusion

A new feature selection method DivF is proposed in this paper. It focuses on evaluating feature's classification ability specific to each target class. This information is constructed as a discriminative information vector. Then, both feature complementarity and feature redundancy are measured by comparing the differences of the new vectors. The process of selecting features is also accomplished in the new vector space. Two dilemmas for the common feature selection methods, finding partially predominant features and measuring multi-feature classification redundancy, are both solved by the new method. Experimental results on low-dimensional and high-dimensional data validate the new method's effectiveness.

Acknowledgments. This work was partially supported by the National Natural Science Foundation of China (61070089) and the Science Foundation of Tianjin (14JCY-BJC15700).

References

1. Guyon, I., Elisseeff, A.: An introduction to variable and feature selection. J. Mach. Learn. Res. **3**, 1157–1182 (2003)
2. Tang, J., Alelyani, S., Liu, H.: Feature selection for classification: a review. In: Aggarwal, C. (ed.) Data Classification: Algorithms and Applications. CRC Press, Chapman (2014)
3. Blum, A.L., Langley, P.: Selection of relevant features and examples in machine learning. Artif. Intell. **97**(1), 245–271 (1997)
4. Kohavi, R., John, G.H.: Wrappers for feature subset selection. Artif. Intell. **97**(1), 273–324 (1997)
5. Inza, I., Larrañaga, P., Blanco, R., Cerrolaza, A.J.: Filter versus wrapper gene selection approaches in DNA microarray domains. Artif. Intell. Med. **31**(2), 91–103 (2004)
6. Peng, H., Long, F., Ding, C.: Feature selection based on mutual information criteria of max-dependency, max-relevance, and min-redundancy. IEEE Trans. Pattern Anal. Mach. Intell. **27**(8), 1226–1238 (2005)
7. Brown, G., Pocock, A., Zhao, M., Luján, M.: Conditional likelihood maximisation: a unifying framework for information theoretic feature selection. J. Mach. Learn. Res. **12**, 27–66 (2012)
8. Hua, J., Tembe, W.D., Dougherty, E.R.: Performance of feature-selection methods in the classification of high-dimension data. Pattern Recognit. **42**(3), 409–424 (2009)
9. Gu, Q., Li, Z., Han, J.: Generalized Fisher Score for feature selection. In: Proceedings of the 27th UAI, pp. 266–273 (2011)
10. He, X., Cai, D., Niyogi, P.: Laplacian score for feature selection. In: Proceedings of NIPS 18, pp. 507–514 (2005)
11. Zhang, Y., Zhou, Z.H.: Multi-label dimensionality reduction via dependence maximization. ACM Trans. Knowl. Discov. Data **4**(3), 1503–1505 (2010)
12. Zhao, Z., Liu, H.: Spectral feature selection for supervised and unsupervised learning. In: Proceedings of the 24th ICML, pp. 1151–1157 (2007)
13. Zhao, Z., Wang, L., Liu, H., Ye, J.: On similarity preserving feature selection. IEEE Trans. Knowl. Data Eng. **25**(3), 619–632 (2013)
14. Quinlan, J.R.: C4.5: Programs for Machine Learning. Morgan Kaufmann, San Mateo (1993)
15. Yu, L., Liu, H.: Efficient feature selection via analysis of relevance and redundancy. J. Mach. Learn. Res. **5**, 1205–1224 (2004)
16. Mitra, P., Murthy, C.A., Pal, S.K.: Unsupervised feature selection using feature similarity. IEEE Trans. Pattern Anal. Mach. Intell. **24**(4), 301–312 (2002)

17. Hall, M.A.: Correlation-based feature subset selection for machine learning. Ph.D. thesis, Dept. Computer Science, Waikato Univ., Hamilton, New Zealand (1999)
18. Fleuret, F.: Fast binary feature selection with conditional mutual information. J. Mach. Learn. Res. **5**, 1531–1555 (2004)
19. Wasikowski, M., Chen, X.W.: Combating the small sample class imbalance problem using feature selection. IEEE Trans. Knowl. Data Eng. **22**(10), 1388–1400 (2010)
20. Kim, H., Drake, B.L., Park, H.: Adaptive nonlinear discriminant analysis by regularized minimum squared errors. IEEE Trans. Knowl. Data Eng. **18**(5), 603–612 (2006)
21. Wei, J.M., Wang, S.Q., Yuan, X.J.: Ensemble rough hypercuboid approach for classifying cancers. IEEE Trans. Knowl. Data Eng. **22**(3), 381–391 (2010)
22. Yu, L., Liu, H.: Feature selection for high-dimensional data: a fast correlation-based filter solution. In: Proceedings of the 20th ICML, pp. 856–863 (2003)
23. Robnik-Šikonja, M., Kononenko, I.: Theoretical and empirical analysis of ReliefF and RReliefF. Mach. Learn. **53**(1–2), 23–69 (2003)
24. Hall, M., Frank, E., Holmes, G., Pfahringer, B., Reutemann, P., Witten, I.H.: The weka data mining software: an update. SIGKDD Explor. **11**(1), 10–18 (2009)

A Label Inference Method Based on Maximal Entropy Random Walk over Graphs

Jing Pan[1,2], Yajun Yang[1,2(✉)], Qinghua Hu[1,2], and Hong Shi[1,2]

[1] School of Computer Science and Technology, Tianjin University, Tianjin, China
{panjing,yjyang,huqinghua,serena}@tju.edu.cn
[2] Tianjin Key Laboratory of Cognitive Computing and Application, Tianjin, China

Abstract. With the rapid development of Internet, graphs have been widely used to model the complex relationships among various entities in real world. However, the labels on the graphs are always incomplete. The accurate label inference is required for many real applications such as personalized service and product recommendation. In this paper, we propose a novel label inference method based on maximal entropy random walk. The main idea is that a small number of vertices in graphs propagate their labels to other unlabeled vertices in a way of random walk with the maximal entropy guidance. We give the algorithm and analyze the time and space complexities. We confirm the effectiveness of our algorithm through conducting experiments on real datasets.

Keywords: Label inference · Random walk · Maximal entropy

1 Introduction

With the rapid development of Internet, graphs have been widely used to model the complex relationships among various entities in real applications including social network, web graph and biology. These graphs are always with labels. For an example of social network, every people is represented by a vertex in the graph. The vertices may be with the labels to indicate the hobbies, occupations or other attributes for their representing people. As another example of protein interaction network, the labels on the vertices indicate the types of the protein. These graphs are called as labeled graphs. However, due to the limitation of information collection and the existence of noise, the labels on graphs are always incomplete, i.e., only a fraction of the vertices are attached with the labels on the graph and the remaining ones are unknown. For the example, in Sina Weibo, which is the largest micro-blog platform in China, only $20\% - 30\%$ users exhibit the labels about their occupations. Therefore, inferring the labels for the vertices in graphs is a critical problem in many applications.

Consider a concrete application of social network, e.g., Sina Weibo, an internet enterprise tries to provide more personalized service for the users according to the users' information such as their hobbies, occupations and so on. However, only a small amount of information or labels are explicitly provided by rarely users themselves due to the privacy or other concerns, then an effective label

© Springer International Publishing Switzerland 2016
F. Li et al. (Eds.): APWeb 2016, Part I, LNCS 9931, pp. 506–518, 2016.
DOI: 10.1007/978-3-319-45814-4_41

inference mechanism is required to infer the missing labels for the remaining users. By the accurate label inference, more personalized services such as product and content recommendation may be provided for the targeting customers. It improves the user experience effectively.

Over the last several years, there have been an increasing interest in label inference from the graphs. The first category of these works focuses on building a classifier to distinguish the vertices with the distinct labels in a network [8,12,14]. For these methods, the sufficient background knowledge is necessary for extracting features from the users [6,18]. Specially, an actual and reliable sub-network is required as a training dataset including all the kinds of labels. However, such sub-network is difficult to find out, even it may be not exist. The other category is a community detection method to identify the group affiliation for the individuals in the network [17]. The individuals in the same group are deem as with the uniform labels. Homophily [13] is a common assumption in these methods to infer user labels, i.e., the similar users tend to interact with each other directly. Hence the users with the similar labels are aggregated naturally to a community by one-hop interactions.

In this paper, we study the problem of identifying the labels for all the vertices in the labeled graphs. The main idea of our label inference mechanism is that a small number of labeled vertices propagate their labels to the entire network in a way of random walk under the guidance by the maximal entropy model. Different to the previous methods, our method does not require too much background knowledge or training sub-networks. On the other hand, homophily assumption is not necessary, that is, it allows the users with the same label appear in the network dispersedly but not in some common community.

The main contributions of this paper are summarized as below. First, we propose a novel label inference model based on the maximal entropy random walk. Second, we give the algorithm to infer labels with our model and analyze the time and space complexities of our algorithm. Our algorithm can handle both undirected and directed graphs. Finally, we confirm the effectiveness of our algorithm through conducting experiments on real datasets by compared with SVM, Homophily, and traditional random walk algorithms.

The rest of the paper is organized as follow. Section 2 defines the problem. Section 3 proposes the random walk model with the maximal entropy guidance. The label inference algorithm is given in Sect. 4. The experimental results on real networks are presented in Sect. 5. The related works are introduced in Sect. 6. Finally, we conclude this paper in Sect. 7.

2 Problem Statement

A labeled graph is a simple directed graph, denoted as $G = (V, E, \Sigma, L)$, where V is the set of vertices and E is the set of edges in G. The total number of nodes is n. Each edge $e \in E$ is represented by (u, v), $u, v \in V$. e is called u's outgoing edge or v's incoming edge and v (or u) is called u (or v)'s outgoing(or incoming) neighbor. Σ is the set of all the labels in G. L is a label function mapping V to Σ. Each $L(u)$ is a label in Σ, i.e. $L(u) \in \Sigma$. For example, in Sina Weibo, a user u

Table 1. Important notations

Notation	Description
$G(V, E)$	structure of the graph
Σ, V_s, L	label set, labeled vertex set, label function mapping V to Σ
$T(u)$	label-weight set on vertex u
$w_x(u)$	the probability that vertex u has the label x
$p_{i,j}^{(k)}$	transition probability at the k-th iteration of maximal entropy random walk
$N^+(v_i)$, $N^-(v_i)$	outgoing neighbor set of v_i, incoming neighbor set of v_i
$H(v_j)$	entropy value of v_j
$w_x(v_i, v_j)$	probability that v_i have the label l_x propagated by its incoming neighbor v_i

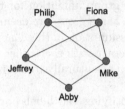

Fig. 1. An example graph from a company

Table 2. Labels on vertices in Fig. 1

ID	Name	Label
1	Fiona	Manager
2	Philip	Programmer
3	Jeffrey	Human Resource
4	Mike	Programmer
5	Abby	Manager

may have an occupation label as "Manager", then L(u) = "Manager". Note that all the labels in Σ are with the same type, e.g., "Manager" and "Programmer" are two labels about the same label type "occupation". In real applications, a user may has several labels with different types. In the following, we first focus on the case that every vertex has a label with the same type and we will discuss how to deal with the general case that every vertex may have the labels with several types in Sect. 4.3. Our work can be easily extended to handle undirected graphs, in which an undirected edge (u, v) is equivalent to two directed edges (u, v) and (v, u) (Table 1).

Figure 1 illustrates an example graph in a social network. There are five users in this graph that are staffs in software company and the labels about their occupations are shown in Table 2. In this example, the occupation of Fiona is "Manager", so she is labeled in "Manager".

Problem Statement: Given a labeled graph $G = (V, E, \Sigma, L)$ and the labeled vertex set V_s where $V_s \subset V$, i.e., only the labels on the vertices in V_s is known, and our work is to infer $L(u)$ for every vertex $u \in V - V_s$.

3 Inference Model Based on Maximal Entropy Random Walk

3.1 Framework

In this section, we introduce the framework of our label inference model as shown in Fig. 2. The input is $G(V, E, \Sigma, L)$ and V_s, which describe the structure of the

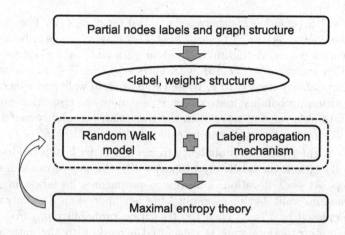

Fig. 2. The framework of label inference model

entire graph $G(V, E)$ and the labeled vertex set respectively. The main idea of our method is propagating the labels from the vertices in V_s to the unlabeled vertices in the way of random walk. It is worth noting that an unlabeled vertex $v \in V - V_s$ may become a vertex with several labels in the random walk process. Σ is the set of all the possible labels in graph G, $\Sigma = \{l_1, \cdots, l_p\}$, then every vertex in G only has a label from Σ. We use a tuple set $T(u) = \{< l_x, w_x(u) > | 1 \leq x \leq p\}$ to denote the labels l_x $(1 \leq x \leq p)$ on vertex u with the corresponding weight $w_x(u)$. $T(u)$ is called label-weight set of u. Here, $w_x(u)$ indicates the probability that vertex u has the label l_x and $\sum_{x=1}^{p} w_x(u) = 1$. For every vertex $u \in V_s$, if it has a label l_{x_1}, then $w_{x_1}(u) = 1$ and $w_x(u) = 0$ for $x \neq x_1$ and $1 \leq x \leq p$. For every vertex $u \in V - V_s$, $w_x(u)$ is initialized as $\frac{1}{p}$ for $1 \leq x \leq p$. It means u may have p labels with the same probability at the beginning. The value of $w_x(u)$ $(1 \leq x \leq p)$ for every vertex is updated iteratively after every one step random walk and $w_x(u)$ becomes larger and larger if u has the label l_x in real world. We discuss how to update $w_x(u)$ in Sect. 3.3. Moreover, the random walk is under the guidance of maximal entropy. After every one step random walk, our algorithm computes the entropy value for every vertex $v \in V - V_s$. For every vertex u in G, the transition probability from u to all its outgoing neighbors in the next one step random walk is calculated according to the entropy values of all the outgoing neighbors of u. Finally, when the convergence is satisfied, the random walk terminates, and the label set $L(u)$ for every vertex $u \in V_s$ can be inferred by its $T(u)$.

3.2   One by One Step Random Walk

A random walk is a mathematical formalization of a path that consists of a succession of random steps. Give a directed graph $G(V, E)$, where the number of the vertex is $n = |V|$. Graph G can be represented as a $n \times n$ symmetric adjacency matrix M, where $m_{i,j}$ is an entry of M at the i-th row and the j-th column.

Then $m_{i,j} = 1$ if $(v_i, v_j) \in E$ otherwise $m_{i,j} = 0$ for $v_i, v_j \in V$. The degree of a vertex v_i is denoted as $d_i = \sum_{j=1}^{n} m_{i,j}$. Let $D = \text{diag}(d_1, \cdots, d_n)$ be a diagonal matrix of vertex degree. A traditional random work on G can be defined utilizing the transition matrix $P = D^{-1}M$ with entries $p_{i,j} = \frac{m_{i,j}}{d_i}$. $p_{i,j}$ is called the transition probability from v_i to v_j by one step random walk and P is called the initial transition probability matrix. Let $P^{(k)}$ denote the transition probability matrix of k-th step random walk, then the transition probability matrix $P^{(k+1)}$ of $(k+1)$-th step random walk can be calculated as $P^{(k+1)} = P^k \times P$. An entry $p_{i,j}^{(k)}$ in $P^{(k)}$ is the transition probability from v_i to v_j by k step random walk.

In our algorithm, the labels are propagated in the way of random walk one by one step. At each iteration, a vertex v_i propagates its labels in $T(v_i)$ by one step random walk. Let v_j is an outgoing neighbor of v_i. The entropy value of v_j is calculated by $T(v_j)$. Then the transition probability $p_{i,j}$ from v_i to v_j by next one step random walk is calculated according to the entropy values of all v_i's outgoing neighbors. Different to the traditional random walk, the transition probability $p_{i,j}$ for one step random walk is not an uniform value in our method, that is, $p_{i,j}$ for the k-th and $(k+1)$-th iteration are different. We use $p_{i,j}^{(k)}$ to denote the transition probability at the k-th iteration. With $p_{i,j}^{(k)}$, all v_j's incoming neighbors propagate their labels to v_j and then $T(v_j)$ is updated. The initial transition probability $p_{i,j}^{(1)}$ from v_i to v_j is given below.

$$p_{i,j}^{(1)} = \frac{I(v_i, v_j)}{\sum_{v_j \in N^+(v_i)} I(v_i, v_j)} \tag{1}$$

Where $N^+(v_i)$ is the outgoing neighbor set of v_i and $I(v_i, v_j)$ is parameter used to reflect the closeness between v_i and v_j. The value of $I(v_i, v_j)$ can be given by the real application requirement. For example, in DBLP dataset, $I(v_i, v_j)$ is the number of the papers in which researcher v_i and v_j are co-author. If there is no requirement about $I(v_i, v_j)$, then $I(v_i, v_j) = 1$ for every pair of vertex v_i and v_j in G.

3.3   Updating Label-Weight Set Under Maximal Entropy Guidance

In this section, we discuss how to compute the transition probability $p_{i,j}^{(k)}$ under maximal entropy guidance and how to update $T(v_j)$ for every vertex $v_j \in V - V_s$. We introduce the maximal entropy for computing the transition probability $p_{i,j}^{(k)}$ which is well known in information theory. Without loss of generality, at k-th iteration, every vertex v_j is with a label-weight set $T(v_j) = \{< l_x, w_x(v_j) > | 1 \leq x \leq p\}$. Then the entropy value $H(v_j)$ of v_j can be calculated as below.

$$H(v_j) = -\sum_{x=1}^{p} w_x(v_j) \times \ln w_x(v_j) \tag{2}$$

Note that $w_x(v_j) \times \ln w_x(v_j) = 0$ if $w_x(v_j) = 0$. $H(v_j)$ indicates the uncertainty about the labels on vertex v_j. The larger $H(v_j)$ results in the more uncertainty on v_j for label inference. Therefore, the more labels from the other vertices are expected to propagate their labels to v_j for improving the label inference on v_j. There is a special case to be handled with carefully, where v_j has no label, i.e., $w_x(v_i) = 0$ for $1 \leq x \leq p$. In this case, $H(v_j)$ is computed as $H(v_j) = -\sum_{x=1}^{p} \frac{1}{p} \times \ln \frac{1}{p}$. It means the uncertainty on v_i is the maximum. If v_j is an outgoing neighbor of v_i, then the transition probability $p_{i,j}^{(k)}$ from v_i to v_j at k-th iteration is computed as follows.

$$p_{i,j}^{(k)} = \frac{H(v_j)}{\sum_{v_j \in N^+(v_i)} H(v_j)} \tag{3}$$

where $N^+(v_i)$ is the outgoing neighbor set of v_i.

Next, we introduce how to update $T(v_j)$ utilizing the transition probability $p_{i,j}^{(k)}$ at k-th iteration. We use $w_x(v_i, v_j)$ to denote the probability that v_j have the label l_x propagated by its incoming neighbor v_i at this iteration, then $w_x(v_i, v_j)$ can be calculated by

$$w_x(v_i, v_j) = w_x(v_i) \times p_{i,j}^{(k)} \tag{4}$$

By $w_x(v_i, v_j)$, the $w_x(v_j)$ updates utilizing the following equation

$$w_x(v_j) = \frac{\sum_{v_i \in N^-(v_j)} w_x(v_i, v_j)}{\sum_{x=1}^{p} \sum_{v_i \in N^-(v_j)} w_x(v_i, v_j)} \tag{5}$$

where $N^-(v_j)$ is the incoming neighbor set of v_j. Thus for every vertex $v_j \in V_s$, the $T(v_j)$ is also updated after updating every $w_x(v_j)$ ($1 \leq x \leq p$).

4   Label Inference Algorithm

4.1   Label Inference Algorithm

The algorithm for label inference is shown in Algorithm 1. The input is a labeled graph G and vertex subset V_s, where the labels on every vertex $v \in V_s$ is given. The output is $T(v_j)$ for every vertex in $v_j \in V_r$, where $V_r = V - V_s$. Initially, Algorithm 1 computes the initial transition probability $p_{u,v}^{(1)}$ by Eq. (1) for every vertex $u \in V_s$ and every u's outgoing neighbor v (line 1 to 2). Line 3 to 13 shows the one by one step random walk under maximal entropy guidance. In each iteration, for every vertex $v_j \in V_r$, algorithm first computes its $H(v_j)$. Next, algorithm computes $p_{i,j}^{(k)}$ for all v_j's incoming neighbor v_i, then $w_x(v_i, v_j)$ and $w_x(v_j)$ can be computed. Finally, $T(v_j)$ can be updated with updating $w_x(v_j)$. The algorithm terminates when the convergence is satisfied. The condition of convergence is given by the following equation.

$$\sum_{v_j \in V_r} \sum_{1 \leq x \leq p} |w_x^k(v_j) - w_x^{k-1}(v_j)| \leq \theta \times p \times |V_r| \tag{6}$$

where $w_x^k(v_j)$ is essentially $w_x(v_j)$ at k-th iteration, and θ is a threshold given by user to control the number of iterations for convergence. As discussion in [10], the convergence can be satisfied for a maximal entropy random walk.

Algorithm 1. LABEL-INFERENCE $(G(V, E, \Sigma, L), V_s)$

Input: $G(V, E, \Sigma, L), V_s$.
Output: $T(v_j)$ for every vertex $v_j \in V_r$ (or $V - V_s$).

1: **for** every vertex $u \in V_s$ **do**
2:    computes $p_{u,v}^{(1)}$ for every $v \in N^+(u)$;
3: **while** $\sum_{v_j \in V_r} \sum_{1 \le x \le p} |w_x^k(v_j) - w_x^{k-1}(v_j)| \ge \theta \times p \times |V_r|$ **do**
4:    **for** every vertex $v_j \in V_r$ **do**
5:       computes $H(v_j)$ by Eq. (2);
6:    **for** $v_j \in V - V_s$ **do**
7:       **for** $v_i \in N^-(v_j)$ **do**
8:          computes $p_{i,j}^{(k)}$ by Eq. (3);
9:          **for** $x = 1$ to p **do**
10:             computes $w_x(v_i, v_j)$ by Eq. (4);
11:       **for** $x = 1$ to p **do**
12:          computes $w_x(v_j)$ by Eq. (5);
13:       updates $T(v_j)$ by $w_x(v_j)$;
14: **return** $T(v_j)$ for every vertex $v_j \in V_r$

4.2   Complexity Analysis

We analyze the time and space complexities of Algorithm 1. Let n be the number of the vertices in G. For the time complexity, at each iteration, algorithm needs to compute $H(v_j)$ for every $v_j \in V_r$, thus the time complexity of this step is $O(|V_r|)$. The complexity of computing $p_{i,j}^{(k)}$ is $O(nd)$, where d is the average out-degree of all the vertices in G. Because there are p labels in G, then complexity of computing $w_x(v_j)$ is $O(|V_r| + pnd) = O(pnd)$. Let the number of iterations be k, then the total complexity of Algorithm 1 is $O(kpnd)$. In the worst case, $d = n$ but d is always far less than n in real applications.

For the space complexity, at each iteration, Algorithm 1 needs to calculate and maintain $w_x(v_i, v_j)$ $(1 \le x \le p)$, where v_i is a vertex in G and v_j is an outgoing neighbor of v_i. Then the space complexity is $O(pnd)$.

4.3   Discussion About Multiply Types of Labels in Graph

In the above discussion, we assume that all the vertices in G have a label with the same type, e.g., all users have the label with the same type "occupation" in a social network. Next, we introduce how to deal with the general case where the vertex may have the labels with several types. For example, an user may have two different label types such as "hobby" and "occupation". Generally, if the graph G has q label types, then the graph is considered as q labeled graphs $G_1, \cdots, G_q$, where every $G_i (1 \le i \le q)$ has the same structure (V, E) as the original graph G but only have one label type. Our label inference method can be used on G_q to infer the labels for the q-th label type.

5   Experiments

In this section, we compare our method with other relevant approaches from different aspects to demonstrate the effectiveness of the model we proposed on label inference. In the following, we introduce the datasets, the baseline methods, and report our results. All the experiments are conducted on a 3.2 GHz Intel Core i5 CPU PC with the 16 GB main memory, running on Windows 7 and the programming language is Python.

5.1   Dataset and Experiment Setup

DBLP is a dataset that describes the researcher cooperation network. We treat the research domains as users' labels. We extract the journal information and conference information from DBLP network, and we label these journals and conferences into seven computer domain, including Artificial Intelligence, Data Mining, Computer Security, Programming Language, Computer Architecture, Theoretical Computer Science, and Human-computer Interaction. We extract 1,000, 3,000, 5,000, 7,000, and 10,000 vertices in the DBLP graph respectively to conduct the experiments. Meanwhile, We extract 4, 5, 6, and 7 research domains to be the labels for researchers by combining some domains. We evaluate the performance of experiments by precision and recall.

In order to demonstrate the effectiveness of the label inference method (LIM), we compare our method against a number of baseline approaches. Since our model considers both the local network structure of individual users and the effects from neighbor influence, we use the following approaches to show the performance of our method from different perspectives:

(1) SVM: We use the traditional SVM classifier as the first baseline.
(2) Homophily: From the paper [13], we know that the label of a user is associated with the user's neighbors. Therefore, we use this method as a baseline which can infer labels for users by majority votes on the user's neighbors. And this method is called Homophily.
(3) TRW: Our method is modified by the traditional random walk, so we apply the traditional random walk as another baseline. We refer to this approach as TRW.

5.2   Experimental Results

Exp1-Impact of the Proportion of Unlabeled Vertices. In Fig. 3, we study the performance of our proposed model LIM and the baselines mentioned in Sect. 5.1 on the different proportion of unlabeled vertices. We conduct this experiment in 5,000 vertices graph and we set the label number as 5 classes. We set the unlabeled scale 20 %, 40 %, 60 %, 80 %, and 90 % respectively. We present the result of precision and recall in Fig. 3(a) and (b) respectively. We can discovery that the decline tendency of our method is much slow than other methods along with the change of the scale of unlabeled vertices. And the precision of our

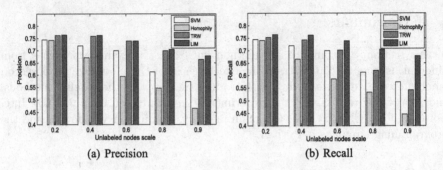

(a) Precision   (b) Recall

Fig. 3. Impact of the proportion of unlabeled vertices

method is almost 70 % at the condition of there are 90 % vertices lack of label. We can get the conclusion that the performance is influenced by the proportion of unlabeled vertices.

Exp2-Impact of Vertex Size. We study the performance of our method and the baselines in different vertex size and the results can be seen in Fig. 4. We conduct this experiment in 1,000, 3,000, 5,000, 7,000, and 10,000 vertices graph respectively. And we set the label number as 4 classes. The Fig. 4(a) and (b) show the results of precision and recall of all the methods. We can analyze the results to get that the precision and recall have no relation with the vertex size in most cases. As the number of vertices changing, the performance has no obvious change in most cases.

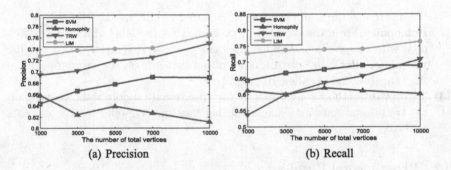

(a) Precision   (b) Recall

Fig. 4. Impact of vertex size

Exp3-Impact of Label Number. We study the impact of the number of vertices' labels in Fig. 5. We show the performance of our method and the baselines on 4, 5, 6, and 7 label classes in the graph which includes 5,000 vertices and the proportion of unlabeled vertices is 80 %, which we can have a intuitive understanding on precision and recall according to the Fig. 5(a) and (b) respectively. It is obvious that the larger the label number is, the lower the precision is because of the unrelated labels have interference for label inference when unlabeled vertices collect labels from labeled vertices. However, the descent rate of performance of our method is slower than other methods.

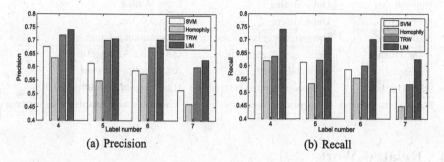

(a) Precision   (b) Recall

Fig. 5. Impact of label number

Exp4-Impact of Threshold θ. We study the impact of threshold θ which has a detailed introduction in Sect. 4.1. The precision and recall value are different in different θ. We show the performance of this experiment in Fig. 6. We conduct the experiment in 7,000 vertices graph and the unlabeled vertices scale is 80 %. The method we compare to is TRW only. We can see the results from Fig. 6. The threshold affects the iteration convergence times. However, the threshold doesn't have an important impact on the experimental performance.

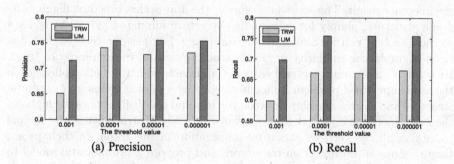

(a) Precision   (b) Recall

Fig. 6. Impact of threshold θ

Exp5-Real Case Study. In Table 3 we give a real case study on DBLP dataset. We can discover that most inference results are accurate on our method. However, The last row which is marked in bold type is wrong. By analyzing the papers of this researcher, both the true label and the inference label are his main domains and the number of published papers are almost same. In this real case study, we should make efforts on this special case to improvement our algorithm in future.

Table 3. Real case study

Name	True label	Inference label
Jiawei Han	Data Mining	Data Mining
Philip S. Yu	Data Mining	Data Mining
Jeffrey Yu	Data Mining	Data Mining
Nate Foster	Programming Language	Programming Language
Onur Mutlu	Computer Architecture	Computer Architecture
Yi Lin	Artificial Intelligence	Artificial Intelligence
Alex Pentland	Human-computer Interaction	Human-computer Interaction
Ronald L. Rivest	**Computer Security**	**Theoretical Computer Science**

6   Related Work

Label inference is an important problem in graphs and there have been an increasing interest in this problem over the last several years. The existing works for label inference problem can be divided into two categories.

The first category of the works [2,20,22] proposes different ideas which based on building feature vectors for vertices mainly from their textual informations and neighbor link informations. By training the feature vectors they build a classifier for inferring labels for vertices. For example, [15] uses the Bayesian classifier and iteration method to update attribute information. These works have a strong dependency on the background knowledge of users. However, the background knowledge isn't sufficient and always outdated which strongly influences the inference result. The second category is the approaches based on digging the graph structure, mainly for surrounding structure mining [4,19]. [5] provides a method on how can we automatically discover role labels for vertices. For vertices in graph, the similarity in graph structure decides the similarity of them. However, it is an unsupervised learning approach which is not applicable on the semi-supervised problem we focus on. [21] uses mining technology on graph structure and the relationship between vertices and their neighbors and combines the two aspects infer social roles for vertices. They abstract five social factors to mining all the vertices' structure information in the graph. In their paper, they propose an optimization framework and propose a probabilistic model to integrate the vertices' structure information and local neighbors' information to infer labels. In [14] studies user attribute inference in university social networks by applying community detection. However, these methods that depend on the surrounding structure maybe give a wrong judge if two users have same label but not in the adjacent structure.

Random Walk model is a good method to mining graph structure [1,9,11,16]. It is well studied in multiclass semi-supervised learning [7,23]. The main idea of these work is to estimate a distribution over the missing labels based on Markov random walk. Meanwhile, there are many works is proposed to improve the performance of random walk which combines with the maximum entropy theory [3,10]. However, the problem they focus on is link prediction or some else, which is different with the problem our paper focus on.

7   Conclusion

In this paper, we study how to infer labels for unlabeled nodes in graphs or social networks. We first define the label inference problem. Second, we propose the maximal entropy random walk inference model to solve this problem. We improve the result precision by the proposed model. Finally, we confirm the effectiveness of our algorithm through conducting experiments on real datasets. In the next, we will do more works about how to deal with multi-attributes graphs and how to optimize our model to save more time and memories.

Acknowledgments. This work is supported by the grant of the National Natural Science Foundation of China No. 61432011, 61402323 and 61502335.

References

1. Azran, A.: The rendezvous algorithm: Multiclass semi-supervised learning with Markov random walks. In: Proceedings of the 24th International Conference on Machine Learning, pp. 49–56. ACM (2007)
2. Bhagat, S., Cormode, G., Muthukrishnan, S.: Node classification in social networks. In: Aggarwal, C.C. (ed.) Social Network Data Analytics, pp. 115–148. Springer, New York (2011)
3. Burda, Z., Duda, J., Luck, J., Waclaw, B.: Localization of the maximal entropy random walk. Phys. Rev. Lett. **102**(16), 160602 (2009)
4. Fortunato, S.: Community detection in graphs. Phys. Rep. **486**(3), 75–174 (2010)
5. Henderson, K., Gallagher, B., Eliassi-Rad, T., Tong, H., Basu, S., Akoglu, L., Koutra, D., Faloutsos, C., Li, L.: Rolx: structural role extraction & mining in large graphs. In: Proceedings of the 18th ACM SIGKDD International Conference on Knowledge Discovery and Data Mining, pp. 1231–1239. ACM (2012)
6. Hu, X., Liu, H.: Social status and role analysis of Palin's email network. In: Proceedings of the 21st International Conference Companion on World Wide Web, pp. 531–532. ACM (2012)
7. Jaakkola, M.S.T., Szummer, M.: Partially labeled classification with Markov random walks. In: Advances in Neural Information Processing Systems (NIPS), vol. 14, pp. 945–952 (2002)
8. Leuski, A.: Email is a stage: discovering people roles from email archives. In: Proceedings of the 27th Annual International ACM SIGIR Conference on Research and Development in Information Retrieval, pp. 502–503. ACM (2004)
9. Li, R.H., Yu, J.X., Huang, X., Cheng, H.: Random-walk domination in large graphs. In: 2014 IEEE 30th International Conference on Data Engineering (ICDE), pp. 736–747. IEEE (2014)
10. Li, R.H., Yu, J.X., Liu, J.: Link prediction: the power of maximal entropy random walk. In: Proceedings of the 20th ACM International Conference on Information and Knowledge Management, pp. 1147–1156. ACM (2011)
11. Lovász, L., et al.: Random walks on graphs: a survey. Comb. Paul Erdos is Eighty **2**, 353–398 (1996)

12. McCallum, A., Wang, X., Corrada-Emmanuel, A.: Topic and role discovery in social networks with experiments on enron and academic email. J. Artif. Intell. Res. **30**, 249–272 (2007)

13. McPherson, M., Smith-Lovin, L., Cook, J.M.: Birds of a feather: homophily in social networks. Ann. Rev. Sociol. **27**, 415–444 (2001)

14. Mislove, A., Viswanath, B., Gummadi, K.P., Druschel, P.: You are who you know: inferring user profiles in online social networks. In: Proceedings of the Third ACM International Conference on Web Search and Data Mining, pp. 251–260. ACM (2010)

15. Neville, J., Jensen, D.: Iterative classification in relational data. In: Proceedings of the AAAI-2000 Workshop on Learning Statistical Models from Relational Data, pp. 13–20 (2000)

16. Ribeiro, B., Wang, P., Murai, F., Towsley, D.: Sampling directed graphs with random walks. In: 2012 Proceedings IEEE INFOCOM, pp. 1692–1700. IEEE (2012)

17. Wang, G., Zhao, Y., Shi, X., Yu, P.S.: Magnet community identification on social networks. In: Proceedings of the 18th ACM SIGKDD International Conference on Knowledge Discovery and Data Mining, pp. 588–596. ACM (2012)

18. Welser, H.T., Cosley, D., Kossinets, G., Lin, A., Dokshin, F., Gay, G., Smith, M.: Finding social roles in Wikipedia. In: Proceedings of the 2011 iConference, pp. 122–129. ACM (2011)

19. Xie, J., Szymanski, B.K.: Community detection using a neighborhood strength driven label propagation algorithm. In: 2011 IEEE Network Science Workshop (NSW), pp. 188–195. IEEE (2011)

20. Zhao, Y., Sundaresan, N., Shen, Z., Yu, P.S.: Anatomy of a web-scale resale market: a data mining approach. In: Proceedings of the 22nd International Conference on World Wide Web, pp. 1533–1544. International World Wide Web Conferences Steering Committee (2013)

21. Zhao, Y., Wang, G., Yu, P.S., Liu, S., Zhang, S.: Inferring social roles and statuses in social networks. In: Proceedings of the 19th ACM SIGKDD International Conference on Knowledge Discovery and Data Mining, pp. 695–703. ACM (2013)

22. Zheleva, E., Getoor, L.: To join or not to join: the illusion of privacy in social networks with mixed public and private user profiles. In: Proceedings of the 18th International Conference on World Wide Web, pp. 531–540. ACM (2009)

23. Zhu, X., Ghahramani, Z., Lafferty, J., et al.: Semi-supervised learning using Gaussian fields and harmonic functions. In: ICML, vol. 3, pp. 912–919 (2003)

An Adaptive kNN Using Listwise Approach for Implicit Feedback

Bu-Xiao Wu[1], Jing Xiao[1(✉)], Jia Zhu[1], and Chen Ding[2]

[1] School of Computer Science, South China Normal University, Guangzhou, China
xiaojing@scnu.edu.cn
[2] Department of Computer Science, Ryerson University, Toronto, Canada

Abstract. Collaborative Filtering is a very popular method in recommendation systems. In item recommendation tasks, a list of items is recommended to users by ranking, but traditional CF methods do not treat it as a ranking problem for implicit feedback datasets. In this paper, we propose MAP-kNN, an adaptive kNN approach using listwise approach for implicit feedback datasets. The similarity matrix is learned by maximizing the Mean Average Precision, which is a well-known measurement in information retrieval for representing the performance of a list of ranked items. An optimization strategy and a new sampling method are proposed to improve the learning efficiency of MAP-kNN. The complexity of our algorithm over each iteration after optimization is lower than other methods that also use listwise approach. Experimental results on two datasets indicate that our approach outperforms other state of the art recommendation approaches.

Keywords: Recommendation systems · Collaborative Filtering · Learning to Rank · Implicit feedback

1 Introduction

Recommender systems are widely used to suggest items that the users may have interest. For example, Last.fm[1] is personalized to recommend a list of music to users that they may like. Collaborative Filtering (CF) method is an effective way in recommender systems. CF can be memory based or model based. K-nearest neighbor (kNN) CF [1,2] is a typical memory based approach. The main idea about memory based CF is to find the similarities among users or items. For instance, item based CF [2] is to compute the similarities of items using users' past behavior. The items with high similarity will be recommended to users. The classical model based approach is Matrix Factorization (MF) [3]. MF learns the latent factors of users and items from observed data. Most CF methods are based on scenarios where users provide explicit feedback (i.e., ratings). Actually, most feedback is implicit in real world scenarios [4–6]. Types of implicit feedback

---

[1] www.last.fm

© Springer International Publishing Switzerland 2016
F. Li et al. (Eds.): APWeb 2016, Part I, LNCS 9931, pp. 519–530, 2016.
DOI: 10.1007/978-3-319-45814-4_42

include buying history, playing song history, browsing history. The common features are binary, positive or negative. Recently, some CF methods are proposed for implicit feedback. WR-MF [7] is an effective CF method using MF using implicit feedback.

In an item recommendation system, there is a personalized ranking for a list of items based on users' buying history and ratings. Accordingly some algorithms that combine learning to rank (LTR) with collaborative filtering are proposed. CoFiRank [8] considered collaborative filtering as a ranking problem and used Maximum Margin Matrix Factorization to optimize ranking. ListRank-MF [9] integrated listwise LTR algorithm with MF and a ranked list of items were obtained by minimizing loss function. But CoFiRank and ListRank-MF only apply to rating datasets. Bayesian Personalized Ranking (BPR) [10] is a state of the art algorithm for implicit feedback datasets. BPR employed a pairwise way to optimize the measure of Area Under the ROC Curve (AUC). It has been applied to two recommender models, MF and adaptive kNN. TFMAP [11] has been recently proposed; it is a context aware recommendation algorithm using tensor factorization by optimizing Mean Average Precision. Both of AUC and MAP are famous measurements in information retrieval. AUC is based on pairwise comparison between positive items and negative items, while MAP is a listwise measure that represents the performance of a ranked item list. CLiMF [12] was another CF method which combined MF with LTR for implicit feedback datasets, where the latent factors of users and items were learned by maximizing the Mean Reciprocal Rank.

The similarity matrix is computed by formulas such as cosine similarity, Pearson correlation, in traditional kNN collaboration filtering. Since users emphasize the top-ranked items in recommender systems, optimizing MAP is a good choice for learning the similarity matrix. What's more, kNN is a classic and effective algorithm which is simple and easy to be implemented. In this paper, we propose MAP-kNN which is an adaptive kNN using listwise approach for implicit feedback by optimizing MAP. Although BPR-kNN is also based on kNN model, MAP-kNN is more effective, because MAP is top-biased while AUC is not. The contributions of this paper can be summarized as follows:

1. We present a new adaptive kNN approach in which the item similarity matrix is learned by maximizing MAP for implicit feedback. The experiments on two datasets presented that our approach outperforms other state-of-the-art approaches.
2. An optimization strategy is proposed to reduce the computational complexity of parameter learning of MAP-kNN. Compared with other methods based on listwise approach, the complexity of MAP-kNN is less expensive and it is linear to the number of positive items that users have.
3. We construct a new sampling distribution to sample top-ranked negative items as much as possible. The experimental results show that the new sampling method can get fast convergence of parameter learning of MAP-kNN.

The rest of the paper is organized as follows: Sect. 2 introduces the related work of collaborative filtering systems using learning to rank; Sect. 3 describes the method of parameter learning; a new sampling method is presented in Sect. 4; Sect. 5 gives the experimental results and discussions. Conclusion is given in Sect. 6.

2   Related Work

Various algorithms have been proposed in recommender systems and collaborative filtering is a well-known effective algorithm for recommender systems. Most of CF methods have been devoted to the context of the rating problem. The top-N recommendation for implicit feedback has not received enough attention. In this section, we review the work which is closely related to existing collaborative filtering approaches that combine CF with learning to rank.

Previously, the most prevalent technique in recommender systems is kNN collaborative filtering. However, the calculation of similarity matrix requires expensive computation to construct by formulas (i.e., cosine similarity and the Pearson correlation) in traditional kNN models [1, 2]. Matrix factorization (MF) techniques have become very popular in collaborative filtering. In early work, MF models have been shown high efficiency for rating prediction problem. Recently, Hu proposed WR-MF [7] that used MF models for implicit dataset. WR-MF used a least square optimization with case weights. The case weights were used to reduce the impact of negative items in parameters learning.

For top-N recommendation task, the ultimate purpose of recommender system is to suggest users a ranked list of items. Learning to rank has attracted much research attention in recommender systems. Rendle proposed a state of the art algorithm called Bayesian Personalized Ranking (BPR) [10]. BPR was applied to implicit feedback scenarios. It presented an optimization framework, and was applied to two models: MF and adaptive kNN. The idea of BRP is based on pairwise comparisons between a small set of positive items and a very large set of negative items. BPR leads to maximize the Area Under the Curve (AUC). There are some other pairwise methods used in recommender systems [13–15].

To alleviate the drawback that pairwise approach ignores the fact that ranking is a prediction task of a list of objects [10], some listwise approaches in recommender systems have been proposed. ListRank-MF [9] was proposed for CF that combines listwise LTR method with MF for explicit feedback. The main idea of ListRank-MF is using listwise LTR method to rank items for users. The latent factors of users and items are learned by minimizing the loss function that uses the cross entropy of top one probabilities [15] in MF. CLiMF [12] is an effective approach in top-N recommendations for implicit feedback. It is a listwise way to learn the latent factors of users and items by maximizing Mean Reciprocal Rank (MRR), which is a famous metric of information retrieval. TFMAP [11] is a top-N context aware recommendation approach based on tensor factorization and it is applied to implicit feedback data with contextual information. The main idea of TFMAP is learning parameter by maximizing MAP. MAP is a listwise measure in information retrieval. Compared with AUC, MAP is a top biased measure that mistakes in items recommendation at the top of the list carry more weights than mistakes at the bottom of the list. The maximization of MAP for a very large data collection could have large computational cost. Therefore, TFMAP has not optimized all items but sampled some negative items and all positive items for each user.

Since the top biased MAP is very important in top-N recommendations, similar to TFMAP, in this paper we propose an adaptive kNN approach that similarity matrix is learned by maximizing MAP in a very fast way. We also introduce a new sampling method to improve the efficiency of parameter learning.

3 Parameter Learning

In this section, we present an adaptive kNN using listwise approach for implicit feedback that the similarity matrix of items is learned by maximizing MAP. First, we introduce the smoothed MAP and the parameter learning by optimizing MAP. Then, we show how to reduce the high computational cost by our optimization approach.

3.1 Smoothing the MAP and Optimization

The definition of MAP can be expressed as:

$$MAP = \frac{1}{|U|} \sum_{i \in U} \frac{1}{|N_i|} \sum_{j \in N_i} \frac{\sum_{k \in T} Y_{ij} I(R_{ik} \le R_{ij}))}{R_{ij}} \tag{1}$$

where N_i is the set of items which are positive to user i; T is the set of all items. U is the set of all users. Y_{ij} is a binary number; if user i has positive behavior to item j, Y_{ij} is 1; if not, Y_{ij} is 0. R_{ij} denotes the rank of item j in the ranked list for user i. $I(x)$ is an indicator function; if x is true, $I(x)$ is 1; if not, $I(x)$ is 0.

Note that MAP is a non smooth function, it cannot be directly maximized for parameter learning. Using logistic function is an effective way to smooth the measure of information retrieval [10–12]. Let $g(x)$ represent the logistic function, it is expressed as following:

$$g(x) = \frac{1}{1 + e^{-x}} \tag{2}$$

The approximation of $I(R_{ik} \le R_{ij})$ is derived by a logistic function and it can be expressed as following:

$$I(R_{ik} \le R_{ij}) \approx g(S_{ik} - S_{ij}) \tag{3}$$

where S_{ik} is the predicted score that user i assigns to item k. For user i, the items are ranked by descending order on the basis of predicted scores. If S_{ik} is much higher than S_{ij}, $g(S_{ik} - S_{ij})$ is close to 1. If S_{ij} is much higher than S_{ik}, $g(S_{ik} - S_{ij})$ is close to 0. We use an adaptive kNN model to compute the predicted scores. For user i to item j, the main idea about predicted scores computed for item-based kNN is depending on the similarity between item j and other items that the user has positive behavior. The model of item based kNN is expressed as follows:

$$S_{ij} = \sum_{k \in N_i} \theta_{jk} \tag{4}$$

where θ is the $N \times N$ symmetric item similarity matrix. θ_{jk} is the similarity between item j and item k. The non smoothed $1/R_{ij}$ can be approximated as:

$$\frac{1}{R_{ij}} = \frac{1}{\sum_{k \in T} I(S_{ik} \leq S_{ij})} \approx \frac{1}{\sum_{k \in T} g(S_{ik} - S_{ij})} \tag{5}$$

Here we have the approximation of $1/R_{ij}$. The main idea of Eq. (5) is that the rank of item j is obtained by comparing item j's score with other items' scores. The approximation of R_{ij} will have a higher value, if the score of item j is greater than other items' scores. On the basis of above mentioned approximation, the objective function for an individual user can be described as:

$$F(\theta) = \sum_{i \in U} \frac{1}{|N_i|} \sum_{j \in N_i} \frac{\sum_{k \in T} Y_{ik} g(\sum_{t \in N_i} \theta_{kt} - \theta_{jt})}{\sum_{k \in T} g(\sum_{t \in N_i} \theta_{kt} - \theta_{jt})} - \frac{\lambda}{2} \|\theta\|^2 \tag{6}$$

The $-\frac{\lambda}{2} \|\theta\|^2$ term is necessary for regularizing model since it can avoid overfitting. In this paper, we use gradient ascent method to maximize the objective function. The gradient of objective function with respect to the similarity matrix of items is calculated as follows:

$$\frac{\partial F}{\partial \theta_{jt}} = \sum_{i \in U} \frac{1}{|N_i|} \Big(\frac{\sum_{k \in T} Y_{ik} g'(S_{ik} - S_{ij}) \sum_{k \in T} g(S_{ik} - S_{ij})}{(\sum_{k \in T} g(S_{ik} - S_{ij}))^2}$$
$$- \frac{\sum_{k \in T} Y_{ik} g(S_{ik} - S_{ij}) \sum_{k \in T} g'(S_{ik} - S_{ij})}{(\sum_{k \in T} g(S_{ik} - S_{ij}))^2} \Big) - \lambda \theta_{jt} \tag{7}$$

$$\frac{\partial F}{\partial \theta_{kt}} = \sum_{i \in U} \frac{1}{|N_i|} \Big(\frac{Y_{ik} g'(S_{ik} - S_{ij}) \sum_{k \in T} g(S_{ik} - S_{ij})}{(\sum_{k \in T} g(S_{ik} - S_{ij}))^2}$$
$$- \frac{g'(S_{ik} - S_{ij}) \sum_{k \in T} Y_{ik} g(S_{ik} - S_{ij})}{(\sum_{k \in T} g(S_{ik} - S_{ij}))^2} \Big) - \lambda \theta_{kt} \tag{8}$$

$$g(S_{ik} - S_{ij}) = g(\sum_{t \in N_i} \theta_{kt} - \theta_{jt}) \tag{9}$$

$$\frac{\partial g(S_{ik} - S_{ij})}{\partial \theta_{jt}} = \frac{e^{-(\sum_{t \in N_i} \theta_{kt} - \theta_{jt})}}{(1 + e^{-(\sum_{t \in N_i} \theta_{kt} - \theta_{jt})})^2} \tag{10}$$

$$\frac{\partial g(S_{ik} - S_{ij})}{\partial \theta_{kt}} = \frac{-e^{-(\sum_{t \in N_i} \theta_{kt} - \theta_{jt})}}{(1 + e^{-(\sum_{t \in N_i} \theta_{kt} - \theta_{jt})})^2} \tag{11}$$

Since k is the number of all items, updating θ_{kt} will cost a large amount of time. Optimizing the parameters of all items is not necessary according to the definition of MAP, thus sampling a relatively small number of negative items and all positive items for each user is an efficient way to solve this problem [11]. We use this idea to reduce the complexity of computation. The gradients of similarity matrices of items can be calculated as follows:

$$\frac{\partial F}{\partial \theta_{jt}} = \sum_{i \in U} \frac{1}{|N_i|} \left(\frac{\sum_{k \in s_i} Y_{ik} g'(S_{ik} - S_{ij}) \sum_{k \in s_i} g(S_{ik} - S_{ij})}{(\sum_{k \in s_i} g(S_{ik} - S_{ij}))^2} \right.$$
$$\left. - \frac{\sum_{k \in s_i} Y_{ik} g(S_{ik} - S_{ij}) \sum_{k \in s_i} g'(S_{ik} - S_{ij})}{(\sum_{k \in s_i} g(S_{ik} - S_{ij}))^2} \right) - \lambda \theta_{jt} \qquad (12)$$

$$\frac{\partial F}{\partial \theta_{jt}} = \sum_{i \in U} \frac{1}{|N_i|} \left(\frac{Y_{ik} g'(S_{ik} - S_{ij}) \sum_{k \in s_i} g(S_{ik} - S_{ij})}{(\sum_{k \in s_i} g(S_{ik} - S_{ij}))^2} \right.$$
$$\left. - \frac{g'(S_{ik} - S_{ij}) \sum_{k \in s_i} Y_{ik} g(S_{ik} - S_{ij})}{(\sum_{k \in s_i} g(S_{ik} - S_{ij}))^2} \right) - \lambda \theta_{kt} \qquad (13)$$

where s_i is the set of items for user i which contains the sampled negative items and all positive items.

3.2 Optimizing the Complexity of Learning

The computational complexity of calculating and updating θ_{jt} is $O(2d^3 U)$, where d is the number of average positive items per user. The complexity of calculating and updating θ_{kt} over each iteration is $O(2d^4 U)$. Although d is a small number, but it still needs a lot of time to calculate and update. According to Eqs. (10) and (11), for each item t in N_i, θ_{jt} is updated with the same value, and θ_{kt} is also updated with the same value in item j. Therefore it can be optimized as shown in Algorithm 1. We calculate the gradients of θ_{jt} and θ_{kt} but do not update them immediately. The gradients of θ_{jt} and θ_{kt} are updated until all the gradients of θ have been calculated. The computational complexity of calculating the gradients of θ_{jt} and θ_{kt} is $O(2d^2 U)$. The computation complexity of updating the gradient θ_{jt} and θ_{kt} is $O(2d^2 U)$. It significantly reduces the computational complexity of parameter learning for one iteration.

The computational complexity of parameter learning is $O(Kd^2 U)$ in CliMF, where K is the number of latent dimension. For TFMAP, the computational complexity of parameter learning is $O(2Kd^2 UC)$, where C is the number of context type. In non-contextual information case, the computational complexity of parameter learning is $O(2Kd^2 U)$. In contrast to CliMF and TFMAP, because K usually values from 10 to 100, our approach learns faster by optimizing the complexity. We let $A = dU$, where A is the number of non zeros in the user item matrix. Note that A is much greater than d, so it achieves a close to linear computational complexity of A. Due to its low complexity, our algorithm is suitable for large scale of data in practice.

4 A New Sampling Distribution

Since top-ranked negative items are the most influential items for MAP optimization [11], sampling the negative items that have the lowest rank in the list is useless to improve MAP. It is important to sample the top-ranked negative items in the list. However it is hard to find top-ranked negative items in the list after several iterations during learning process (positive items usually have higher rank in MAP optimizing process). In order to improve the efficiency of parameter learning and get faster convergence rate, our target is to sample top-ranked negative items as much as possible.

Since our approach is based on the similarity matrix of items, it means that users will probably like the items which are similar to the items that users have liked previously. For example, if user i buys both item k and item j, it means item k and item j may have some similar features. According to this idea, we construct a new sampling distribution to sample negative items in the higher ranking positions as much as possible for all negative items. We define C_{jk} to represent the correlation of item j and k; the value of C_{jk} is the number of users who have positive activities on both item j and k. In order to calculate faster, we build the user-item list to store positive items for each user. The computational complexity of C_{jk} is $O(d^2 U)$. We define X_{ij} as:

$$X_{ij} = \sum_{k \in P_i} C_{jk} \tag{14}$$

where X_{ij} represents the possibility that item j has higher rank in the list of user i. If X_{ij} has higher values, it means item j could have higher rank in the list of user i. The sampling distribution is:

$$P_{ij} \propto log(X_{ij} + z) \tag{15}$$

where z is the parameter of smoothing Eq. (14) and $z \in [1, +\infty]$. Note that the distribution of sampling negative items does not change after the correlation of items C are obtained. The cost of sampling a negative item from the distribution is $O(1)$. So, sampling from Eq. (14) does not increase the computational complexity of learning for similarity matrices of items.

To validate the new distribution in which sampling negative items can improve the efficiency of parameter learning and get faster convergence rate, we present an experimental study that compares MAPs using new sampling distribution and uniform distribution. The dataset used is Last.fm and the details are shown in the next section. As shown in Fig. 1, the uniform distribution needs about 200 iterations to achieve convergence. The new distribution only needs 50 iterations to achieve convergence and shows better prediction quality than uniform distribution. The results indicate the new distribution is effective and can improve the efficiency of parameter learning.

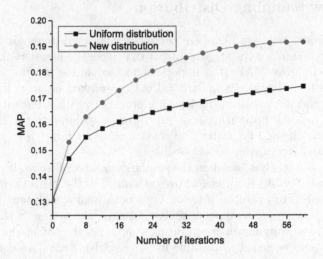

Fig. 1. MAP values of Last.fm dataset during the learning process

---

Algorithm 1. Learning similarity matrix

Input: The set of positive items by user i $N(i)$, the set of negative items by user i
   $NI(i)$, the set of users U, learning rate α, regularization parameter λ, the maximal
   number of iterations $itermax$.

output: The similarity matrix θ.

1: Initialize θ with random values and $it = 0$.
2: **repeat**
3:    **for all** $i \in U$ **do**
4:        $NI(i) \leftarrow SampleNegativeItems(i)$
5:        $updateV \leftarrow 0$
6:        **for all** $t \in N(i)$ **do**
7:            **for all** $j \in N(i)$ **do**
8:                $updateV_j \leftarrow updateV_j + \alpha \frac{\partial F}{\partial \theta_{jt}}$ based on Eq. (11)
9:            **end for**
10:           **for all** $k \in N(i) \bigcup NI(i)$ **do**
11:               $updateV_k \leftarrow updateV_k + \alpha \frac{\partial F}{\partial \theta_{kt}}$ based on Eq. (12)
12:           **end for**
13:       **end for**
14:       **for all** $j \in N(i) \bigcup NI(i)$ **do**
15:           **for all** $t \in N(i)$ **do**
16:               $\theta_{jt} \leftarrow \theta_{jt} + updateV_j$
17:           **end for**
18:       **end for**
19:   **end for**
20:   $it \leftarrow it + 1$
21: **until** $it > itermax$

Algorithm 2. Sampling method for negative items

Input: The parameter of smoothing z, The matrix of possibility about higher rank X, The number of positive items by user i I_i.

output: The set of sampling negative items NI.

1: $q \leftarrow 0$
2: **repeat**
3:     Draw j form $P_{ij} \propto log(X_{ij} + z)$
4:     $NI_q \leftarrow j$
5:     $q \leftarrow q + 1$
6: **until** $q > I_i$

Table 1. Statistics of two datasets

Dataset	lastfm-2k	Epinions
Number of users	1892	4718
Number of items	17632	49288
Number of non zeros	92834	346035
Avg of items per user	49.07	73.44
Sparseness	99.72	99.85

5 Experimental Evaluation

5.1 Data Sets

We evaluate the performance of algorithm on two different real datasets, namely lastfm-2k and Epinions[2] [12]. lastfm-2k is a subset of data crawled from a famous music website Last.fm and lastfm-2k contains 1892 users, 17632 artists, 92834 user-listened artist relations. In lastfm-2k, artists are treated as items; the task is to predict the artist that users may have interest. Epinions dataset is a social network dataset from Epinions; it contain 4718 users, 49288 trustees and 346035 trustees relationships. The trustees are treated as items, the task is to predict the trustees for users. The statistics of these two datasets are shown in Table 1.

5.2 Evaluation Methodology

To evaluate the performance of MAP-kNN, two experimental methodologies are employed. First experimental methodology is 5-fold-Leave-One-Out-Cross-Validation. It is used to validate the new distribution in which sampling negative items can improve the efficiency of parameter learning and get faster convergence rate. The result is shown in Fig. 1 and the discussion is in previous section. The other experiment methodology is to separate each dataset into training set and test set under different number of items that users have positive action.

---

[2] www.epinions.com.

Table 2. Comparison among approaches for Last.fm

	Given 10				Given 15				Given 20			
	MAP	P@1	P@5	P@10	MAP	P@1	P@5	P@10	MAP	P@1	P@5	P@10
Cosine-kNN	0.091	0.160	0.149	0.141	0.092	0.160	0.146	0.143	0.191	0.467	0.374	0.313
BPR-kNN	0.165	0.454	0.383	0.325	0.172	0.466	0.385	0.338	0.205	0.529	0.423	0.354
TFMAP-noC	0.129	0.283	0.238	0.208	0.113	0.266	0.231	0.208	0.111	0.268	0.190	0.171
CliMF	0.117	0.302	0.235	0.207	0.118	0.295	0.243	0.213	0.106	0.276	0.199	0.179
MAP-kNN	0.171	0.461	0.386	0.332	0.175	0.470	0.389	0.337	0.213	0.553	0.439	0.364

Table 3. Comparison among approaches for Epinions

	Given 10				Given 15				Given 20			
	MAP	P@1	P@5	P@10	MAP	P@1	P@5	P@10	MAP	P@1	P@5	P@10
Cosine-kNN	0.054	0.075	0.069	0.683	0.092	0.161	0.147	0.141	0.125	0.271	0.229	0.209
BPR-kNN	0.078	0.326	0.253	0.218	0.134	0.382	0.304	0.266	0.142	0.372	0.298	0.260
TFMAP-noC	0.068	0.259	0.169	0.144	0.061	0.239	0.142	0.132	0.057	0.208	0.134	0.117
CliMF	0.067	0.257	0.168	0.142	0.063	0.233	0.153	0.131	0.059	0.207	0.140	0.120
MAP-kNN	0.100	0.329	0.260	0.227	0.140	0.395	0.319	0.276	0.150	0.389	0.312	0.274

For instance, "Given 10" represents that it will randomly select 10 out of users positive items as training set for each user, the other positive items will be treated as test set. This methodology is used to compare with other methods under different conditions of users' positive items. The learning rate α in Algorithm 1 is set to 0.1 and the parameter of smoothing z in Algorithm 2 is set to 1.1.

In order to evaluate the performance of recommendation, we choose two evaluation metrics. The first evaluation metric used in the experiment is MAP, which is defined as Eq. (1). The other evaluate metrics is Precision which is a common metric in Top-N recommendation. Precision is defined as:

$$Precision = \frac{\sum_{u \in U} |RL(u) \cap TL(u)|}{\sum_{u \in U} |RL(u)|} \tag{16}$$

where U is a set of all users; $RL(u)$ is the list of items recommended to user u. $TL(u)$ is the list of items for user u in test data. In order to measure the ranked list of size-N items for each user in detail, we have different values of N, such as precision at top-1(P@1) or precision at top-5(P@5).

5.3 Compared Algorithms

In order to show the effectiveness of our proposed recommendation approach, we compare the recommendation results with the following methods:

1. Cosine-kNN: Cosine-kNN is a k-nearest neighbor method where the item similarity matrix is computed by the formula of cosine similarity.

2. BPR-kNN: this method is proposed by Rendle, which is a adaptive kNN using pairwise approach to learn item similarity matrix by optimize AUC. Both MAP-kNN and BPR-kNN are adaptive kNN methods.
3. TFMAP-noC: TFMAP is a context aware recommendation algorithm using tensor factorization by optimize MAP. Because the datasets we used do not contain contextual information, we use TFMAP-noC which is a variant of the TFMAP that is context free method. Both MAP-kNN and TFMAP-noC are learning parameter by optimizing MAP.
4. CliMF: CliMF is an MF method using listwise approach to learn user and item factors by optimizing MRR.

5.4 Results and Discussion

Table 2 shows that the recommendation performance on Lastfm dataset under different conditions of users' positive items. First of all, we can see that MAP-kNN outperforms all other methods in prediction quality. Table 3 summarizes the overall performance of different methods in epinions dataset under different conditions of users' positive items.

Compared with Cosine-kNN, Tables 2 and 3 show the results that the item similarity by maximizing MAP is more precise than the item similarity by computing Cosine similarity. Since MAP-kNN and BPR-kNN are both based on kNN method, our experiments can indicate that it is better to optimize MAP than AUC due to the top biased feature of MAP.

Compared with TFMAP-noC and CliMF, although they also optimize top biased metric(MAP and MRR), MAP-kNN has better performance. A most important reason is that CliMF and TFMAP-noC directly approximate the rank of item j in the list of user i R_{ij} is $R_{ij} \approx 1/g(S_{ij})$. It is based on the idea that the higher the item rank, the greater the predicted relevance score is. But in fact, it would be better to compare the score of the item with the scores of other items.

The approximation of R_{ij} in CliMF and TFMAP-noC does not compare with the scores of other items. The more positive items users have, the more imprecise the approximation of R_{ij} is. The performance of Given 15 and Give 20 are worse than Given 10 in CliMF and TFMAP-onC, because more positive items may bring noise to the approximation of R_{ij}. MAP-kNN considers this situation; R_{ij} is approximated by comparing with other items' scores, and the performance of Given 15 and Given 20 are better than Given 10 in MAP-kNN.

6   Conclusions

In this paper, we have introduced the MAP-kNN which is an adaptive kNN method for implicit feedback dataset, where the item similarity is obtained by optimizing MAP. To improve the efficiency of parameter learning and get faster convergence rate, we construct a new sampling distribution to sample negative items. We also analyze the computational complexity of MAP-kNN, in contrast to other listwise methods for recommender systems, MAP-kNN has less computational complexity of parameter learning and we find it to be linear to the number of positive items users have. Because MAP-kNN is computationally inexpensive, it is suitable for large scale of data in practice. Experimental results for two datasets verify MAP-kNN has better performance than other state of the art methods in terms of prediction quality.

Acknowledgement. This work was partially supported by the National Natural Science Foundation of China (NSFC) projects No. 61202296, No. 61370229, No. 61370178, the S&T Projects of Guangdong Province No. 2013B090800024, No. 2014B010103004, No. 2014B010117007, No. 2015A030401087, No. 2015B010110002, GDUPS(2015), the Natural Science Foundation of Guangdong Province project No. S2012030006242 and the Science and Technology Program of Guangzhou project No. 201508010067.

References

1. Koren, Y.: Factorization meets the neighborhood: a multifaceted collaborative filtering model. In: Proceedings of SIGKDD, pp. 426–434 (2008)
2. Deshpande, M., Karypis, G.: Item-based top-N recommendation algorithms. ACM Trans. Inf. Syst. **22**(1), 143–177 (2004)
3. Sarwar, B., Karypis, G., Konstan, J., et al.: Incremental singular value decomposition algorithms for highly scalable recommender systems. In: Fifth International Conference on Computer and Information Science, pp. 27–28 (2002)
4. Ostuni, V., Noia, T., Sciascio, E., et al.: Top-N recommendations from implicit feedback leveraging linked open data. In: Proceedings of RecSys, pp. 85–92 (2013)
5. Aiolli, F.: Convex AUC optimization for top-N recommendation with implicit feedback. In: Proceedings of RecSys, pp. 293–296 (2014)
6. Lim, D., McAuley, J., Lanckriet, G.: Top-N recommendation with missing implicit feedback. In: Proceedings of RecSys, pp. 309–312 (2015)
7. Hu, Y., Koren, Y., Volinsky, C.: Collaborative filtering for implicit feedback datasets. In: Proceedings of ICDM, pp. 263–272 (2008)
8. Weimer, M., Karatzoglou, A., Le, Q.V., et al.: CoFi rank-maximum margin matrix factorization for collaborative ranking. In: Proceedings of NIPS, pp. 1–8 (2007)
9. Shi, Y., Larson, M., Hanjalic, A.: List-wise learning to rank with matrix factorization for collaborative filtering. In: Proceedings of RecSys, pp. 269–272 (2010)
10. Rendle, S., Freudenthaler, C., Gantner, Z., et al.: BPR: Bayesian personalized ranking from implicit feedback. In: Proceedings of UAI, pp. 452–461 (2009)
11. Shi, Y., Karatzoglou, A., Baltrunas, L., et al.: TFMAP: optimizing MAP for top-N context-aware recommendation. In: Proceedings of SIGIR, pp. 155–164 (2012)
12. Shi, Y., Karatzoglou, A., Baltrunas, L., et al.: CLiMF: learning to maximize reciprocal rank with collaborative less-is-more filtering. In: Proceedings of RecSys, pp. 139–146 (2012)
13. Rendle, S., Freudenthaler, C.: Improving pairwise learning for item recommendation from implicit feedback. In: Proceedings of WSDM, pp. 273–282 (2014)
14. Zhong, H., Pan, W., Xu, C., et al.: Adaptive pairwise preference learning for collaborative recommendation with implicit feedbacks. In: Proceedings of CIKM, pp. 1999–2002 (2014)
15. Cao, Z., Qin, T., Liu, T.Y., et al.: Learning to rank: from pairwise approach to listwise approach. In: Proceedings of ICML, pp. 129–136 (2007)

Quantifying the Effect of Sentiment on Topic Evolution in Chinese Microblog

Peng Fu[1,2], Zheng Lin[1(✉)], Hailun Lin[1], Fengcheng Yuan[1], Weiping Wang[1], and Dan Meng[1]

[1] Institute of Information Engineering, Chinese Academy of Sciences, Beijing, China
{fupeng,linzheng,linhailun,yuanfengcheng,wangweiping,mengdan}@iie.ac.cn
[2] University of Chinese Academy of Sciences, Beijing, China

Abstract. The role of sentiment on topic evolution in social media is an interesting problem and has not been fully investigated. Quantifying the effect of sentiment on topic evolution can help people understand the relationship between sentiment and information diffusion. In this paper, we propose a method to identify the stages of topic evolution and introduce a new metric called popularity strength to measure their popularity. We also classify topics into four categories and quantify the effect of sentiment on different classes. Our findings show that "Good news illumines widely, and bad news flies quickly", and sentiment has complex dynamics towards topic evolution.

Keywords: Sentiment analysis · Topic evolution · Information diffusion

1 Introduction

Social media has recently become a vivid service where users are more likely to share their information and opinions with others. Information diffusion in social media has been a research hotspot with many scenarios, including detecting popular topics [1,3,4], modeling information diffusion [5,14] and social contagion [10,12,16]. Sentiment analysis, also known as opinion mining, aims to identify the sentiment polarity of texts, which helps us understand the perspectives and preferences of users [6,7] or study publishers' attitudes toward social events [8,21]. Both measuring topic evolution and sentiment analysis have attracted a lot of attention in data mining and natural language processing. However, research on the role of sentiment on topic evolution has not been fully investigated.

In this paper, we concentrate on quantifying the effect of sentiment on different stages of topics in Chinese social media world. Although recent studies found that many facts of the content may affect information diffusion [3,15,17], little work has been devoted to measure whether sentiment has impact on topic evolution.

Therefore, we take Weibo, the most popular microblog service in China, as the research object, and explore the characteristics of sentiment on messages diffusion and the influence of sentiment on topic evolution. We first investigate

© Springer International Publishing Switzerland 2016
F. Li et al. (Eds.): APWeb 2016, Part I, LNCS 9931, pp. 531–542, 2016.
DOI: 10.1007/978-3-319-45814-4_43

the role of sentiment in Weibo's messages diffusion. Then, we formulate topic evolution into three stages and quantify the sentiments in every stage. Finally, we classify topics into four categories by their evolutionary patterns and quantify the shift of sentiments in each class of topics. Our study finds that sentiment has complex dynamics to messages diffusion in Chinese microblog. For instance, "Good news illumines widely, and bad news flies quickly"; Topics which reach the peak stage in a short time are emotionally positive; On the contrary, topics which fade out quickly usually shift of sentiment toward negativeness over topic evolution. The main contributions of this paper can be summarized as follows:

- We propose a method to identify the stages of topic evolution, which make it easy to extract topic features.
- We define the popularity strength for topic evolution, which can measure the popularity of different periods within the current topic.
- We further use a softmax function to classify topics and study the effect of sentiment on each topic class.
- We make the first attempt to quantify the effect of sentiment on topic evolution of Chinese microblog world.

The rest of this paper is organized as follows. Section 2 introduces related work. Section 3 describes the methods for quantifying the effect of sentiment on topic evolution. We present our datasets, experimental setup and results in Sect. 4. Finally, the paper is concluded in Sect. 5.

2   Related Work

Information diffusion, affected by users, topics and graphics topology, has become an important research subfield of social media. There are many researches on information diffusion focused on modeling and predicting pathways of diffusion of information in social networks. Liben-Nowell et al. traced the information spreading processes at a person-by-person level [22]. Sun et al. studied the diffusion patterns on the Facebook [23]. In another work, Lin et al. tracked the evolution of an arbitrary topic and revealed the latent diffusion paths of that topic in a social community [24]. Jain et al. investigated the temporal aspects of user behavior and related it to the evolution of topics on Twitter [25]. Besides, it is shown that many characteristics of the content may affect information diffusion [15,17,19].

Since sentiment is one of the important characteristics of content. It may affect information diffusion in some ways. Therefore, a number of researches studied the effect of sentiment on information diffusion for social media like Twitter. De Choudhury M et al. gave the study that human moods are passed via online interactions [2]. Stieglitz and Linh found that sentiment-charged tweets are more likely to be retweeted than neutral ones [13]. Roy Ka-Wei Lee and Ee-Peng Lim gave a research on measuring latent user factors in sentiment diffusion [11]. In addition, some studies have shown that positive sentiment-charged content could acquire more attention [9,13]. Dang-Xuan and Stieglitz gave the result that positive and negative sentiments tend to receive significantly more replies compared to sentiment-neutral contents [20]. Emilio Ferrara and Zayao Yang explored the complex dynamics intertwining sentiment and information diffusion in Twitter [18]. However, these studies have not revealed the relationship between sentiment and topic evolution. Therefore, we aims at quantifying the effect of sentiment on topic evolution in Chinese microblog world.

3   Methodology

In this section, we present the proposed methods for sentiment analysis, topic evolution modeling, topic classification and quantifying the effect of sentiment on topic evolution. Before providing the details of these models, we first briefly introduce the problem definition and notations that will be used in sentiment analysis, topic evolution modeling and topic classification.

3.1   Problem Definition

Given a corpus of Weibo $D = \{d_1, d_2, \ldots, d_n\}$, each d_i consists of the posting time, the content, the user id and the hashtag if it contains. Sentiment analysis predicts the polarity of the content. We use $s = \{-1, 0, +1\}$ to represent the sentiment polarity of d_i. If d_i is a negative message, $s = -1$; if d_i is a positive message, $s = +1$ and $s = 0$ when the polarity is neutral. We regard topic evolution modeling problem as observing the number of messages on the topic with the temporal growth. The messages of a topic are divided into day slots by the posting date. Then, we divide the topic's day slots into three stages: developing stage, peak stage and declining stage. Given a topic, which is identified by the hashtag, $T = \{y_1, y_2, \ldots, y_k\}$ where k is the number of different posting date by day, and y_k is the number of Weibo messages in the kth day. We use label $L = \{L_a, L_b, L_c\}$ to represent the developing, peak and declining stages respectively. Topic evolution model can be expressed as a function $f : T \to T_L$, where T_L is a sequence consists of label L.

The popularity strength of a topic evolution stage is defined as the sum of the number of retweets, replies, favors and the number of messages in the stage. The popularity strength score is calculated by:

$$ps = reteets + replies + favors + mn \tag{1}$$

where $retweets$, $replies$ and $favors$ are the number of total retweets, replies and favors in the stage respectively, and mn is the number of messages in this stage.

3.2   Sentiment Analysis

The purpose of sentiment analysis is to determine whether a given Weibo message d_i is positive (+1), neutral (0) or negative (−1). We trained a sentiment classifier using Support Vector Machine [27] on the training data.

Unlike English words in a sentence are separated by a blank space. A Chinese sentence consists of a string of continuous words. Therefore, we adopt ICTCLAS [26], a famous Chinese word segmentation software, to segment a Weibo content into a set of words $w = \{w_1, w_2, \ldots, w_i\}$. We calculate TF-IDF of each word w_i as feature values.

3.3 Modeling Topic Evolution

To explore the topic evolution, we consider the Weibo messages with a hashtag. Weibo allows users tag the content with a topic by using hashtag which is a short sentence inserted into a pair of hash (#) symbols. We regard topic evolution problem as a function $f : T \to T_L$, where T is a numerical vector and T_L is a label sequence.

More precisely, we firstly use Lagrange polynomial interpolation to formulate topic evolution function $L(x)$:

$$L(x) = \sum_{j=1}^{k} y_j \prod_{i=1, j \neq i}^{k} \frac{x - x_i}{x_j - x_i} \tag{2}$$

where x_i and x_j denote distinct day points and y_j denotes the normalized messages number in topic T. The derivative of function $L(x)$ at given point (x_j, y_j) represents the evolution tendency of a topic. Hence, we calculate the derivative of $L(x)$:

$$L'(x) = \sum_{j=1}^{k} y_j \prod_{i=1, j \neq i}^{k} \left(\frac{x - x_i}{x_j - x_i} \sum_{l=1, l \neq j}^{k} \frac{1}{x - x_l} \right) \tag{3}$$

where (x_j, y_j) is the given point. If $L'(x_j) > 0$, it indicates that the given point is at the developing stage. Otherwise, the given point is at declining stage. If y_i is the maximum number, the point (x_j, y_j) is at the peek stage. By using this method, it is easy to get label sequences L_T.

3.4 Topic Classification

In order to further study the relationship between sentiment and topic evolution, we classified the topics into different categories according to their evolutionary patterns by a softmax function. Inspired by previous studies that identified topic clusters by an optimal Gaussian mixture model [18,28], we followed their way classifying the topics into four categories: (1) Topics which gather a large popularity strength, but maintain a very short period of time. (2) Topics that get a large popularity strength before the peak, and quickly fade out. (3) Topics reach the peak suddenly, and have most of the popularity strength after the peak stage. (4) Topics with a balanced popularity strength before, during and after the peak stage. Topic evolution feature function $f_e(T, T_L)$ takes topic T and topic label sequence T_L as input, and outputs an 3-dimensional feature vector describing topic evolution. Hence, we define,

$$p(y = j | f_e(T, T_L); \theta) = \frac{\exp(\theta_j^T f_e(T, T_L))}{\sum\limits_{j=1}^{k} \exp(\theta_j^T f_e(T, T_L))} \tag{4}$$

where θ is the parameter of the function. The output 3-dimensional features of $f_e(T, T_L)$ include: (1) the proportion of the developing stage by the popularity strength of the developing stage; (2) the relative location of the peak stage in the entire topic period, multiplied by the popularity strength of the peak stage; (3) the proportion of the declining stage multiplied by the popularity strength of the declining stage. The cost function is:

$$J(\theta) = -\frac{1}{m} \left[\sum_{i=1}^{m} \sum_{k=1}^{K} I\{y^{(i)} = k\} \log \frac{\exp(\theta^{(k)T} f_e(T, T_L)^{(i)})}{\sum\limits_{j=1}^{K} \exp(\theta^{(j)T} f_e(T, T_L)^{(i)})} \right] \tag{5}$$

where $I\{\cdot\}$ is the indicator function, that is $I\{$a true statement$\} = 1$ and $I\{$a false statement$\} = 0$. K is the number of classes and m is the size of training set. We use gradient descent to optimize it.

4   Experiments and Results

4.1   Experimental Setups

Datasets. We quantify the effect of sentiment on topic evolution on two corpora: the Weibo messages posted during November 12 to 19, 2015 as Dataset$A$ and the Weibo messages posted during January 8 to 15, 2016 as Dataset$B$. Each message consists of textual content, user id, the date it was posted, whether or not it was retweeted from another Weibo message and some additional metadata. The training set for sentiment analysis contains 10,000 messages for each sentiment.

We extracted all Weibo messages in Chinese that do not contain media content and remove all duplicates. Thus, we get 7,890,279 Weibo messages in dataset$A$, and 8,692,295 Weibo messages in dataset$B$. More details are presented in Table 1

Table 1. General statistics of the datasets

Items	dataset$A$	dataset$B$
Number of messages	7,890,279	8,692,295
Messages with hashtag	1,244,727	1,707,352
Distinct hashtags	11,603	169,539
Distinct users	59,684	2,776,731

Parameter settings. For sentiment analysis, we use liner kernel and L2-regularized L2-loss support vector for SVM training. The accuracy of sentiment classification is about 87 %, which is acquired by 5-fold cross-validation.

4.2 The Characteristics of Sentiment on Message Diffusion

Experiments in this section are designed to investigate the relation between
public opinion and message diffusion.

The distribution of the two datasets on sentiment polarity is presented in
Fig. 1. It can be found that most of the messages in both datasets are neutral.
The positive messages are twice more than the negative ones. When combining
datasetA and datasetB together, there are 56.31 % of the messages are neutral,
30.96 % are positive and 12.73 % are negative on average.

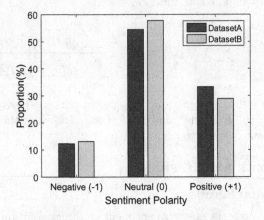

Fig. 1. Distribution of sentiments

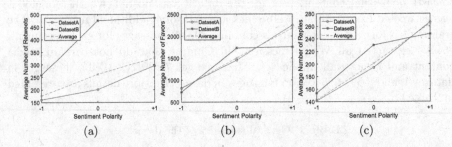

Fig. 2. The effect of sentiment on retweet, favor and reply

Figure 2 presents the average number of retweets, favors and replies in different polarities. Here, we focus on the messages that have been retweeted at
least once, including 2,215,703 messages in datasetA and 532,579 messages in
datasetB. In Fig. 2(a), we can observe that negative messages have the smallest
number of retweets on average, compared with the neutral and positive ones.
Meanwhile, positive messages have the largest number of reteets on average. In
Fig. 2(b) and (c), we can see the similar phenomenon: the positive messages still
get the most favors and replies and the negative ones get the fewest.

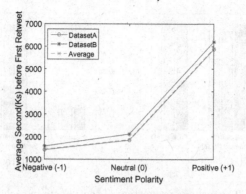

Fig. 3. The effect of sentiment on reply and diffusion speed

Figure 3 presents the speed of messages diffusion which is measured by the average time interval between the original Weibo message and the first retweet. It can be seen that negative messages attract first retweet faster than the neutral and the positive ones, and even twice faster than the positive messages.

In general, according to Figs. 2 and 3, we can conclude that: (1) positive messages spread wider than the neutral and negative ones, and collect more favors and replies. It suggests that users tend to favor, discuss and rebroadcast positive and neutral information. (2) Negative messages spread much faster than the positive ones and a little faster than neutral ones. For positive information, it requires more time to be rebroadcasted.

4.3   Sentiment in Topic Evolution

In this part, we divide a topic evolution into three stages, the developing stage, the peak stage and the declining stage, by the method proposed in Sect. 3.3. We analyze the topics which have more than 100 messages both in dataset A and dataset B. Table 2 presents the dataset used in this experiment.

Table 2. The data used in topic evolution

Items	dataset A	dataset B
Number of messages	1,035,463	908,312
Distinct hashtags	1,613	1,982
Number of day slots	8	8

Since the messages of each topic are divided into day slots by the posting date, we get 2,380 developing day slots, 1,613 peak day slots and 3,308 declining day slots in dataset A and 5,491 developing day slots, 1,982 peak day slots and 5,175 declining day slots in dataset B. On average, developing stage lasts 1.48 days in

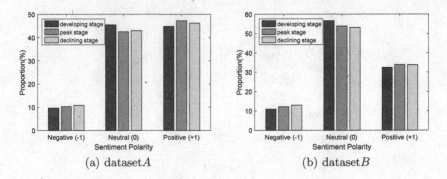

Fig. 4. The average proportion of sentiments in each stage

datasetA and 2.77 days in datasetB; declining stage lasts 2.05 days in datasetA and 2.61 days in datasetB. Peak stage lasts 1 day in both datasets. We quantify the sentiment in each stage by calculating the proportion of messages of the three different sentiment polarities for each topic.

Figure 4 presents the average proportion of sentiments at each stage of the topics in both datasets. The proportions of both negative and positive sentiments grow up steadily. The proportion of the negative messages goes up over 2 % from the developing stage to the declining stage and the proportion of the positive messages grows up about 1 % toward the end in both datasets. Meanwhile, the proportion of the neutral messages drops by 3 % in both datasets. It can be concluded that the proportion of the negative and positive messages keep a little increasing in general before the demise of the entire topic period. This suggests that, there is a general shift emotions toward more clear sentiment polarity.

4.4   The Effect of Sentiment on Different Topics

To further study the effect of sentiment on topics, we use the same data as Sect. 4.3 and classify the topics into four categories. The class_1 topics only appear in one day slot and gather a large popularity strength. The class_2 topics get most of popularity strengths before the peak stage and fade out quickly. Contrarily, the class_3 topics reach the peak stage in a short time and gather most of popularity strengths later. The class_4 topics have the balanced popularity strength in the developing, peak and declining stages. The numbers of topics for each class are shown in Table 3.

Figure 5 demonstrates the proportion of sentiments at each evolution stage in class_1. Both figures of datasetA and datasetB only present the proportion of sentiments in peak stage, because the topics in class_1 last one day. The proportion of the negative sentiment in datasetA is below the average, while above in datasetB. The proportion of the neutral sentiment is less than the average in both datasets. At the same time, the proportion of positive sentiment is higher than the average in both datasets, 8 % and 11 % higher respectively. This suggests that, in general, the transient topics will end up with more positive sentiments.

Table 3. The number of topics in each class

Class	dataset A	dataset B
Class_1	209	91
Class_2	291	459
Class_3	295	541
Class_4	818	891

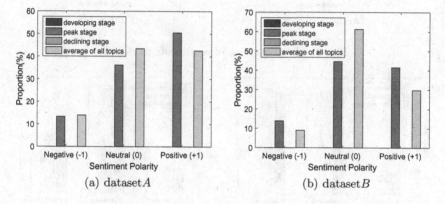

(a) dataset A (b) dataset B

Fig. 5. The proportion of sentiments in class_1

Figure 6 demonstrates the distribution of sentiments in class_2. The amount of the negative sentiment grows a bit during the topic evolution stages in both datasets, while the positive sentiment drops more than 3 % throughout the evolution stages. Meanwhile, the proportion of the neutral sentiment increases by 2 % in both datasets. It indicates that topics which get the most popularity strengths before the peak stage and fade out quickly usually shift of sentiment toward negativeness over time.

The quantifying changes of sentiments in class_3 can be seen in Fig. 7. In contrast to class_2, the proportion of the negative sentiment drops a bit during the topic evolution stages in both datasets, while the positive sentiment grows up throughout the evolution stages. Notably, the proportion of positive messages closes to the proportion of neutral messages in dataset A. This suggests that topics which reach the peak stage in a short time and gather most of popularity strengths later are emotionally positive.

Figure 8 presents the distribution of different sentiments on topics with balanced popularity strength in the developing, peak and declining stages. The proportions of both negative and positive sentiments grow a bit in the two datasets. At the same time, the amount of the neutral sentiment decreases in both datasets. It indicates that the sentiment of topics in class_4 usually shift neutral polarity to negative or positive polarities with the evolution of topics.

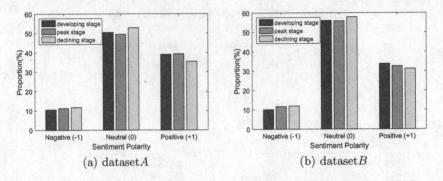

Fig. 6. The proportion of sentiments in class_2

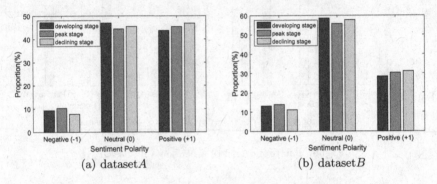

Fig. 7. The proportion of sentiments in class_3

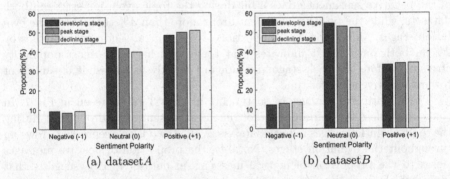

Fig. 8. The proportion of sentiments in class_4

5   Conclusions

In this paper, we quantify the effect of sentiments on topic evolution in Chinese microblog. In order to investigate the topic evolution, we propose a method to identify different stages of a topic period. We also assign a popularity strength to each topic evolution stage to measure the degree of user participation.

In this way, we are able to extract topic evolution features. Finally, we use a softmax function to classify topics into four categories, and analyse the effect of sentiments on the topic evolution for each category respectively. Through extensive experiments, we find that the negative sentiment messages spread much faster than the positive ones, but the positive sentiment messages spread broader than the negative ones. Besides, in terms of different topic stages and different topic categories, the sentiment has effect on the topic evolution more or less.

Acknowledgement. This work was supported by National Natural Science Foundation of China (No. 61502478, No.61402464), National High-Tech Research and Development Program of China (2013AA013204) and National HeGaoJi Key Project (2013ZX01039-002-001-001).

References

1. Bakshy, E., Rosenn, I., Marlow, C., Adamic, L.: The role of social networks in information diffusion. In: Proceedings of the 21st International Conference on World Wide Web, pp. 519–528. ACM, April 2012
2. De Choudhury, M., Counts, S., Gamon, M.: Not all moods are created equal! exploring human emotional states in social media. In: Sixth International AAAI Conference on Weblogs and Social Media, May 2012
3. Mathioudakis, M., Koudas, N.: TwitterMonitor: trend detection over the Twitter stream. In: Proceedings of the 2010 ACM SIGMOD International Conference on Management of data, pp. 1155–1158. ACM, June 2010
4. Dong, G., Zou, X., Wang, W., Hu, Y., Shen, G., Orkphol, K., Yang, W.: Measurement and analysis of burst topic in microblog. Int. J. Database Theory Appl. **8**(2), 145–156 (2015)
5. Gui, H., Sun, Y., Han, J., Brova, G.: Modeling topic diffusion in multi-relational bibliographic information networks. In: Proceedings of the 23rd ACM International Conference on Conference on Information and Knowledge Management, pp. 649–658. ACM, November 2014
6. Singh, V.K., Mukherjee, M., Mehta, G.K.: Combining a content filtering heuristic and sentiment analysis for movie recommendations. In: Venugopal, K.R., Patnaik, L.M. (eds.) ICIP 2011. CCIS, vol. 157, pp. 659–664. Springer, Heidelberg (2011)
7. Liang, T.P., Li, X., Yang, C.T., Wang, M.: What in consumer reviews affects the sales of mobile apps: a multifacet sentiment analysis approach. Int. J. Electron. Commer. **20**(2), 236–260 (2015)
8. Zhou, X., Tao, X., Yong, J., Yang, Z.: Sentiment analysis on tweets for social events. In: 2013 IEEE 17th International Conference on Computer Supported Cooperative Work in Design (CSCWD), pp. 557–562. IEEE, June 2013
9. Bayer, M., Sommer, W., Schacht, A.: Font size matters-emotion and attention in cortical responses to written words. PLoS One **7**(5), e36042 (2012)
10. Myers, S.A., Zhu, C., Leskovec, J.: Information diffusion and external influence in networks. In: Proceedings of the 18th ACM SIGKDD International Conference on Knowledge Discovery and Data Mining, pp. 33–41. ACM, August 2012
11. Lee, R.K.-W., Lim, E.-P.: Measuring user influence, susceptibility and cynicalness in sentiment diffusion. In: Kazai, G., Rauber, A., Fuhr, N., Hanbury, A. (eds.) ECIR 2015. LNCS, vol. 9022, pp. 411–422. Springer, Heidelberg (2015)

12. Lerman, K., Ghosh, R.: Information contagion: an empirical study of the spread of news on digg and Twitter social networks. In: ICWSM 2010, pp. 90–97 (2010)
13. Stieglitz, S., Dang-Xuan, L.: Emotions and information diffusion in social media sentiment of microblogs and sharing behavior. J. Manag. Inf. Syst. **29**(4), 217–248 (2013)
14. Wang, Y., Agichtein, E., Benzi, M.: TM-LDA: efficient online modeling of latent topic transitions in social media. In: Proceedings of the 18th ACM SIGKDD International Conference on Knowledge Discovery and Data Mining, pp. 123–131. ACM, August 2012
15. Suh, B., Hong, L., Pirolli, P., Chi, E.H.: Want to be retweeted? Large scale analytics on factors impacting retweet in Twitter network. In: 2010 IEEE Second International Conference on Social Computing (SocialCom), pp. 177–184. IEEE, August 2010
16. Weng, L., Menczer, F., Ahn, Y.Y.: Predicting successful memes using network and community structure. In: Eighth International AAAI Conference on Weblogs and Social Media, May 2014
17. Recuero, R., Araujo, R., Zago, G.: How does social capital affect Retweets? In: Fifth International AAAI Conference on Weblogs and Social Media, July 2011
18. Ferrara, E., Yang, Z.: Quantifying the effect of sentiment on information diffusion in social media. PeerJ Comput. Sci. **1**, e26 (2015)
19. Nagarajan, M., Purohit, H., Sheth, A.P.: A qualitative examination of topical tweet and retweet practices. In: ICWSM 2010, pp. 295–298 (2010)
20. Dang-Xuan, L., Stieglitz, S.: Impact and diffusion of sentiment in political communication-an empirical analysis of political weblogs. In: ICWSM, May 2012
21. Kagan, V., Stevens, A., Subrahmanian, V.S.: Using Twitter sentiment to forecast the 2013 Pakistani election and the 2014 indian election. IEEE Intell. Syst. **1**, 2–5 (2015)
22. Liben-Nowell, D., Kleinberg, J.: Tracing information flow on a global scale using Internet chain-letter data. Proc. Natl. Acad. Sci. **105**(12), 4633–4638 (2008)
23. Sun, E., Rosenn, I., Marlow, C., Lento, T.M.: Gesundheit! modeling contagion through Facebook news feed. In: ICWSM, May 2009
24. Lin, C.X., Mei, Q., Han, J., Jiang, Y., Danilevsky, M.: The joint inference of topic diffusion and evolution in social communities. In: 2011 IEEE 11th International Conference on Data Mining (ICDM), pp. 378–387. IEEE, December 2011
25. Jain, M., Rajyalakshmi, S., Tripathy, R.M., Bagchi, A.: Temporal analysis of user behavior and topic evolution on Twitter. In: Bhatnagar, V., Srinivasa, S. (eds.) BDA 2013. LNCS, vol. 8302, pp. 22–36. Springer, Heidelberg (2013)
26. Zhang, H.P., Yu, H.K., Xiong, D.Y., Liu, Q.: HHMM-based Chinese lexical analyzer ICTCLAS. In: Proceedings of the Second SIGHAN Workshop on Chinese Language Processing, vol. 17, pp. 184–187. Association for Computational Linguistics, July 2003
27. Fan, R.E., Chang, K.W., Hsieh, C.J., Wang, X.R., Lin, C.J.: LIBLINEAR: a library for large linear classification. J. Mach. Learn. Res. **9**, 1871–1874 (2008)
28. Lehmann, J., Gonalves, B., Ramasco, J.J., Cattuto, C.: Dynamical classes of collective attention in Twitter. In: Proceedings of the 21st International Conference on World Wide Web, pp. 251–260. ACM, April 2012

Measuring Directional Semantic Similarity
with Multi-features

Bo Liu, Xuanhua Shi, and Hai Jin[(⊠)]

Services Computing Technology and System Lab, Cluster and Grid Computing Lab,
Big Data Technology and System Lab, School of Computer Science and Technology,
Huazhong University of Science and Technology, Wuhan 430074, China
{borenaliu,xhshi,hjin}@hust.edu.cn

Abstract. Semantic similarity measures between linguistic terms are
essential in many *Natural Language Processing* (NLP) applications. Term
similarity is most conventionally perceived as a symmetric relation. However, semantic directional (asymmetric) relations exist in lexical semantics and make symmetric similarity measures less suitable for their identification. Furthermore, directional similarity actually represents even
more general conditions and is more practical in some specific NLP
applications than symmetric similarity. As the footstone of similarity
measures, current semantic features cannot efficiently represent large
scale web text collections. Hence, we propose a new directional similarity method, considering feature representations both in linguistic and
extra linguistic dimensions. We evaluate our approach on standard word
similarity, reporting state-of-the-art performance on multiple datasets.
Experiments show that our directional method handles both symmetric
and directional semantic relations and leads to clear improvements in
entity search and query expansion.

Keywords: Semantic similarity · Directional relation · Multi-features

1   Introduction

Semantic similarity is widely exploited in the identification of semantically similar terms from text data, and it lies at a core of NLP applications like *Information
Retrieval* (IR) and *Question Answering* (QA). Most current semantic similarity measures are prominently aiming at resolving symmetric similarity relations
in linguistics. However, quite a few semantic relations are directional in those
applications mentioned above. For instance, a user looking for 'baby food' will
be satisfied with results like 'baby pap' or 'baby juice'; but when she is looking
for 'frozen juice', as a candidate result, 'frozen food' will not be suitable. To
illustrate the directional situation 'pap → food' and 'juice → food', we follow
the lexical denotation [1]: most of the ontological word-interrelationships, like
hyponym/hypernym ('skyscraper → building'), meronym/holonym 'window →
building'), and value/attribute ('slow → velocity') are all directional relations,
signified as '$u \rightarrow v$', which means that the latter term v is expanded by the

© Springer International Publishing Switzerland 2016
F. Li et al. (Eds.): APWeb 2016, Part I, LNCS 9931, pp. 543–554, 2016.
DOI: 10.1007/978-3-319-45814-4_44

former term u. Given the potential advantages of directional semantic relations, it is necessary to identify these relations in order to improve the retrieval performance and catch users' query intention.

However, little research touched upon such directional relations when dealing with similarity relations. Hence, how to measure directional similarity is an inevitable challenge. Moreover, directional similarity actually represents even more general conditions: the degree of symmetric similarity for word pair can be measured with directional similarity method, but not the vice versa. It means that if a word pair (u, v) behaves as a directional relation, the directional similarity scores of '$u \rightarrow v$' and '$v \rightarrow u$' tend to be different; while when these two scores are equal in fact, it can be transformed into a symmetric relation. From this point of view, directional similarity is more practical than symmetric similarity.

Another challenge for semantic similarity is how to represent semantic features for large scale web text collections such as Wikipedia, providing not only sufficient linguistic corpus but also rich structured knowledge including hyperlink text and category information. As the first step to model semantic similarity, proper feature representations are required. Lexical terms' feature sets will be compared and calculated by corresponding similarity measures. Although there are latest feature representation groups [2–4] widely used in semantic similarity, almost all models have significant limitations that directly affect the performance of measures. In this work, we analyze the models' properties, and find out that it is worth to exploit the possibility of complement of these models.

In this paper, we propose a directional semantic similarity measurement. We advance the feature contrast model to emphasis on the unique features of terms. Meanwhile, we present directional similarity approaches with two views of feature representation, vector-based view and graph-based view, then compute similarity scores by using the term pair's vectors, supplemented by the structural relatedness between them.

We evaluate our approach on standard word similarity task, and it achieves better performance than other state-of-the-art symmetric measures on multiple datasets. The experiments show that our approach designed to capture directional similarity relations, can reflect a general blend of the symmetric and directional similarity. We also demonstrate the advantages of multi-feature representations, which lead to clear improvements in application-oriented tasks such as entity search and query expansion.

In summary, the contributions of this paper are as follows:

- In order to present the directional relation of term pair, we propose a directional similarity model by conducting a parametric function of their common features and unique features (Sect. 2).
- Aiming to comprehensively represent the web text's features, we exploit feature representations both in linguistic and extra linguistic dimensions, and measure directional similarity with multiple features (Sect. 3).
- We demonstrate the superiority of our directional methods through extensive experiments (Sect. 4).

In addition, we review related work in Sect. 5 and conclude the paper in Sect. 6.

2   Directional Similarity Model

This paper presents a new model for computing directional semantic similarity. After providing some preliminary definitions, we introduce the Tversky's feature contrast theory [5], which serves as a basis to motivate the present directional model.

2.1   Preliminaries

In this section, we define and formalize different term features gathered from encyclopedic knowledge, especially refer to Wikipedia in our research.

Vector-based Feature. Here we first define vector-based feature. We refer to any vector space used to represent the entire Wikipedia plain text and any vector feature corresponds to every unique term (maybe a concept or named entity, etc., for simplicity, we uniformly denote them as terms) in the corpus. We leverage the automatically generated corpus in [6], comprising text collections from various domains and topics and provides a suitable word coverage. We follow the vector representations in [7], to obtain continuous term vectors of 300 dimensions, that are constructed by NASARI approach[1] with the Word2Vec [2] toolkit.

Definition 1. *We define vector space F^m as a real-valued $M * L$ matrix space, and define set of term $T = \{t_1, t_2, \ldots, t_i, \ldots\}$, $t_i \subseteq F^m$, $i \in 1, 2, \ldots, L$, where F^m is the whole distribution of feature vector space, $\overrightarrow{t_i}$ represents vector of term t_i and $\overrightarrow{t_i} = [v_1^i, v_2^i, \ldots, v_m^i,]$ is M-dimensional ($M = 300$) as mentioned above.*

Graph-based Feature. We then define graph-based feature of Wikipedia corpus. We derive structural features based on Wikipedia by using one of its most important properties – the linking of terms within articles. These links are regarded as structural features within the text that helps to define and compare its terms.

Definition 2. *We define a knowledge graph as an undirected graph $G(V, E)$, where V represents the set of all vertices (that we call terms or anchor texts), E represents the set of all edges (that we also call linked relations), connecting vertices in V.*

For further explanation, we comply this rule: term which matches an article title must have a set of links in its article, the anchor text of a certain link must match another known term from Wikipedia, and the graph should provide these terms as nodes and these linked relations as edges.

Definition 3. *We define a path P through the graph $G(V, E)$ is a shortest sequence of nodes and linked relations $n_1 \rightarrow n_2 \rightarrow \ldots \rightarrow n_k$ such that every two nodes have one or more shortest paths reach each other.*

---

[1] http://lcl.uniroma1.it/nasari/.

Definition 4. *We define k-hop neighbor node sets $\Gamma(n_i)$ and $\Gamma(n_j)$ respectively for node pair (n_i, n_j) according to a k-level traversal algorithm (when their path length is equal to k), such that each node pair in graph $G(V, E)$ have corresponding extended node set.*

2.2   Feature Contrast Model for Directional Similarity

Using these definitions, we now introduce the underlying theory of our directional model. Remarkably, from a feature contrast perspective posed by Tversky [5], the presence of directional similarity can be quantified by terms' features comparison processing: a parameter-regulated function of both common features and distinctive features. In more details, according to feature contrast model, the similarity of a term t_i to a term t_j is a function of the features common to t_i and t_j, those in t_i but not in t_j and those in t_j but not in t_i. Figure 1 provides an abstract representation of the contrast model.

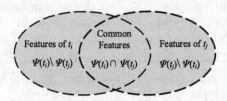

Fig. 1. The feature contrast model

If we admit a function $\psi(t)$ that yields the set of features relevant to term t, which we have already defined briefly before, the *feature-based contrast* (FC) model can be adapted by the following equation:

$$Sim_{FC}(t_i, t_j) = \frac{F(\psi(t_i) \cap \psi(t_j))}{F(\psi(t_i) \cap \psi(t_j)) + \alpha F(\psi(t_i) \setminus \psi(t_j)) + \beta F(\psi(t_j) \setminus \psi(t_i))} \quad (1)$$

where $F()$ is referred to some functions that quantified the salience of a set of features in different view and measurement standards.

We observe that in the case $\alpha \neq \beta$, which assesses the degree to which term t_i is similar to term t_j, $Sim_{FC}(t_i, t_j)$ is not equal to $Sim_{FC}(t_j, t_i)$, means that, the similarity is intended to be directional, or also called asymmetric. Our motivation in this paper are consistent with this case that we tend to emphasize on the unique features in t_i rather than the features unique to t_j. We advance this FC model and consider $\alpha = 0, \beta = 1$ or $\alpha = 1, \beta = 0$, to emphasize asymmetrically one of the objects. Thus, a more useful definition of directional feature-based similarity can be shown in the following:

$$DiSim(t_i \rightarrow t_j) = \frac{F(\psi(t_i) \cap \psi(t_j))}{F(\psi(t_i) \cap \psi(t_j)) + F(\psi(t_i) \setminus \psi(t_j))} \quad (2)$$

$$DiSim(t_j \to t_i) = \frac{F(\psi(t_i) \cap \psi(t_j))}{F(\psi(t_i) \cap \psi(t_j)) + F(\psi(t_j) \setminus \psi(t_i))} \qquad (3)$$

We consider the directional approach as $DiSim(t_i \to t_j)$ where term t_i is the subject and t_j is the referent. Let T_i and T_j denote the sets of features associated with the terms t_i, t_j, respectively. In this way, the function $F()$ is refined by the set operations, and the equations can be presented as below:

$$DiSim(t_i \to t_j) = \frac{|T_i \cap T_j|}{|T_i \cap T_j| + |T_i - T_j|} = \frac{|T_i \cap T_j|}{|T_i|} \qquad (4)$$

$$DiSim(t_j \to t_i) = \frac{|T_i \cap T_j|}{|T_i \cap T_j| + |T_j - T_i|} = \frac{|T_i \cap T_j|}{|T_j|} \qquad (5)$$

where $T_i \cap T_j$ denotes the common features of two objects $\psi(t_i) \cap \psi(t_j)$; $T_i - T_j$ and $T_j - T_i$ denote the distinct features of one object $\psi(t_i) \setminus \psi(t_j)$ and $\psi(t_j) \setminus \psi(t_i)$.

Note that this general model enables to rewrite several existing similarity measures, which can be augmented to compute relatedness score between term pairs [8]. We argue that an implement may be established that will lead to a new similarity function, so we specifically adapt some classical similarity measures on the directional model in the next section.

3   Directional Similarities with Multi-features

After briefing the directional model on the basis of feature contrast theory, we introduce how we implement it with multi-view feature representations of terms. Our opinion is that it cannot exist a single similarity measure that can fit all possible semantic feature forms. To bridge this gap and try to fit different problems, we consider a parametric family of similarity measures obtained by instantiating the directional model both for the vector-based and graph-based feature settings.

3.1   Vector-Based Directional Similarity

We obtain vector-based feature for entire corpus followed by the definitions in preliminaries. In this section, we will discuss how to implement the directional similarity model with vector-based feature of terms.

Undoubtedly, as a measure of correlation between two vectors, Cosine similarity is the most popular in many NLP tasks. As a result, we will always get Cosine similarity score of one term pair. Obviously, a typical characteristic of Cosine is symmetry which is limitable for general conditions. Now, we generalize the directional model DiSim to the similarity methods DiCosine, aiming to suggest a nice interpretation for measuring the real-valued vectors.

Note that, in the Eq. (4), $\frac{|T_i \cap T_j|}{|T_i|}$ can be simply defined as a vector multiplication which means dot product of the vectors, these equations are developed by below:

$$DiSim(t_i \to t_j) = \frac{|T_i \cap T_j|}{|T_i|} = \frac{|T_i \cap T_j|}{|T_i \cap T_i|} = \frac{\vec{t_i} \cdot \vec{t_j}}{\vec{t_i} \cdot \vec{t_i}} \qquad (6)$$

$$DiSim(t_j \rightarrow t_i) = \frac{|T_i \cap T_j|}{|T_j|} = \frac{|T_i \cap T_j|}{|T_j \cap T_j|} = \frac{\vec{t_i} \cdot \vec{t_j}}{\vec{t_j} \cdot \vec{t_j}} \tag{7}$$

where $\vec{t_i} \cdot \vec{t_j}$ denotes the common features of two terms, and $\vec{t_i} \cdot \vec{t_i} = ||\vec{t_i}||^2$ denotes the unique features of one term from the term pair.

Nicely, with the equations above, the Cosine similarity function can be seen as a product of the square roots of the two directional similarities:

$$Cos(\vec{t_i}, \vec{t_j}) = \frac{\vec{t_i} \cdot \vec{t_j}}{||\vec{t_i}|| \cdot ||\vec{t_j}||} = DiSim(t_i \rightarrow t_j)^{\frac{1}{2}} DiSim(t_j \rightarrow t_i)^{\frac{1}{2}} \tag{8}$$

The idea behind the DiCosine is to give asymmetric weights to the directional similarities in the formula above:

$$DiCos_\lambda(\vec{t_i}, \vec{t_j}) = DiSim(t_i \rightarrow t_j)^\lambda DiSim(t_j \rightarrow t_i)^{1-\lambda}$$
$$= \frac{\vec{t_i} \cdot \vec{t_j}}{||\vec{t_i}||^{2\lambda} \cdot ||\vec{t_j}||^{2(1-\lambda)}} \tag{9}$$

where $0 \leq \lambda \leq 1$, $\lambda \in [0,1]$ is a weighting parameter to tune.

Finally, note that Eq. (9) still holds for real valued vector representations. Clearly in this case, the strict spatial geometric interpretation cannot be applied but we can still consider DiCosine as a function indicating how much information we can get from two feature vectors.

3.2 Graph-Based Directional Similarity

Following the Definitions 4, the graph G allows us to determine the connection relations of terms inside the corpus. These relations are regarded as salient features within the text that helps to define and compare its terms. Hence, we put forward a weighted directional Jaccard similarity (DiJaccard) as a proper implementation for the directional model in Eqs. (4) and (5).

$$DiJaccard(n_i \rightarrow n_j) = \frac{\sum_{p=1}^{K} weight(p) \sum_{q=K-p+1}^{K} |\Gamma_p(n_i) \cap \Gamma_q(n_j)|}{\sum_{p=1}^{K} weight(p) |\Gamma_p(n_i)|} \tag{10}$$

Algorithm 1 shows details of our graph-based DiSim method. The first phrase is to obtain the path length and neighbor node sets for node pair (n_i, n_j), which are defined in Sect. 2.1. We illuminate an I/O efficient level-traversal algorithm for generating the neighbor set of node in G. The second phrase is to accomplish DiJaccard similarity as Eq. (10) under two nodes' neighbor sets. To distinguishingly rank the similarity scores, we additionally use weight function: all nodes in same level receive the same weight, and the weight would get decrease with the increase of level's number. With regard to the settings of weight value and feasible path length range, we denote that the weight function $weight(i)$ is then calculated as $1/i$ ($i = 1, 2, \ldots, 15$). Namely, when the path length equals to 1, the weighted DiJaccard will reverse to naive one which only considers the former node's degree; and when the path length exceeds 15, this path is regarded as infeasible, and the similarity score will be set to 0.

Algorithm 1. Graph-based DiSim Method

Input: G, A, B
Output: $DiJaccard(A, B)$
1: $plength \leftarrow PLength(A, B)$, $weight(i) \leftarrow 1/i$, $\Gamma\_A(i)$ represents the i-th level adjacent list;
2: **for** each $i \in [1, 15]$ **do**
3:     **if** $(B \in \Gamma\_A(i))$ **then**
4:         $plength = i$; break; //function $PLength(A, B)$
5: **if** $(plength == 1)$ **then**
6:     **return** $DiJaccard(A, B) \leftarrow \frac{1}{|\Gamma\_A(1)|}$;
7: **else if** $(plength == 15)$ **then**
8:     **return** $DiJaccard(A, B) \leftarrow 0$;
9: **else**
10:     $tmp\_A\_B \leftarrow \sum_{i=1}^{plength}(weight(i)(\sum_{j=plength-i+1}^{plength} |\Gamma\_A(i) \cap \Gamma\_B(j)|))$;
11:     $tmp\_A \leftarrow \sum_{i=1}^{plength} |\Gamma\_A(i)| * weight(i)$;
12:     $DiJaccard(A, B) \leftarrow \frac{tmp\_A\_B}{tmp\_A}$;
13:     **return** $DiJaccard(A, B)$;

3.3 Similarity Scores with Multi-features

As we proposed previously, even though proved efficient mostly, these two features still have respective limitations. After analyzing the two feature models, we find it suitable to complement each other: the property of the vector-based feature that leads similarity relations to fall into local sub-optimal context can be offset by the global information of graph-based feature. In addition, the drawbacks of the graph-based model that lacks explicit linguistic information are also avoided properly. For a conclusion, we make a linear combination score and regard the weighted DiJaccard score of term pair as the complement for the DiCosine score.

$$Score(t_i \rightarrow t_j) = Score(DiCos_\lambda(\overrightarrow{t_i}, \overrightarrow{t_j})) + Score(DiJaccard(n_i \rightarrow n_j)) \quad (11)$$

Such a combination shows improvements over single measure due to the fact that different measures capture different text characteristics. This rare complement of scaled corpus and structural information of Web sources such as Wikipedia makes our works (and for other NLP applications) initiative.

4   Experimental Study

In the following we detail the experiments in two parts: at first part, we validate our model for identification of directional relations in term pairs, afterwards in order to be compatible with previous similarity works, we experiment with three ground-truth datasets; and at second part we demonstrate how the directional model can be utilized in two IR task, entity search and query expansion.

4.1   Similarity Measures Analysis

Datasets. We take three standard datasets for our directionality validation test and word similarity experiments, which are widely used in the literature: WordSim-353 [9], SimLex-999 [10], and MEN-3000 [11]. Although they tend to differ from genuine or relatedness similarity in their focus, we distinguish no longer in our experiments, as we mainly address the similarity study with human judgments and concrete IR task.

Directionality Validation

Extended Dataset. To validate our model's capacity for identifying the directional relations in term pairs, we extend standard datasets into two reversed subsets: the form of word pair (u, v) in two subsets can be expressed as $u \rightarrow v$ and $v \rightarrow u$ respectively. We assign two annotators to judge the word pairs through the lexical denotations mentioned above, and they are required to figure out whether the directional relations are existing in the word pairs. In another words, the directional degree of two opposite word pairs could be distinguished, for instance, for the pairs *daffodil $\rightarrow$ flower* and *flower $\rightarrow$ daffodil, daffodil $\rightarrow$ flower* should be assigned relatively a higher similarity value, and these two pairs would be tagged positive $(+)$ directional polarity; these pairs *daffodil $\rightarrow$ rose* and *rose $\rightarrow$ daffodil* should be marked the approximately same scores, and tagged negative $(-)$ non-polarity. After the raw judgements, we select almost 300 word pairs possess directional polarity from standard datasets, which are taken as polarity judgments in the test.

Table 1. Accuracy of polarity-annotated word Pairs in different datasets

Directional-tagged dataset	Correct rate
Subset_1(50 pairs from WordSim-353)	85 %
Subset_2(250 pairs from MEN-3000)	91 %

With polarity judgments, we have nicely checked validity of directional approaches. Briefly, we use the DiCosine method to compute word pairs' scores in the extended subsets. After tuning the parameter, we also judge the polarity of our results: the positive $(+)$ would be set for the two word pairs if the score of DiCosine(u, v) is higher than DiCosine(v, u). Table 1 shows that most of the word pairs' polarity are correctly predicted (only 9 % is incorrect in MEN-3000).

Similarity Measures Analysis

Baselines. Callback that our approach has multi-views of feature representation, vector-based, graph-based and combined, for each view we compare performance with corresponding baselines. We use (1) Cosine which is strong baseline for embedded vectors comparison in prevailing Word2Vec-likeness models. For graph-based techniques, we use (2) SimRank [12] which is a classic method, iteratively computes nodes' scores as the sum of similarities between their neighbors in the graph. (3) M&W method [4] has a similar framework as our method. It measures entity relatedness using a variation of Jaccard similarity on Wikipedia page interlinks. (4) EntityRel [13] is a recent approach which learns strong relevance contexts by constraining the length of path and a weighted composite similarity method for path selection. We use (5) Weighted Overlap method [14] as the baseline for combined method, which considers both lexical and structural information for the corpus.

Evaluation. For evaluation of similarity methods proposed, we follow: the score assigned to each pair of terms is computed by a similarity method m, and m's quality and availability is evaluated by achieving the Spearman correlation coefficient score (ρ) between the ranking derived from m's scores and the one derived from the human scores, a setting in which the absolute similarity scores do not play a role and it is solely their ranking that matters.

Table 2. The table shows Spearman's ρ scores for different measures, which are implemented in the best tuning parameters, for our method, the typical value of λ is 0.65; for the SimRank, the damping factor set is 0.8 as a common value. We adapt them on the same graph G for a fair comparison.

View	Method	Datasets			Average
		WS-Sim-353	SimLex-999	MEN-3000	
Combined	DiCosine+DiJaccard	0.856	0.650	0.761	0.756
	Weighted Overlap	0.740	0.653	0.605	0.673
Vector-based	DiCosine	0.791	0.430	0.756	0.659
	Cosine	0.750	0.462	0.680	0.631
Graph-based	DiJaccard	0.453	0.448	0.485	0.462
	M&W	0.410	0.460	0.461	0.444
	EntityRel	0.472	0.455	0.430	0.453
	SimRank	0.353	0.333	0.355	0.347

From the overall results in Table 2, we can see that: by jointly viewing and modeling the contextual occurrence and the linked relationship, our combined directional method can outperform most of competitors according to the Spearman correlation. As shown in Table 2, a notable performance deficiency is attributed to the graph-based method group, that because interlink graph model differs most from the ground truth when the items being compared cannot be resolved to suitable linkages in Wikipedia or web page [4]. For example, there is no article for the concept *defeat*; the anchor directs only to specific military encounters and other reference information. These cases are common in the above datasets, for this reason, we consider to utilize the graph-based method as a complement for the combined methods.

4.2 Downstream Tasks

One primary goal of measuring semantic similarity is to improve downstream NLP tasks. We implement our method to two practical applications.

Entity Search. Unlike traditional web search engine that finds disorganized web pages, entity search retrieves knowledge directly by generating relevant entities in response to a query. In this task, we evaluate our similarity model through INEX09 query set[2], in which almost 35 queries contain a list of relevant entities each of which corresponds to a Wikipedia page. We first extract the underlying entities mentioned in query q through entity linking technique, and then rank each candidate entity by its summed similarity scores to the entities in q.

Table 3. Precision@k of different similarity approaches to complete entity search

Methods	Precision		
	Precision@5	Precision@10	Precision@20
Ours	0.705	0.667	0.580
LSCR	0.655	0.564	0.447
Cosine	0.430	0.367	0.322

Our comparisons include the LSCR model [15] which has outstanding performance in entity search task, and the benchmarking Cosine model. LSCR model develops a hybrid solution by formalizing both context and category matching, this model is based on generative language modeling techniques which is similar to our vector-based method. For evaluating, we adopt the common metric of precision@k, i.e., the percentage of relevant entities in the top-k results (we set $k = 5, 10, 20$). Table 3 lists the entity search results of the competitors. Our method shows potential advantage over the previous models. LSCR model performs well for combining both long and short-context, however, our method leverages the high-quality embedding features and rich link structures for entity of Wikipedia, which leads to accurate query results.

Query Expansion. As a well-known method to improve effectiveness of IR systems, query expansion task is to augment a given textual input (specifically a query) with expansion terms, which are obtained in our case by the directional similarity measures.

We pick up the representative directional model balAPinc [1] as the reference for term expansion with some common words. As shown in Table 4, compared with the balAPinc model, which uses the counting-based word-word features, our work replaces its feature construction with semantic enrichment step, such as learning multi-features combining with a learned representations and structural information.

---

[2] http://www.inex.otago.ac.nz/tracks/entityranking/entity-ranking.asp.

Table 4. Top expansion terms scored by BalAPinc and our method

word	balAPinc	Ours
computer	network, program, telephone, science, machine, industry,	CPU, laptop, hardware, pc, operating system, Microsoft, keyboard, software
university	highschool, college, seminary, campus, educational institution, higher education, institute	college, student, school, institute, undergraduate education, academic, major, research, course, faculty
airport	gatwick, airfield, london, terminal, port, air base	aircraft, air station, airport shuttle, airline, flight, international airport

5 Related Work

Semantic similarity techniques have been widely studied in recent years. There are some conventional but strong methods such as Cosine, SimRank and Jaccard, and other up-to-date methods such as Weighted Overlap [14], semantic SimRank (SR#) [16]. However, most of them only focus on identifying symmetric semantic relation and ignore the directional relation. It also appears that a few approaches can tackle this situation [1,17], but they still rely on relatively stale feature constructions, which need statistical information with considerable manual efforts, and make the entire model impracticable for large scale text data.

As the basis of similarity measure, latest feature representation approaches are divided into two main families: (1) *neural network language models* (NNLM), like the skip-gram model and GloVe model [2,3] and (2) graph structural models, such as *Wikipedia link-based measure* (WLM) [4]. Even though vector-based models efficiently generate explicit vector features for all individual terms in corpus, they do relatively be influenced by noise (e.g., bounded surrounding context). Models based on graph may do better on the representation for global structural knowledge, but they easily confront inadequate taxonomical relationships or hyperlink connections, which finally leads to low results quality. Therefore, it is necessary to propose a new method to compute similarity based on terms and their relations, considering feature representations in multiple dimensions.

6 Conclusion

Semantic similarity is a basic issue in most NLP applications. However, most current methods cannot identify both symmetric and directional semantic relations between textual terms. Therefore, we propose and demonstrate a new directional similarity method to capture almost semantic relations over massive web text. The experiments show that, our method achieves excellent effectiveness and lead to clear improvements in IR tasks. In the future, we plan to explore better feature construction approach to make higher efficiency.

Acknowledgments. This paper is partly supported by the NSFC under grants No. 61433019 and No. 61370104, International Science and Technology Cooperation Program of China under grant No. 2015DFE12860, National 863 Hi-Tech Research and Development Program under grant 2014AA01A301, and Chinese Universities Scientific Fund under grant No. 2015MS077.

References

1. Kotlerman, L., Dagan, I., Szpektor, I., Zhitomirsky-Geffet, M.: Directional distributional similarity for lexical inference. Nat. Lang. Eng. **16**(4), 359–389 (2010)
2. Mikolov, T., Chen, K., Corrado, G., Dean, J.: Efficient Estimation of Word Representations in Vector Space (2013). CoRR abs/1301.3781
3. Pennington, J., Socher, R., Manning, C.D.: GloVe: global vectors for word representation. In: Proceedings of ACL, Baltimore (2014)
4. Milne, D., Witten, I.H.: An Effective, Low-cost measure of semantic relatedness obtained from Wikipedia links. In: Proceedings of AAAI, Chicago (2008)
5. Tversky, A.: Features of similarity. Psychol. Rev. **84**(4), 327–352 (1977)
6. Iacobacci, I., Pilehvar, M.T., Navigli, R.: SensEmbed: learning sense embeddings for word and relational similarity. In: Proceedings of ACL, Beijing (2015)
7. Camacho-Collados, J., Pilehvar, M.T., Navigli, R.: NASARI: a novel approach to a semantically-aware representation of items. In: Proceedings of ACL, Beijing (2015)
8. Pirró, G., Euzenat, J.: A feature and information theoretic framework for semantic similarity and relatedness. In: Patel-Schneider, P.F., Pan, Y., Hitzler, P., Mika, P., Zhang, L., Pan, J.Z., Horrocks, I., Glimm, B. (eds.) ISWC 2010, Part I. LNCS, vol. 6496, pp. 615–630. Springer, Heidelberg (2010)
9. Finkelstein, L., Gabrilovich, E., Matias, Y., Rivlin, E., Solan, Z., Wolfman, G., Ruppin, E.: Placing search in context: the concept revisited. ACM Trans. Inf. Syst. **20**(1), 116–131 (2002)
10. Hill, F., Reichart, R., Korhonen, A.: SimLex-999: evaluating semantic models with similarity estimation. Comput. Linguist. **41**(4), 665–695 (2015)
11. Bruni, E., Tran, K.N., Baroni, M.: Multimodal distributional semantics. J. Artif. Intell. Res. **49**(1), 1–47 (2014)
12. Jeh, G., Widom, J.: SimRank: a measure of structural-context similarity. In: Proceedings of SIGKDD, Alberta, pp. 538–543 (2002)
13. Ji, M., He, Q., Han, J., Spangler, S.: Mining strong relevance between heterogeneous entities from unstructured biomedical data. Data Min. Knowl. Disc. **29**(4), 976–998 (2015)
14. Pilehvar, M.T., Jurgens, D., Navigli, R.: Align, disambiguate and walk: a unified approach for measuring semantic similarity. In: Proceedings of ACL, Sofia (2013)
15. Chen, Y., Gao, L., Shi, S., Du, X., Wen, J.: Improving context and category matching for entity search. In: Proceedings of ACL, Québec (2014)
16. Yu, W., McCann, J.A.: High quality graph-based similarity search. In: Proceedings of SIGIR, Santiago, pp. 83–92 (2015)
17. Turney, P.D., Mohammad, S.M.: Experiments with three approaches to recognizing lexical entailment. Nat. Lang. Eng. **21**(3), 437–476 (2015)

Finding Latest Influential Research Papers Through Modeling Two Views of Citation Links

Lu Huang[1], Hongyan Liu[2(✉)], Jun He[1(✉)], and Xiaoyong Du[1(✉)]

[1] Key Labs of Data Engineering and Knowledge Engineering,
Ministry of Education, China School of Information,
Renmin University of China, Beijing 100872, China
{huanglu_annie,hejun,duyong}@ruc.edu.cn
[2] Department of Management Science and Engineering,
Tsinghua University, Beijing 100084, China
liuhy@sem.tsinghua.edu.cn

Abstract. Finding hidden topics and latest topic influential papers in a corpus can help researchers get a quick overview and recent development of a scientific research field. Existing work focused on finding milestone papers which are usually published many years ago. Finding latest influential papers is a more challenging problem due to lack of enough citation information of newly published papers. In this paper, we study this problem and propose a novel way of modeling citation links with a probabilistic generative model. The key idea is to consider two views of citation, both citing and being cited of each paper. Through this idea, we can not only model topic dependence between cited and citing papers but also incorporate latest papers into our model. Based on these ideas, we jointly model the two views with an extension of topic model, *Bi-Citation-LDA* model, which can not only find previous important papers but also find newly published influential papers in each topic. Experiments on real dataset and comparison with existing methods indicate that our model can effectively find latest topic influential papers.

1 Introduction

When we reach a new research area, we all want to understand the whole picture of the area quickly by reading a short list of important papers. Therefore, how to find such a reading list and find hidden topics is a very important problem. Pervious works studying this problem focused on finding milestone papers which are usually papers published years ago. Although topic milestone papers can somehow help researchers understand a research area, researchers also expect to know the latest progress in the field and read latest important work to find research trend. Therefore, in this paper we study how to find *influential papers* which not only are representative papers but also reflect latest progress of a research area. In the meantime, we also want to analyze why a paper is influential and find its potential influential aspects. Finding latest influential papers is more challenging compared with finding milestone papers, as newly published papers are usually hardly cited. Thus, we need to develop effective approach for this problem.

© Springer International Publishing Switzerland 2016
F. Li et al. (Eds.): APWeb 2016, Part I, LNCS 9931, pp. 555–566, 2016.
DOI: 10.1007/978-3-319-45814-4_45

Topic modeling is a popular approach to finding hidden topics in a corpus and has been extensively used in the field of research paper analysis. Existing approaches model paper citations as a research paper's attributes and get citation distribution within each topic to find representative papers. However, the representative papers (milestone papers) found are typically works published years ago and can hardly reflect the latest progress in the scientific field. Wang et al. [1] studied topic milestone paper discovery and developed a model called Citation-LDA for topic evolution analysis. They modeled each paper as "a bag of citation" and found milestone papers based on co-citation relation. This is a very innovative idea. However, there is still room for improving. First of all, based on this model, research paper is modeled as "a bag of citation", which could be biased against newly published papers and rarely cited papers. Second, using only references as paper representation may not produce the expected results. For example, two paper may have the same reference list, but they may describe very different approaches to solving a problem.

In this paper, we propose a novel topic model to find *influential papers* and analyze the influential aspects of each paper. In our model, we jointly model the content and link information of each paper. More importantly, we not only consider the content and references of each paper, but also take the set of citing papers into account. We model each research paper as two bags of information: one containing paper contents and references and the other containing citing papers (a set of papers which cited the target paper). The content and references bag sub-model is similar to the link-LDA model [3] which captures the notion that documents share the same words and references tend to discuss the same topic, while the citing paper bag sub-model can represent the document's influential aspects. We propose this model based on the following intuitions:

First, the set of citing papers for a paper can reflect its most influential topics. By incorporating citing papers, we can model the flow of topics among papers and thereby newly published papers can be considered.

Second, a bag of citing papers can be a supplement to the co-citation relation. In our model, co-citation (two papers are cited together by other papers) and bibliographic coupling (two works cited a common third work) are considered simultaneously to measure the similarity relationship between papers.

Based on the above intuitions, we propose a novel topic model to find the hidden topics and topic influential papers in a corpus. Unlike previous models, our model can find both milestone papers published years ago and newly published papers as topic influential papers, and can find the attractive aspects of these papers at the same time.

The rest of the paper is organized as follows. In Sect. 2 we discussed related work. Then, in Sect. 3, we formulate the research problem and describe our proposed probabilistic model. Based on this model, we explain how to find influential papers and influential aspects in Sect. 4. Experiment results and analysis are given in Sect. 5. Finally we conclude the paper in Sect. 6.

2 Related Work

Early studies related to our work are about citation analysis and topic models. Citation analysis usually models paper citations in a graph and measures paper similarities and impacts using graph algorithms such as page-rank. Bibliometric measures such as co-citation analysis [4] and bibliographic coupling [5] are proposed for citation analysis and have been widely used in many digital library projects for similarity measure of publications. A recent work [6] compared citation analysis, bibliographic coupling, and direct citation for representing the research front and found that among the three pure citation-based approaches, bibliographic coupling outperforms the other two methods. Regarding citation analysis in our work, we measure paper impacts based on mutual influence of co-citation relation and bibliographic coupling. Bibliographic coupling has been used for representing the research front in our model which accords with the results of [6].

Probabilistic topic models have been widely used to discover the hidden topics in a corpus. Topic modeling over text [7, 8] which models documents as "a bag of words" has been thoroughly studied. These models find hidden topics and topic descriptions while ignore the document impact on each topic. Recent works including [1–3, 9] integrate citations which is only the reference information into topic model and model paper impact based on co-citation relations. Wang et al. [1] models each document as "a bag of citation" and discovers milestone papers in each topic. Nallapati et al. [2] jointly models paper text and citations and proposes Link-PLSA-LDA for citation prediction task. Erosheva et al. [3] adopt the assumption that a document is a mixed member of "a bag of words" and "a bag of references". While Kataria et al. [9] incorporates the context in which a document links to another document to improve link prediction and log-likelihood estimation on unseen content. All these four papers incorporate citations into their models, which can find representative papers for each topic. But these representative papers are sampled from paper citations (references of papers in corpus), which eliminate the newly published papers and rarely cited papers.

To our knowledge, there is no existing approach that describes document through both "a bag of references" and "a bag of citing papers". We make full use of the two kinds of link information in a corpus and model the topic dependence between cited and citing papers for topic influential paper discovery.

3 Probabilistic Model of Research Papers

3.1 Problem Formulation

Given a corpus with M documents with each being a research paper, let V denote the set of the word vocabulary occurred in the corpus. Each document d is represented by both its content and citation information. From one point of view, each paper is represented by its content consisting of N_d word tokens and a set of L_d papers cited by document d, which are called cited paper set. From another point of view, we can regard a document as a set of L'_d papers that cited the paper d, which are called citing paper set. Both cited paper and citing paper are from the corpus.

From the first point of view, we assume that each document d is generated with a topic distribution θ_d. From the second point of view, we assume that the set of papers citing document d has a mixed topic distribution δ_d. We think that each paper in the citing paper set is attracted by some aspects of document d and choose to cite it. Therefore, we use I_d to represent the probabilistic distribution over different aspects for which a paper chooses to cite document d. In this way, attracted topics propagate from the cited paper to citing papers.

Let K denote the number of topics hidden in the corpus of M documents and there exists a $K \times |V|$ topic-word distribution matrix ϕ representing a probabilistic distribution of topics over words. Similarly, a $K \times M$ topic-citation distribution matrix ψ represents a probabilistic distribution of topics over citations including cited papers and citing papers.

In this paper, we want to find the K hidden topics in a corpus and under each topic we want to discover the influential papers which could best represent the topic. Topic influential papers are a list of ranked papers. The meaning of "influential" can be explained in the following two aspects: First, the influential papers should contain the most important previous works in a topic which is similar to the milestone papers studied in previous work [1]. Second, the influential papers should contain some newly published papers which can best reflect the latest development trend in a topic. Existing work focused on finding milestone papers, which can hardly find latest important papers.

3.2   Probabilistic Model of Two Views of Citations

To find hidden topics and influential papers of each topic, we jointly model the two views of a document by a generative model depicted by the graph shown in Fig. 1.

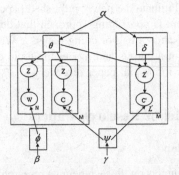

Fig. 1. Graphical representation of Bi-Citation-LDA model

As shown in Fig. 1, we model each document d in the corpus as two parts, the left part of the graph represents a mixed bag of word content and references, and the right part represents a bag of citing papers that cited document d. For each document d in M, we assume that the author choose to mix d with a topic distribution θ_d and for each word in d, a topic is assigned to it with a multinomial distribution $Z \sim Mult(\cdot|\theta_d)$ and each word is chosen from topic-word distribution ϕ based on the assigned topic. Similarly, for each reference in d, when the topic is assigned, each reference is chosen from topic-citation distribution ψ. While for the right part, there exists a topic distribution δ_d within the bag of citing papers. Papers in this bag can jointly cite more than one papers besides document d. Thus δ_d can be used to explain the reference behavior of the citing papers. Further, the bag of citing papers chooses to cite document d to reference some influential topics in d and the influential topics have a distribution of I_d which can be obtained based on θ_d and δ_d. In this way, we can incorporate latest papers into this model and enable influential topics to flow from the cited part to the citing part. Therefore, for each paper citing document d, an influential topic is assigned to it with a multinomial distribution $Z \sim Mult(\cdot|\theta_d, \delta_d)$ and each citing paper is chosen from topic-citation distribution ψ based on the topic.

The generative steps of a document based on the above described model, Bi-Citation-LDA, are given below, where α, β and γ are Dirichlet smoothing parameters.

Model Bi-Citation-LDA

For each document d
  Generate $\theta_d \sim Dir(\cdot|\alpha)$
    for each word position $i = 1, 2 \cdots, N_d$
      Generate topic $Z_i \sim Mult(\cdot|\theta_d)$
      Generate word $w_i \sim Mult(\cdot|Z_i, \phi_{Z_i})$, where $\phi_{Z_i} \sim Dir(\cdot|\beta)$
    for each reference $c_i, i = 1, 2 \cdots, L_d$
      Generate topic $Z_i \sim Mult(\cdot|\theta_d)$
    Generate reference $c_i \sim Mult(\cdot|Z_i, \psi_{Z_i})$, where $\psi_{Z_i} \sim Dir(\cdot|\gamma)$
If d has been cited
  Generate $\delta_d \sim Dir(\cdot|\alpha)$
    for each citing paper c'_j
      Generate topic $Z \sim Mult(\cdot|\theta_d, \delta_d)$
      Generate paper $c'_j \sim Mult(\cdot|Z_j, \psi_{Z_j})$, where $\psi_{Z_j} \sim Dir(\cdot|\gamma)$

3.3 Inference of Bi-Citation-LDA

Given a corpus of documents and its corresponding citing papers, we can learn the set of parameters ψ and ϕ through Gibbs sampling [11]. Let w be a word of a document d, c be a reference paper (cited paper), and c' be a citing paper of document d. Formally, given the model parameters α, β and γ, based on the Bi-Citation-LDA model, the joint probability distribution of the latent and the observed variables(marked in bold) can be factorized as follows:

$$p(\boldsymbol{w}, \boldsymbol{c}, \boldsymbol{c'}|\alpha, \beta, \gamma) = p(\boldsymbol{w}|z, \beta)p(\boldsymbol{c}|z, \gamma)p(\boldsymbol{c'}|z', \gamma)p(z|\alpha)p(z'|\alpha) \tag{1}$$

The updating rules during Gibbs sampling for the topic assignment of word w_i, reference paper c_i and citing paper c_i' in the ith position to topic k are shown in Table 1, where symbol $\neg i$ means that all other element (word, reference or citing paper) are considered except the current word, reference paper and citing paper. Due to the space limit, we omit the details of the deduction of these formulas.

Table 1. Gibbs updates for Bi-Citation-LDA

$$p(Z_i = k|w_i = t, w_{\neg i}, Z_{\neg i}) \propto \frac{n_{k,\neg i}^{(t)} + \beta}{\sum_{t=1}^{V}(n_{k,\neg i}^{(t)} + \beta)} \frac{n_{d,\neg i}^{(k)} + \alpha}{\sum_{k=1}^{K}(n_{d,\neg i}^{(k)} + \alpha)}$$

$$p(Z_i = k|c_i = t, c_{\neg i}, Z_{\neg i}) \propto \frac{n_{k,\neg i}^{(t)} + \gamma}{\sum_{t=1}^{M}(n_{k,\neg i}^{(t)} + \gamma)} \frac{n_{d,\neg i}^{(k)} + \alpha}{\sum_{k=1}^{K}(n_{d,\neg i}^{(k)} + \alpha)}$$

$$p(Z_i' = k|c_i' = t, c_{\neg i}', Z_{\neg i}') \propto \frac{n_{k,\neg i}^{(t)} + \gamma}{\sum_{t=1}^{M}(n_{k,\neg i}^{(t)} + \gamma)} \frac{n_{d,\neg i}^{(k)} + \alpha}{\sum_{k=1}^{K}(n_{d,\neg i}^{(k)} + \alpha)} \frac{n_{d',\neg i}^{(k)} + \alpha}{\sum_{k=1}^{K}(n_{d',\neg i}^{(k)} + \alpha)}$$

Where $n_{k,i}^{(t)}$ denotes the number of times element t (word, reference or citing paper) is sampled from topic k, $n_{d,i}^{(k)}$ denotes the number of times topic k is observed in document d in the content and reference bag while $n_{d',i}^{(k)}$ denotes the number of times topic k is observed in the citing paper bag denoted by d' to distinguish from document d.

After the Gibbs sampling process, we can estimate the posterior distribution ψ (topic-citation distribution) and ϕ (topic-word distribution) by following formulas:

$$\psi(d_i, Z_k) = \psi(d_i = t, Z_i = k) = \frac{n_k^{(t)} + \gamma}{\sum_{t=1}^{M}(n_k^{(t)} + \gamma)} \tag{2}$$

$$\phi(w_i, Z_k) = \phi(w_i = t, Z_i = k) = \frac{n_k^{(t)} + \beta}{\sum_{t=1}^{V}(n_k^{(t)} + \beta)} \tag{3}$$

4 Finding Influential Papers

4.1 Describing Each Topic and Finding Topic Influential Papers

For each topic, we can choose keywords to summarize it based on the posterior distribution ϕ. Within each topic Z_k, $\phi(w_i, Z_k)$ indicates the importance of each word w_i to the topic. Thus we can choose the top ranked words in each topic to describe the topic. In our paper, we use the title of each paper to represent the document's content.

Based on our proposed Bi-Citation-LDA model, we can derive a posterior distribution ψ. Within each topic Z_k, $\psi(d_i, Z_k)$ indicates the importance of each paper d_i (d_i can be cited paper or citing paper). Each $\psi(d_i, Z_k)$ can be computed using formula (2). Documents can be ranked based on $\psi(d_i, Z_k)$ and those ranking at top of each topic Z_k can be considered as topic influential papers. These influential papers not only contain important papers but also cover latest papers, which is because that we not only model documents from the content and reference point of view, but also model documents from the being cited point of view, reflecting the latest development of the topics hidden in the document.

4.2 Influence Aspect Analysis

In previous works, the set of papers citing a document d has not been fully considered to find latest influential papers. In this paper, our proposed Bi-Citation-LDA model considers document d's citing papers as a bag which can reflect d's influential aspects, and thereby reflect d's importance in terms of being influential to a topic's development.

In the document's view, we can derive a doc-topic distribution θ_d to represent the main topics in each document d. δ_d represents the topic distribution among each bag of citing papers and for each document who has been cited, there exist a I_d doc-topic distribution to reflect the influential topics in d, which can be calculated by normalizing the product of θ_d and δ_d. Thus, we use $I_d(Z_k)$ to denote that how much each hidden topic Z_k of document d can influence other papers, which can be computed as follows:

$$I_d(Z_k) = \frac{\dfrac{n_d^{(k)} + \alpha}{\sum_{i=1}^{K}(n_d^{(i)} + \alpha)} \dfrac{n_{d'}^{(k)} + \alpha}{\sum_{i=1}^{K}(n_{d'}^{(i)} + \alpha)}}{\sum_{j=1}^{K} \dfrac{n_d^{(j)} + \alpha}{\sum_{i=1}^{K}(n_d^{(i)} + \alpha)} \dfrac{n_{d'}^{(j)} + \alpha}{\sum_{i=1}^{K}(n_{d'}^{(i)} + \alpha)}} \tag{4}$$

Where $n_d^{(k)}$ denotes the number of times topic Z_k is observed in document d in the content and reference bag while $n_{d'}^{(k)}$ denotes the number of times topic Z_k is observed in the citing paper bag d'.

Therefore, those top ranked topics with high $I_d(Z_k)$ can indicate the influential aspects of the document d.

5   Experiments

In this section, we evaluate the performance of our proposed model and method for latest influential paper mining. We conducted experiments on a real research paper dataset. Through experiments, we want to answer following three questions:

Q1: Can Bi-Citation-LDA find newly published papers in each topic?
Q2: How's the quality of influential papers in each topic found? Can they best represent each topic?

Q3: Can Bi-Citation-LDA find the influential aspects of each paper?

We compare the experimental results with that of the Citation-LDA model [1] to demonstrate the effectiveness of our work. In our experiments, we choose $K = 100$ as our topic number which has a relatively low perplexity of the model.

5.1   Datasets

We use the ACL Anthology Network (ANN) [10] as the dataset in our experiments. The ACL Anthology is a digital archive of research papers in natural language processing (NLP). Major conference and journal papers in the NLP field can be found in this dataset. ANN has several releases from year 2002 to year 2013. The latest 2013 release contains 21,212 papers (including citing papers and cited papers) with 110,930 citations. We use paper title as paper content and each paper has a bag of reference paper IDs and citing paper IDs as inputs.

5.2   Finding Latest Influential Papers

To answer the first question, Q1, we compare Bi-Citation-LDA with Citation-LDA using the top 15 papers found in topic "sentiment analysis", the same one analyzed in [1]. We run the model with datasets of different years (from year 2002 to year 2013). We choose these four years because *sentiment analysis* was put forward in year 2002[1] and during these four years, a large number of papers related to *sentiment analysis* has been published. Tables 2 and 3 show the top 15 influential and milestone papers based on Bi-Citation-LDA and Citation-LDA respectively.

In the two tables, the two digit numbers following the first character in each *paper ID* represent the year when the paper was published. For example, C94-2174 was published in year 1994. Column *weight* represents $\psi(d_i, Z_k)$, which means how much a paper d_i can represent topic Z_k Papers marked in bold are newly published papers. Take the dataset of year 2002 for example. We can't find paper published in 2002 as milestone papers using model Citation-LDA, while our model discovers three papers. Similar things happen for years from 2003 to 2013, which means that our model Bi-Citation-LDA can discover newly published papers in each topic while citation-LDA can only find early published papers as milestone papers.

5.3   Influence Analysis

To answer the second question, Q2, we use the milestone papers found by Citation-LDA from 2013 ANN dataset as baseline. We think that if a paper is influential on a topic, it will gradually recognized and become a milestone paper for the topic. Therefore, we compare the influential papers mined by our model with the milestone paper found by Citation-LDA. Table 4 shows the top 15 high milestone papers found by citation-LDA in topic "sentiment analysis" with the 2013 ANN dataset.

---

[1] http://en.wikipedia.org/wiki/Sentiment_analysis.

Table 2. The top-15 influential papers in "sentiment analysis" using Bi-Citation-LDA

2002		2003		2004		2005	
PaperID	weight	PaperID	weight	Paper ID	weight	PaperID	weight
C94-2174	0.0463	C88-1016	0.046	P97-1023	0.0451	P02-1053	0.0515
P98-1112	0.0463	P99-1032	0.043	W02-1011	0.0412	W02-101	0.0481
P99-1032	0.0463	C94-2174	0.041	P99-1032	0.0393	P97-1023	0.0415
P97-1005	0.0432	P97-1005	0.039	P97-1005	0.0373	P99-1032	0.0315
W01-1626	0.0278	W02-1011	0.025	C94-2174	0.0373	P97-1005	0.0299
J93-1001	0.0247	W01-1626	0.023	P02-1053	0.0353	W03-1017	0.0299
P97-1023	0.0216	W03-1014	0.023	J93-2001	0.0294	W03-1014	0.0282
W02-1011	0.0216	P02-1053	0.021	W01-1626	0.0255	P04-1035	0.0282
W02-2034	0.0216	P84-1063	0.021	W03-1014	0.0196	C94-2174	0.0249
C00-1044	0.0185	P97-1023	0.018	W03-0404	0.0196	W01-1626	0.0216
C94-1092	0.0185	P98-1112	0.018	J04-3002	0.0177	W03-0404	0.0216
C86-1129	0.0185	W03-0404	0.018	W03-1017	0.0157	J04-3002	0.0183
H92-1041	0.0155	C00-1044	0.016	C00-1044	0.0157	H05-1073	0.0149
P88-1003	0.0155	P82-1005	0.016	P04-1035	0.0138	H05-1044	0.0149
P02-1053	0.0155	W03-1017	0.014	J00-4001	0.0118	C00-1044	0.0133

Table 3. The top-15 milestone papers in "sentiment analysis" using citation-LDA

2002		2003		2004		2005	
Paper ID	Weight	Paper ID	Weight	Paper ID	Weight	Paper ID	Weight
P97-1023	0.0759	P97-1005	0.0909	P97-1023	0.0649	P02-1053	0.0821
P98-2127	0.0633	C94-2174	0.0788	P97-1005	0.0573	P97-1023	0.0699
J94-2004	0.0506	P99-1032	0.0424	P02-1053	0.0496	W02-1011	0.0669
P99-1032	0.0506	P97-1023	0.0424	W02-1011	0.0458	P97-1005	0.0517
P97-1005	0.0506	C88-1026	0.0364	C94-2174	0.0458	C94-2174	0.0486
J93-2004	0.0506	J93-2001	0.0364	J93-1003	0.042	P04-1035	0.0365
J93-1003	0.0506	P02-1053	0.0303	P98-2127	0.0305	W03-1017	0.0304
W00-1302	0.038	P90-1014	0.0303	J93-2004	0.0305	J93-1003	0.0304
P99-1042	0.038	C88-1060	0.0303	W02-2001	0.0267	J94-2004	0.0243
H91-1061	0.038	P89-1032	0.0242	P99-1032	0.0267	W03-1014	0.0213
W98-1126	0.0253	E95-1002	0.0242	W03-1812	0.0191	C00-1044	0.0213
W01-1626	0.0253	C90-3022	0.0242	C00-1044	0.0191	P99-1032	0.0213
C00-1044	0.0253	W02-1011	0.0182	W03-1810	0.0191	W03-0404	0.0182
W01-1605	0.0253	C00-1044	0.0182	W03-1809	0.0191	J96-2004	0.0182
M91-1020	0.0253	A97-1015	0.0182	J94-2004	0.0191	P02-1022	0.0152

Table 4. Top-15 High Impact papers in Topic "Sentiment Analysis" with citation-LDA (2013)

PaperID	Weight	Paper title
W02-1011	0.0681	Thumbs Up? Sentiment Classification Using Machine Learning Techniques
P02-1053	0.0559	Thumbs Up Or Thumbs Down? Semantic Orientation Applied To Unsupervised Classification Of Reviews
H05-1044	0.0455	Recognizing Contextual Polarity In Phrase-Level Sentiment Analysis
P04-1035	0.0377	A Sentimental Education: Sentiment Analysis Using Subjectivity Summarization Based On Minimum Cuts
P97-1023	0.0319	Predicting The Semantic Orientation Of Adjectives
C04-1200	0.0264	Determining The Sentiment Of Opinions
H05-1043	0.0255	Extracting Product Features And Opinions From Reviews
W03-1014	0.0238	Learning Extraction Patterns For Subjective Expressions
W03-1017	0.0232	Towards Answering Opinion Questions: Separating Facts From Opinions And Identifying The Polarity Of Opinion Sentences
P07-1056	0.0154	Biographies Bollywood Boom-boxes and Blenders: Domain Adaptation for Sentiment Classification
P05-1015	0.0148	Seeing Stars: Exploiting Class Relationships For Sentiment Categorization With Respect To Rating Scales
J04-3002	0.0133	Learning Subjective Language
W03-0404	0.0122	Learning Subjective Nouns Using Extraction Pattern Bootstrapping
W06-0301	0.0116	Extracting Opinions Opinion Holders And Topics Expressed In Online News Media Text
C00-1044	0.0107	Effects Of Adjective Orientation And Gradability On Sentence Subjectivity

First, we compare the top-15 topic influential papers about "sentiment analysis" for year 2002 to 2005 found by Bi-Citation-LDA (as shown in Table 2) with the milestone papers given in Table 4. From the results we can see that the highlighted papers found by our model in Table 2 appear in Table 4 and the two newly published paper which marked in bold but not highlighted in Table 2 appear in the top-30 papers in 2013 Citation-LDA results. These results demonstrate that our Bi-Citation-LDA can find newly published papers that are influential papers and can be milestone papers several years later.

In the meantime, we want to demonstrate that our model can find the influential papers that can cover most of the milestone papers found by Citation-LDA. We compute the coverage proportion, the proportion of the top-10 milestone papers found by Citation-LDA appearing in the top-15 influential papers found by Bi-Citation-LDA. Figure 2 shows the coverage proportion. From the figure we can see that influential papers can cover most of the milestone papers. While in year 2004, the coverage proportion only reaches 60 percent. This is because that *Sentiment analysis* just emerged in year 2004. Papers published in this year had hardly been cited. Hence, the top-10 milestone papers found from this dataset is not very accurate, and some of them just slightly related to *sentiment analysis*.

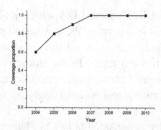

Fig. 2. Coverage proportion from year 2004 to 2010

5.4   Influential Aspect Analysis of Each Paper

To answer the third question, Q3, as we discussed in Sect. 4, for each document, we can get a doc-topic distribution I_d to represent the influential aspects of each paper. Table 5 shows some examples which are related to sentiment analysis. The number before each topic denotes the topic number in our results and the number in the bracket represents the topic weight in this document. Take paper *Convolution Kernels for Opinion Holder Extraction* published in 2011 for example. The top 2 topics of this paper are topic 14 *"extraction, relation"* and topic 37 *"sentiment analysis"*, while the most influential topic (most paper cited this paper for the topic) is topic 37 *"sentiment analysis"*. From the results, we can demonstrate that our Bi-Citation-LDA can find some influential aspects of each paper in the corpus.

Table 5. Results of influential aspect analysis

Year	Paper title	
2011	**Twitter Catches The Flu: Detecting Influenza Epidemics using Twitter**	
main topics	37 (0.0574) Sentiment analysis	66 (0.0410) Social media ` twitter
influential topics	66 (0.0410) Social media twitter	40 (0.0007) detection medical texts
2010	**Convolution Kernels for Opinion Holder Extraction**	
main topics	14 (0.1136) extraction ` relation	37 (0.0682) Sentiment analysis
influential topics	37 (0.0056) Sentiment analysis	14 (0.0031) extraction relation
2009	**Contrastive Summarization: An Experiment with Consumer Reviews**	
main topics	37 (0.0446)Sentiment analysis	98 (0.0268)document summarization
influential topics	98 (0.0017) document summarization	37 (0.0012)Sentiment analysis

6   Conclusion and Future Work

Influential paper mining is very important and challenging. Existing work about milestone paper mining can only find papers published many years ago. In this paper, we propose a novel approach for latest topic influential paper discovery and present a probabilistic generative model Bi-Citation-LDA which jointly considers the cited and citing papers of each document in a corpus. We conducted experiments on real datasets and results show that our model can find not only milestone papers but also newly published influential papers in each topic. In the future, we will further improve the model by introducing time decay factor. Also, our model can be applied to graph data for network clustering and ranking and we will study its application to this area.

Acknowledgement. This work was supported in part by National Natural Science Foundation of China under grant No. 71272029, 71432004, 71490724 and 61472426, the 863 program under grant No. 2014AA015204, and the Beijing Municipal Natural Science Foundation under grant No. 4152026.

References

1. Wang, X., Zhai, C., Roth, D.: Understanding evolution of research themes: a probabilistic generative model for citations. In: Proceedings of the 19th ACM SIGKDD International Conference on Knowledge Discovery and Data Mining, pp. 1115–1123. ACM (2013)
2. Nallapati, R.M., Ahmed, A., Xing, E.P., Cohen, W.W.: Joint latent topic models for text and citations. In: Proceedings of the 14th ACM SIGKDD International Conference on Knowledge Discovery and Data Mining, pp. 542–550. ACM (2008)
3. Erosheva, E., Fienberg, S., Lafferty, J.: Mixed-membership models of scientific publications. Proc. Nat. Acad. Sci. USA **101**(Suppl. 1), 5220–5227 (2004)
4. Eugene, G.: Citation analysis as a tool in journal evaluation. Am. Assoc. Advance. Sci. **178** (4060), 471–479 (1972)
5. Mirton, K.M.: Bibliographic coupling between scientific papers. Am. Documentation **14**(1), 10–25 (1963)
6. Kevin, B.W., Richard, K.: Co-citation analysis, bibliographic coupling, and direct citation: which citation approach represents the research front most accurately? J. Am. Soc. Inf. Sci. Technol. **61**(12), 2389–2404 (2010)
7. David, M.B., Ng, A.Y., Jordan, M.I.: Latent dirichlet allocation. J. Mach. Learn. Res. **3**, 993–1022 (2003)
8. Griffiths, T.L., Steyvers, M.: Finding scientific topics. Proc. Nat. Acad. Sci. USA **101** (Supp. 1), 5228–5235 (2004)
9. Saurabh, K., Mitra, P., Bhatia, S.: Utilizing context in generative bayesian models for linked corpus. In: AAAI, vol. 10 (2010)
10. Dragomir, R.R., Muthukrishnan, P., Qazvinian, V.: The ACL anthology network corpus. In: Proceedings of the 2009 Workshop on Text and Citation Analysis for Scholarly Digital Libraries, pp. 54–61. Association for Computational Linguistics. (2009)
11. Limin, Y., Mimno, D., McCallum, A.: Efficient methods for topic model inference on streaming document collections. In: Proceedings of the 15th ACM SIGKDD International Conference on Knowledge Discovery and Data Mining, pp. 937–946. ACM (2009)

Multi-label Chinese Microblog Emotion Classification via Convolutional Neural Network

Yaqi Wang[1], Shi Feng[1,2(✉)], Daling Wang[1,2], Ge Yu[1,2],
and Yifei Zhang[1,2]

[1] School of Computer Science and Engineering, Northeastern University,
Shenyang, China
wyq20100695@163.com, {fengshi,wangdaling,yuge,
zhangyifei}@cse.neu.edu.cn
[2] Key Laboratory of Medical Image Computing, Ministry of Education,
Northeastern University, Shenyang 110819, China

Abstract. Recently, analyzing people's sentiments in microblogs has attracted more and more attentions from both academic and industrial communities. The traditional methods usually treat the sentiment analysis as a kind of single-label supervised learning problem that classifies the microblog according to sentiment orientation or single-labeled emotion. However, in fact multiple fine-grained emotions may be coexisting in just one tweet or even one sentence of the microblog. In this paper, we regard the emotion detection in microblogs as a multi-label classification problem. We leverage the skip-gram language model to learn distributed word representations as input features, and utilize a Convolutional Neural Network (CNN) based method to solve multi-label emotion classification problem in the Chinese microblog sentences without any manually designed features. Extensive experiments are conducted on two public short text datasets. The experimental results demonstrate that the proposed method outperforms strong baselines by a large margin and achieves excellent performance in terms of multi-label classification metrics.

Keywords: Sentiment analysis · Emotion classification · Multi-label classification · Convolutional Neural Network

1 Introduction

With the fast development of social media, more and more people are willing to express their feelings and emotions via microblogging platform. The microblog has aggregated huge number of tweets that containing people's opinions about personal lives, celebrities, social events and so on. Therefore, analyzing people's sentiments in microblogs has attracted more and more attentions from both academic and industrial communities.

The traditional sentiment analysis problems usually classify the microblog into positive and negative orientations or identify people's fine-grained emotions in microblogs, such as *happiness, sadness, like, anger, disgust, fear, surprise* and so on.

© Springer International Publishing Switzerland 2016
F. Li et al. (Eds.): APWeb 2016, Part I, LNCS 9931, pp. 567–580, 2016.
DOI: 10.1007/978-3-319-45814-4_46

Table 1. The examples of coexisting multiple emotions in microblogs

Microblogs	Emotions
The movie tonight is fantastic, but the dinner sucks!	*happiness, like, disgust, anger*
The ending of this novel is so miserable and terrible, and I think I won't read it anymore.	*disgust, fear*
What an unbelievable farewell performance by Kobe Bryant! It's a pity that I can't see him in NBA anymore!	*surprise, like, sadness*

In these methods, each microblog is associated with a single sentiment orientation or emotion label. This is commonly referred as a single-label classification problem. However, in fact multiple emotions may be coexisting in just one tweet or even one sentence of the microblog, as shown in the following examples of Table 1.

Different from the traditional single-label sentiment classification problem, there are critical new challenges for detecting the emotions in microblogs. Firstly, the microblog usually has a length limitation of 140 characters, which leads to extremely sparse vectors for the learning algorithms. The free and informal writing styles of users also set obstacles for the emotion detection in microblogs. Secondly, although the text is short, multiple emotions may be coexisting in one microblog. As shown in Table 1, the second tweet expresses a *disgust* emotion as well as a *fear* emotion simultaneously. Therefore, for the emotion detection task, each microblog can be associated with multiple fine-grained emotion labels, which is rarely studied in the previous literature.

To tackle these challenges, in this paper, we regard the emotion detection in microblogs as a multi-label classification problem. We utilize word distributed representation to alleviate the sparse vector problem and leverage a Convolutional Neural Network (CNN) based method to detect the coexisting mixed emotions in the short text of microblogs.

Multi-label classification has been extensively studied in many applications [1]. However, only limited literature are available for analyzing the coexisting emotions in text. Ye et al. employed multi-label classification algorithm to classify the input online news articles into categories according to different readers' emotions [2]. Bhowmick et al. presented a novel method for classifying news into multiple emotion categories using an ensemble based multi-label classification technique called RAKEL [3]. Read et al. used restricted Boltzmann machines to develop better feature-space representations which could make labels less interdependent and easier to predict for multi-label classification [4]. Although the above two studies have achieved promising results, both of them focused on the online long news articles with dense vector space. Liu et al. gave detailed empirical study of different classic multi-label learning methods for the multiple emotion sentiment classification task of Chinese microblogs [5]. The goal of Liu's work is similar to ours. However, they conducted the experiments on two very small datasets that only had a few hundreds of microblogs and the classic learning methods they used were heavily dependent on the quality of the lexicons and manually designed features.

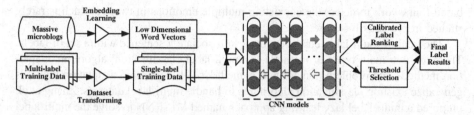

Fig. 1. The details of our proposed framework

In this paper, we utilize a Convolutional Neural Network (CNN) based method to solve multi-label emotion classification problem without any manually designed features in the Chinese microblogs. The details of our proposed framework are shown in Fig. 1. (1) We preprocess the microblogs and transform the multi-label training dataset into single-label dataset. (2) We utilize skip-gram language model and extremely large corpus to train the distributed representation of each word in the single-label dataset. The dense vector representations of the microblog sentences are generated by the word vector with appropriate semantic composition method. (3) We use a CNN based method to train one multi-class classifier and k binary classifiers to get the probability values of the k emotion labels. (4) We leverage a multi-label learning method based on Calibrated Label Ranking (CLR) to get the final emotion labels of each microblog. As a powerful deep learning algorithm, CNN has achieved remarkable performance in computer vision and speech recognition. However, as far as we know, CNN has never before been reported for solving the multi-label text classification task.

The main contributions of this paper are summarized as follows: (1) To the best of our knowledge, this is the first attempt to explore the feasibility and effectiveness of CNN model with the help of word embedding to detect the coexisting emotions in the microblog short text. (2) We utilize skip-gram language model to get word distributed representation instead of the TFIDF based extremely sparse vectors. Meanwhile, the input features are learned by CNN automatically rather than being designed manually. (3) We conduct extensive experiments on two public short text datasets. The experimental results demonstrate that the proposed method achieves excellent performance in terms of multi-label classification metrics.

2 Related Work

Analyzing the sentiments and emotions in microblogs has attracted increasing research interest in recent years. The methods used for microblog sentiment analysis generally can be classified into lexicon-based [6] and learning-based [7] categories. The performance of many existing learning based methods are heavily dependent on the choice of feature representation of tweets. For example, Mohammad et al. designed a number of hand-crafted features to implement a twitter sentiment classification system that achieved the best performance of SemEval 2013 task [8]. Unlike the previous studies, on one hand the CNN model used in this paper could automatically learn discriminative features for the classification task. On the other hand, we utilize the multi-label learning based framework to detect the coexisting multiple emotions in tweets, which is rarely studied in previous literature.

The goal of multi-label classification is to associate a sentence with a set of labels. There are two main types of methods for multi-label classification: algorithm adaptation methods and problem transformation methods [9]. Algorithm adaptation methods generalize existing classification algorithms to handle multi-label datasets. Zhang et al. proposed a multi-label lazy learning approach named ML-KNN to solve the multi-label problem [10]. Problem transformation methods transform a multi-label classification problem into single-label problems. Wang et al. proposed a calibrated label ranking based framework for detecting the multi-label fine-grained emotions in the Chinese microblogs [11]. According to the detailed empirical study of different multi-label classification methods by [5] and our previous work, we utilize the Calibrated Label Ranking (CLR) as the learning method for the multi-label microblog emotion classification task.

As a powerful deep learning algorithm, Convolutional Neural Network could decrease the number of features by capturing local semantic features and is to use end-to-end automatically trainable systems which do not rely on human-designed heuristics. CNN has been successfully used in image recognition, speech recognition [12], and machine translation [13]. Wei et al. propose a flexible deep CNN infrastructure, called Hypotheses-CNN-Pooling (HCP), to solve the multi-label problem of images [14]. In recent years, CNN has achieved remarkable performances for various challenging NLP tasks, such as part-of-speech tagging, chunking, named entity recognition and semantic role labeling [15]. Santos et al. proposed a new deep CNN that exploited from character- to sentence-level information to perform sentiment analysis of short text [16]. Kim et al. proposed a simple CNN architecture to allow for the use of both task-specific and static vectors for sentence-level classification tasks [17]. Although a lot of papers have been published for CNN based text classification, no literature has been reported for exploiting the feasibility and effectiveness of CNN model for the multi-label microblog emotion classification task. In this paper, we introduce a CNN framework to solve multi-label emotion classification problem in the Chinese microblogs without any manually designed features.

3   Microblog Modeling Using Convolutional Neural Network

In this section, we propose a CNN framework to solve multi-label sentiment classification problem in the Chinese microblogs. Suppose $Y = \{y_1, y_2, \ldots, y_k\}$ represents the emotion label set. Giving the training microblog dataset $D = \{(x_i, Y_i) | 1 \leq i \leq p\}$ where p is the number of sentences in the D, x_i is a sentence, Y_i represents the emotion label subset, $Y_i \subseteq Y$. Our aim is to use a CNN classifier to predict the emotion label set $h(x) \subseteq Y$ of a sentence x in test dataset T.

3.1   Preprocessing of Datasets and Training Dataset Transformation

The raw microblog data is inconsistency and redundancy. We remove all the URLs, @ symbols, stop words and English tweets. Then, we use Jieba[1] python package for Chinese text segmentation which break the microblogs into words.

Because we use Calibrated Label Ranking (CLR) as method for multi-label classification, we need to transform the multi-label training microblog dataset into single-label dataset. The goal of problem transformation method is to transform the multi-label problem into other well-established learning scenario. For the given training dataset, each sentence in the microblog may contain one or more relevant emotion labels, therefore we need to transform the multi-label training dataset into the single-label train dataset so that we can use the CNN learning algorithm to address the multi-label learning problem. The dataset transformation method is shown in Algorithm 1.

Algorithm 1. Training dataset transformation

Input: The multi-label training microblog sentence dataset D;

Output: The single-label training dataset D' after transformed;

Description:

1. FOR every microblog sentence x_i, where Y_i is the relevant emotion label set of the sentence x_i and $D = \{(x_i, Y_i) | 1 \le i \le p\}$, where p is the number of microblog sentences in dataset D.

2. FOR every label y_j in the label set Y, where $Y = \{like, happiness, sadness, fear, surprise, disgust, anger\}$ or $Y = \{anger, anxiety, expect, hate, joy, love, sorrow, surprise\}$.

3. If $y_j \in Y_i$, put the microblog sentence (x_i, y_j) into D'.

In Algorithm 1, we associate each microblog sentence with only one emotion label, which means that a sentence x_i in D may appear multiple times in D'.

3.2   Convolutional Neural Network Model

The CNN model is a variant of neural network, which includes the input layer, convolutional layer, pooling layer and fully connected layer. The CNN model used in this paper for modeling the microblog sentences is shown in Fig. 2 below.

The neurons in CNN are usually arranged in three dimensions: depth, width, and height. The size of input layer, convolutional layer, pooling layer and output layer are depth $\times$ width $\times$ height. For example, when a sentence has 70 words and each word is represented by 300 dimensional vector, the size of input layer is $1 \times 70 \times 300$.

Input Layer. This layer contains the input data. Let $s_i \in R^k$ be the k-dimensional word vector corresponding to the i-th word in the sentence. The max sentence length in the datasets is n. If the length is less than n, the sentence is padded necessarily. Therefore, every sentence is represented as

---

[1] https://github.com/fxsjy/jieba.

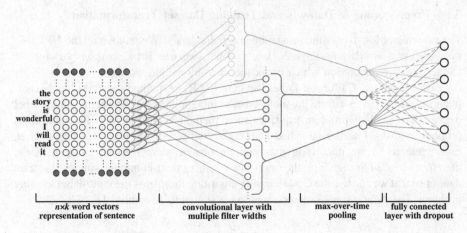

Fig. 2. Convolutional approach to sentence-level classification with different filter widths

$$s_{1:n} = s_1 \oplus s_2 \oplus \ldots \oplus s_n \tag{1}$$

where $\oplus$ is the concatenation operator.

Convolutional Layer. In this layer, local semantic features can be captured and the connection between two connected neuron represents the layer of convolutional and input. Here a convolution operation involves a filter $w \in \mathbb{R}^{hk}$, which is applied to a window of h words to produce a new feature. The filter w can also be considered as the weight matrix in the convolution layer. We can calculate the value of h words to generate new feature z_i:

$$z_i = f(w \cdot s_{i:i+h-1} + b) \tag{2}$$

where f is an activation function such as rectified linear units and $b \in \mathbb{R}$ is a bias term. This filter is applied to the sentence $\{s_{1:h}, s_{2:h+1}, \ldots, s_{n-h+1:n}\}$ to produce a new feature map $z = [z_1, z_2, \ldots, z_{n-h+1}]$, with $z \in \mathbb{R}^{n-h+1}$.

Pooling Layer. In this layer, the max-over-time pooling layer selects global semantic features and attempts to capture the most important feature with the highest value for each feature map. The pooling layer is aimed at sampling with common statistics methods, such as mean value, max value and L2-norm. By doing this, the problem of overfitting is addressed to a certain degree. We then apply max-over-time pooling operation over the new feature map by convolution operation and take maximum value $z_{max} = \max(z)$ as the feature corresponding to the filter. Because the model employs several parallel convolutional layers with different window sizes of words, we obtain multiple features so that we can get the input vectors for classifier.

Fully Connected Layer. In this layer, the dropout regularization method is used to summarize the features. The neurons in fully connected layers have full connections with all neurons in the previous layer. The structure of a fully connected layer is the same as that of the layer in the ordinary neural network. In order to avoid overfitting and approximate a way to solve the exponentially neural network computational architectures, we apply dropout on the fully connected layers. When training the model, the neurons which is dropped out has a probability p to be temporarily removed from the network, as shown in Fig. 2. When calculating input and output both in the forward propagation and the back propagation, the dropout neurons are ignored. That means the dropout will prevent neurons from co-adapting too much.

Softmax Layer. Finally, these features are passed to the softmax layer and the output of softmax layer is the probability distribution over labels.

Input Word Vectors. The input of the CNN model is the distributed representation of the words, namely word vectors, in the microblog sentences. These vectors could be initialized randomly during CNN model training. But a popular method to improve the performance of CNN is to initialize word vectors by unsupervised neural language model. In this paper, we utilize the skip-gram model [18] to pre-train word vectors from extremely large corpus that has more than 30 million Chinese microblogs. Throughout the training process of CNN, one strategy is to keep the pre-trained word vectors static, another strategy is to fine-tune the vectors via back propagation.

4 Calibrated Label Ranking Based Multi-label Classification

Based on the transformed single label microblog dataset and the CNN classification algorithm, in this section we leverage the calibrated label ranking method to detect the multiple emotions in the microblog short text.

4.1 Ranking Function

A single label multi-class classifier $f(\cdot)$ is built, so that we can get the confidence of the emotion labels for a given microblog sentence. We use $f(x_i, y_j)$ represents the confidence that sentence x_i has the probability of emotion label y_j. The $f(x_i, y_j)$ can be transformed to a rank function $rank_f(x_i, y_j)$ subsequently and we use the Formula 3 to get the rank value.

$$rank_f(x_i, y_j) = \sum_{k=1, j\neq k}^{q} [\![f(x_i, y_j) > f(x_i, y_k)]\!] \tag{3}$$

where $y_j \in Y$, Y is the label set, q is the number of labels, $\delta = 1$ if the predicate δ holds, and 0 otherwise.

4.2 Threshold Selection and Label Set Result

After we get the rank values of all labels by the rank function, the key issue is to gain a threshold to divide the rank value list into relevant label set and irrelevant label set. In order to determine the threshold, the processing procedure is as follows:

(1) Transform the multi-class single label training dataset into q separate datasets and each dataset is associated with a binary label (with/without an emotion label).
(2) For q binary labels training datasets, we construct q binary CNN classifiers $g_j(x_i)$, $j \in \{1, 2, \ldots, q\}$ and x_i represents a sentence. We use the Formula 4 to construct the training dataset of the q binary classifiers.

$$D_j = \left\{ (x_i, \phi(Y_i, y_j)) | 1 \leq i \leq p \right\}$$

$$\phi(Y_i, y_j) = \begin{cases} +1, y_j \in Y_i \\ -1, otherwise \end{cases} \tag{4}$$

where D_j is the train dataset, $j \in \{1, 2, \ldots, q\}$, y_j represent the emotion label, p is the number of sentences in the training dataset and Y_i is the relevant label set of the sentence x_i. If the label y_j in the labels set Y_i, the $\phi(Y_i, y_j) = +1$, otherwise $\phi(Y_i, y_j) = -1$.

The q binary classifiers $g_j(x_i)$ trained by the separate training datasets classify every sentence in the testing dataset. If x_i is corresponding with y_j, then $g_j(x_i) = +1$, else $g_j(x_i) = -1$.

(3) Apply the output results of the q binary classifiers to product the threshold and the rank value for every microblog sentence in the testing dataset. The rank value $rank_f(x_i, y_j)$ of label y_j, and the threshold value $rank_f^*(x_i, y_v)$ is updated as follow.

$$rank_f^*(x_i, y_j) = rank_f(x_i, y_j) + [\![g_j(x_i) > 0]\!] \tag{5}$$

$$rank_f^*(x_i, y_v) = \sum_{j=1}^{q} [\![g_j(x_i) < 0]\!] \tag{6}$$

Finally, we compare the rank value with the threshold value so that we can get the relevant label set and irrelevant label set. If the rank value of a label is bigger than the threshold, namely $rank_f^*(x_i, y_v)$, we put the label into the relevant label set. Otherwise we put it into the irrelevant label set. We regard the relevant label set as the final result of multi-label classification. The formula of multi-label emotion classifier $h(x_i)$ is:

$$h(x_i) = \left\{ y_j | rank_f^*(x_i, y_j) > rank_f^*(x_i, y_v), y_j \in Y \right\} \tag{7}$$

In summary, for a sentence in test dataset, the process of predicting its multi-label emotion set is shown as Algorithm 2.

Algorithm 2. Predicting Multi-label Emotion Set for a Sentence in Test Dataset

Input: a sentence x, word vectors, dataset D' from Algorithm 1;

Output: emotion set of x;

Description:

    1. Transform x and D' into the feature representation of word vectors.

    2. Put the word vectors of x and D' into CNN model.

    3. Get multi-class probability of emotion label y_j and multiple binary labels.

    4. Apply Calibrated Label Ranking algorithm on the multi-class probability of emotion label y_j to get the rank value:

$$rank_f(x_i, y_j) = \sum_{k=1, j \neq k}^{q} [\![f(x_i, y_j) > f(x_i, y_k)]\!]$$

    5. Apply threshold selection on multiple binary labels to update rank value and get threshold value:

$$rank_f^*(x_i, y_j) = rank_f(x_i, y_j) + [\![g_j(x_i) > 0]\!]$$
$$rank_f^*(x_i, y_v) = \sum_{j=1}^{q} [\![g_j(x_i) < 0]\!]$$

    6. Get final result of multi-label classification, that is emotion set of x:

$$h(x_i) = \{ y_j | rank_f^*(x_i, y_j) > rank_f^*(x_i, y_v), y_j \in Y \}$$

    7. Return the emotion set of x.

---

5 Experiment

5.1 Experiment Setup and Datasets

We conduct our experiment on the NLPCC 2014 Emotion Analysis in Chinese Weibo Text (EACWT) task[2] and Ren_CECps [19] dataset respectively. In the EACWT, the training dataset contains 15,677 microblog sentences, the testing dataset contains 3,816 microblog sentences, the fine-grained emotion label set contains seven basic emotions, i.e. Y = {*like, happiness, sadness, fear, surprise, disgust, anger*} and every sentence has less than two emotion labels. In the Ren_CECps, the training dataset contains 26,000 Chinese blog sentences, the testing dataset contains 6,153 Chinese blog sentences, the fine-grained emotion label set contains eight basic emotions, i.e. Y = {*anger, anxiety, expect, hate, joy, love, sorrow, surprise*} and every sentence has less than three emotion label. Note that Ren_CECps is a Chinese blog dataset which contains documents with different topics. However, the average length of the sentences is similar to that of the microblog dataset EACWT. Therefore, we can utilize the sentences in Ren_CECps as well as EACWT to evaluate the performance of our proposed short text multi-label emotion classification algorithm.

Table 2 show the statistics information of the number of sentences in the datasets with different number of emotion labels. In Table 2, the Ren_CECps dataset has much more sentences with two emotion labels. We argue that using these two datasets with different emotion label distributions, we can better evaluate the feasibility and effectiveness of our proposed model and method.

---

[2] http://tcci.ccf.org.cn/conference/2014/pages/page04_eva.html.

Table 2. The number of sentences in the datasets with different number of emotion labels

# of emotion labels in sentence	# of sentences in the dataset	
	EACWT	Ren_CECps
1	17,545	18,812
2	1,948	11,416
3	0	1,824

5.2 Hyper-parameters Settings

For the CNN model we use: rectified linear units, filter windows (h) of 2, 3, 4 with 100 feature maps each, dropout rate (p) of 0.5, l_2 constraint (s) of 3, mini-batch size of 50 based on stochastic gradient descent. The Adadelta update rule is used. The max sentence length n is 70 and 90 in the EACWT and Ren_CECps respectively.

We utilize the well-known word embedding tool word2vec to train the word vectors on a crawled 30 million Chinese microblog corpus. The word2vec hyper-parameters of our pre-trained word vectors includes: the selected algorithm is skip-gram, the size of dimensionality is 50, 100, 200 and 300 respectively, the size of window is 8, the value of sample is 1e-4, and the times of iteration is 15.

5.3 Model Variations and Baselines

In this paper, we conduct the experiments with several variants of the CNN model. The SVM classifier that has achieved excellent performance in many previous sentiment classification tasks is viewed as the baseline model.

SVM-tfidf: We use the tfidf to generate the unigram feature vectors, and apply SVM classifier with CLR algorithm to solve the microblog multi-label classification problem. In the previous work, the SVM-tfidf method has achieved the best performance in the EACWT evaluation task [11]. We regard it as a strong baseline.

SVM-word2vec: We use the vector semantic composition based degeneration algorithm for modeling the microblog sentence. The vectors of words in the sentence are added together to form the distributed representation of the microblog sentence.

CNN-static: The inputs of CNN are the pre-trained vectors of words in microblog sentences from word2vec. If the word is not in the set of pre-trained word vectors, the unknown one is randomly initialized. Here "-static" means that the word vectors are kept static during training and only other parameters of the model are learned.

CNN-non-static: It is the same as the CNN-static but the pre-trained vectors are fine-tuned during training for different datasets.

CNN-rand: The input of all words vector are randomly initialized and modified during training.

Evaluation metrics. The classic multi-label learning evaluation metrics Ranking Loss (RL), Hamming Loss (HL), One-Error (OE), Subset Accuracy (SA), macro F1 (maF), micro F1 (miF) and Average Precision (AP) are selected for evaluating the performance of the proposed model and algorithm. The seven metrics can be grouped into label-based metrics, example-based metrics and ranking-based metrics. For more details about these metrics, please refer to Madjarov et al. [20] and Zhang et al. [9].

5.4 Experimental Results

We show the best result of every model in the Table 3 and Table 4. "↓" means the smaller values are better. "↑" means the larger values are better. Meanwhile, the SVM-word2vec-300, CNN-static-300, CNN-non-static-300 and CNN-rand-300, "-300" represent that the input word vectors have the dimensionality of 300 in the corresponding models. We can clearly see that in EACWT dataset the model CNN-non-static-300 has achieved the best performances under the every evaluation metric comparing with the other models in the Table 3. In Ren_CECps dataset, CNN-static-300 is better than other baselines and CNN models under all the metrics except Hamming Loss in the Table 4.

Based on the observations from above two tables, we have following discussions.

(1) With appropriate parameter settings, the CNN based models outperform the SVM algorithms. We argue that the CNN models can automatically learn the higher level of feature representation, so it is the better than the SVM based methods.

(2) The performance of CNN models can be significantly improved by using pre-trained word vectors. This is because that compared with randomly initialized vectors, pre-trained vectors could learn the semantic information of the word from extremely large corpus. Moreover, pre-trained vectors can be fine-tuned on the dataset in "non-static" model, so that vectors indeed reflect the word semantic in the dataset. We can see that CNN-non-static model has achieved best performances in EACWT dataset.

(3) The "non-static" model does not perform best as we expected in the Ren_CECps dataset. This may because that Ren_CECps contains various topics and has more scattered emotion labels. Different documents have the different topics. So it is more difficult for learning the effective feature representations and semantic vectors. The CNN-static model has achieved the best performances in Ren_CECps dataset. A similar observation of "static" model is reported for a single-label text classification task in [17]. We think the performance of "static" and "non-static" CNN models are related to the topic and label distribution for the text classification.

Finally, we demonstrate the effects of word vector dimension for EACWT dataset in Fig. 3. Generally, we find that when the word dimension is bigger, a better performance for the CNN-non-static model could be achieved. Similar performances can be observed in Ren_CECps dataset. We omit the results of micro F1 metric in EACWT and all the metrics in Ren_CECps due to length limitation.

Table 3. Microblog multi-label emotion classification results on the EACWT dataset

Algorithm	RL↓	HL↓	OE↓	SA↑	maF↑	miF↑	AP↑
SVM-tfidf	0.1879	0.2045	0.3160	0.6014	0.5532	0.6014	0.7250
SVM-word2vec-300	0.1805	0.2130	0.3052	0.5767	0.5373	0.5767	0.7257
CNN-static-300	0.1807	0.2201	0.2942	0.5655	0.5401	0.5655	0.7273
CNN-non-static-300	**0.1645**	**0.1968**	**0.2628**	**0.6132**	**0.6005**	**0.6132**	**0.7556**
CNN-rand-300	0.2098	0.2336	0.3511	0.5482	0.5173	0.5482	0.6926

Table 4. Microblog multi-label emotion classification results on the Ren_CECps dataset

Algorithm	RL↓	HL↓	OE↓	SA↑	maF↑	miF↑	AP↑
SVM-tfidf	0.1914	**0.2888**	0.4965	0.2665	0.3095	0.2665	0.6125
SVM-word2vec-300	0.1917	0.3043	0.4687	0.2790	0.3491	0.2790	0.6226
CNN-static-300	**0.1882**	0.3164	**0.4352**	**0.2889**	**0.3868**	**0.2889**	**0.6320**
CNN-non-static-300	0.1893	0.3269	0.4493	0.2668	0.3682	0.2668	0.6219
CNN-rand-300	0.1996	0.3778	0.5264	0.2192	0.2780	0.2192	0.5782

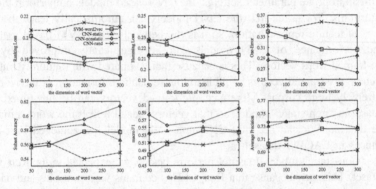

Fig. 3. The experiment results for different word dimensions for the EACWT dataset

6   Conclusions and Future Work

Different from the traditional sentiment orientation classification problem, multiple fine-grained emotions can co-exist in the microblog sentences, which can be regarded as a multi-label learning problem. In this paper, we propose a Convolutional Neural Network framework using Calibrated Label Ranking to detect the multiple emotions in the Chinese microblog sentences. Extensive experiments on two dataset show that static and non-static CNN methods with the help of word embedding could achieve better performances than other strong baseline algorithms by a large margin.

In the future, on one hand we want to design new problem transformation methods for the multiple emotion detection task. On the other hand, we intend to revise the architecture of CNN models for multi-label learning problem and incorporate other deep learning models, such as RNN, to further improve the performance of the task.

Acknowledgements. The work is supported by National Natural Science Foundation of China (61370074, 61402091), the Fundamental Research Funds for the Central Universities of China under Grant N140404012.

References

1. Tsoumakas, G., Katakis, I., Vlahavas, I.P.: Mining multi-label data. In: Maimon, O., Rokach, L. (eds.) Data Mining and Knowledge Discovery Handbook, pp. 667–685. Springer, Heidelberg (2010)
2. Ye, L., Xu, R., Xu, J.: Emotion prediction of news articles from reader's perspective based on multi-label classification. In: Proceedings of ICMLC, pp. 2019–2024 (2012)
3. Bhowmick, P.K.: Reader perspective emotion analysis in text through ensemble based multi-label classification framework. Comput. Inf. Sci. (CCSECIS) **2**(4), 64–74 (2009)
4. Read, J., Pérez-Cruz, F.: Deep learning for multi-label classification. CoRR abs/1502.05988 (2015)
5. Liu, S., Chen, J.: A multi-label classification based approach for sentiment classification. Expert Syst. Appl. **42**(3), 1083–1093 (2015)
6. Tang, D., Wei, F., Qin, B., Zhou, M., Liu, T.: Building large-scale Twitter-specific sentiment lexicon: a representation learning approach. In: Proceedings of COLING, pp. 172–182 (2014)
7. Liu, K., Li, W., Guo, M.: Emoticon smoothed language models for Twitter sentiment analysis. In: Proceedings of AAAI (2012)
8. Mohammad, S.M., Kiritchenko, S., Zhu, X.: NRC-Canada: building the state-of-the-art in sentiment analysis of Tweets. CoRR abs/1308.6242 (2013)
9. Zhang, M., Zhou, Z.: A review on multi-label learning algorithms. IEEE Trans. Knowl. Data Eng. **26**(8), 1819–1837 (2014)
10. Zhang, M., Zhou, Z.: ML-KNN: a lazy learning approach to multi-label learning. Pattern Recogn. (PR) **40**(7), 2038–2048 (2007)
11. Wang, M., Liu, M., Feng, S., Wang, D., Zhang, Y.: A novel calibrated label ranking based method for multiple emotions detection in Chinese microblogs. In: Zong, C., Nie, J.-Y., Zhao, D., Feng, Y. (eds.) NLPCC 2014. CCIS, vol. 496, pp. 238–250. Springer, Heidelberg (2014)
12. Abdel-Hamid, O., Deng, L., Yu, D.: Exploring convolutional neural network structures and optimization techniques for speech recognition. In: Proceedings of INTERSPEECH, pp. 3366–3370 (2013)
13. Meng, F., Lu, Z., Wang, M., Li, H., Jiang, W., Liu, Q.: Encoding source language with convolutional neural network for machine translation. ACL (1), pp. 20–30 (2015)
14. Wei, Y., Xia, W., Huang, J., Ni, B., Dong, J., Zhao, Y., Yan, S.: CNN: single-label to multi-label. CoRR abs/1406.5726 (2014)
15. Collobert, R., Weston, J.: A unified architecture for natural language processing: deep neural networks with multitask learning. In: Proceedings of ICML pp. 160–167 (2008)
16. Santos, C., Gatti, M.: Deep convolutional neural networks for sentiment analysis of short texts. In: Proceedings of COLING pp. 69–78 (2014)
17. Kim, Y.: Convolutional neural networks for sentence classification. In: Proceedings of EMNLP pp. 1746–1751 (2014)

18. Mikolov, T., Sutskever, I., Chen, K., Corrado, G., Dean. J.: Distributed representations of words and phrases and their compositionality. In: Proceedings of NIPS (2013)
19. Quan, C., Ren, F.: A blog emotion corpus for emotional expression analysis in Chinese. Comput. Speech Lang. (CSL) 24(4), 726–749 (2010)
20. Madjarov, G., Kocev, D., Gjorgjevikj, D., Dzeroski, S.: An extensive experimental comparison of methods for multi-label learning. Pattern Recogn. (PR) 45(9), 3084–3104 (2012)

Mining Recent High Expected Weighted Itemsets from Uncertain Databases

Wensheng Gan[1], Jerry Chun-Wei Lin[1(✉)], Philippe Fournier-Viger[2], and Han-Chieh Chao[1,3]

[1] School of Computer Science and Technology, Shenzhen, China
wsgan001@gmail.com, jerrylin@ieee.org, hcc@ndhu.edu.tw
[2] School of Natural Sciences and Humanities Harbin Institute of Technology
Shenzhen Graduate School, Shenzhen, China
philfv@hitsz.edu.cn
[3] Department of Computer Science and Information Engineering,
National Dong Hwa University, Hualien, Taiwan

Abstract. Weighted Frequent Itemset Mining (WFIM) has been proposed as an alternative to frequent itemset mining that considers not only the frequency of items but also their relative importance. However, some limitations of WFIM make it unrealistic in many real-world applications. In this paper, we present a new type of knowledge called Recent High Expected Weighted Itemset (RHEWI) to consider the recency, weight and uncertainty of desired patterns, thus more up-to-date and relevant results can be provided to the users. A projection-based algorithm named RHEWI-P is presented to mine RHEWIs based on a novel *upper-bound downward closure (UBDC)* property. An improved algorithm named RHEWI-PS is further proposed to introduce a *sorted upper-bound downward closure (SUBDC)* property for pruning unpromising candidates. An experimental evaluation against the state-of-the-art HEWI-Uapriori algorithm is carried on both real-world and synthetic datasets, and the results show that the proposed algorithms are highly efficient and acceptable to mine the required information.

Keywords: Weighted frequent itemset · Recency constraint · Uncertian data · Upper bound · *SUBDC* strategy

1 Introduction

In the field of data mining, pattern mining is an important topic consisting of discovering interesting patterns such as association rules [3,9], sequential patterns [6], among others [11,17]. Agrawal et al. proposed the Apriori algorithm [4] to mine association rules by firstly discovering the set of frequent itemsets. Frequent Itemset Mining (FIM) has become an important data mining task having a wide range of real-world applications. However, an important limitation of FIM [3,9] is that it considers the frequencies of items/itemsets in a transactional database but the other implicit factors, such as their weight, interest,

© Springer International Publishing Switzerland 2016
F. Li et al. (Eds.): APWeb 2016, Part I, LNCS 9931, pp. 581–593, 2016.
DOI: 10.1007/978-3-319-45814-4_47

risk or profit. To address this issue, the problem of Weighted Frequent Itemset Mining (WFIM) was proposed by considering both the weight (importance) and frequency of patterns [7, 8, 11, 15–17].

Cai et al. [7] first defined a weighted-support model. Wang et al. [16] assigned different weights to various items and mined Weighted Association Rules (WAR). Tao et al. [15] developed the WARM (Weighted Association Rule Mining) algorithm and designed the *weighted downward closure* property to mine weighted frequent itemsets (WFIs). In recent years, many interesting issues in WFIM have been extensively studied, and several extensions have been proposed in many other fields, such as mining weighted association rules without pre-assigned weights [14], infrequent weighted itemset mining [8], multilingual summarizer on WFIs [10], and so on. Although WFIM can reveal more useful information than FIM, WFIM still suffers from several limitations. For example, it does not consider how recent the discovered patterns are. As a result, algorithms of WFIM may discover many WFIs that only appeared in the far past and irrelevant to the up-to-date decision-making. It is unfair to measure the interestingness of patterns without considering their recency. Lin et al. thus proposed the RWFIM algorithm to discover the recent WFIs [11]. In addition, the huge amount of the collected data may inaccurate, imprecise, or incomplete from wireless sensor network [2], which reveal the data uncertainty property. Thus, it is a crucial challenge to design an effective algorithm and solve the above limitations.

To the best of our knowledge, no algorithm has been yet proposed to address the problem for mining up-to-date weighted itemsets in uncertain databases. The only related work is the HEWI-Uapriori algorithm [12], which was proposed to discover weighted itemsets in uncertain databases by using the level-wisely time-consuming approach. It may, however, return a huge amount of useless out-of-date weighted itemsets without considering the time-sensitive constraint. In this study, we address the issue to mine the interesting patterns from uncertain databases with weight, uncertainty and recency constraints. A novel type of pattern called Recent High Expected Weighted Itemsets (RHEWIs) is proposed to reveal more useful and meaningful HEWIs by considering both weight, probability, and recency constraints of the itemsets. Besides, a time-decay strategy is defined to automatically assign recency to transactions, which can be adjusted according to a user's preferences. Moreover, the projection-based RHEWI-P algorithm is proposed to efficiently mine RHEWIs by utilizing the *upper-bound downward closure* (*UBDC*) property. An improved version named RHEWI-PS is further proposed. It relies on a new *downward closure* property with sorted strategy to early prune unpromising candidates. Substantial experiments are conducted on both real-life and synthetic datasets to show that the presented algorithms have good performance to discover the set of complete RHEWIs in uncertain databases.

2   Preliminaries and Problem Statement

Let $I = \{i_1, i_2, \ldots, i_m\}$ be a finite set of m distinct items appearing in a transactional database $D = \{T_1, T_2, \ldots, T_n\}$, where each transaction $T_q \in D$ is a subset

Table 1. An uncertain database

TID	Occurred time	Transaction (*item, prob.*)
T_1	2015/1/08, 09:10	a:0.3, b:0.8, c:1.0
T_2	2015/1/09, 11:20	d:1.0, f:0.5
T_3	2015/1/11, 08:20	b:0.6, c:0.7, d:0.9, e:1.0, f:0.7
T_4	2015/1/12, 09:15	a:0.5, c:0.45, f:1.0
T_5	2015/1/12, 15:20	c:0.9, d:1.0, e:0.7
T_6	2015/1/14, 08:30	b:0.7, d:0.3
T_7	2015/1/14, 15:25	a:0.8, b:0.4, c:0.9, d:1.0, e:0.85
T_8	2015/1/15, 09:10	c:0.9, d:0.5, f:1.0
T_9	2015/1/16, 08:30	a:0.5, e:0.4
T_{10}	2015/1/18, 09:00	b:1.0, c:0.9, d:0.7, e:1.0, f:1.0

Table 2. Derived RHEWIs

Itemset	*expWSup*	*Recency*
(c)	5.750	3.7097
(d)	2.970	3.8954
(e)	3.160	3.2284
(f)	2.940	2.6927
(bc)	1.736	2.1663
(cd)	2.720	3.1009
(ce)	2.695	2.3784
(cf)	2.329	2.4202
(de)	2.126	2.3784
(cde)	2.080	2.3784

of I, and has a unique identifier called its *TID*. Based on the attribute uncertainty database model [2], a unique existence probability $p(i_j, T_q)$ is assigned to each item i_j in each transaction T_q. It indicates that an item i_j exists in T_q with a probability $p(i_j, T_q)$. A unique existence weight $w(i_j)$ is assigned to each item $i_j \in I$, which represents its importance (e.g., weight, interest, risk). Existence weights for all items are stored in a weight table $wtable = \{w(i_1), w(i_2), \ldots, w(i_m)\}$. An itemset $X = \{i_1, i_2, \ldots, i_k\}$ with k distinct items is called a k-itemset, X is said to be contained in a transaction T_q if $X \subseteq T_q$, and $TIDs(X)$ denotes the *TIDs* of all transactions in D containing X. As a running example, Table 1 shows an uncertain database containing 10 transactions, which are sorted by purchased time.

Definition 1. The weight of an item i_j in D is denoted as $w(i_j)$, and represents the importance of this item to the user ($w(i_j) \in (0, 1]$).

For example, a user could define the weight table of items in Table 1 as $wtable = \{w(a): 0.3, w(b): 0.4, w(c): 1.0, w(d): 0.55, w(e): 0.8, w(f): 0.7\}$.

Definition 2. The weight of an item i_j in T_q is defined as the weight of i_j in D. Thus, $w(i_j, T_q) = w(i_j), 1 \leq q \leq |D|$.

Definition 3. The weight of an itemset X in D is denoted as $w(X)$ and defined as the sum of the weights of all items in X divided by the number of items in X: $w(X) = \frac{\sum_{i_j \in X} w(i_j)}{|X|}$, where $|X|$ is the number of items in X.

For example, the weight of (b) in T_1 is $w(b, T_1) = w(b) = 0.4$, and the weight of (bcd) is calculated as $w(bcd) = (w(b) + w(c) + w(d))/3 = (0.4 + 1.0 + 0.55)/3 = 0.650$.

Definition 4. The weight of an itemset X in T_q is defined as the weight of the itemset X in D, that is: $w(X, T_q) = w(X)$.

For example, the weight of (bcd) in T_3 is $w(bcd, T_3) = w(bcd) = 0.650$.

Definition 5. The weighted support of an itemset X in D w.r.t. $wsup(X)$ is defined as: $wsup(X) = \sum_{X \subseteq T_q \land T_q \in D} w(X, T_q) = w(X) \times sup(X)$, where $sup(X)$ is the support of X in D.

For example, the (bcd) appears in transactions T_3, T_7 and T_{10}, the weighted support of (bcd) is calculated as $wsup(bcd) = \{w(bcd, T_3) + w(bcd, T_7) + w(bcd, T_{10})\} = w(bcd) \times 3 = (0.4 + 1.0 + 0.55)/3 \times 3 = 1.950$.

Definition 6. Assume that the transactions are sorted by purchase time, and the *Recency* of T_q is denoted as $R(T_q) = (1 - \delta)^{T_{current} - T_q}$, where δ is a user-specified time-decay factor ($\delta \in (0,1]$), $T_{current}$ is the current timestamp which is equal to the number of transactions in D, and T_q is the currently processed transaction.

Thus, a high *Recency* value is assigned to the most recent transactions. For example, assume that the user set the time-decay factor δ to 0.15. The *Recency* of T_1 and T_9 are respectively calculated as $R(T_1) = (1\text{-}0.15)^{(10-1)} = 0.2316$ and $R(T_9) = (1\text{-}0.15)^{(10-9)} = 0.85$.

Definition 7. The *Recency* of an itemset X in T_q is $R(X, T_q) = R(T_q)$.

Definition 8. The *Recency* of an itemset X in D is denoted as $R(X)$, and defined as: $R(X) = \sum_{X \subseteq T_q \land T_q \in D} R(X, T_q)$.

For example, the *Recency* of (c) in T_1 is $R(c, T_1) = R(T_1) = 0.2316$. The *Recency* of (bcd) in D is calculated as $R(bcd) = R(bcd, T_3) + R(bcd, T_7) + R(bcd, T_{10}) = 0.3206 + 0.6141 + 1.000 = 1.9347$.

Definition 9. The probability of an itemset X in T_q is denoted as $p(X, T_q)$, and defined as $p(X, T_q) = \prod_{i_j \in X} p(i_j, T_q)$, where j is the j-th item in X.

Definition 10. The expected support of an itemset X in the uncertain database is denoted as $expSup(X)$, which can be defined as the sum of the expected probabilities of X in the transactions as:

$$expSup(X) = \sum_{X \subseteq T_q \land T_q \in D} p(X, T_q) = \sum_{X \subseteq T_q \land T_q \in D} \left(\prod_{i_j \in X} p(i_j, T_q) \right).$$

Definition 11. The expected weighted support of an itemset X in an uncertain database D is denoted as $expWSup(X)$, which can be defined as the weight of X multiplied by the expected support of X as:

$$expWSup(X) = w(X) \times expSup(X) = w(X) \times \sum_{X \subseteq T_q \land T_q \in D} p(X, T_q).$$

Definition 12. An itemset X is defined as a high expected weighted itemset (HEWI) in an uncertain database D if its expected weighted support in D is no less than the minimum expected weighted support count as: $expWSup(X) \geq \alpha \times |D|$, where $|D|$ is the number of transactions in D and α is the minimum expected weighted support threshold.

Definition 13. An itemset X in an uncertain database D is said to be a recent high expected weighted itemset ($RHEWI$) if it satisfies the following two conditions: $RHEWI \leftarrow \{X | expWSup(X) \geq \alpha \times |D| \wedge R(X) \geq \beta\}$, where α is the minimum expected weighted support threshold and β is the minimum recency threshold.

For example, if we set $\beta = 2.0$ and $\alpha = 15\%$, respectively. The full set of RHEWIs of the running example database is shown in Table 2.

Definition 14. Given an uncertain database D, a weight table $wtable$ containing the weight of each item, the user-specified minimum expected weighted support threshold α and a minimum recency threshold β. The problem of high expected weighted itemset mining (abbreviated as RHEWIM) is to find the set of complete RHEWIs, and each pattern X satisfies the following two conditions as: (1) $expWSup(X) \geq \alpha \times |D|$; (2) $R(X) \geq \beta$.

3   Proposed RHEWI-P and RHEWI-PS Algorithms

3.1   Proposed Upper-Bound Downward Closure Property

A major challenge of the proposed RHEWI is that the well-known *downward closure* (*DC*) property used in FIM and WFIM does not hold in RHEWIM, that is the subsets of a RHEWI may or may not be RHEWIs. For example, consider the RHEWIs shown in Table 2 for the running example. The item (b) is not a RHEWI but its superset (bc) is a RHEWI. Thus, traditional approaches used in FIM and WFIM cannot be directly applied to solve this problem. Therefore, we introduce the *upper-bound downward closure (UBDC)* property to reduce the size of the search space for mining RHEWIs.

Definition 15 (Transactional upper-bound weight, *tubw*). The transactional upper-bound weight of a transaction T_q ($tubw(T_q)$) is defined as:

$$tubw(T_q) = max\{w(i_1, T_q), w(i_2, T_q), \ldots, w(i_j, T_q)\},$$

where j is the number of items in T_q.

For example, $tubw(T_1) = max\{w(a, T_1), w(b, T_1), w(c, T_1)\} = max\{0.3, 0.4, 1.0\} = 1.0$.

Theorem 1. *The weight of any itemset X in a transaction T_q is always smaller or equal to the transactional upper-bound weight of T_q. Thus, $w(X, T_q) \leq tubw(T_q)$.*

Proof. Since $tubw(T_q) = max\{w(i_1, T_q), w(i_2, T_q), \ldots, w(i_j, T_q)\}$, thus:

$$w(X, T_q) = \frac{\sum_{i_j \in X \wedge X \subseteq T_q} w(i_j, T_q)}{|X|} \leq \frac{max\{w(i_j, T_q)\} \times |X|}{|X|} = tubw(T_q).$$

Definition 16 (Transaction upper-bound probability, *tubp*). The transaction upper-bound probability of a transaction T_q is denoted as $tubp(T_q)$, and defined as:

$$tubp(T_q) = max\{p(i_1, T_q), p(i_2, T_q), \ldots, p(i_j, T_q)\},$$

in which j is the number of items in T_q.

Theorem 2. *The probability of any itemset X in a transaction T_q is always smaller or equal to the transaction upper-bound probability of T_q. Thus, $p(X, T_q) \leq tubp(T_q)$.*

Proof. Since $tubp(T_q) = max\{p(i_1, T_q), p(i_2, T_q), \ldots, p(i_j, T_q)\}$ and $p(i_j, T_q) \in [0, 1)$, $p(X, T_q) = \prod_{i_j \in X} p(i_j, T_q) \leq p(i_j, T_q) \leq (max\{p(i_j, T_q)\} = tubp(T_q))$. Therefore, it can be obtained that $p(X, T_q) \leq tubp(T_q)$.

Definition 17 (Transaction upper-bound weighted probability, *tubwp*). The transaction upper-bound weighted probability of a transaction T_q is denoted as $tubwp(T_q)$, and defined as: $tubwp(T_q) = tubw(T_q) \times tubp(T_q)$.

Definition 18 (Transactional accumulation upper-bound weighted probability, *taubwp*). The transactional accumulation upper-bound weighted probability of an itemset X in D is denoted as $taubwp(X)$ and defined as: $taubwp(X) = \sum_{X \subseteq T_q \land T_q \in D} tubwp(T_q)$.

For example, the *taubwp* of (bcd) is calculated as $taubwp(bcd) = tubwp(T_3) + tubwp(T_7) + tubwp(T_{10}) = 1.0 + 1.0 + 1.0 = 3.0$.

Definition 19. An itemset X in D is called a recent high expected weighted upper-bound itemset ($RHEWUBI$) if it satisfies the following two conditions as: $RHEWUBI \leftarrow \{X | taubwp(X) \geq \alpha \times |D| \land R(X) \geq \beta\}$.

For example, (b) is a RHEWUBI since $taubwp(b) = 4.385 > 1.5$, $R(b) = 2.6883 > 2.0$; but (bcd) is not a RHEWUBI since $taubwp(bcd) = 3.0 > 1.5$, $R(bc) = 1.9347 < 2.0$.

Theorem 3 (Upper-bound downward closure property of RHEWUBI, UBDC property). *Let X^k be a k-itemset. If X^k is a RHEWUBI, then any subset X^{k-1} of X^k is also a RHEWUBI, that is $taubwp(X^k) \leq taubw(X^{k-1})$ and $R(X^k) \leq R(X^{k-1})$.*

Proof. Since $X^{k-1} \subseteq X^k$, $TIDs(X^k) \subseteq TIDs(X^{k-1})$. Thus:

$$taubwp(X^k) = \sum_{X^k \subseteq T_q \land T_q \in D} tubwp(T_q) \leq \sum_{X^{k-1} \subseteq T_q \land T_q \in D} tubwp(T_q) = taubwp(X^{k-1}),$$
$$R(X^k) = \sum_{X^k \subseteq T_q \land T_q \in D} R(X^k, T_q) \leq \sum_{X^{k-1} \subseteq T_q \land T_q \in D} R(X^{k-1}, T_q) = R(X^{k-1}).$$

It thus can be concluded that if X^k is a RHEWUBI, any subset X^{k-1} is also a RHEWUBI. Thus the *UBDC* property holds and can be used to prune unpromising candidates w.r.t. the minimum expected weighted support threshold and minimum recency threshold during the search procedure for mining the desired RHEWIs.

Theorem 4 (RHEWIs $\subseteq$ RHEWUBIs). *Let RHEWUBIs be the set of recent high expected weighted upper-bound itemsets in the database D, and RHEWIs be the set of recent high expected weighted itemsets in D. The proposed upper-bound downward closure (UBDC) property implies that RHEWIs $\subseteq$ RHEWUBIs. If a pattern is not a RHEWUBI, it will not be a RHEWI.*

Proof. $\forall X$ in D, $w(X) = w(X, T_q)$. From Theorems 1 and 3, it can be obtained that $w(X, T_q) \leq tubw(T_q)$ and $R(X^k) \leq R(X^{k-1})$. Thus:

$expWSup(X) = w(X) \times \sum_{X \subseteq T_q \wedge T_q \in D} p(X, T_q) = \sum_{X \subseteq T_q \wedge T_q \in D} (w(X) \times p(X, T_q))$
$= \sum_{X \subseteq T_q \wedge T_q \in D} (w(X, T_q) \times p(X, T_q)) \leq \sum_{X \subseteq T_q \wedge T_q \in D} (tubw(T_q) \times tubp(T_q))$
$= \sum_{X \subseteq T_q \wedge T_q \in D} tubwp(T_q) = taubwp(X).$
$\Rightarrow expWSup(X) \leq taubwp(X).$

Thus, if an itemset X is not a RHEWUBI, it is either not a RHEWI in the uncertain database D.

3.2 Procedure of RHEWI-P Algorithm

The proposed RHEWI-P algorithm works as follows. It first scans the original database to discover the set of recent high expected weighted upper-bound 1-itemsets ($RHEWUBI^1$) and recent high expected weighted 1-itemsets ($RHEWI^1$). Then, the designed **Mining-RHEWI**$(X, db_{|X}, k)$ function is recursively called to mine the set of complete RHEWIs. The RHEWI-P algorithm and the **Mining-RHEWI**$(X, db_{|X}, k)$ procedure are respectively given in Algorithm 1 and Algorithm 2. In these pseudo-code, note that the projection of a database db by an itemset X is defined as the set of transactions from db containing X and denoted as $db_{|X}$.

3.3 Procedure of RHEWI-PS Algorithm

Lemma 1. *Let X^k be a k-itemset and X^{k-1} be any of subset of X^{k-1}. Assume that all the 1-items are sorted in the total descending order $\prec$ of item weights, i.e. $w(i_1) \geq w(i_2) \geq \cdots \geq w(i_k) > 0$. It has $w(X^k) \leq w(X^{k-1})$.*

Proof. According to [13] and Definition 3, the lemma $w(X^k) \leq w(X^{k-1})$ holds.

For example, assume that the transactions (sub-projected database) are sorted in descending order of the 1-items weight (w.r.t. $c \prec e \prec f \prec d \prec b \prec a$), the weight of (cd) is always no less than that of its supersets (cdb), (cda) and $(cdba)$, which are respectively calculated as $w(cd) = (1.0 + 0.55)/2 = 0.775$, $w(cdb) = (1.0 + 0.55 + 0.4)/3 = 0.650$, $w(cda) = (1.0 + 0.5 + 0.3)/3 = 0.600$, and $w(cdba) = (1.0 + 0.55 + 0.4 + 0.3)/4 = 0.5625$. Thus, the weights of (cdb), (cda) and $(cdba)$ are all less than the weight of (cd).

Lemma 2. *The concept of expSup always has the anti-monotonicity.*

Proof. Let X^{k-1} be a $(k-1)$ itemset and X^k be any of its supersets. Based on [12], it holds $expSup(X^{k-1}) \geq expSup(X^k)$.

Input: D, an uncertain database; *wtable*, a predefined weight table; δ, the
time-decay threshold; α, the minimum expected weighted support
threshold; β, the minimum recency threshold.

Output: The set of recent high expected weighted itemsets (RHEWIs).

1 **for** *each* $T_q \in D$ **do**

2 $\quad$ calculate $R(T_q)$, $tubw(T_q)$, $tubp(T_q)$, $tubwp(T_q)$;

3 **for** *each 1-item* $i_j \in D$ **do**

4 $\quad$ calculate $R(i_j)$, $taubwp(i_j)$;

5 **for** *each 1-item* $i_j \in D$ **do**

6 $\quad$ **if** $taubwp(i_j) \geq \alpha \times |D| \wedge R(i_j) \geq \beta$ **then**

7 $\quad\quad$ $RHEWUBI^1 \leftarrow RHEWUBI^1 \cup \{i_j\}$;

8 $\quad\quad$ calculate $expWSup(i_j)$;

9 $\quad\quad$ **if** $expWSup(i_j) \geq \alpha \times |D|$ **then**

10 $\quad\quad\quad$ $RHEWI^1 \leftarrow RHEWI^1 \cup \{i_j\}$;

11 sort $RHEWUBI^1$ in lexicographical order, and set $k = 1$;

12 **for** *each* $i_j \in RHEWUBI^1$ **do**

13 $\quad$ scan D to project all related transactions within i_j into the sub-database
$db_{|i_j}$ of i_j;

14 $\quad$ call **Mining-RHEWI**$(i_j, db_{|i_j}, k)$;

15 **return** *RHEWIs*.

Algorithm 1. RHEWI-P

Input: X, a prefix itemset; $db_{|X}$; the projected database by X; k, the length of
X.

Output: The set of *RHEWIs* with the prefix X.

1 generate $PC^{k+1} \leftarrow \{X'|X' = X \cup \{y\} \wedge y \in RHEWUBI^1 \wedge y$ is greater than all
items in X according to the lexicographical order$\}$;

2 **for** *each (k+1)-itemset* $X' \in PC^{k+1}$ **do**

3 $\quad$ **for** *each* $X' \subseteq T_q \wedge T_q \in db_{|X'}$ **do**

4 $\quad\quad$ obtain the projected sub-database $db_{|X'}$ of X';

5 $\quad\quad$ calculate $R(T_q)$, $tubw(T_q)$, $tubp(T_q)$, $tubwp(T_q)$;

6 $\quad$ calculate $R(X')$, $taubwp(X')$, $expSup(X')$;

7 $\quad$ **if** $taubwp(X') \geq \alpha \times |D| \wedge R(X') \geq \beta$ **then**

8 $\quad\quad$ calculate $expWSup(X') = w(X') \times expSup(X')$;

9 $\quad\quad$ **if** $expWSup(X') \geq \alpha \times |D|$ **then**

10 $\quad\quad\quad$ $RHEWI^{k+1} \leftarrow RHEWI^{k+1} \cup X'$;

11 $\quad\quad$ $RHEWUBI^{k+1} \leftarrow RHEWUBI^{k+1} \cup X'$;

12 $\quad\quad$ call **Mining-RHEWI**$(X', db_{|X'}, k+1)$;

13 $RHEWIs \leftarrow \bigcup_k RHEWIs^{k+1}$;

14 **return** *RHEWIs* with the prefix X.

Algorithm 2. Mining-RHEWI$(X, db_{|X}, k)$

Input: X, a prefix itemset; $db_{|X}$; the projected database by X; k, the length of X.

Output: The set of $RHEWIs$ with a prefix X.

1   generate $PC^{k+1} \leftarrow \{X'|X' = X \cup \{y\} \wedge y \in RHEWUBI^1 \wedge y$ is greater than all items in X according to the descending order of weight value$\}$;

2   **for** *each (k+1)-itemset* $X' \in PC^{k+1}$ **do**

3   scan $db_{|X}$ to generate projected sub-database $db_{|X'}$ of X';

4   **for** *each* $T_q \in db_{|X'}$ **do**

5   calculate $R(T_q)$;

6   calculate $R(X')$ and $expSup(X')$;

7   calculate $expWSup(X') = w(X') \times expSup(X')$;

8   **if** $expWSup(X') \geq \alpha \times |D| \wedge R(X') \geq \beta$ **then**

9        $RHEWI^{k+1} \leftarrow RHEWI^{k+1} \cup X'$;

10   call **Mining-RHEWI'**(X', $db_{|X'}$, $k+1$);

11   $RHEWIs \leftarrow \bigcup_k RHEWIs^{k+1}$;

12   **return** $RHEWIs$ with the prefix X.

Algorithm 3. Mining-RHEWI'(X, $db_{|X}$, k)

Lemma 3. *Assume that all the 1-items are sorted in the total order $\prec$ of items their weight values w.r.t. the descending order, i.e., $w(i_1) \geq w(i_2) \geq \cdots \geq w(i_k) > 0$. The expected weighted support of an itemset X is no less than that of any of its supersets (extension projected-itemsets).*

Proof. Let X^k denotes any supersets of a $(k\text{-}1)$-itemset X^{k-1}. Based on Lemmas 1 and 2, the two following conditions are obtained: (1) $w(X^{k-1}) \geq w(X^k)$; (2) $expSup(X^{k-1}) \geq expSup(X^k)$. Thus, $w(X^{k-1}) \times expSup(X^{k-1}) \geq w(X^k) \times expSup(X^k)$, $expWSup(X^{k-1}) \geq expWSup(X^k)$ holds.

Theorem 5 (Sorted upper-bound downward closure property, $SUBDC$ property). *When all the 1-items are sorted in the total descending order $\prec$ of item weights, let X^k be a k-itemset. The sorted upper-bound downward closure property ensures that: $expWSup(X^k) \leq expWSup(X^{k-1})$ and $R(X^k) \leq R(X^{k-1})$.*

By utilizing the $SUBDC$ property, an improved algorithm named projection-based and sorted-based RHEWI-PS algorithm is proposed. It is similar to the RHEWI-P algorithm, the differences are that (1) in Algorithm 1 Line 11, sort $RHEWUBI^1$ and the processed transactions in descending order of weight value before generating the initial sub-databases; (2) the mining procedure. Moreover, the mining procedure for deriving RHEWIs is modified to verify the new pruning property with sorted strategy for each generated itemset. Due to the page limitation, the pseudo-code of the modified **Mining-RHEWIs'** procedure is provided in Algorithm 3.

4   Experiments

We performed extensive experiments to evaluate the proposed algorithms. Since this is the first paper to mine interesting patterns from uncertain databases with weight and recency constraints, none algorithm was available for comparison. We thus used a state-of-the-art HEWI-Uapriori [12] algorithm as a baseline for further evaluation. All the compared algorithms were implemented in Java language and performed on a personal computer with an Intel Core i5-3460 dual-core processor and 4 GB of RAM, running on the 32-bit Microsoft Windows 7 operating system. Experiments are conducted on two datasets including one real-world dataset (foodmart [1]) and one synthetic dataset (T10I4D100K [5]). The weight of each item was randomly selected in (0,1] similar to the setting in [8,12]. In addition, due to the attribute uncertainty model [2], a unique probability value in the range of 0.5 to 1.0 is assigned to each item in every transaction in the used datasets.

4.1   Runtime

The runtime of the three algorithms are first compared under varied minimum expected weighted support thresholds (α) with a fixed β, and under varied minimum recency thresholds (β) with a fixed α. Results are shown in Fig. 1.

In Fig. 1, it can be seen that RHEWI-P and RHEWI-PS perform very well compared to HEWI-Uapriori on most datasets, and RHEWI-PS has the best performance. This result is reasonable since both the proposed RHEWI-P and

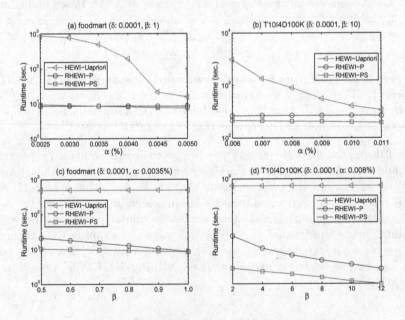

Fig. 1. Runtime under various α and β.

RHEWI-PS algorithms consider recency patterns. By considering recency as an additional constraint, a larger part of the search space can be pruned and less patterns are discovered. Differing from the apriori-like candidate generation-and-test mechanism, the projection-based mechanism can generally reduce the cost of database scans since transactions will become smaller as larger itemsets are explored, so the two projection-based approaches perform better than HEWI-Uapriori. RHEWI-PS is faster than RHEWI-P because it adopts the novel *sorted upper-bound downward closure (SUBDC)* property to eliminate itemsets early. This allows RHEWI-PS to avoid performing the projection operation for forming the sub-database of each pruned itemset. When α was fixed and β was varied, it is interesting to find that the runtime of the RHEWI-P and RHEWI-PS algorithms were significantly decreased when β is increased, while that of HEWI-Uapriori was unchanged. The reason is that the proposed two algorithms discover the RHEWIs by using recency, weight and uncertainty constraints, while HEWI-Uapriori does not consider the recency constraint. Therefore, the parameter β does not effect the performance of HEWI-Uapriori.

4.2   Patterns Analysis

With the same parameter settings as above, the number patterns are further compared under varied α with a fixed β, and varied β with a fixed α, respectively. Results are shown in Fig. 2. Note that HEWI-Uapriori discovers HEWIs, and the RHEWI-P and RHEWI-PS algorithms generate the desired RHEWIs.

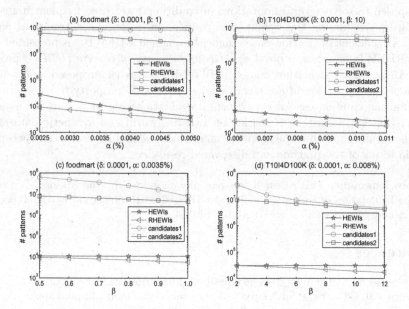

Fig. 2. Number of generated patterns under various α and β.

The generated candidates of RHEWI-P and RHEWI-PS are respectively denoted as candidate1 and candidate2 in the experiments.

From Fig. 2, it can be seen that the number of discovered RHEWIs is always smaller than the number of HEWIs, and candidate2 is much less than candidate1 under varied α and β on all datasets. It indicates that (1) numerous HEWIs are discovered but few of them are the desired up-to-date patterns; fewer RHEWIs are produced when considering the recency of transactions; (2) the search space in RHEWI-PS is more compact than that of RHEWI-P; the *SUBDC* property plays an active role to early prune the unpromising candidates. These situations frequently happen especially when α or β is set to small value since numerous irrelevant HEWIs or RHEWIs are pruned. From Fig. 2, it can also be seen that the number of generated four types of patterns decrease as α is increased. However, the number of RHEWIs, candidate1 and candidate2 dramatically decreases as β increases, while the number of HEWIs remains stable, as shown in Fig. 2(c) and (d). The reason is the same as mentioned in runtime analysis. In particular, less RHEWIs are found but they are more valuable for real-life applications compared to HEWIs since they represent recent trends. Thus, the recency constraint has significant influence on the mining results.

5   Conclusion

In real-world, each object has its distinct importance or interestingness, and the huge amount of the collected data may contain uncertainty. In this paper, a novel knowledge named recent high expected weighted itemsets (RHEWIs) is proposed to solve several limitations of traditional weighted frequent itemset mining by considering the weight, uncertainty and recency constraints of patterns. An efficient projection-based algorithm named RHEWI-P is presented to mine RHEWIs based on a novel *upper-bound downward closure (UBDC)* property. An improved algorithm named RHEWI-PS is further proposed, and relies on the *sorted upper-bound downward closure (SUBDC)* property to early prune unpromising candidates. An experimental evaluation against a state-of-the-art HEWI-Uapriori algorithm is carried on two real-world and synthetic datasets, and the results show that the proposed algorithms are highly efficient and acceptable in terms of the runtime and discovered results.

Acknowledgment. This research was partially supported by the National Natural Science Foundation of China (NSFC) under Grant No.61503092, and by the Tencent Project under grant CCF-TencentRAGR20140114.

References

1. Frequent itemset mining dataset repository. http://fimi.ua.ac.be/data/
2. Aggarwal, C.C., Yu, P.S.: A survey of uncertain data algorithms and applications. IEEE Trans. Knowl. Data Eng. **21**(5), 609–623 (2009)
3. Agrawal, R., Imielinski, T., Swami, A.: Database mining: a performance perspective. IEEE Trans. Knowl. Data Eng. **5**, 914–925 (1993)

4. Agrawal, R., Srikant, R.: Fast algorithms for mining association rules in large databases. In: The International Conference on Very Large Data, Bases, pp. 487–499 (1994)
5. Agrawal, R., Srikant, R.: Quest synthetic data generator. http://www.Almaden.ibm.com/cs/quest/syndata.html
6. Agrawal, R., Srikant, R.: Mining sequential patterns. In: The International Conference on Data, Engineering, pp. 3–14 (1995)
7. Cai, C.H., Fu, A.W.C., Kwong, W.W.: Mining association rules with weighted items. In: International Database Engineering and Applications Symposium, pp. 68–77 (1998)
8. Cagliero, L., Garza, P.: Infrequent weighted itemset mining using frequent pattern growth. IEEE Trans. Knowl. Data Eng. **26**(4), 903–915 (2014)
9. Chen, M.S., Han, J., Yu, P.S.: Data mining: an overview from a database perspective. IEEE Trans. Knowl. Data Eng. **8**(6), 866–883 (1996)
10. Baralis, E., Cagliero, L., Fiori, A., Garza, P.: MWI-Sum: a multilingual summarizer based on frequent weighted itemsets. ACM Trans. Inf. Syst. **34**(1), 5 (2015)
11. Lin, J.C.W., Gan, W., Fournier-Viger, P., Hong, T.P.: RWFIM: recent weighted-frequent itemsets mining. Eng. Appl. Artif. Intell. **45**, 18–32 (2015)
12. Lin, J.C.W., Gan, W., Fournier-Viger, P., Hong, T.P., Tseng, V.S.: Weighted frequent itemset mining over uncertain databases. Appl. Intell. **41**(1), 232–250 (2016)
13. Lin, J.C.-W., Gan, W., Fournier-Viger, P., Hong, T.-P.: Efficient mining of weighted frequent itemsets in uncertain databases. In: Perner, P. (ed.) MLDM 2016. LNCS, vol. 9729, pp. 236–250. Springer, Heidelberg (2016). doi:10.1007/978-3-319-41920-6_18
14. Sun, K., Bai, F.: Mining weighted association rules without preassigned weights. IEEE Trans. Knowl. Data Eng. **20**(4), 489–495 (2008)
15. Tao, F., Murtagh, F., Farid, M.: Weighted association rule mining using weighted support and significance framework. In: ACM SIGKDD International Conference on Knowledge Discovery and Data Mining, pp. 661–666 (2003)
16. Wang, W., Yang, J., Yu, P.S.: Efficient mining of weighted association rules (WAR). In: ACM SIGKDD Intern. Conf. on Knowledge Discovery and Data Mining, pp. 270–274 (2000)
17. Yun, U., Leggett, J.: WFIM: weighted frequent itemset mining with a weight range and a minimum weight. In: SIAM International Conference on Data Mining, pp. 636–640 (2005)

Context-Aware Chinese Microblog Sentiment Classification with Bidirectional LSTM

Yang Wang[1], Shi Feng[1,2(✉)], Daling Wang[1,2], Yifei Zhang[1,2], and Ge Yu[1,2]

[1] School of Computer Science and Engineering,
Northeastern University, Shenyang, China
wangyang@stumail.neu.edu.cn,
{fengshi,wangdaling,zhangyifei,yuge}@cse.neu.edu.cn
[2] Key Laboratory of Medical Image Computing (Northeastern University),
Ministry of Education, Shenyang 110819, China

Abstract. Recently, with the fast development of the microblog, analyzing the sentiment orientations of the tweets has become a hot research topic for both academic and industrial communities. Most of the existing methods treat each microblog as an independent training instance. However, the sentiments embedded in tweets are usually ambiguous and context-aware. Even a non-sentiment word might convey a clear emotional tendency in the microblog conversations. In this paper, we regard the microblog conversation as sequence, and leverage bidirectional Long Short-Term Memory (BLSTM) models to incorporate preceding tweets for context-aware sentiment classification. Our proposed method could not only alleviate the sparsity problem in the feature space, but also capture the long distance sentiment dependency in the microblog conversations. Extensive experiments on a benchmark dataset show that the bidirectional LSTM models with context information could outperform other strong baseline algorithms.

Keywords: Context-aware sentiment · Recurrent neural networks · Bidirectional long short-term memory · Sentiment classification

1 Introduction

Sentiment analysis, also known as opinion mining, is a fundamental task in natural language processing and computational linguistics. Analyzing the sentiment orientations of the user generated content on the Web has become a hot research topic for both academic and industrial communities in recent years. A crucial problem for sentiment analysis task is that the same word may express quite opposite orientations in different context. For example, the adjective "unpredictable" indicates a negative polarity in the phrase "unpredictable steering" from a car review; On the other hand, it could also expresses a positive orientation in the phrase "unpredictable plot" from a movie review [1]. A lot of literature have been published for building context-sensitive sentiment lexicons

© Springer International Publishing Switzerland 2016
F. Li et al. (Eds.): APWeb 2016, Part I, LNCS 9931, pp. 594–606, 2016.
DOI: 10.1007/978-3-319-45814-4_48

and classifiers [2–5]. However, most of these existing methods need manually designed word or phrase level contextual features or specific topic domains.

With the popular microblogging services such as Twitter and Weibo, users can conveniently express their personal feelings and opinions about all kinds of issues in real time. The user relationships in microblog may be asymmetrical, and every user is able to follow other users, comment on and retweet the microblogs. Microblog could evolve into widespread real-time conversation streams because of the informal, colloquial, asymmetrical and fragmented characteristics. The microblog data is length limited, sparse, and lack of enough context. Therefore, the microblog has more implicit context of the sentiment related features. The context of a current microblog is usually implicit in the preceding microblogs or the information during the interactions. Let us consider the following tweets from two microblog users:

Allen: @**Thomas** *Exactly!*

Bill: @**Thomas** *You're right! I can't agree with you more!!*

Obviously, these two tweets are replies to the preceding ones. We can infer that these tweets may contain strong sentiment orientations. However, it is really difficult for the traditional machine learning based methods to classify the orientations in these sparse tweets that have no explicit lexical nor syntactic sentiment features. The Fig. 1 show the entire conversations of these tweets.

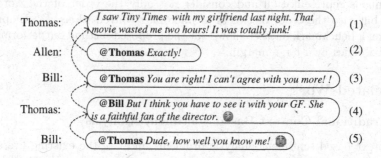

Thomas: *I saw Tiny Times with my girlfriend last night. That movie wasted me two hours! It was totally junk!* (1)

Allen: @**Thomas** *Exactly!* (2)

Bill: @**Thomas** *You are right! I can't agree with you more! !* (3)

Thomas: @**Bill** *But I think you have to see it with your GF. She is a faithful fan of the director.* (4)

Bill: @**Thomas** *Dude, how well you know me!* (5)

Fig. 1. An illustration of the microblog conversations between three users.

In Fig. 1, Thomas expresses a clear negative sentiment toward the movie "Tiny Times" in tweet (1). Then Allen and Bill agree with Thomas in tweet (2) and (3). Both of these two microblogs are discussing the same movie, but they omit the topic words and express consistent orientations with the preceding one. Only with the help of the context tweets, we can infer that the tweet (2) and (3) express negative sentiments. From this example, we can find that how to merge the context knowledge, and enrich the features of microblog are critical problems in the context-aware sentiment analysis task for microblog conversations.

Different from the traditional context-aware sentiment analysis problem whose solutions rely on explicit semantic features and domain-sensitive words, the context-aware sentiment analysis task in microblogs has brought in several new challenges: **Sparsity**. Users have very limited space to express their feelings in microblogs. Therefore, the vectors formed by tweets are extremely sparse, which set obstacles for directly building classifiers from vector space model; **Context-aware sentiments**. The sentiments embedded in microblog conversations are usually implicit and context-aware. Even a single non-sentiment word can express obvious sentiments given certain context; **Long distance dependency**. The microblog conversations may be quite long. So the background topics and sentiment indicators could be long distance away from the target tweet. The traditional bag-of-words based machine learning methods could not distinguish the implicit or hidden dependency in the long conversations.

To tackle these challenges, in this paper we regard the microblog conversations as sequences and leverage the Recurrent Neural Network (RNN) models to build context-aware sentiment classifiers for Chinese microblogs. The RNN based model with its variants have achieved promising results for many sentiment analysis tasks. However, to the best of our knowledge, little literature has studied the context-aware sentiment classification problem by using the RNN based models. The main contributions of this work are as follows: (1) We model the microblog conversations as sequences and utilize preceding tweets to enrich the content of the target tweets, which could alleviate the sparsity problem and incorporate context information. (2) We develop improved RNN based model bidirectional Long Short-Term Memory (BLSTM), which could give microblogs a continuous representation and consider not only the order of the words in tweets, but also the order of tweets in microblog conversations. (3) Empirical results on a benchmark dataset show that the proposed method outperforms the strong baselines by a large margin.

2   Related Work

2.1   Traditional Context Based Sentiment Analysis

For context based sentiment analysis problems, most of the existing literature are based on the traditional user generated content, such as reviews and blogs. McDonald et al. [2] treated the sentiment labels on sentences as the sequence tagging problem, and utilized CRFs model to score each sentence in the product reviews. Katz et al. [3] used the information retrieval technology to identify the keywords with the emotional information, and then used these keywords and the context to construct the corresponding features. Yang et al. [4] considered both local and global context information to model the complex sentence. They put the vocabulary and discourse knowledge into CRFs model. To deal with the problem that the same sentiment words might have different polarities in the different context, Lu et al. [5] proposed a method for combining information from different sources to learn context-aware sentiment lexicon.

Most of these existing methods relied on explicit semantic features or topic-focused domains. However, the sentiments embedded in microblogs are usually more abbreviated and obscure without obvious sentiment words. Furthermore, the context sentiment related information in microblog conversations could be long distance dependent with each other and usually have scattered topics, which bring new challenges for the sentiment analysis task.

2.2 Sentiment Analysis Based on Microblog Context

Recently, some researches have already treated the microblog content with the dialogue relationship as the context information. Vanzo et al. [6] regarded the context based sentiment classification problem as the sequence labeling problem of the Twitter stream. They used the bag of words model, word meanings and sentiment information as features and used the SVM^{hmm} as the classifier. Mukherjee et al. [7] combined the conjunctions with dialogue information to improve the accuracy of classification results, and discussed the impact of negative words. Wu et al. [8] proposed a new framework to extract context knowledge from unlabeled data for the microblog classification problem, then they defined the relationship between words and words and the relationship between words and emotions. Li [9] classified the microblog text by using the sentiment dictionary, and considered the influence of the user who made comments. Wu et al. [10] utilized the social context information, such as users' historical tweets and friend relations, to tackle the shortness and noise problem of the microblog data.

The previous literature have already proposed the idea of modeling the microblogs as sequences. However, most of these methods usually need to manually design context based features and resorted to the user profiles and social relations to enrich the feature set. In this paper, we leverage the RNN based models to classify the context-aware sentiments in microblogs. Although the RNN based methods have achieved promising results for many sentiment analysis tasks [11,12], to the best of our knowledge, they have never before been reported for context-aware sentiment classification. Our proposed algorithm considers not only the order of words in tweets but also the order of the tweets in conversations, which is capable of learning continuous representations without feature engineering and meanwhile capturing context dependency.

3   Preliminary

The traditional microblog sentiment classification methods treat each post equally in the dataset and do not care about the orders of the posts. However, we can see from the examples in Fig. 1 that without context information we can hardly detect the sentiments embedded in microblog (2) and (3), whose polarities are highly rely on the preceding microblog (1). In this paper, we model the incoming microblog stream as a sequence. Suppose that we have a labeled training set D with microblogs from *Positive, Neutral, Negative* categories. There are several threads in D, and each thread contains sequential microblogs that forming a conversation tree (shown below in Fig. 2) with an original tweet as root and a number of branches as conversations.

Given the training dataset $D = \{(m_1; y_1) \cdots (m_n; y_n)\}$, the sentiment classification problem can be defined as learning a mapping function $f : M \rightarrow Y$

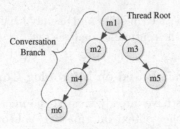

Fig. 2. An example of conversation tree in the training dataset D. Each node is representing a microblog post and the link in the tree denotes the retweet or the '@' actions in microblog stream. There is one original tweet acting as the thread root and the retweet and reference actions have formed the conversation branches in the tree.

where the microblog $m \in M$, the sentiment polarity label $y \in Y$ and $Y = \{Positive, Neutral, Negative\}$. As aforementioned, the conversation branches in D are chain microblogs regarded as sequences. To capture the context information for each m and tackle the sparsity problem, we incorporate the preceding tweets of m to enrich the microblog representations. Therefore, each instance in D is transformed to $(pre(m), m; y)$ where $pre(m)$ denotes the preceding tweets of m in the conversation branch. We aim to learn a context-aware sentiment classification model with the help of the expanded representation as context to detect the polarity of target microblog in the testing set T. Note that we do not consider the sentiment labels of preceding tweets in the conversation.

4   Context-Aware Sentiment Analysis for Microblogs

4.1   Modeling Context-Aware Sentiments in Chinese Microblog

Based on our problem formulation, one key problem is how to model the sequence of the microblog conversation. The traditional language models such as N-Gram can only take into account the limited number of words around the current word, so they cannot model the long term dependency relationships well in the long microblog conversations. On the other hand, the bag-of-words model does not consider the order of the words in the conversation sequence. In this section, we first provide a brief overview of the basic recurrent neural network model for the context-aware sentiment analysis task and introduce the notations used in the paper. Afterwards, we leverage the LSTM and bidirectional LSTM to model the conversation branches in the sequential microblog stream.

Recurrent Neural Network. RNN [14] is a kind of neural network model which is suitable for modeling the sequence information, and it can maintain a memory based on history information, which enables the model to predict the current output conditioned on long distance features. At each time step t, the state of the model's hidden layer is not only related to the current input, but also related to the state of hidden layer at the previous time step $t - 1$. Given the expanded representation $(pre(m), m; y)$ of a training instance in D, we concatenate the words in $pre(m)$ and m as a sequence $x_1 x_2 \cdots x_n$, which could reflect not only the order of the words in sentences, but also the order of the tweets in conversations. The word x is the input of the recurrent neural network and the calculation formula of hidden state $h_t \in \mathbb{R}^m$ at time step t is as follows:

$$h_t = f(W x_t + U h_{t-1} + b) \tag{1}$$

where W, U and b are the parameters of the model, x_t is the embedding distributed representation of word x at step t. The historical information is stored in the variable h_{t-1}. RNN has demonstrated promising results in some sequence labeling task. However, due to the existence of gradient vanishing problem, learning long-range dependencies with a vanilla RNN is very difficult.

Long Short Term Memory. LSTM [13] model solves the gradient vanishing problem by introducing the memory cells c_t at each time step t and three gate structures: input gate i_t, forget gate f_t and output gate o_t:

$$
\begin{aligned}
i_t &= \sigma(W^i x_t + U^i h_{t-1} + b^i) \\
f_t &= \sigma(W^f x_t + U^f h_{t-1} + b^f) \\
o_t &= \sigma(W^o x_t + U^o h_{t-1} + b^o) \\
g_t &= \tanh(W^g x_t + U^g h_{t-1} + b^g) \\
c_t &= f_t \odot c_{t-1} + i_t \odot g_t \\
h_t &= o_t \odot \tanh(c_t)
\end{aligned}
\tag{2}
$$

Here $\sigma(\cdot)$ is the sigmod function and $\tanh(\cdot)$ is the hyperbolic tangent function. $\odot$ is the element-wise multiplication operator. The memory cell c_t is computed additively, so the error derivatives can flow in another path, thus avoid the gradient vanishing problem. Parameters of the LSTM are W^j, U^j, b^j for $j \in \{i, f, o, g\}$. Similar with RNN model, the input of LSTM is the embedding representation of the word x in the conversation sequence. LSTM is capable of mapping word sequences with variable length to a fixed-length vector by recursively transforming current word vector x_t with the output vector of the previous step h_{t-1}. With the help of the powerful cell and the three gates, the LSTM is able to model the long dependency in the word sequence. During training, the dropout technique that randomly drop units from the neural network is used to avoid overfitting problem and epochs measure how many times every example has been seen during training.

Bidirectional Long Short Term Memory. Considering we need to use the history information of microblogs, we apply the bidirectional LSTM(**BLSTM**) model for making use of the past and future context information. The basic idea of BLSTM is to connect two LSTM hidden layers of opposite directions to the same output. Bidirectional LSTM model is a 2 layers LSTM network, which is composed of forward hidden sequence $\overrightarrow{h}$ and backward hidden sequence $\overleftarrow{h}$ of

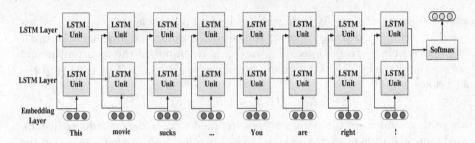

Fig. 3. The bidirectional LSTM model. The input of the model is the word embedding corresponding to each word. Softmax function is used to predict the sentiment polarity.

LSTM models, both of which are connected to the output layer as shown in Fig. 3. The details of the equation of BLSTM model are discussed in Sect. 4.2. Because the bidirectional LSTM model looks at the sequence twice, from left to right and right to left, the output layer can get information from past and future states and the context information can be handled well. The bidirectional LSTM models are trained by back propagation through time(BPTT) [14].

4.2   Context-Aware Sentiment Classification

In our model, each word corresponds to a specific index, then we turn indexes into dense word vectors of fixed size, and each microblog with its preceding tweets is encoded as a sequence of word embeddings. The input of the model at time t is the input word embedding x_t, which is one column of the embedding matrix X. The bidirectional RNN model can be represented as the following equations:

$$\overrightarrow{h_t} = f(W_{x\overrightarrow{h}}x_t + U_{\overrightarrow{h}\overrightarrow{h}}\overrightarrow{h}_{t-1} + b_{\overrightarrow{h}})$$
$$\overleftarrow{h_t} = f(W_{x\overleftarrow{h}}x_t + U_{\overleftarrow{h}\overleftarrow{h}}\overleftarrow{h}_{t-1} + b_{\overleftarrow{h}})$$
$$P(y = j|w_{1:t}) = \frac{exp(\overrightarrow{h_t} \cdot \overrightarrow{p}^j + \overleftarrow{h_t} \cdot \overleftarrow{p}^j + q^j)}{\Sigma_{j\in\{-1,0,1\}}exp(\overrightarrow{h_t} \cdot \overrightarrow{p}^{j'} + \overleftarrow{h_t} \cdot \overleftarrow{p}^{j'} + q^{j'})} \tag{3}$$

where p^j is the j-th column of $P \in \mathbb{R}^{m \times 3}$ and q^j is the j-th element of $q \in \mathbb{R}^3$. P,q,X are parameters of the model to be learned during training strategy. $\{-1, 0, +1\}$ represents categories $\{Negative, Neutral, Positive\}$ respectively. The final orientation of the microblog is predicted by the softmax function at the top layer (as shown in Fig. 3). By replacing the hidden states in the bidirectional RNN with LSTM memory blocks, we can get the aforementioned BLSTM models as the context-aware sentiment representation learning layer in our proposed framework.

5   Experiment

5.1   Datasets

Our experiments were performed on the COAE-2015 Task-1 dataset[1]. The training set that crawled from Weibo contains 2,800 Chinese microblogs with *Positive*, *Negative* and *Neutral* labels. The organizers of the evaluation task have provided a benchmark testing set. However, they do not disclose the gold standard labels of the microblog in the set. We ask two graduate students who majoring in opinion mining to manually label part of the test data. There are about 65 % of the results having the consistent sentiment labels, and we remove the remaining microblogs with the inconsistent annotations. Finally we have 4,248 labeled Chinese microblogs, which belong to 555 threads formed by retweet and reference '@' relationship. Among the data set, there are 1,571 positive, 1,030 negative, and 1,647 neutral microblog respectively. The statistics information of the conversation length are shown in Table 1. We can see that most of the sequential conversations have the length of two or three. But there are still quite a number of conversations contain more than 3 sequential microblogs.

Table 1. The statistics of the conversation branch length in the dataset

	Length 2	Length 3	Length 3+
Percentage	57.8 %	25.1 %	17.1 %

The statistics information of the sentiment polarity drift between the root microblog and the retweet microblogs are shown in Table 2. The horizontal header represents the source microblog polarity information and the vertical header represents the retweet microblog polarity information.

Table 2. The sentiment drift information of the source and retweet microblog

Retweet	Source		
	Positive	*Neutral*	*Negative*
Positive	725	629	52
Neutral	287	908	141
Negative	126	554	217

We can see that a lot of microblogs retain the source microblogs polarities, and there are still quite a number of microblogs have the reverse orientations. These polarity drifts can be indicated by features such as contrastive connectives

---

[1] http://www.ccir2015.com/.

or context dependencies, which could be captured by our LSTM models. Finally we select 4/5 of the whole dataset as the training set and the remaining as the validation set. The dataset used in this paper has been made public[2].

5.2 Experiment Setup

The input of LSTM models are 50 or 100 dimensions word embedding vectors that constructed by selecting the 10,000 most frequent features in the dataset. We choose the Adam algorithm as the optimizer and choose the softmax as the top layer to predict the final sentiment labels. For regularization, early stopping and dropout [13] are used during the training procedure. In the LSTM model, we compared the impact of the dropout and weight regularize mechanism on the neural network structures. We set the dropout rate to 0.5 on the input-to-hidden layers and hidden-to-output layer and use the L2 regularization.

5.3 Comparison Methods

Bag-of-words and SVM. We build a Support Vector Machine (SVM) based classifier on the training data provided. There are two kinds of feature space: (1) unigram, bigram and trigram (1 to 3-gram); (2) unigram, bigram, trigram, 4-gram and 5-gram (1 to 5-gram). For implementation, we utilize the TFID-FVectorizer function provided by Scikit-learn[3] to get the document-term matrix and calculate the TFIDF weight respectively. The SVM is implemented by LinearSVC function in Scikit-learn.

Conditional Random Fields. CRFs [15] are a class of statistical modelling method often applied in pattern recognition and machine learning, where they are used for structured prediction. One of the most common use-cases for structured prediction is chain-structured outputs. These occur naturally in sequence labeling tasks, such as Part-of-Speech tagging or segmentation in natural language processing.

We regard the conversation microblog with the retweet relationship as the input sequence, and the sentiment of each microblog as the output sequence. We use the linear chain CRF model to solve the context-aware sentiment classification problem. Each microblog is a node in our chain, and the microblogs with the retweet relationship are connected with an edge. Therefore, the conversation branch described in Sect. 3 has formed the chain in CRF model. The length of the chain varies with the number of microblogs with the retweet relationship, just the same as the conversation branches in Fig. 1. The CRF model with linear chain structure is implemented by Pystruct[4] package.

---

[2] http://github.com/leotywy/coae2015.

[3] http://scikit-learn.org/stable/.

[4] http://pystruct.github.io/.

Gated Recurrent Units (GRU). In order to make a comparison, we utilize an alternative RNN model called GRU [14] to train a sentiment classifier. GRU is designed to have more persistent memory thus making it easier to capture long-term dependencies. GRUs have fewer parameters (W and U are smaller) to train and the training time could be less than the basic LSTM model.

5.4   Experiment Results

We first study the effect of some hyperparameters on the one layer LSTM model with dropout and L2 regularize. We fix the number of hidden unit as 64, batch size as 10, and compare the impact of the dropout rate and epoch of the model.

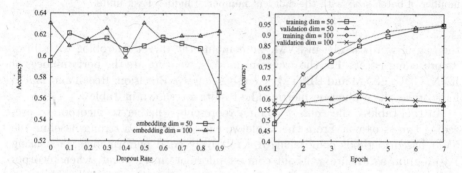

Fig. 4. The performance of 1 layer LSTM model with different dropout rate and epoch.

Figure 4 shows the higher dimensional word embedding (100) generally performs better. From the results, we find that setting the dropout rate to 0.5 and training the model with 4 or 5 epoch can get the good performance with the larger word embedding dimensions. When the epoch is larger than 5, model may be over fitting. Therefore using the early stopping method may be a good choice on this small dataset. Then we fix the dropout rate to 0.5 and explore the impact of the number of hidden units and the batch size as shown in Fig. 5.

In Fig. 5, with different number of hidden units, the impact of the dimension of the word embedding on the results is not very obvious. 100 dimensional word embedding is slightly better than 50 dimensional word embedding. For both of the smaller and larger word embedding, the batch size of training data for 8 or 16 is good enough. In the following experiments, we choose batch size as 8. For the smaller word embedding, the number of neural units in the hidden layer for 16 performs well. For the larger word embedding, the number of neural units in the hidden layer for 32 or 128 performance better than the smaller model by using the smaller batch size, but the model is not as stable as the smaller word embedding model when the batch size changes.

Then we compare the results of the traditional methods and recurrent neural network models. We perform hyperparameters tuning on all these models. For

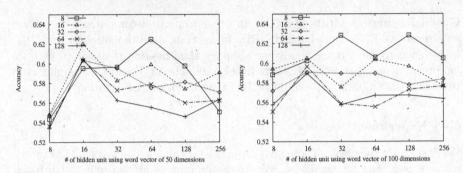

Fig. 5. The performance of BLSTM model trained with different dimension of word embedding, number of hidden units and batch size. Different lines represent different number of batch sizes with the different number of hidden layer units.

the bag of words model in SVM, we combine the target microblog with all of its preceding tweets. For the recurrent models, we compare the performances of RNN, GRU, LSTM and BLSTM. We use Accuracy, Precision, Recall and Macro F1 score as the evaluation metric. The results are shown in Table 3.

In the Table 3, the *context* column represents whether to incorporate preceding tweets or not. From this table we can see that as a strong baseline the SVM based methods could achieve a relatively good performance when using 1 to 3-gram as features without context information. However, when incorporating the preceding tweets, the performances decrease dramatically. This may because that bag of words based SVM does not consider word and sentence orders. Including preceding tweets has brought in more noise instead of useful features. The performance of CRF based method is dragged down by the sparse feature space and limited training instances.

All of the recurrent neutral network based methods could achieve better results when incorporating context information, which indicates that RNN based methods are good at modeling the context-aware sentiments in the microblog conversations. Note that in Table 3 the vanilla RNN does not have a context-aware results because when considering preceding tweets, the training procedure does not converge for vanilla RNN. We do not report some of the context-free classification results due to space limitations.

A good classification performance is achieved when using one layer LSTM with dropout and regularization. Dropout is a technique that can randomly drop units from the neural network during training, which can prevent the neural network overfitting effectively. We argue that the input word embedding could alleviate the sparsity problem of the feature space. The GRU and LSTM model could map the long word sequences into fixed-length word vectors and capture the context-aware sentiments in microblogs with long distance dependencies. The performance of GRU model and LSTM is similar, but slightly worse than that of the BLSTM. We find that the bidirectional LSTM achieve a better performance than the unidirectional LSTM at the context aware sentiment analysis task.

Table 3. Experiment results of the shallow learning methods and RNN models

Model	Features/structures	Context	Accuracy	Precision	Recall	MacroF1
SVM	1 to 3 gram	No	0.6235	0.6134	0.6079	0.6082
		Yes	0.5847	0.5699	0.5677	0.5683
	1 to 5 gram	No	0.6176	0.6089	0.6027	0.6042
		Yes	0.5965	0.5798	0.5777	0.5779
CRF		Yes	0.4264	0.4327	0.4187	0.4113
RNN	1 layer+dropout+L2	No	0.5800	0.5722	0.5718	0.5702
		Yes	-	-	-	-
GRU	1 layer+dropout+L2	No	0.5894	0.5959	0.5839	0.5794
		Yes	0.6188	0.6059	0.6012	0.6019
LSTM	1 layer	No	0.5965	0.5940	0.5760	0.5804
		Yes	0.6165	0.6145	0.6113	0.6093
	1 layer+dropout+L2	No	0.6153	0.6100	0.5900	0.5815
		Yes	0.6205	0.6145	0.6058	0.6057
	BLSTM	No	0.6024	0.5839	0.5776	0.5757
		Yes	0.6223	0.6114	0.5996	0.6008
	BLSTM+dropout+L2	No	0.6118	0.5921	0.5855	0.5802
		Yes	**0.6282**	**0.6196**	**0.6189**	**0.6183**

The BLSTM provides the output with both the past and the future context information for every timestep in the input sequence.

6   Conclusion

In this paper, we utilize the different LSTM models for the context-aware Chinese microblog sentiment analysis task. The extensive experiments on a benchmark dataset demonstrate that the LSTM based models can be more effective at representing the sequential microblog conversations and modeling the context-aware sentiments. The best performance is achieved when using bidirectional LSTM networks with dropout and regularization.

In the future, we intend to use the pretrained word embedding with larger corpus as the initial input to the recurrent network. We will also consider attention-based RNN models for further improving the sequential representation of the microblog conversations. Our proposed method is language-independent, and we will conduct more experiments to evaluate the models on Twitter corpus.

Acknowledgments. The work is supported by National Natural Science Foundation of China (61370074, 61402091), the Fundamental Research Funds for the Central Universities of China under Grant N140404012.

References

1. Turney, P.D.: Thumbs up or thumbs down? semantic orientation applied to unsupervised classification of reviews. In: Proceedings of ACL, pp. 417–424 (2002)
2. McDonald, R., Hannan, K., Neylon, T., Wells, M., Reynar, J.: Structured models for fine-to-coarse sentiment analysis. In: Proceedings of ACL, pp. 432–439 (2007)
3. Katz, G., Ofek, N., Shapira, B.: ConSent: context-based sentiment analysis. Knowl.-Based Syst. **84**, 162–178 (2015)
4. Yang, B., Cardie, C.: Context-aware learning for sentence-level sentiment analysis with posterior regularization. In: Proceedings of ACL, pp. 325–335 (2014)
5. Lu, Y., Castellanos, M., Dayal, U., Zhai, C.: Automatic construction of a context-aware sentiment Lexicon: an optimization approach. In: Proceedings of WWW, pp. 347–356 (2011)
6. Vanzo, A., Croce, D., Basili, R.: A context-based model for sentiment analysis in twitter. In: Proceedings of COLING, pp. 2345–2354 (2014)
7. Mukherjee, S., Bhattacharyya, P.: Sentiment analysis in twitter with lightweight discourse analysis. In: Proceedings of COLING, pp. 1847–1864 (2013)
8. Wu, F., Song, Y., Huang, Y.: Microblog sentiment classification with with contextual knowledge regularization. In: Proceedings of AAAI, pp. 2332–2338 (2015)
9. Li, D., Shuai, X., Sun, G., Tang, J., Ding, Y., Luo, Z.: Mining topic-level opinion influence in microblog. In: Proceedings of CIKM, pp. 1562–1566 (2012)
10. Wu, F., Huang, Y., Song, Y.: Structured microblog sentiment classification via social context regularization. Neurocomputing **175**, 599–609 (2016)
11. Tang, D., Qin, B., Liu, T.: Document modeling with gated recurrent neural network for sentiment classification. In: Proceedings of EMNLP, pp. 1422–1432 (2015)
12. Wang, X., Liu, Y., Sun, C.: Predicting polarities of tweets by composing word embeddings with long short-term memory. In: Proceedings of ACL, pp. 1343–1353 (2015)
13. Hinton, G.E., Srivastava, N., Krizhevsky, A., Sutskever, I., Salakhutdinov, R.R.: Improving Neural Networks by Preventing Co-adaptation of Feature Detectors (2012). arXiv preprint. arXiv:1207.0580
14. Greff, K., Srivastava, R.K., Koutnk J., Steunebrink B.R., Schmidhuber J.: LSTM: A Search Space Odyssey (2015). arXiv preprint. arXiv:1503.04069
15. Lafferty, J., McCallum, A.: Conditional random fields: probabilistic models for segmenting and labeling sequence data. In: Proceeding of ICML, pp. 282–289 (2001)

Author Index

Printed in the United States
By Bookmasters